Angiogenesis Assays

Angiogenesis Assays
A critical appraisal of current techniques

Carolyn A. Staton
Academic Unit of Surgical Oncology
University of Sheffield Medical School
Sheffield, UK

Claire Lewis
Academic Unit of Pathology
University of Sheffield Medical School
Sheffield, UK

Roy Bicknell
Division of Immunity and Infection
University of Birmingham
Birmingham, UK

John Wiley & Sons, Ltd

Other Wiley Editorial Offices

John Wiley & Sons Inc., 111 River Street, Hoboken, NJ 07030, USA

Jossey-Bass, 989 Market Street, San Francisco, CA 94103-1741, USA

Wiley-VCH Verlag GmbH, Boschstr. 12, D-69469 Weinheim, Germany

John Wiley & Sons Australia Ltd, 42 McDougall Street, Milton, Queensland 4064, Australia

John Wiley & Sons (Asia) Pte Ltd, 2 Clementi Loop #02-01, Jin Xing Distripark, Singapore 129809

John Wiley & Sons Canada Ltd, 6045 Freemont Blvd, Mississauga, Ontario, L5R 4J3

Wiley also publishes its books in a variety of electronic formats. Some content that appears in print may
not be available in electronic books.

Library of Congress Cataloging-in-Publication Data
Angiogenesis assays : a critical appraisal of current techniques /
  [edited by] Carolyn Staton, Claire Lewis, Roy Bicknell.
       p. ; cm.
   Includes bibliographical references and index.
   ISBN-13: 978-0-470-01600-8 (alk. paper)
   ISBN-10: 0-470-01600-0 (alk. paper)

  1. Blood-vessels–Growth. 2. Neovascularization. I. Staton, Carolyn.
II. Lewis, Claire E. III. Bicknell, R. J. (R. John)
   [DNLM: 1. Neovascularization, Pathologic–physiopathology. 2.
Biological Assay–methods. 3. Models, Cardiovascular. 4.
Neovascularization, Physiologic–physiology. WG 500 A58425 2007]

   QP102.A544 2007
   612.1'3–dc22   2006031246

British Library Cataloguing in Publication Data

A catalogue record for this book is available from the British Library

ISBN-10 0 470 01600 0   ISBN-13 978-0-470-01600-8

Typeset in 10½/12½pt Times by Thomson Digital
Printed and bound in Great Britain by Antony Rowe Ltd., Chippenham, Wilts
This book is printed on acid-free paper responsibly manufactured from sustainable forestry
in which at least two trees are planted for each one used for paper production.

Cover images provided by Carolyn A. Staton

Contents

Foreword

Heterogeneity of angiogenesis in disease: need for diverse approaches to study blood vessel growth and regression

The growth of new blood vessels from existing vessels, familiar to most as angiogenesis or neovascularization, has acquired an importance that would have been difficult to imagine a few years ago. The proven efficacy of recently approved drugs that block angiogenesis in tumours and age-related macular degeneration has heightened the visibility and relevance of research on blood vessel growth and regression. The promise of factors that stimulate functional revascularization of organs, starved of their blood supply by ischaemic vascular disease or other conditions, is also increasing. Success in the clinic has shifted into high gear the search for even more efficacious drugs. How can these agents be identified, screened and tested? How can their mechanism of action be determined? The seemingly ideal approach for evaluating agents would be through pre-clinical models of the targeted diseases. However, few pre-clinical models faithfully mimic human disease. Therefore, the search continues for faster, easier, more relevant ways of assessing agents that stimulate or inhibit angiogenesis.

The use of 'angiogenesis' as a generic term to describe vascular proliferation suggests a single process. Yet, angiogenesis occurs under many different conditions. Blood vessel growth is driven by multiple factors and occurs in varied settings. Are newly formed blood vessels the same regardless of the driving stimulus and environmental conditions in which they grow? Does each condition produce a homogeneous population of new blood vessels? Is angiogenesis in tumours the same as in eye disease, inflammation, or wound healing?

The answer to these questions is clearly, no. Examples of the heterogeneity of new blood vessels are accumulating at an increasing rate. The amount of angiogenesis driven by vascular endothelial growth factor (VEGF) is dose-dependent, and the phenotype of the new blood vessels is governed by local concentration. Greater concentrations of VEGF produce more abundant blood vessels and exaggerate vessel abnormalities. Lower concentrations of VEGF drive less angiogenesis and promote a more normal vessel phenotype. Most blood vessels in tumours have multiple, sometimes bizarre, abnormalities, and the types and severity of the abnormalities vary within each tumour and among tumours of different varieties. Blood vessels at sites of inflammation are leaky because tiny gaps form reversibly between endothelial cells, just as they do after a mosquito bite. In contrast, tumour vessels leak because of structural defects in the endothelium, which may have multiple layers in some regions and an incomplete monolayer in others. Unlike new or remodelled blood vessels at sites of inflammation, which support high blood flow and robust influx of inflammatory cells, tumour vessel abnormalities may impair blood flow and repel entry of immune cells. When local blood flow in tumours is less than required for cell viability, necrosis results. Loss of blood flow to necrotic regions of tumour may redirect flow to adjacent viable regions that then grow even faster.

An important measure of blood vessel diversity stems from the chamaeleon-like characteristics of endothelial cells and mural cells (pericytes or smooth muscle cells), the two cell types that form the vasculature. In normal blood vessels, both types of cells adapt structurally and functionally to their environment to provide organ-specific features, such as impermeability of the blood–brain barrier, lymphocyte trafficking of high-endothelial venules, and efficient plasma filtration of renal glomeruli. The same chamaeleon-like, adaptive properties of endothelial cells and mural cells underlie the growth, remodelling, and heterogeneity of blood vessels at sites of disease.

Blood vessels that grow or undergo remodelling in disease reflect the integrated action of multiple angiogenic growth factors and inhibitors, substances that augment or limit blood flow, changing composition of the extracellular matrix, and other dynamic environmental conditions. Because of the diversity of conditions that influence angiogenesis in health and disease, newly formed blood vessels are heterogeneous. As a result, no single *in vitro* assay or *in vivo* model can simulate all forms of angiogenesis. Only a broad range of experimental models can mimic the spectrum of conditions that new blood vessels experience under different pathological circumstances.

The 19 chapters of this book review the attributes and limitations of *in vitro* assays, *in vivo* models and clinical settings for studying angiogenesis. Varied approaches are used to observe, characterize, compare, stimulate or block angiogenesis under different conditions. *In vitro* methods make it possible to examine endothelial cell proliferation, migration and tube formation, as well as to investigate effects of fluid shear stress and flow, membrane and intracellular signalling events, and – in co-culture experiments – interactions of mural cells with endothelial cells. Powerful *in vivo* models have been developed to study angiogenesis in normal and disease settings. Models range from the chick chorioallantoic membrane, mammalian cornea, implanted Matrigel plugs, subcutaneous air sacs and transgenic mouse models of cancer to real-time viewing of sprouting endothelial cells in transparent developing zebrafish or in tumours growing in subcutaneous windowed chambers. *In vivo* models also provide approaches for assessing the contribution of bone marrow-derived cells to growing blood vessels. In concert with clues from pre-clinical models, clinical research is searching for better ways to monitor the action of angiogenesis inhibitors or promoters in patients.

Given the momentum of research on angiogenesis and the broad-based development of agents to manipulate blood vessel growth, the future is exciting and promising. But in moving forward, an ongoing challenge is to determine how to link properties of angiogenic blood vessels identified in pre-clinical assays to those in human disease. VEGF and multiple other factors clearly stimulate angiogenesis in the cornea and Matrigel plugs, but how do the new vessels compare with those at sites of angiogenesis in human cancer? Which assays provide the most meaningful information about angiogenesis in cancer, eye disease or inflammation? Which assays give a reliable fingerprint of blood vessel growth and remodelling driven *in vivo* by VEGF, PDGF, angiopoietins, ephrins, sphingolipids or chemokines, alone or in various combinations? Can data from *in vitro* or *in vivo* assays predict response to angiogenesis inhibitors in human disease? What biomarkers identified in pre-clinical assays can serve as meaningful readouts for actions of angiogenesis-related drugs in humans?

Better understanding of the process of angiogenesis and properties of newly formed blood vessels will lead to even more informative assays and biomarkers. These in turn will help in screening and evaluating of new, more efficacious drugs and other novel tools in vascular biology. Together, these advances will further the exploitation of vascular abnormalities as targets for drug delivery and the control of blood vessel growth and regression in health and disease.

Donald M. McDonald
San Francisco, California, USA
4 May 2006

Preface

Angiogenesis, the development of new blood vessels from the existing vasculature, is essential in normal developmental processes and in numerous pathological conditions such as diabetic retinopathy, rheumatoid arthritis, psoriasis and cancer. This process is a multi-factorial and highly structured sequence of cellular events comprising vascular initiation, formation, maturation, remodelling and regression which, under physiological conditions, are controlled and modulated to meet the tissue requirements. However, under pathological conditions this tight regulation is lost. As angiogenesis is a key player in over 70 different disease states there is a need to study this process in great detail for the development of future therapeutic strategies.

One of the most important technical challenges in studies of angiogenesis is selection of the appropriate assay. The ideal angiogenesis assay would be fast, easy, robust, with reliable readouts, automated computational analysis, multi-parameter assessment, including positive and negative controls and should relate directly to results seen in the clinic. Sadly, such a 'gold-standard' angiogenesis assay has yet to be developed. Endothelial cells whose migration, proliferation, differentiation and structural rearrangement is central to the angiogenesis process are commonly studied in *in vitro* assays, but they are not the only cell type involved in angiogenesis. Therefore the most translatable assays would include the supporting cells (e.g. pericytes, smooth muscle cells and fibroblasts), the extracellular matrix and/or basement membrane and the circulating blood. However, no *in vitro* assays exist which fully model all the components of this complex process. While *in vivo* assays have the components present, these are limited by species used, organ sites and lack of quantitative analysis.

Due to these technical challenges and the variety of assays being used between different laboratories, there is a need to highlight the details and limitations of each assay currently in use. In this book, therefore, we have invited experts in the use of a diverse range of assays to outline the key components and give a critical appraisal of the strengths and weaknesses of these assays. This book aims to provide the information to enable researchers in this field to make informed choices about the type of assays to use for their research and to recognise the limitations of these assays. As anti-angiogenic agents are now in clinical use a critical analysis of the biological end-points currently being used in clinical trials to assess the efficacy of these drugs is included and the book finishes with a discussion of the direction future studies may take.

Carolyn A. Staton
Claire Lewis
Roy Bicknell

List of Contributors

Andrade, Silvia P.
Departments of Physiology and Biophysics
Institute of Biological Sciences
Federal University of Minas
  Gerais-Av. Antonio Carlos 6627
Campus Pampulha
Cx Post 486 - CEP 31270-901
Belo Horizonte/MG, Brazil

Asai, Tomohiro
Department of Medical Biochemistry
School of Pharmaceutical Sciences
University of Shizuoka
Yada, Suruga-ku
Shizuoka 422-8526, Japan

Auerbach, Robert
Department of Zoology
University of Wisconsin
1117 West Johnson Street
Madison, WI 53706, USA

van Beijnum, Judy R.
Angiogenesis Laboratory
Research Institute for Growth and
  Development (GROW)
Department of Pathology
Maastricht University and University
  Hospital Maastricht
Maastricht, The Netherlands

Castoldi, Gianluigi
Haematology Section
Department of Biomedical Sciences
University of Ferrara
S. Anna Hospital
Corso Giovecca
203 44100 Ferrara, Italy

Chau, Cindy H.
Molecular Pharmacology Section
Medical Oncology Branch
Centre for Cancer Research
National Cancer Institute
National Institutes of Health
Bethesda, MD, 20892, USA

Cuneo, Antonio
Haematology Section
Department of Biomedical
  Sciences
University of Ferrara
S. Anna Hospital
Corso Giovecca
203 44100 Ferrara, Italy

Dewhirst, Mark W.
Department of Radiation Oncology
Duke University School of Medicine
Durham, NC 27710, USA

Egginton, Stuart
Centre for Cardiovascular
  Sciences and Department
  of Physiology
The Medical School
The University of Birmingham
Birmingham B15 2TT, UK

Fan, Tai-Ping
Angiogenesis and TCM Laboratory
Department of Pharmacology
University of Cambridge
Tennis Court Road
Cambridge CB2 1PD, UK

Ferreira, Monica A. N. D.
General Pathology
Institute of Biological Sciences
Federal University of Minas
    Gerais-Av. Antonio Carlos 6627
Campus Pampulha
Cx Post 486 - CEP 31270-901
Belo Horizonte/MG, Brazil

Figg, William D.
Molecular Pharmacology Section
Medical Oncology Branch
Centre for Cancer Research
National Cancer Institute
National Institutes of Health
Bethesda, MD 20892, USA

Griffioen, Arjan W.
Angiogenesis Laboratory
Research Institute for Growth and
    Development (GROW)
Department of Pathology
Maastricht University and
    University Hospital Maastricht
Maastricht, The Netherlands

Griffoni, Cristiana
Department of Experimental Biology
University of Bologna
Via Selmi 3
40126 Bologna, Italy

Hatziapostolou, Maria
Laboratory of Molecular Pharmacology
Department of Pharmacy
University of Patras
GR26504 Patras, Greece

Hillen, Femke
Angiogenesis Laboratory
Research Institute for Growth and
    Development (GROW)
Department of Pathology
Maastricht University and University
    Hospital Maastricht
Maastricht, The Netherlands

Hirschi, Karen K.
Baylor College of Medicine
One Baylor Plaza
Houston, TX 77030, USA

Hoffman, Robert M.
AntiCancer, Inc.
7917 Ostrow Street
San Diego, CA 92111, USA

Iorio, Rosa Anna
Department of Experimental Biology
University of Bologna
Via Selmi 3
40126 Bologna, Italy

Lang, Sven A.
Department of Surgery
University of Regensburg
93042 Regensburg, Germany

Laschke, Matthias W.
Institute for Clinical and Experimental
    Surgery
University of Saarland
D-66421 Homburg/Saar, Germany

Lee, Wen-Sen
Department of Physiology
Taipei Medical University
Taipei 110, Taiwan

Lenaz, Patrizia
Department of Experimental Biology
University of Bologna
Via Selmi 3
40126 Bologna, Italy

Melotte, Veerle
Angiogenesis Laboratory
Research Institute for Growth and
    Development (GROW)
Department of Pathology
Maastricht University and
    University Hospital Maastricht
Maastricht, The Netherlands

Menger, Michael D.
Institute for Clinical and Experimental
  Surgery
University of Saarland
D-66421 Homburg/Saar, Germany

Mikolon, David
Moores UCSD Cancer Center
University of California
San Diego
3855 Health Science Drive
MC0803, USA

Mills Shaw, Kenna R.
National Institute of Child Health and
  Human Development
National Institutes of Health
9000 Rockville Pike, Bethesda
Maryland 20892, USA

Mousa, Shaker A.
Pharmaceutical Research Institute
  at Albany College of Pharmacy
106 New Scotland Avenue
Albany, NY 12208, USA

Nash, Gerard B.
Centre for Cardiovascular
  Sciences and Department
  of Physiology
The Medical School
The University of Birmingham
Birmingham B15 2TT, UK

Nix, Melissa K.
Baylor College of Medicine
One Baylor Plaza
Houston, TX 77030, USA

Oku, Naoto
Department of Medical Biochemistry
School of Pharmaceutical Sciences
University of Shizuoka
Yada, Suruga-ku
Shizuoka 422-8526, Japan

Papadimitriou, Evangelia
Laboratory of Molecular
  Pharmacology
Department of Pharmacy
University of Patras
GR26504 Patras, Greece

Polytarchou, Christos
Laboratory of Molecular
  Pharmacology
Department of Pharmacy
University of Patras
GR26504 Patras, Greece

Rigolin, Gian Matteo
Haematology Section
Department of Biomedical Sciences
University of Ferrara
S. Anna Hospital
Corso Giovecca
203 44100 Ferrara, Italy

Santi, Spartaco
Department of Experimental
  Biology
University of Bologna
Via Selmi 3
40126 Bologna, Italy

Shan, Siqing
Department of Radiation Oncology
Duke University School of Medicine
Durham, NC 27710, USA

Smiley, Shannon
Roswell Park Cancer Institute
Elm and Carlton Sreets
Buffalo
New York 14263, USA

Smith, Ewen J.
Tumour Targeting Group
Academic Unit of Pathology
University of Sheffield
Sheffield, S10 2RX, UK

Spisni, Enzo
Department of Experimental Biology
University of Bologna
Via Selmi 3
40126 Bologna, Italy

Staton, Carolyn A.
Academic Unit of Surgical Oncology
K-Floor, Royal Hallamshire Hospital
Glossop Road, Sheffield S10 2JF

Stoeltzing, Oliver
Department of Surgery
University of Regensburg
93042 Regensburg, Germany

Strillaci, Antonio.
Department of Experimental Biology
University of Bologna
Via Selmi 3
40126 Bologna, Italy

Stupack, Dwayne G.
Department of Pathology & Moores UCSD
Cancer Center
University of California
San Diego
3855 Health Science Drive
MC0803, USA

Tomasi, Vittorio
Department of Experimental Biology
University of Bologna
Via Selmi 3
40126 Bologna, Italy

Vollmar, Brigitte
Department of Experimental Surgery
University of Rostock
Schillingallee 70
D-18055 Rostock, Germany

Weinstein, Brant, M.
National Institute of Child Health and
  Human Development
National Institutes of Health
9000 Rockville Pike
Bethesda
Maryland 20892, USA

Wong, Michael K. K.
Roswell Park Cancer Institute
Elm and Carlton Streets
Buffalo
New York 14263, USA

Yonezawa, Sei
Department of Medical Biochemistry
School of Pharmaceutical Sciences
University of Shizuoka
Yada, Suruga-ku
Shizuoka 422-8526, Japan

Zijlstra, Andries
Department of Cell Biology
The Scripps Research Institute
10550 North Torrey Pines Road
La Jolla, CA 92037, USA

1

Endothelial cell biology

Femke Hillen, Veerle Melotte, Judy R. van Beijnum and Arjan W. Griffioen

Abstract

Vascular endothelial cells are organized as a thin layer on the interior surface of all vessels and are known to function in a variety of important physiological processes. The interactions of endothelial cells with other cells and with the extracellular matrix are crucial in endothelial cell functions such as the initiation of coagulation, leukocyte adhesion and the selection of a leukocyte infiltrate, the angiogenesis cascade, and transport of molecules through the vessel wall by active or passive mechanisms. This chapter highlights these processes and describes endothelial cells, their heterogeneity, various isolation techniques and their use in *in vitro* models.

Keywords

endothelial cell morphology; endothelial cell functions; angiogenesis; isolation and culture; heterogeneity

1.1   Introduction

In 1661, Marcello Malpighi described for the first time the existence of capillaries in the mesenterium and the lung of a frog. The anatomical research of blood vessels was greatly advanced and stimulated by contributions of the pioneers in the development of microscopy, Antonie von Leeuwenhoek (1632–1723) and Jan Swammerdam (1637–1680), who developed with Friedrich Ruysch (1638–1731) the injection techniques for coloured solutions into the vessel lumen. Friedrich Gustav Jacob Henle introduced the expression 'epithelium' in 1837. Between 1841 and 1859, Henle, von Koelliker and Frey showed that the capillaries have their own wall, like a structureless skin with nuclei. A forceful discussion started about the origin, development and functions of endothelial cells, lasting until around

Angiogenesis Assays   Edited by Carolyn A. Staton, Claire Lewis and Roy Bicknell
© 2006 John Wiley & Sons, Ltd

1930. For many years the endothelium was thought of as an inert single layer of cells that passively allowed the passage of water and small molecules across the vessel wall. In the 1920s and 1930s a new area began when Lewis and Shibuya published their first results on cultivation of endothelial cells. Between 1884 and 1950, 135 papers were published dealing with various cultivation techniques for endothelial cells (Thilo-Korner and Heinrich, 1983).

1.2   Morphology of the endothelium

As a monolayer lining the entire circulatory system, the endothelial cell surface consists of about 1 to 6 × 10^{13} cells and weighs approximately 1 kg (Cines *et al.*, 1998; Sumpio *et al.*, 2002). The whole circulatory system has a common basic structure and consists of three different layers: the *tunica intima* constitutes endothelium supported by a basement membrane and delicate collagenous tissue, an intermediate muscular layer which is named the *tunica media* and an outer supporting tissue layer called the *tunica adventitia* (Gallagher, 2005).

It is currently widely recognized that endothelial cells show a remarkable heterogeneity along the vascular tree, as a biological adaptation to local needs. This heterogeneity is most obvious at the morphological level. Based on the endothelium, vessel phenotype can be classified as continuous, fenestrated or discontinuous. These phenotypes relate to the differences in permeability displayed by various vascular beds. In *continuous capillaries* endothelial cells line the full surface of the vascular wall. This vessel type is found in most tissues. In *fenestrated capillaries* the endothelial cells have small openings, called fenestrae, about 80–100 nm in diameter. Their permeability is much greater than that of continuous endothelium type capillaries and they are found in the small intestine, endocrine glands and the kidney. Fenestrae are sheltered by a small, non-membranous, permeable diaphragm, and allow the rapid passage of macromole-cules. The basement membrane of endothelial cells in fenestrated vessels is continuous over the fenestrae. *Discontinuous capillaries*, also called sinusoids, have a large lumen, many fenestrations with no diaphragm and a discontinuous or even absent basal lamina. Such vessels are found in the liver, spleen, lymph nodes, bone marrow and some endocrine glands (Cleaver and Melton, 2003; Ghitescu and Robert, 2002). Broad modulations even exist within each type of endothelium, for example, within the continuous endothelium, the extremes are the brain capillaries (with very few plasmalemmal vesicles) and the heart capillaries (rich in such vesicles) (Renkin, 1988). Beside this traditional classification, other distinguishing features are used, such as endothelial cell size or shape, orientation with respect to the direction of blood flow, complexity of inter-endothelial junctions, presence or absence of diaphragms on fenestrations and of plasmalemmal bodies, and composi-tion of the vessel wall (Cleaver and Melton, 2003; Ghitescu and Robert, 2002).

In addition to morphological heterogeneity, there is also functional heterogeneity of endothelial cells, including roles in control of vasoconstriction and vasodilatation,

blood coagulation and anticoagulation, fibrinolysis, leukocyte homing, acute inflammation and wound healing, atherogenesis, antigen presentation and catabolism of
lipoproteins.. Structural and functional diversity of endothelial cells is, as might be
expected, the result of molecular differences between endothelial cell populations.
These differences have been investigated between various populations of endothelial
cells, such as those of arteries and veins (Lawson *et al.*, 2001; Wang *et al.*, 1998;
Zhong *et al.*, 2000), large and small vessels (Muller *et al.*, 2002) and normal and
tumour vessels (Carson-Walter *et al.*, 2001; St Croix *et al.*, 2000).

In the mature vascular system, the endothelium is supported by mural cells that
express characteristics specific to their localization. The arteries and veins are
surrounded by single or multiple layers of vascular smooth muscle cells, whereas
the smallest capillaries are partially covered by solitary cells referred to as
pericytes (Gerhardt and Betsholtz, 2003). Smooth muscle cells maintain the
integrity of the vessel and provide support for the endothelium. They control
blood flow by contracting or dilating in response to specific stimuli.

Smooth muscle cells synthesize the connective tissue matrix of the vessel wall,
which is composed of elastin, collagen and proteoglycans. Like endothelial cells,
smooth muscle cells show a very low level of proliferation in the normal artery but
proliferate in response to vessel injury.

Pericytes are associated with capillaries and post-capillary venules. They provide
structural support to the endothelial cells and mediate endothelial cell function.
Pericytes constitute a heterogeneous population of cells and their ontogeny is not
well understood. Differences in pericyte morphology and distribution among
vascular beds suggest tissue-specific functions. The number of pericytes also varies
among different tissues and among vessels at different sites. Pericytes are plastic and
have the capacity to differentiate into other mesenchymal cell types, such as smooth
muscle cells, fibroblasts and osteoblasts (Jung *et al.*, 2002).

Arteries and veins

A well-known anatomical and physiological distinction between vessels is that of
arteries and veins (Carmeliet, 2003). Arterial vessels carry afferent circulation and
are exposed to the highest pressure and flow and are characteristically surrounded
by a thick medial layer consisting mostly of vascular smooth muscle cells. In
contrast, venous vessels carry efferent circulation with low pressure, have less
surrounding smooth muscle, and possess specialized structures, such as valves, to
ensure blood flow in a single direction. Although differences in fluid dynamics
within the circulation play an important role in determining the characteristic
structure of an artery or vein, recent evidence suggests that the identity of
endothelial cells lining these vessels is established before the onset of circulation
by genetic mechanisms during embryonic development (Lawson *et al.*, 2002;
Wang *et al.*, 1998). Several breakthrough discoveries have led to our current
understanding of the molecular difference between arterial and venous endothelial

cells. In 1998, the group of Anderson showed that EphrinB2 and EphB4 were specific markers for arterial and venous endothelial cells, respectively (Sato, 2003), which showed for the first time that the arterial–venous distinction had a genetic basis. Consequently, in the cardiovascular system, EphrinB2 expression is restricted to the arteries, smooth muscle cells, pericytes and mesenchyme that surround sites of vascular remodelling. EphB4 is expressed predominantly on venous and lymphatic endothelial cells (Harvey and Oliver, 2004). The difference between arteries and veins is also guided by *gridlock* (grl), an artery-restricted gene that is expressed in the lateral posterior mesoderm and acts downstream of the notch signalling pathway (see below). The gridlock gene was first described by Weinstein and Fishman in 1995 (Weinstein *et al.*, 1995; Zhong *et al.*, 2000). In 2001, the same researchers observed that the Notch signalling pathway is regulated by the earlier described gridlock gene. In mammals, four different Notch receptors (Notch 1–4) have been cloned and characterized and these receptors bind to five ligands (Jagged 1 and 2 and Delta-like 1, 3 and 4). The Notch pathway is activated when endothelial cells adopt a venous phenotype but when this pathway is inhibited by the gridlock gene, endothelial cells assume the arterial fate. Among the potential molecules that may act upstream of the Notch pathway to induce arterial differentiation is vascular endothelial growth factor (VEGF). Most recently, three independent groups discovered that VEGF act as an inducer of the arterial fate of endothelial cells (Harvey and Oliver, 2004). In zebrafish it was discovered that the sonic Hedgehog pathway, which lies upstream of VEGF, also functions in regulating the arterial fate of endothelial cells (Sato, 2003). Since their isolation in the early 1990s, members of the Hedgehog family of intercellular signalling proteins have been recognized as key mediators of many fundamental processes of embryonic development. Several studies suggest an important role for sonic Hedgehog, in particular, during blood vessel development. Recent work has shown that sonic Hedgehog can promote angiogenic blood vessel growth in part by inducing the expression of vascular endothelium growth factor, and as well as angiopoietin-1 and -2. These observations suggest that sonic Hedgehog may cooperate with vascular-specific growth factors during the development of the embryonic vasculature.

1.3   Endothelial cell adhesion and interactions

Endothelial cells have an important function in the interaction with each other and with a large variety of other cells, among which are pericytes, smooth muscle cells and leukocytes, as well as with the extracellular matrix. To accomplish these functions endothelial cells are equipped with a variety of different adhesion molecules.

Endothelial cell–cell interactions

Cell–cell-interactions are important for the regulation of tissue integrity, and the generation of barriers between different tissues and body compartments. Individual cells are anchored together through adhesion junctions, organized in three categories: tight junctions, adherens junctions and gap junctions (Bazzoni and Dejana, 2004; Dejana, 2004). The adhesion molecules that function in these structures as well as several other molecules important in cell–cell adhesion will be discussed. The intercellular interactions, mediated by these adhesion receptors, are important in the regulation of intracellular signalling.

Adherens and tight junctions both share the same binding feature. In both types of junctions, adhesion is mediated by transmembrane proteins that promote homophilic interactions and form a zipper-like structure along the cell border (Figure 1.1).

Tight junctions are responsible for regulating paracellular permeability and play a role in maintaining cell polarity by subdividing the plasma membrane into an

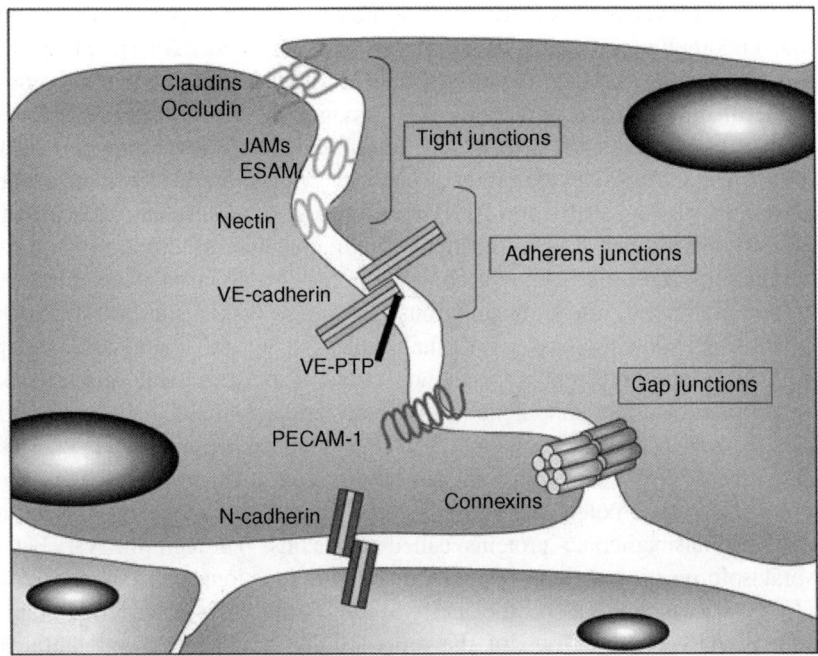

Figure 1.1 Endothelial cell–cell junctions transmembrane adhesive proteins between endothelial cells are organized in three classes. Members of the tight junctions are claudins, occludin, junctional adhesion molecules (JAMs) and endothelial cell selective adhesion molecule (ESAM). The adherens junctions are represented with adhesion molecules like vascular endothelial cadherin (VE-cadherin), which, through its extracellular domain, is associated with vascular endothelial protein tyrosine phosphatase (VE-PTP). Nectin has a role in the organization of both tight junctions and adherens junctions. Outside these junctional zipper-like molecules, platelet endothelial cell adhesion molecule (PECAM) participates to endothelial cell–cell adhesion. In endothelial cells, neuronal cadherin (N-cadherin) is connecting endothelial cells to pericytes and smooth muscle cells. Gap junctions are composed of arrays of small channels that permit small molecules to shuttle from one cell to another and thus directly link the interior of adjacent cells. Adapted from a figure by Dejana (2004). (A colour reproduction of this figure can be viewed in the colour section towards the centre of the book).

apical and a basolateral side. These structures are located at the boundary between apical and basolateral domains. The main function of tight junctions is their barrier function. The adhesion molecules that form these stuctures have a molecular architecture that is highly complex. Zonula-occludens-1 (ZO-1) was first discovered in 1986 and is perhaps the most extensively studied tight junction molecule (Dejana, 2004). Other important tight junction proteins are occludins, claudins (Schneeberger and Lynch, 2004), junctional adhesion molecules (JAMs; Keiper *et al.*, 2005) and endothelial cell selective adhesion molecule (ESAM; Hirata *et al.*, 2001).

Adherens junctions are important in the regulation of contact inhibition of cell growth, transendothelial migration of leukocytes and solutes, and in the organization of new vessels during angiogenesis (Bazzoni and Dejana, 2004). They are distributed in all blood and lymphatic vessels. Endothelial cells express an important key player in these structures, a member of the cadherin family, called vascular endothelial cadherin (VE-cadherin; Vincent *et al.*, 2004). VE-cadherin forms dimers that then undertake a second head-to-head dimerization with another VE-cadherin dimer on an adjacent cell. Through its extracellular domain, VE-cadherin is associated with a vascular endothelial protein tyrosine phosphatase (VE-PTP). The latter molecule binds through its cytoplasmatic tail to components like β-catenin, plakoglobulin and P120, that through signalling mediate cell shape and polarity. Nectin and its cytoplasmatic binding partner afadin are also present on endothelium, but little is known about their specific function. They carry out a role in both adherence and tight junctions (Takai and Nakanishi, 2003).

Gap junctions allow the passage of small molecular weight solutes and ions from cell to cell. These intercellular junctions allow direct electrical and metabolic communication between endothelial cells, between endothelial cells and smooth muscle cells and between endothelial cells and lymphocytes or monocytes (Nilius and Droogmans, 2001). Because ions can flow through them, gap junctions permit changes in membrane potential to pass from cell to cell which are constructed as a hexamer of transmembrane proteins called connexins. Through the variable use of several isoforms of connexins, there is variability in functional cell–cell interactions.

Endothelial cells have also other cell-specific homophilic adhesion proteins at the intercellular contacts. Two of the most studied are platelet endothelial cell adhesion molecule-1 (PECAM-1) and S-endo-1, both members of the immunoglobulin superfamily. The amino-terminal immunoglobulin-like domain of PECAM-1 is involved in homophilic binding on adjacent cells. Other domains of this molecule are involved in heterophilic adhesive interactions with several ligands such as $\alpha_v\beta_3$, CD38 and several proteoglycans (Jackson, 2003). S-endo-1 (also termed CD146, Mel-CAM, MCAM, MUC18 or A32 antigen) is a membrane glycoprotein involved in homophilic cell–cell interactions, but its binding partner is still unknown (Bardin *et al.*, 2001).

Another member of the cadherin family, N-cadherin, with the same type of dimerization, can be found at comparable levels to VE-cadherin in most endothelial cells. In contrast to VE-cadherin, N-cadherin is localized at the basal side of endothelial cells and is in contact with pericytes or smooth muscle cells.

Endothelial cell–matrix interactions

Maintenance of the integrity of the vessel wall is one of the most important functions accomplished through interactions between the vascular endothelium and the surrounding matrix. The sub-endothelium, a protein rich matrix underneath the endothelial cells, is crucial in the preservation of optimal endothelial cell functioning. Specific matrix ligands and receptors on the membrane contribute to the maintenance of an intact endothelial cell layer. The extracellular matrix (ECM) is organized in two layers, one of which is composed of a vascular basement membrane or basal lamina and smooth muscle cells, and the other is composed of interstitial matrix. The basement membrane consists of a network of molecules such as collagen IV, laminin, heparin sulphate proteoglycans and nidogen/entactin (Kalluri, 2003), whereas typical components of the interstitial matrix are fibrillar collagens and glycoproteins such as fibronectin (Iivanainen *et al.*, 2003). The extracellular matrix not only has a mechanical role in supporting and maintaining tissue architecture but can also be described as a dynamic structure that regulates migration, proliferation and differentiation of endothelial cells. Under normal physiological conditions in resting tissues, endothelial cells have a slow turnover and adhesive interactions with the extracellular matrix are stable. When angiogenic stimuli are present, one of the first events to occur is the production of specific proteases (matrix metalloproteinases) by endothelial cells that are capable of degrading matrix components. This causes specific molecular interactions between vascular endothelial cells and the surrounding microenvironment to change, paving the way for the formation of new blood vessels.

These interactions with the extracellular matrix occur mainly through integrins and heparan sulphate proteoglycans. Integrins are heterodimeric transmembrane proteins that consist of an α and a β subunit. There are 18 known α and eight known β subunits which form at least 24 different heterodimers in mammals. These molecules recognize ECM components and are expressed by all adhesive cells (Iivanainen *et al.*, 2003). Integrin-mediated cellular adhesion to ECM leads to intracellular signalling and modulates endothelial cell adhesion by targeting matrix degrading enzymes to the site of sprouting. For example, integrin $\alpha_v\beta_3$, the integrin that is the best characterized for its role in angiogenesis, interacts directly with MMP-2 (Brooks *et al.*, 1996). Another function of integrins is the regulation of the activity of a number of angiogenic and antiangiogenic factors, for example, $\alpha_v\beta_3$ directly associates with, and regulates the signalling of, vascular endothelial growth factor (VEGF) receptor 2. In addition, $\alpha_v\beta_3$ is induced in endothelial cells by angiogenic growth factors such as VEGF and bFGF. Other antiangiogenic molecules, such as endostatin, angiostatin and thrombospondin, that are natural components of the ECM, can also bind to $\alpha_v\beta_3$ and disrupt the endothelial cell–extracellular matrix interactions. Finally, it is known that many signalling pathways activated by integrins are also directly or indirectly activated by growth factors (Li *et al.*, 2003; Stupack and Cheresh, 2004).

A second group of endothelial receptors are the cell surface heparan sulphate proteoglycans (HSPGs) (Iivanainen *et al.*, 2003). Many matrix components have heparin binding motifs that mediate the interaction with cell surface HSPGs. This group of cell surface adhesion molecules consist of a core protein that is covalently linked to heparin sulphate-type glycosaminoglycan side-chains. There are two main HSPG gene families that are present in the membrane of cells: the syndecans and glypicans. Syndecans are transmembrane molecules that signal through various pathways by their cytoplasmic tail. Glypicans do not have a hydrophobic transmembrane or cytoplasmic domain and are anchored to the cell surface at the extracellular site by a glycosyl-phosphatidyl-inositol (GPI) anchor. This anchor gives glypicans the potential to participate in intracellular signalling. HSPGs can also contribute to signalling by interaction with other matrix receptors that anchor directly to the cytoskeleton and serve as an integrin co-receptor. Endothelial cells express syndecan-1, syndecan-4, glypican-1 and glypican-4. Other membrane glycoproteins, which carry heparan sulphate side-chains and are present on endothelial cells, are betaglycan and CD44. Syndecan-1 and 4 are known to be induced during neovascularization during wound repair (Gallo *et al.*, 1996).

In normal physiological conditions endothelial cells are quiescent and bound to the ECM. The structure of the ECM is complex and highly cross-linked, and only certain domains of the matrix components can bind to endothelial cells. Due to an angiogenic response, induced by VEGF, bFGF, PDGF and several chemokines, pericytes are detached, endothelial cells are dislodged from the blood vessels by degrading and invading through the ECM and detach from the adhesive components. The proteolytic degradation of the ECM is mediated by matrix proteinases. Their role in physiological and tumour-associated angiogenesis has been widely investigated. The best characterized enzymes, important in the degradation of both the vascular basement membrane and the underlying ECM, are the matrix metalloproteinases (MMPs) (Iivanainen *et al.*, 2003). MMPs are a family consisting of 22 members of zinc-dependent endopeptidases that can degrade ECM, cytokines, chemokines and their receptors. Based on their structure and substrate specificity, they are classified into several subgroups: collagenases, stromelysins, matrilysins, gelatinases and membrane type MMPs. Most of them are secreted as zymogens that will be activated by other MMPs or serine proteinases. After detachment of endothelial cells, MMPs can promote migration and proliferation of endothelial cells. In the initial step of angiogenesis a fibrin gel, a provisional matrix generated from fibrinogen leakage, is polymerized and endothelial cells attach to these provisional matrix components including fibrin, vitronectin, fibronectin, collagen I and thrombin.

Pro-angiogenic factors like VEGF and bFGF, produced by macrophages and tumour cells, are captured in the ECM and require matrix metalloproteinases such as MMP2 and MMP9 for mobilization of the growth factors and the initiation of angiogenesis. MMPs are predominantly secreted by stromal and immune cells. MMP-mediated degradation can be a positive and negative regulator of tumour angiogenensis (Sottile, 2004). At early degradation specific domains of matrix components like collagen, laminin and fibronectin provide pro-angiogenic signals. When degradation reaches completion, fragments like endostatin, arrestin, canstatin, tumstatin and other collagen fragments exert anti-angiogenic properties.

Endothelial cell–leukocyte interactions

Endothelial cells are also known to be of critical importance to the selection of white blood cells forming an inflammatory infiltrate. At the time of inflammation, due to either tissue injury or infection, several processes occur to facilitate the infiltration of leukocytes into the tissue. Among these are vasodilatation, increased blood flow and release of histamine and inflammatory cytokines. Due to these processes the endothelial cells become activated and interactions with leukocytes occur. The different steps in leukocyte sequestration into the surrounding tissue are tethering, rolling of the leukocyte along the vessel wall, firm adhesion to the endothelial cells, and transmigration though the vascular wall. All these sequential steps in the adhesion cascade are mediated through the intricately regulated expression of adhesion molecules (Figure 1.2) (Yadav *et al.*, 2003).

A group of adhesion molecules, important in tethering and rolling of leukocytes along the venular wall, are the selectins (Bevilacqua, 1993; Elangbam *et al.*, 1997; Gonlugur and Efeoglu, 2004; Kaila and Thomas, 2002). The term selectin was proposed to highlight the amino terminal lectin domain and the selective function and expression of these molecules. The first member of this family is E-selectin

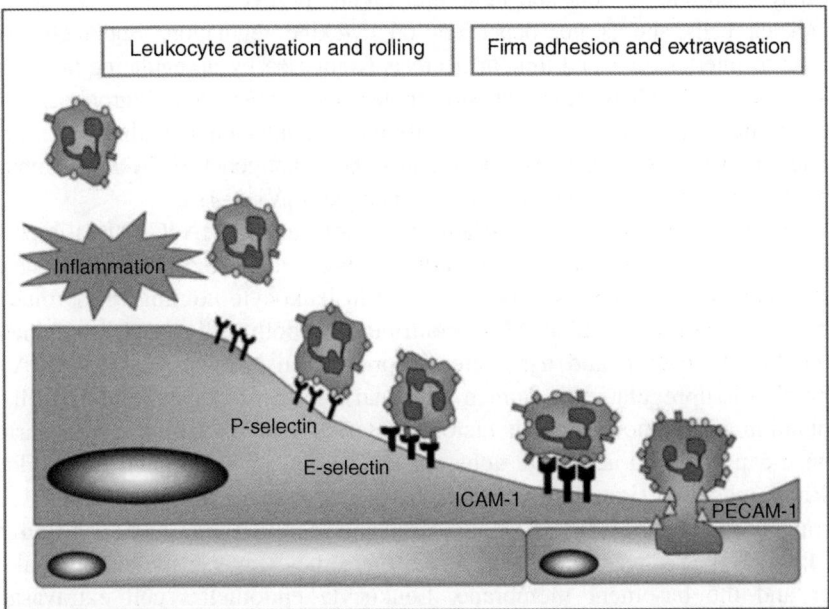

Figure 1.2 Leukocyte vessel wall interactions during vascular inflammation at the initial stages of inflammation endothelial cells become activated and the expression of selectin is upregulated. Cytokines at the site of inflammation activate leukocytes and the rolling of leukocytes occurs via interactions with endothelial cell P-selectin and E-selectin. Leukocyte VLA-4 and LFA-1 interact in the stage of firm adhesion with ICAM-1. The transmigration of leukocytes is mediated by endothelial cell PECAM-1. Adapted from a figure by Kakkar and Lefer (2004). (A colour reproduction of this figure can be viewed in the colour section towards the centre of the book).

which is only present on activated endothelium and inducible, over several hours, by interleukin-1, tumour necrosis factor and endotoxin (bacterial lipopolysaccharides). E-selectin participates in rolling of leukocytes along the endothelial surface. P-selectin is found in α-granules of platelets and Weibel-Palade bodies of endothelial cells. It is activated by thrombin, histamine and platelet-activating factor and can be rapidly distributed to the cell surface by exocytosis. It plays a role, like E-selectin, in rolling of leukocytes during immune surveillance. The last member, L-selectin, is constitutively expressed mainly on the surface of leukocytes, such as monocytes and neutrophils. E-, P- and L-selectin bind to one or more types of carbohydrate, such as sialyl Lewis, P-selectin glycoprotein ligand-1 and CD34, sulphated polysaccharides and phosphated mono- and polysaccharides (Elangbam *et al.*, 1997).

After a leukocyte has rolled along the venular surface, it adheres firmly to it. This process occurs through immunoglobulin-like cell adhesion molecules on the endothelial surface (Bevilacqua, 1993; Elangbam *et al.*, 1997; Gonlugur and Efeoglu, 2004). Intercellular adhesion molecule-1 and -2 (ICAM-1 and ICAM-2) and vascular cell adhesion molecule-1 (VCAM-1) are the key players in adhesion of leukocytes. Like E-selectin, ICAM-1 can be upregulated quickly by interleukin-1 and tumour necrosis factor over a period of several hours. ICAM-1 is expressed on a variety of cell types and may contribute to adhesion in several events. ICAM-2 is mainly expressed by endothelial cells and is not dependent on cytokine regulation. Endothelial cells respond to interleukin-1 and tumour necrosis factor also by upregulating the expression of VCAM-1. These three adhesion molecules interact with neutrophils, monocytes, lymphocytes and natural killer cells through interaction with the β2-family of integrins such as leukocyte function associated antigen-1 (LFA-1), membrane-activated complex-1 (Mac-1) and very late antigen-4 (VLA-4).

Mucosal vascular addressin cell adhesion molecule-1 (MAdCAM-1) is preferentially expressed on endothelial cells in intestinal mucosa, submucosa and Peyer's patches and plays an important role in leukocyte homing to the mucosal immune compartment. MAdCAM-1 contributes to both rolling and firm adhesion by binding L-selectin and $\alpha_4\beta_7$ integrin present on lymphocytes. MAdCAM-1 expression is upregulated on human intestinal microvascular endothelial cells by stimulation with tumour necrosis factor-α, interleukin 1-β, or lipopolysaccharides.

Once captured and activated, adherent leukocytes are stimulated to de-adhere, by RhoA-mediated signalling delivered by the cytoplasmatic tail of ICAM-1, and migrate through the vessel wall (Nourshargh and Marelli-Berg, 2005; van Buul and Hordijk, 2004; Yadav *et al.*, 2003). From there they traverse the endothelial cell layer and the basement membrane. Leukocyte endothelial cell extravasation mostly takes place through small gaps at intercellular endothelial junctions, the paracellular route, and sometimes through the endothelial cells themselves, the transcellular pathway, a mechanism which is still obscure (Engelhardt and Wolburg, 2004). In the paracellular route, leukocytes prefer to migrate through sites where the junctional complexes are less tight. Molecules like VE-cadherin, platelet-endothelial cell adhesion molecule-1, CD99, ICAM-2 and members of the junctional adhesion molecule (JAM) family (Ebnet *et al.*, 2004; Muller, 2003), of which the precise involvement is still unknown, guide the leukocytes through the endothelial barrier. The process of transmigration seems to prepare the leukocytes for migration in the extravascular tissue (Yadav *et al.*, 2003).

1.4 Coagulation and haemostasis

Since endothelial cells are in direct contact with the blood, they play an important role in haemostasis and blood coagulation (Guyton, 2005). Antithrombotic, prothrombotic and fibrinolytic activities of endothelial cells are discussed here.

Prothrombotic activities of endothelial cells

The initial event in wound healing after vascular injury is the attraction, adherence and aggregation of circulating platelets. Fibrinogen and vWF are the major adhesive components that form bridges between different platelets. P-selectin, stored in the α-granules of platelets and the Weible-Palade bodies of endothelial cells, is rapidly expressed, by exocytosis, on the cell surface of platelets and endothelial cells and will contribute to adhesion of platelets to endothelial cells and leukocytes (Geng, 2003). Once aggregated in the platelet plug, platelets release stored growth factors such as VEGF, PDGF, TGF-β and such cytokines as PF4 and CD40L that induce proliferation, differentiation and migration of endothelial cells and smooth muscle cells (Anitua *et al.*, 2004; Rhee *et al.*, 2004). The contribution of platelets to angiogenesis *in vitro* is due to VEGF and not due to bFGF (Verheul *et al.*, 2000).

Tissue factor (TF), expressed by endothelial cells, is known as the primary cellular initiator of blood coagulation (Figure 1.3). It can upregulate the balance between proangiogenic and antiangiogenic factors by playing a key role in the regulation of VEGF and thrombospondin (Zhang *et al.*, 1994). The expression of TF can be upregulated by bFGF (Kaneko *et al.*, 2003).

Antithrombotic activities of endothelial cells

Thrombomodulin (TM) is a high affinity thrombin receptor that is produced by endothelial cells and is the key player in initiating processes of anticoagulation (Figure 1.3). TM converts thrombin from a procoagulant protease into an anticoagulant and slows down the clotting process. The thrombin/TM complex plays a role as a cofactor in the activation of the zymogen protein C (PC) into activated protein C (APC) (Sadler, 1997). This leads to inactivation of factor Va and thrombin, and results in fibrinolysis induced by plasmin. TM is also expressed by other cells types such as keratinocytes, osteoblasts and macrophages (Boffa and Karmochkine, 1998).

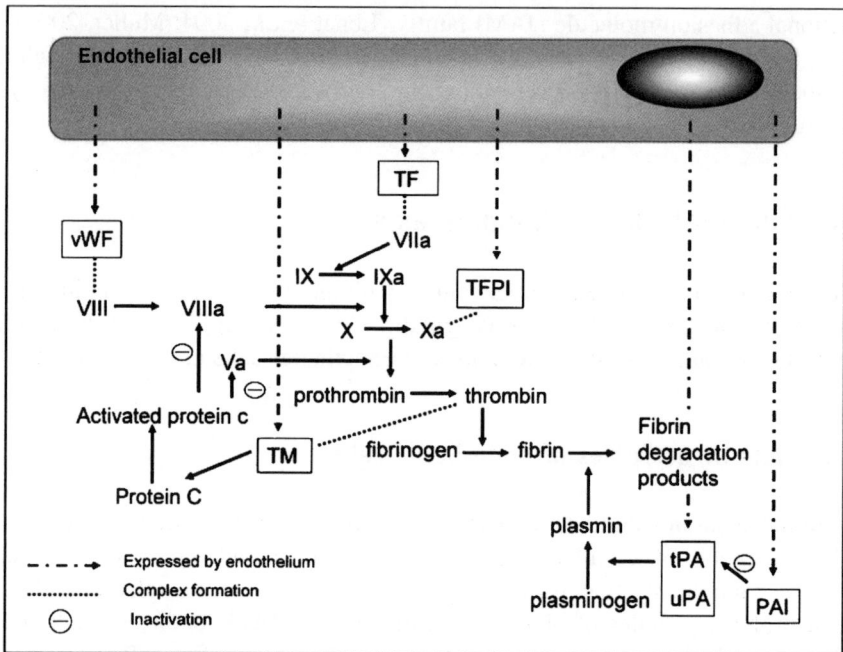

Figure 1.3 Simplified representation of coagulation and fibrinolytic processes endothelial cells produce different molecules important in the pathway of coagulation (white part in the schema) and fibrinolysis (highllighted part in the schema). Tissue factor (TF), tissue factor pathway inhibitor (TFPI) and von Willebrand factor (vWF) are important in coagulation, while thrombomodulin (TM), tissue-type plasminogen activator (tPA), urokinase-type plasminogen activator (uPA) and plasminogen activator inhibitor (PAI) play a role in fibrinolysis.

The inhibitor of TF, tissue factor pathway inhibitor, exists in two forms in humans, TFPI-1 and TFPI-2. These are expressed by endothelial cells and have a major physiological inhibitor function of the coagulant pathway (Price *et al.*, 2004).

Fibrinolytic activity of endothelial cells

A fibrin matrix is formed after wounding of a blood vessel and after leakage of plasma from blood vessels in areas of inflammation and tumourigenesis. The fibrin matrix acts as a barrier preventing further blood leakage and it provides a structure for new microvessels to infiltrate in damaged or activated tissue (Dvorak, 2000). Whenever a blood clot is formed, a large amount of plasminogen is trapped in the clot along with other plasma proteins (Guyton, 2005). Endothelial cells excrete certain serine proteases, plasminogen activators, which convert the zymogen plasminogen into an active protease plasmin, which in turn degrades fibrin into soluble fibrin degradation products. Endothelial cells are able to produce and control two different types of plasminogen activators, tissue-type plasminogen activator (t-PA) and urokinase-type plaminogen activator (u-PA). Both serine proteases are involved in the conversion of the inactive plasminogen to the active plasmin. Plasminogen activators do not only contribute to the formation of new microvessels by degradation of fibrin, but also have a crucial role in the prevention of thrombosis. To prevent fibrin clots in the lumen of a blood vessel, and subsequently ischaemia and eventually death in surrounding tissue, endothelial cells produce anticoagulant molecules. Inhibition of the fibrinolytic system may occur either at the level of the PA, by specific plasminogen activator inhibitors (PAI-1 and PAI-2), or at the level of plasmin by α_2-antiplasmin. Inhibition of platelet aggregation and stimulation of fibrinolysis, however, should be area and time limited, to prevent recurrent bleeding. Molecular interactions between the fibrinolytic system and matrix metalloproteinases (MMPs) systems have been recognized at different levels, suggesting that both systems play a crucial role in extracellular matrix degradation or remodelling.

1.5 Transport

Endothelial cells play an important role in the transport of molecules across the vascular wall. Several pathways for molecular transport are described, among which are transcellular transport (through cells) via caveolae and their generated channels and paracellular transport (between cells) via intercellular (inter-endothelial cell) junctions (Simionescu *et al.*, 2002).

Caveolae

Endothelial cell vesicles were first described by Palade in 1953 in the capillaries of the heart, after which they have been identified in a variety of cells and tissues. Endothelial cells are among the richest in caveolae. There are clear variations in their surface density between the known morphological types of endothelium; the largest population of caveolae are present in the continuous endothelium, while their numbers are much lower in the fenestrated endothelium and sporadic in discontinuous endothelium (Stan, 2002). These plasmalemmal vesicles take part in a number of functions; the bulk transport of large (and even small) solutes across the endothelial barrier by transcytosis, endocytosis, potocytosis, signal transduction and control of cholesterol trafficking. The specific marker and major component of caveolae is caveolin-1, an integral membrane protein (20–22 kDa). It is a member of a multigene family of caveolin-related proteins, which show similarities in structure but differ in properties and distribution. In addition, single caveolae from opposite fronts could fuse and form occasional transendothelial channels that allow the passage of small molecules across the cell (Simionescu *et al.*, 1975).

It was postulated that caveolae could also fuse and form complex clusters that are open at both fronts of the endothelial cell, the so-called vesiculo-vacuolar organelles or VVOs (Feng *et al.*, 1996; Dvorak and Feng, 2001; Feng *et al.*, 2002).

VVOs span the entire thickness of vascular endothelium thereby providing a potential trans-endothelial connection between the vascular lumen and the extravascular space. They are large collections of vesicles and vacuoles focally expressed in the microvasculature that accompanies tumours, in venules associated with allergic inflammation and in the endothelia of normal venules, often in a parajunctional location. Structures of endothelial junctions vary along the vascular segments. Intercellular junctions are specialized regions of the plasma membrane that are organized when two cells come into contact. In the endothelium, junctional complexes comprise tight junctions, adherens junctions and gap junctions (Bazzoni and Dejana, 2004). Gap functions are communications structures, which allow the passage of small molecular weight solutes between neighbouring cells (Bazzoni and Dejana, 2004).

It has been postulated that vesicles, although morphologically identical, characterize at least two functionally distinct entities: endocytic vesicles and transcytotic vesicles. Endocytosis is the uptake of plasma proteins and molecules by the endothelial cell for its own use and transcytosis refers to the transport of plasma proteins to the subjacent cells and tissues.

Endocytosis in endothelial cells occurs either by a specific mechanism (receptor-mediated) or by a non-specific process (fluid phase or adsorptive). Fluid phase endocytosis is a process in which molecules ingested in bulk by caveolae are delivered to endosomes, multivescular bodies and lysosomes where the degradation occurs. Adsorptive endocytosis is specific for molecules that bind by electrostatic forces to the cell surface and their microdomains. Receptor mediated endocytosis involves internalization by specific binding sites localized in coated pits and coated vesicles that direct the molecules to endosomes, to be further degraded in lysosomes or to reach multivesicular bodies.

The transcellular transfer of molecules between compartments is named transcytosis and is a basic process shared by epithelial cells and endothelial cells (Simionescu *et al.*, 1975). Transcytosis can also occur in two kind of processes; non-specific (fluid phase and non-specific adsorptive transcytosis) and specific (receptor-mediated). In fluid phase transcytosis, a fraction of plasmalemmal vesicles take up plasma and then transfer through the cytoplasm to reach the abluminal front where they apparently discharge their content. Non-specific adsorptive transyctosis implies an electrostatic attachment of permeant molecule to a vesicle carrier, which translocates the probe across the endothelial cell. For this kind of transcytosis, the number and affinity of binding sites is the deciding factor. Some molecules make use of the fluid phase and adsorptive transcytosis in one process (Simionescu *et al.*, 2002).

1.6 Angiogenesis

Angiogenesis is the process of new blood vessel formation from pre-existing vasculature. Growth of blood vessels during embryogenesis occurs mainly through vasculogenesis and angiogenesis (Carmeliet, 2003). Vasculogenesis involves the *de novo* differentiation of endothelial cells from stem cells, called angioblasts, which assemble into a primary capillary plexus. This primitive early network differentiates by angiogenesis, where new capillaries arise through sprouting, branching and intussusceptive growth from pre-existing capillaries. Arteriogenesis refers to subsequent stabilization of these sprouts by mural cells and remodelling into larger blood vessels.

In adults, angiogenesis is stimulated during many physiological processes, such as wound healing, tissue remodelling/regeneration and the female reproductive cycle. Angiogenesis is also of critical importance in a large variety of pathologies such as rheumatoid arthritis, atherosclerosis and cancer. The field of angiogenesis started to gain real attention when it was hypothesized that 1) tumours are most vulnerable at the level of their blood supply and 2) because tumours are dependent on angiogenesis, blocking angiogenesis would be a tool for therapeutic intervention (Folkman, 1971).

The process of angiogenesis is tightly regulated by positive and negative regulators. The balance between both determines the level of ongoing angiogenesis. A number of angiogenic growth factors and inhibitors are discussed below.

Pro-angiogenic factors

The first growth factor, important in angiogenesis, was discovered in the 1980s. The discovery of basic fibroblast growth factor (bFGF), also known as FGF-2, confirmed Folkman's theory that tumours are dependent on angiogenesis (Folkman and Shing, 1992; Kerbel, 2000). It is a pleiotropic mitogen for growth and differentiation, affecting various mammalian cells, among which are endothelial cells, and organ systems. bFGF stimulates all major steps in the angiogenesis cascade and is involved in wound healing, differentiation, proliferation, haematopoiesis, motility and apoptosis, neurothropic activity, vasculogenesis, migration, adhesion, metastasis and tumour angiogenesis. bFGF belongs to a larger family of 23 members and is one of the main stimulators in angiogenesis (Botta *et al.*, 2000; Okada-Ban *et al.*, 2000; Reuss and von Bohlen, 2003). bFGF signalling occurs through four high affinity tyrosine kinase transmembrane receptors (FGFR1-4). It is produced by many cells, including macrophages and tumour cells, and subsequently secreted into the extracellular matrix. At the start of angiogenesis, it can be released from the matrix. bFGF expression can be downregulated by interferon-α and -β. There is an intricate interaction with other growth factors, such as VEGF, that results in a synergistic action in many endothelial cell functions.

VEGF is another important player in the stimulation of angiogenesis. VEGF, or vascular permeability factor, induces vasodilatation of the existing vessels and increases permeability of the vessel wall. VEGF is also a general activator of endothelial cell proliferation and mobility (Ferrara *et al.*, 2003). The main member of the VEGF family of growth factors is VEGF-A. Exon splicing of the human

VEGF-A gene results in seven isoforms that have different properties. The three major isoforms, named after their molecular size, are VEGF121, VEGF165 and VEGF189.

The VEGF family also includes VEGF-B, VEGF-C, VEGF-D and placental growth factor (PlGF). VEGF-B is an angiogenic protein related to VEGF-A, although it acts on a different set of tyrosine kinase receptors. VEGF-C and VEGF-D have been shown to act as lymphangiogenic growth factors. PlGF stimulates angiogenesis and collateral growth in ischaemic heart and limb with a comparable efficiency to VEGF (Autiero *et al.*, 2003). VEGF signals through two main tyrosine kinase receptors, VEGFR1 or Flt-1, and VEGFR2 or KDR (McColl *et al.*, 2004). It has been demonstrated that stimulation of angiogenesis is mainly mediated by signalling through VEGFR2. Endothelial cells also express cell surface VEGF binding sites that can mediate angiogenesis, neuropilin 1 and neuropilin 2, different from the two tyrosine kinase receptors. Like VEGF, PlGF has a role in angiogenic switch by interacting with VEGFR1 and in that way synergizes with VEGF (Autiero *et al.*, 2003).

The angiopoietin (Ang) family includes four members that all bind to the endothelial tyrosine kinase receptor Tie-2 (Jones, 2003). The most remarkable characteristic of this family is the opposing effect of the different ligands binding to the receptor. Ang-1 and Ang-4 can activate the Tie-2 receptor while Ang-2 inhibits the Ang-1-induced Tie-2 phosphorylation. Ang-1, via phosphorylation of Tie-2, is chemotactic for endothelial cells and is involved in embryonic vascular remodelling and promotes cell survival, sprouting and tube formation and reduces inflammation. In adults, it is involved in maturation and stabilization of mature vessels. Ang-2 is involved in postnatal angiogenic and vascular remodelling events and in detaching smooth muscle cells and loosening underlying matrix, thereby allowing endothelial cells to migrate as inter-endothelial cell contacts are relaxed (Thurston, 2003). Ang-2, in concert with VEGF, is also angiogenic, although in the absence of VEGF, Ang-2 may actually induce vessel regression.

Transforming growth factor (TGF) is a multifunctional cytokine involved in proliferation, differentiation, migration and survival of many cell types (Sun, 2004). Three isoforms exist with overlapping and distinct functions. Signalling occurs via one type II receptor and two type I receptors (TGF receptors). Overexpression of TGF-β results in fibrotic conditions, since TGF-β promotes accumulation of extracellular matrix molecules. At the initial stage of tumour growth, TGF-β has a tumour suppressor function due to its inhibition of cell growth. In later stages, it enhances tumour growth by suppressing the immune system and enhancing the cell cycle thereby promoting angiogenesis.

The Eph-B receptors and their ligands are another example of pro-angiogenic molecules. Eprhins can be divided into two subclasses: EprhinA ligands (ErphinA1-A5) are tethered to the cell surface via a GPI-anchor, whereas EphrinB ligands (ephrinB1-B3) are inserted into the plasma membrane via a transmembrane region followed by a conserved cytoplasmic domain. These ligands and their receptors (Eph tyrosine kinase receptors) regulate axon guidance and bundling in the development of the brain, control cell migration and adhesion. The Ephrin-Eph system functions in cell-to-cell rather than long range communications, because Eph receptors and all known Ephrin ligands are attached to the plasma membrane. As described earlier, the Eprhin-B2 ligand was located strictly on arterial endothelial cells, while the corresponding Eph-B4 receptor marked only venous endothelial cells (Patan, 2000).

Different interactions between VEGF, angiopoietin and Eph/ephrin have been described. The most recently identified genes that contribute to these processes are retinaldehyde dehydrogenase 2 (Raldh2), Norrin, Frizzled-4 (Fz4) and Noge-B (Harvey and Oliver, 2004). Generation of Raldh2-null mice indicate that retinaldehyde (RA) plays a crucial role in mammalian vascular development and that it is required to control endothelial cell proliferation and vascular remodelling during vasculogenesis. The Raldh-2 null mice were unable to produce active RA in the embryo, failed to remodel the primary vascular plexus and did not recruit supporting mural cells to developing vessels (Bohnsack *et al.*, 2004; Lai *et al.*, 2003). The Nogo isoforms A, B and C are members of the reticulon family of proteins. Nogo-A and Nogo C are highly expressed in the central nervous system, whereas Nogo B is found in most tissues. Sessa and co-workers described Nogo-B as a regulator of vascular function. Their study shows that NogoB protein is highly enriched in blood vessels, and that the genetic loss of Nogo-A/B expression markedly augments the injury response in femoral arteries (Acevedo *et al.*, 2004). Xu and colleagues defined a Norrin and Fz4 signalling system that plays a central role in vascular development in the eye and ear (Xu *et al.*, 2004).

Inhibitors of angiogenesis

Angiogenesis is also regulated by endogenous inhibitors. Without trying to be exhaustive, several of these molecules are reviewed here. Platelet factor-4 and interferon-α are among the first endogenous molecules described to display inhibitory activity on endothelial cells (Kolber *et al.*, 1995; Gupta *et al.*, 1995). In later years, more endogenous molecules with angiostatic activity were described. Among these were thrombospondin-1 (Rastinejad *et al.*, 1989; Good *et al.*, 1990; Grossfeld *et al.*, 1997), interferon-γ inducible protein-10 (Luster *et al.*, 1995). Other members of this class of endogenously produced anti-angiogenic proteins are angiostatin (O'Reilly *et al.*, 1994), endostatin (O'Reilly *et al.*, 1997), and Bactericidal Permeability Increasing protein (van der Schaft *et al.*, 2002). It is interesting to note that many of these molecules are proteolytic fragments of collagens and other macromolecules. The scientific interest in these molecules is enormous since it is recognized that these molecules may have application in the treatment of diseases that are dependent on angiogenesis. Although receptors have been described for several angiogenesis inhibitors, detailed mechanisms of action have not yet been elucidated.

Angiogenesis in different physiological and pathological processes

The most highly investigated physiological processes in which angiogenesis is crucial are wound healing and the female reproductive cycle. The dynamic mechanism of angiogenesis, in response to tissue injury and subsequent wound healing, is regulated by serum and extracellular matrix components. The most important angiogenic cytokines in wound repair are VEGF, angiopoietins, bFGF and TFG-β (Li et al., 2003). The role of angiogenesis in the reproductive cycle is extensively reviewed in a special issue of *Angiogenesis* (volume 8, issue 2, 2005).

In diabetes mellitus altered regulation of angiogenesis is the cause of many pathologies (Martin et al., 2003). Vasculopathies in the retina and kidney are found, and situations like impaired wound healing, increased risk of rejection of transplanted organs and impaired formation of coronary collaterals can be observed in these patients. VEGF is the main player in diabetic retinopathy and its expression can be increased by several factors including bFGF, PlGF, TNF, TGF-β, IL1. In diabetic nephropathy with aberrant angiogenesis, abnormal local levels of VEGF were detected and angiotensin II expression was increased.

In rheumatoid arthritis and other inflammatory disease, the formation of new vessels plays an important role (Bodolay et al., 2002; Koch, 2003). The synovium of rheumatoid arthritis is known to have a highly vascularized phenotype. At the site of inflammation, leukocytes will undergo extravasation and attack the synovium of various joints. The delivery of an excess of angiogenic factors will cause synovitis. As in the previously described pathologies, the same angiogenic factors (VEGF, bFGF, TGF-β, TNF-α, IL1, angiopoetins) have been described to play an important role in this pathology.

Another research field that has attracted interest is the therapeutic neovascularization in cardiovascular ischaemic regions (Cao et al., 2005). During hypoxia or ischaemia cascade starts with the activation of the HIF-pathway that in turn will increase erythropoietin and VEGF expression which will lead, finally, to an increased blood flow and oxygen delivery.

Tumour angiogenesis

Although many pathologies are associated with changes in angiogenesis regulation (in diabetes, Martin et al., 2003; rheumatoid arthritis, Bodolay et al., 2002; Koch, 2003; and cardiovascular ischaemic conditions, Cao et al., 2005), the field of angiogenesis really developed from the oncological arena. It was realized that tumours can only grow to a volume of $1-2$ mm^3 without the attraction/formation of new blood vessels (Folkman, 1971; Folkman, 1995). Tumours of $1-2$ mm^3 subsequently switch to an angiogenic phenotype and recruit blood vessels from the surrounding stroma. Angiogenesis is therefore pivotal to tumour formation, and can serve as a prognostic factor for cancer.

Several mechanisms of neovascularization have been identified. Among these are sprouting angiogenesis which involves the proliferation and migration of endothelial cells from pre-existing blood vessels and the organization into tubular vascular structures. Intussusceptive angiogenesis (Burri *et al.*, 2004), the formation of transvascular tissue pillars into the lumen of a pre-existing vessel and subsequent 'splitting' of the vessels in two new vessels, was first observed in postnatal remodelling of capillaries in lungs. New vessels can also grow through the use of circulating endothelial progenitor cells (EPCs). Extensive data now support the existence of these cells and their contribution to the formation of new blood vessels (Hristov and Weber, 2004; Ribatti, 2004). Currently, work is also focusing on an alternative tumour blood supply known as vasculogenic mimicry. Vasculogenic mimicry is the process in which tumour cells provide themselves a secondary circulation system of vasculogenic structures lined by the tumour cells themselves. This process is independent of angiogenesis. It is hypothesized that tumour cells acquire the capacity to form a tubular circulatory system by de-differentiation and acquisition of endothelial characteristics. The concept of vasculogenic mimicry was first described in melanomas (Maniotis *et al.*, 1999). The critics on this first publication were abundant (Bissell, 1999; McDonald and Foss, 2000; McDonald *et al.*, 2000; Shubik and Warren, 2000) because evidence that these patterned networks contributed to tumour blood flow appeared to be lacking. Nowadays several research groups are seeking the molecular mechanisms behind vasculogenic mimicry and proof of its contribution to blood flow.

Endothelial cells also contribute to the formation of lymph vessels. Lymphatic vasculature develops shortly after blood vessels during embryogenesis and might have the same origin (Al-Rawi *et al.*, 2005; Alitalo and Carmeliet, 2002). It is hypothesized that venous endothelial cells become responsive to lymphatic signals, differentiate and make lymphatic sprouts. However, the existence of lymphangioblasts or precursors has also been proven (Schneider *et al.*, 1999). Lymphangiogenesis is regulated by VEGF-C and VEGF-D and its receptor VEGFR3. Since the availability of antibodies specifically recognizing lymph vessels, such as antibodies to Prox-1 (Wigle *et al.*, 2002; Wilting *et al.*, 2002), VEGFR-3 (Makinen *et al.*, 2001; Jeltsch *et al.*, 1997; Saharinen and Petrova, 2004), podoplanin (Breiteneder-Geleff *et al.*, 1999) and LYVE-1 (Banerji *et al.*, 1999), this area of research has been accelerated. The involvement of lymphan-giogenesis in tumour growth and metastasis formation (Alitalo and Carmeliet, 2002; Saharinen *et al.*, 2004) is currently under investigation.

1.7 Isolation and culture of endothelial cells

Endothelial cells are extensively used in culture to study diverse aspects of endothelial cell biology and angiogenesis. Among the most extensively used endothelial cells of human origin are human umbilical vein endothelial cells (HUVEC) that can easily be isolated by perfusion of the umbilical vein with trypsin and have been successfully cultured since 1973 (Jaffe *et al.*, 1973). However, a major drawback of primary endothelial cells is their limited lifespan in culture before they enter senescence. By immortalization, a number of endothelial cell lines have been obtained that can be cultured for prolonged periods of time, such as HMEC-1 (reviewed by Bouis *et al.*, 2001). It has been shown that endothelial cells isolated from fresh tissues retain their phenotype in culture for some time (Bussolati *et al.*, 2003; Folkman *et al.*, 1979; Griffioen, 1997; Kallmann *et al.*, 2002), allowing for detailed analysis of their functional properties. However, over time their phenotype may change as a consequence of the altered microenvironment in culture (Favre *et al.*, 2003). Therefore, depending on the research goal, it may be desirable to use endothelial cells isolated directly from the relevant tissue. For example, gene expression profiles of specific subsets of endothelial cells, such as tumour endothelial cells, were obtained from pure populations of endothelial cells (Madden *et al.*, 2004; van Beijnum and Griffioen, 2005).

It has been recognized that there is considerable heterogeneity in endothelial cells derived from different locations in the body, which is related to the microenvironmental factors acting in each different organ and the specialized function for the endothelium therein (McCarthy *et al.*, 1991; Zetter, 1988). Therefore, the use of capillary endothelial cells derived from the condition under study is preferred over the use of more distant endothelial cell types. Isolating capillary endothelial cells from tissues generates challenges related to the relative limited percentage of endothelial cells present in tissues and the impossibility of canulation of the vessels, unlike larger vessels.

A number of different techniques can be employed to isolate endothelial cells from tissues for subsequent culture or analysis. In general, the isolation of capillary endothelial cells starts with the mechanical disruption of the tissue, usually followed by enzymatic digestion to detach cells from each other and the matrix. Depending on the type of tissue, enzyme concentrations and incubation times may vary (Manconi *et al.*, 2000; Scott and Bicknell, 1993). The most commonly used enzymes are collagenase (Bussolati *et al.*, 2003; Fawcett *et al.*, 1991; Folkman *et al.*, 1979), trypsin and dispase or combinations thereof (Marelli-Berg *et al.*, 2000). Merely plating the thus created single-cell suspensions will yield many impurities that in time may overgrow the endothelial cells. Hence, purification of the endothelial cells is generally required. First, this can be performed manually by weeding the culture to remove contaminant colonies, e.g. using Pasteur pipettes or cell scrapers (Chung-Welch *et al.*, 1988; Folkman *et al.*, 1979). In addition, nylon mesh filters can be applied to minimize the number of contaminant cells seeded initially (Richard *et al.*, 1998). Furthermore, gradient density centrifugation may be applied to separate microvascular segments from non-endothelial cells such as fibroblasts and epithelial cells (Grimwood *et al.*, 1995).

Based on the expression of specific markers, endothelial cells can be separated from contaminating cell types by using immunomagnetic separation techniques. This technique uses antibodies coupled to paramagnetic particles or beads that are incubated with the heterogeneous cell samples to capture the endothelial cells. By placing the samples in a magnetic field, the captured endothelial cells can be separated from contaminating cells. Alternatively, endothelial cells can be purified by using fluorescence activated cell sorting (FACS) following antibody or ligand binding. The most commonly used antigens are CD31 (Demeule *et al.*, 2001; McGinn *et al.*, 2004), factor VIII-related antigen (Patel *et al.*, 2003), CD36 (Giordano and Mitola, 2000), CD105 (Bussolati *et al.*, 2003), CD133 (Gussin *et al.*, 2004), VCAM-1 (Gerritsen *et al.*, 1995) and CD146 (St Croix *et al.*, 2000). Also frequently used are beads coated with the lectin *Ulex europaeus* agglutinin-1 (UEA-1) (Hewett and Murray, 1993) that binds specific glycoproteins on endothelial cells.

A number of properties can be used to characterize the endothelial cells after isolation and purification. The expression of endothelial markers such as CD31, CD34, factor VIII-related antigen, angiotensin-converting enzyme (ACE), as well as binding to UEA-1 can be examined. Furthermore, the response of endothelial cells to various cytokines, such as tumour necrosis factor-α, can be assessed to validate the culture (Hewett and Murray, 1993). Another commonly used method is monitoring the ability of endothelial cells to take up a fluorescent derivative of acetylated low density lipoprotein, Dil-Ac-LDL by scavenger receptor (Fawcett *et al.*, 1991).

Establishing a long-term culture of endothelial cells isolated from tissues remains a challenge. Cells derived from different tissues may require different culture conditions and supplementary growth factors (Relou *et al.*, 1998; Scott and Bicknell, 1993). These endothelial cells usually require the presence of human serum for survival. In addition, an artificial matrix, such as collagen, fibronectin or gelatin, needs to be provided in the culture vessel to which the endothelial cells may adhere (Terramani *et al.*, 2000; Relou *et al.*, 1998; Folkman *et al.*, 1979). To date, diverse cultures of endothelial cells have been established that can be used to study endothelial cell biology related to, for example, angiogenesis, thrombosis and haemostasis.

1.8   Endothelial cell heterogeneity and organ specificity

The endothelium exhibits a remarkable diversity of cellular properties that are adapted to the needs of the underlying tissue (Aird, 2003). Even within one organ, endothelial cells display heterogeneity.

At the structural level, endothelial cells at different locations in the vasculature exhibit differences in size, shape, thickness and endothelial cell junctions (Aird, 2003). Between tissues, there is also considerable difference in continuity of the endothelial lining as described above in the adhesion section. Endothelial cell heterogeneity may be explained by variations in the microenvironment ('nurture') as well as by genetic predisposition ('nature') (Aird, 2004). Interactions between endothelial cells and their precise microenvironments are thought to play an important role in determining the phenotype of the endothelial cell. Depending on the tissue or organ, different soluble mediators may be present that act on the endothelial cells and endothelial cells may be in contact with different cell types or matrix components. In addition, haemodynamic factors may also govern changes in endothelium.

Embryonic endothelial cells are very 'plastic' and most of the highly specialized characteristics of endothelial cells are induced during development, whereas adult endothelium is more stable, though still capable of responding to extrinsic factors.

Endothelial cells isolated from particular sites in the body that are subsequently cultured may lose some of their tissue-specific characteristics (Ribatti *et al.*, 2002). For example, blood–brain barrier endothelial cells that specifically express P-glycoprotein lose this expression *in vitro*. When cultured in the presence of brain matrix or astrocytes, the expression is restored. However, matrix derived from other organs is not capable of stimulating this expression (Aird, 2003). Aortic endothelial cells, when cultured on lung extracellular matrix components, express lung-specific endothelial adhesion molecule (Zhu *et al.*, 1991).

Though highly influenced by its surroundings, the endothelium may retain some of its original properties *in vitro*, in the absence of its normal microenvironment (Kallmann *et al.*, 2002). Endothelial markers exist that show constitutive expression throughout the body, such as CD31 and von Willebrand factor.

The molecular diversity of the endothelium has been explored by using phage display technology to identify organ- and disease-specific protein expression on the surface of endothelial cells. This approach has provided clues of a vascular address system that can be used for organ-specific targeting of normal blood vessels or disease related targeting of blood vessels, such as in tumours (Pasqualini *et al.*, 2002). Differences exist between endothelial cells present in tumours compared with endothelial cells in healthy tissue counterparts (Parker *et al.*, 2004). Furthermore, additional heterogeneity may exist between different tumours and even within tumours (Pasqualini *et al.*, 2002).

In *vivo* phage display offers an unbiased approach to identify surface markers on endothelial cells. Random peptides displayed on filamentous phage have been injected in mice and recovered from different organs and malignancies where they interacted with endothelial surface molecules. Peptides specific for brain vasculature have been identified as well as peptides that home to lung, pancreas, skin, intestine, uterus, adrenal gland, retina and prostate (Rajotte *et al.*, 1998) and tumour vasculature (Pasqualini *et al.*, 2002).

The identification of the molecule targeted by the peptide displayed on phage remains a challenging task, however. The famous RGD sequence targets αV integrins in tumour vessels (Pasqualini *et al.*, 1997) and the NGR sequence was shown to target aminopeptidase N or CD13 (Pasqualini *et al.*, 2000). Diverse preclinical studies have shown the targeting potential of these peptides. Using these peptides as targeting moieties shows promise for therapeutic applications, in particular in vascular targeting of tumours. Coupling of drugs to tumour vasculature specific peptides can increase the selective effect of the drug and reduce side effects. Even without conjugation the peptide motifs may be active *in vivo*. A peptide inhibitor of hyaluronan inhibited inflammation-induced leukocyte migration in mice (Mummert *et al.*, 2000). Similarly, selective peptide inhibitors of MMP-2 and MMP-9 localized to mouse tumours and prevented tumour growth (Koivunen *et al.*, 1999).

Overall, mapping the zip codes of the vasculature is a daunting task, but will offer great potential, both for therapeutic applications and for a more general understanding of the endothelial heterogeneity. In addition, it may shed light on preferred metastasis of tumour cells to specific organs.

1.9 Gene expression in endothelial cells

Endothelial cells in a particular microenvironment interact with various types of cells, such as fibroblasts, pericytes, immune cells, tumour cells or organ cells. In addition, endothelial cells can interact with various ECM components and are exposed to different growth factors and cytokines released by its neighbours (Jung *et al.*, 2002). As a result, endothelial cells lining different vessels in different tissues exhibit morphological and functional specializations that are reflected in their global gene expression profiles (Ho *et al.*, 2003; Chi *et al.*, 2003). A gene expression comparison of cerebral endothelial cells (HCEC), part of the blood–brain barrier, with HUVEC revealed that HCEC are characterized by the expression of genes associated with neuroprotection and growth support. Marked differences in growth factor protein release between HCEC and HUVEC in culture were observed that supported the gene expression data (Kallmann *et al.*, 2002).

Endothelial specific gene expression

It has long been recognized that specific growth factors and cytokines act on endothelial cells by interaction with specific receptors on the endothelium. There are different vascular endothelial growth factor (VEGF) isoforms as well receptor isoforms (reviewed by Ferrara, 2004), that are expressed in endothelial cells. More recently, a tissue-specific angiogenic growth factor, endocrine gland derived vascular endothelial growth factor (EG-VEGF), was identified that is selectively expressed in steroidogenic glands and promotes growth of the endocrine gland endothelium (LeCouter *et al.*, 2001).

Large-scale gene expression surveys have shed light on transcriptional differences between diverse types of cells. In addition, public data sets are available to mine expression data generated by various methods such as SAGE and microarray analysis. Using a combination of these approaches, endothelial cell specific genes were identified (Ho *et al.*, 2003). A total of 64 genes were identified that were differentially expressed in endothelial cells compared with non-endothelial cells with a three- to 55-fold difference in expression level, of which 44 were identified by 'virtual subtraction'. Among these were the known endothelial markers CD31, von Willebrand factor, thrombospondin 1, KDR and VE-cadherin. In addition, other novel expressed sequence tags (ESTs) were identified to be differentially expressed in endothelial cells compared with non-endothelial cells.

By using an in silico approach, Huminiecki and colleagues (Huminiecki and Bicknell, 2000) identified four different endothelial specific transcripts. One of the sequences, magic roundabout, showed homology to ROBO1, an axonal guidance molecule, and was termed ROBO4 (Huminiecki and Bicknell, 2000). Later the protein was associated with sites of active angiogenesis, most notably tumour angiogenesis, and was shown to be induced by hypoxia (Huminiecki et al., 2002). Three additional genes were identified, endothelial cell specific molecule-1, 2 and 3 (ECSM1-3). ECSM2 has a predicted transmembrane domain and is possibly an adhesion molecule and ECSM3 has putative functions in extracellular matrix remodelling (Huminiecki and Bicknell, 2000).

Protein disulfide isomerase (PDI) is a ubiquitously expressed protein with important functions in protein folding (Freedman et al., 1994). A novel PDI-like protein was identified to be highly expressed in endothelial cells by comparison with SAGE maps, and named EndoPDI (Sullivan et al., 2003). In contrast to PDI, which is required for endothelial cell survival under resting as well as under stress conditions, EndoPDI exerts its protective role only under conditions of stress (Sullivan et al., 2003).

Gene expression in *in vitro* models of angiogenesis

Gene expression profiling has been applied to endothelial cells treated with different pro-angiogenic cytokines in various models to elucidate genes involved in angiogenesis. Different studies describing gene expression analysis of endothelial cells stimulated with various cytokines *in vitro* identified components of the MAPK signalling pathway (Jun, myc, MAP2K3) and cytokine signalling components (Flt1, IL-8, IL6ST, CCL2, TEK), as well as cell cycle mediators (MCM2, CDC20, CCND2) (Abe and Sato, 2001; Gerritsen et al., 2003b; Jih et al., 2001). However, few genes emerge that are consequently regulated under the influence of, for example, VEGF (Abe and Sato, 2001; Gerritsen et al., 2003b; Jih et al., 2001). Variations in experimental set-ups will have contributed to the observed differences. An interesting observation comes from Gerritsen and colleagues (2003b), who studied the effects of HGF and VEGF on gene expression in endothelial cell culture. Only limited overlap was observed between the gene expression changes induced by HGF or VEGF, although the combination of HGF and VEGF showed moderate overlap with that of VEGF alone, suggesting an additive effect of HGF on gene expression changes. In all treatment groups however, a clear 'cell cycle signature' was observed, with the upregulation of several cyclins, CDCs and mitosis regulators being dominant. In general, most studies describing gene expression analysis of endothelial cells stimulated with growth factors and cytokines show a predominant upregulation of genes involved in cell cycle, apoptosis and metabolism, suggestive of early events that mediate the activation of endothelial cells (Figure 1.4a) (van Beijnum and Griffioen, 2005).

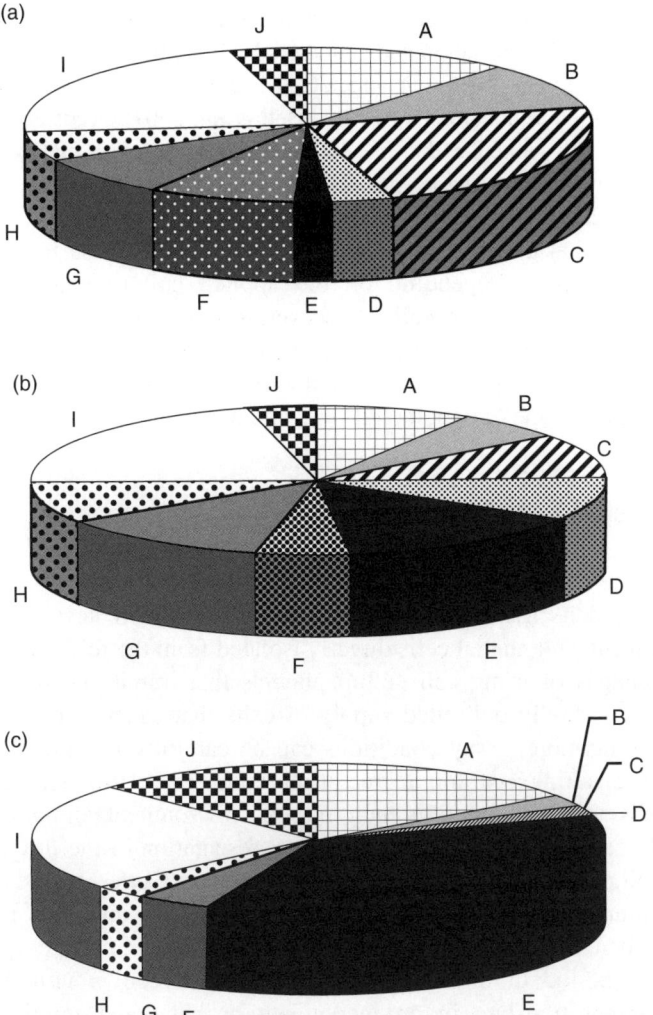

Figure 1.4   Functional classification of annotated genes associated with (Tumour) angiogenesis
(a) Genes upregulated by growth factor stimulation of endothelial cells *in vitro* were classified
according to their functions ($N = 464$) (Abe and Sato, 2001; Gerritsen *et al.*, 2003b; Jih *et al.*,
2001; van Beijnum and Griffioen, 2005; Wang *et al.*, 2000; Zhang *et al.*, 1999). (b) Genes
associated with capillary tube formation of endothelial cells *in vitro* ($N = 385$) (Aitkenhead *et al.*,
2002; Bell *et al.*, 2001; Gerritsen *et al.*, 2003a; Kahn *et al.*, 2000; Wary *et al.*, 2003). (c) Putative
tumour endothelial markers overexpressed in endothelial cells isolated from tumour tissue and
compared with non-tumour endothelial cells ($N = 69$) (Madden *et al.*, 2004; Parker *et al.*, 2004;
St Croix *et al.*, 2000). Gene function was associated with signal transduction (A, squares),
transcription (B, light grey), cell cycle regulation or apoptosis (C, diagonal lines), cytoskeleton
(D, dotted), extracellular matrix remodelling (E, black), growth factor, hormone or cytokine action
(F, dotted grey), metabolism (G, dark grey), protein turnover, modification and transport
(H, dotted lines), cell surface molecules, antigens, receptors and adhesion molecules (I, white), or
unknown function (J, ckeckerboard) (van Beijnum and Griffioen, 2005).

Alternative angiogenesis models use three-dimensional matrices such as Matrigel or type I collagen for tube formation and endothelial differentiation *in vitro*. Integrins and adhesion molecules, e.g. NRCAM and VCAM, are commonly upregulated during tube formation of endothelial cells *in vitro* (Bell *et al.*, 2001; Gerritsen *et al.*, 2003a; Glienke *et al.*, 2000; Aitkenhead *et al.*, 2002). Genes involved in matrix remodelling, such as matrix metalloproteinases, are also induced during this process (Figure 1.4b) (Glienke *et al.*, 2000; van Beijnum and Griffioen, 2005).

It is apparent that different models of angiogenesis give rise to different gene expression signatures. Depending on the applied culture conditions, specific responses are induced in the cells. However, stimuli applied *in vitro* might be incomplete for a perfect mimic of a microenvironment *in vivo* containing additional cells such as pericytes, stromal cells and epithelial cells that strongly influence the phenotype of endothelial cells and hence the gene expression profile.

Gene expression in endothelial cells *in vivo*

Although models of angiogenesis are valuable for initial analysis of endothelial cell biology and its associated gene expression, they will never reach the full potential of using endothelial cells directly isolated from the relevant sources. One of the advantages of using cell culture models that mimic an *in vivo* situation *in vitro* is the virtually unlimited supply of cells that can be obtained for these purposes. Furthermore, assay conditions can be carefully monitored and can be kept constant ensuring a certain degree of reproducibility. However, in cell culture systems, cells are no longer in their natural environment and might respond aberrantly to certain stimuli, giving a false representation of the *in vivo* situation. Also, the culture process itself may induce changes (Favre *et al.*, 2003). When regarding tumour angiogenesis, endothelial cells have generally resided in a tumour microenvironment for months to years, whereas culture systems only cover a time period of days. This discrepancy between *in vitro* and *in vivo* conditions might translate in an incomplete or hampered mimic of *in vivo* conditions.

Relatively few studies have used freshly isolated tumour endothelial cells for gene expression analysis. Pioneering work from St Croix and co-workers (2000) resulted in the description of several tumour endothelial markers (TEMs), genes that showed overexpression in endothelial cells isolated from a fresh colon tumour tissue sample compared with endothelial cells isolated from a normal colon mucosa sample. Serial analysis of gene expression (SAGE) was applied to RNA isolated from endothelial cells that had been selected from the tissues using P1H12 antibody-coupled magnetic beads after enzymatic digestion. Interestingly, TEMs showed a strong bias towards genes functioning in extracellular matrix turnover such as collagens Iα1, Iα2, IIIα1, IVαI, VIα3 and XIIα1, and matrix metalloproteinases MMP-11 and MMP-2, stressing the importance of ECM remodelling during angiogenesis *in vivo*. Furthermore, many TEMs were involved in adhesion and cytoskeletal remodelling, already shown to be involved in tube formation. Genes thought to play a role in the initiation of angiogenesis, e.g. nuclear signalling molecules, and genes involved in cell cycle regulation, metabolism and proliferation were almost completely absent (St Croix *et al.*, 2000).

More recently, SAGE profiles have been published representing malignant brain endothelium and invasive breast carcinoma endothelium (Parker *et al.*, 2004; Madden *et al.*, 2004). Collagens Iα1, IVαI, IVα2 and VIα1 were shown to be overexpressed in glioma endothelial cells, as were a number of matrix metallo-proteinases (Madden *et al.*, 2004). A similar profile is apparent in breast tumour endothelial cells compared with normal breast endothelial cells (Parker *et al.*, 2004). Combining the information in these three SAGE data sets, it is readily apparent that genes involved in extracellular matrix remodelling and cell–cell or cell–matrix contact represent the majority of upregulated genes during tumour angiogenesis in different types of tumours (Figure 1.4c). Furthermore, it has emerged that global gene expression in endothelial cells in *in vitro* models of angiogenesis differs considerably from that of endothelial cells *in vivo* (van Beijnum and Griffioen, 2005).

Gene expression in endothelial cells of different origins

Gene expression is tightly controlled during vascular development. Eph receptors and their ligands play an important role in both angiogenesis and blood vessel develop-ment (Wang *et al.*, 1998). During vessel development, guidance of endothelial cells is tightly regulated, similar to neuronal guidance. Diverse receptor–ligand pairs direct the growth of endothelial cells. Interestingly, many molecules originally identified to play roles during neurogenesis have been associated with angiogenesis, such as the semaphorin receptors neuropilin-1 and neuropilin-2 (Yuan *et al.*, 2002), plexin D1 (Torres-Vazquez *et al.*, 2004), ephrinB2 and its receptor EphB4 (Wang *et al.*, 1998). Very recently, Unc5b, a netrin receptor, has been related to vascular development and guidance of endothelial cells (Lu *et al.*, 2004).

Differences in gene expression have been observed between arterial and venous endothelium. These differences are established even before circulation begins (Lawson and Weinstein, 2002). Venous-specific genes, identified by large-scale gene expression analysis of endothelial cells of diverse origin include EphB4, and a number of genes critically involved in left–right asymmetry such as smoothened, lefty-1, lefty-2 and growth differentiation factor (Chi *et al.*, 2003). Interestingly, defects in left–right symmetry are frequently accompanied by vascular malforma-tions (Meno *et al.*, 1998). Among the arterial endothelial specific genes are the cell surface proteins Notch4, EVA1, CD44 and Ephrin-B1, metabolic enzymes such as alcohol dehydrogenase A1 and endothelial lipase, keratin 7 and the basic helix-loop-helix transcription factor Hey2 that is induced by Notch signalling. Addi-tional evidence for the involvement of Hey2 in arterial differentiation was provided by transduction of HUVEC with a retroviral construct containing the

Hey2 gene (Chi *et al.*, 2003). Hey2 transduction induces the upregulation of different artery-specific genes such as EVA1 and keratin 7. Hey2 induced follistatin expression may act to antagonize venous endothelial cell expressed genes such as lefty-1 and lefty-2 (Chi *et al.*, 2003).

Clear differential expression is also evident between large vessel derived endothelium and microvascular endothelium. Genes involved in biosynthesis and remodelling of extracellular matrix, such as fibronectin, osteonectin and collagens Vα1 and Vα2 are overexpressed in large vessel endothelium. Basement membrane proteins such as collagens IVα1 and IVα2, and its interaction partners such as integrins α4, α1, α9 and β4 are overexpressed in microvascular endothelial cells. In addition, large vessel endothelial cells express genes associated with neuronal guidance such as those discussed above. In contrast microvascular endothelial cells express receptors for paracrine signals of neuroglial cells such as IL-1 receptor, IL-6 receptor, PDGFR and endothelin receptor 1. Furthermore, microvascular endothelial cells express secreted factors that promote survival and differentiation of neuroglial cells (Chi *et al.*, 2003).

Tissue-specific endothelial gene expression is likely to reflect the specialization of the particular vascular bed. Peptide phage display has previously identified a number of peptides specifically reacting with proteins on the endothelial surface in defined locations in the body (Pasqualini *et al.*, 2002). Additional tissue-specific genes were identified by Chi *et al*, who showed SIX3 to be specific for nasal endothelium. Osteoglycin, sFRP and PLA2G12 were specific for lung endothelium, whereas galanin and the calcitonin receptor emerged as uterus specific. Skin endothelium expressed many genes involved in cholesterol biosynthesis, such as stearoyl-CoA desaturase (SCD), fatty acid desaturase (FADS2) and 3-hydroxy-3-methylglutaryl-CoA synthase 1 (HMGCS1), pointing towards active roles played by the endothelium in the organ physiology (Chi *et al.*, 2003).

1.10   Conclusions

In conclusion, endothelial cells from different blood vessels and originating in different tissues have diverse expression profiles that relate to tissue physiology and may actively be involved in disease-related processes. However, data generated using cultured endothelial cells may be biased due to phenotypic drift of endothelial cells in culture. In this regard, gene expression data generated directly from isolated endothelial cells or by using *in vivo* approaches may prove superior.

References

Abe, M. and Sato, Y. (2001) 'cDNA microarray analysis of the gene expression profile of VEGF-activated human umbilical vein endothelial cells', *Angiogenesis*, **4** (4), pp. 289–298.

Acevedo, L., Yu, J., Erdjument-Bromage, H., Miao, R. Q. *et al.* (2004) 'A new role for Nogo as a regulator of vascular remodeling', *Nat. Med.*, **10**(4), pp. 382–388.

Aird, W. C. (2003) 'Endothelial cell heterogeneity', *Crit. Care Med.*, **31**(4) (Suppl), pp. S221–S230.

Aird, W. C. (2004) 'Endothelial cell heterogeity: a case for nature and nurture', *Blood*, **103**(11), pp. 3994–3995.

Aitkenhead, M., Wang, S. J., Nakatsu, M. N., Mestas, J. *et al.* (2002) 'Identification of endothelial cell genes expressed in an *in vitro* model of angiogenesis: induction of ESM-1, (beta)ig-h3, and NrCAM', *Microvasc. Res.*, **63**(2), pp. 159–171.

Al-Rawi, M. A., Mansel, R. E. and Jiang, W. G. (2005) 'Molecular and cellular mechanisms of lymphangiogenesis', *Eur. J. Surg. Oncol.*, **31**(2), pp. 117–121.

Alitalo, K. and Carmeliet, P. (2002) 'Molecular mechanisms of lymphangiogenesis in health and disease', *Cancer Cell*, **1**(3), pp. 219–227.

Anitua, E., Andia, I., Ardanza, B., Nurden, P. and Nurden, A. T. (2004) 'Autologous platelets as a source of proteins for healing and tissue regeneration', *Thromb. Haemost.*, **91**(1), pp. 4–15.

Autiero, M., Luttun, A., Tjwa, M. and Carmeliet, P. (2003) 'Placental growth factor and its receptor, vascular endothelial growth factor receptor-1: novel targets for stimulation of ischaemic tissue revascularization and inhibition of angiogenic and inflammatory disorders', *J. Thromb. Haemost.*, **1**(7), pp. 1356–1370.

Banerji, S., Ni, J., Wang, S. X., Clasper, S. *et al.* (1999) 'LYVE-1, a new homologue of the CD44 glycoprotein, is a lymph-specific receptor for hyaluronan', *J. Cell Biol.*, **144**(4), pp. 789–801.

Bardin, N., Anfosso, F., Masse, J. M., Cramer, E. *et al.* (2001) 'Identification of CD146 as a component of the endothelial junction involved in the control of cell–cell cohesion', *Blood*, **98**(13), pp. 3677–3684.

Bazzoni, G. and Dejana, E. (2004) 'Endothelial cell-to-cell junctions: molecular organization and role in vascular homeostasis', *Physiol. Rev.*, **84**(3), pp. 869–901.

Bell, S. E., Mavila, A., Salazar, R., Bayless, K. J. *et al.* (2001) 'Differential gene expression during capillary morphogenesis in 3D collagen matrices: regulated expression of genes involved in basement membrane matrix assembly, cell cycle progression, cellular differentiation and G-protein signaling', *J. Cell Sci.*, **114**(15), pp. 2755–2773.

Bevilacqua, M. P. (1993) 'Endothelial–leukocyte adhesion molecules', *Annu. Rev. Immunol.*, **11**, pp. 767–804.

Bissell, M. J. (1999) 'Tumour plasticity allows vasculogenic mimicry, a novel form of angiogenic switch. A rose by any other name?', *Am. J. Pathol.*, **155**(3), pp. 675–679.

Bodolay, E., Koch, A. E., Kim, J., Szegedi, G. and Szekanecz, Z. (2002) 'Angiogenesis and chemokines in rheumatoid arthritis and other systemic inflammatory rheumatic diseases', *J. Cell Mol. Med.*, **6**(3), pp. 357–376.

Boffa, M. C. and Karmochkine, M. (1998) 'Thrombomodulin: an overview and potential implications in vascular disorders', *Lupus*, **7** (Suppl. 2), pp. S120–S125.

Bohnsack, B. L., Lai, L., Dolle, P. and Hirschi, K. K. (2004) 'Signaling hierarchy downstream of retinoic acid that independently regulates vascular remodeling and endothelial cell proliferation', *Genes Dev.*, **18**(11), pp. 1345–1358.

Botta, M., Manetti, F. and Corelli, F. (2000) 'Fibroblast growth factors and their inhibitors', *Curr. Pharm. Des*, **6**(18), pp. 1897–1924.

Bouis, D., Hospers, G. A., Meijer, C., Molema, G. and Mulder, N. H. (2001) 'Endothelium *in vitro*: a review of human vascular endothelial cell lines for blood vessel-related research', *Angiogenesis*, **4**(2), pp. 91–102.

Breiteneder-Geleff, S., Soleiman, A., Kowalski, H., Horvat, R. *et al.* (1999) 'Angiosarcomas express mixed endothelial phenotypes of blood and lymphatic capillaries: podoplanin as a specific marker for lymphatic endothelium', *Am. J. Pathol.*, **154**(2), pp. 385–394.

Brooks, P. C., Stromblad, S., Sanders, L. C., von Schalscha, T. L. *et al.* (1996) 'Localization of matrix metalloproteinase MMP-2 to the surface of invasive cells by interaction with integrin alpha v beta', *Cell*, **85**(5), pp. 683–693.

Burri, P. H., Hlushchuk, R. and Djonov, V. (2004) 'Intussusceptive angiogenesis: its emergence, its characteristics, and its significance', *Dev. Dyn.*, **231**(3), pp. 474–488.

Bussolati, B., Deambrosis, I., Russo, S., Deregibus, M. C. and Camussi, G. (2003) 'Altered angiogenesis and survival in human tumour-derived endothelial cells', *FASEB J.*, **17**(9), pp. 1159–1161.

Cao, Y., Hong, A., Schulten, H. and Post, M. J. (2005) 'Update on therapeutic neovascularization', *Cardiovasc. Res.*, **65**(3), pp. 639–648.

Carmeliet, P. (2003) 'Angiogenesis in health and disease', *Nat. Med.*, **9**(6), pp. 653–660.

Carson-Walter, E. B., Watkins, D. N., Nanda, A., Vogelstein, B. *et al.* (2001) 'Cell surface tumour endothelial markers are conserved in mice and humans', *Cancer Res.*, **61**(18), pp. 6649–6655.

Chi, J. T., Chang, H. Y., Haraldsen, G., Jahnsen, F. L. *et al.* (2003) 'Endothelial cell diversity revealed by global expression profiling', *Proc. Natl. Acad. Sci. USA*, **100**(19), pp. 10623–10628.

Chung-Welch, N., Shepro, D., Dunham, B. and Hechtman, H. B. (1988) 'Prostacyclin and prostaglandin E2 secretions by bovine pulmonary microvessel endothelial cells are altered by changes in culture conditions', *J. Cell Physiol.*, **135**( )2, pp. 224–234.

Cines, D. B., Pollak, E. S., Buck, C. A., Loscalzo, J. *et al.* (1998) 'Endothelial cells in physiology and in the pathophysiology of vascular disorders', *Blood*, **91**(10), pp. 3527–3561.

Cleaver, O. and Melton, D. A. (2003) 'Endothelial signaling during development', *Nat. Med.*, **9**(6), pp. 661–668.

Dejana, E. (2004) 'Endothelial cell–cell junctions: happy together', *Nat. Rev. Mol. Cell Biol.*, **5**(4), pp. 261–270.

Demeule, M., Labelle, M., Regina, A., Berthelet, F. and Beliveau, R. (2001) 'Isolation of endothelial cells from brain, lung, and kidney: expression of the multidrug resistance P-glycoprotein isoforms', *Biochem. Biophys. Res. Commun.*, **281**(3), pp. 827–834.

Dvorak, A. M. and Feng, D. (2001) 'The vesiculo-vacuolar organelle (VVO). A new endothelial cell permeability organelle', *J. Histochem. Cytochem.*, **49**(4), pp. 419–432.

Dvorak, H. F. (2000) 'VPF/VEGF and the angiogenic response', *Semin. Perinatol.*, **24**(1), pp. 75–78.

Ebnet, K., Suzuki, A., Ohno, S. and Vestweber, D. (2004) 'Junctional adhesion molecules (JAMs): more molecules with dual functions?', *J. Cell Sci.*, **117**(1), pp. 19–29.

Elangbam, C. S., Qualls, C. W., Jr and Dahlgren, R. R. (1997) 'Cell adhesion molecules–update', *Vet. Pathol.*, **34**(1), pp. 61–73.

Engelhardt, B. and Wolburg, H. (2004) 'Mini-review: Transendothelial migration of leukocytes: through the front door or around the side of the house?', *Eur. J. Immunol.*, **34**(11), pp. 2955–2963.

Favre, C. J., Mancuso, M., Maas, K., McLean, J. W. *et al.* (2003) 'Expression of genes involved in vascular development and angiogenesis in endothelial cells of adult lung', *Am. J. Physiol. Heart Circ. Physiol.*, **285**(5), pp. H1917–H1938.

Fawcett, J., Harris, A. L. and Bicknell, R. (1991) 'Isolation and properties in culture of human adrenal capillary endothelial cells', *Biochem. Biophys. Res. Commun.*, **174**(2), pp. 903–908.

Feng, D., Nagy, J. A., Hipp, J., Dvorak, H. F. and Dvorak, A. M. (1996) 'Vesiculo-vacuolar organelles and the regulation of venule permeability to macromolecules by vascular permeability factor, histamine, and serotonin', *J. Exp. Med.*, **183**(5), pp. 1981–1986.

Feng, D., Nagy, J. A., Dvorak, H. F. and Dvorak, A. M. (2002) 'Ultrastructural studies define soluble macromolecular, particulate, and cellular transendothelial cell pathways in venules, lymphatic vessels, and tumour-associated microvessels in man and animals', *Microsc. Res. Tech.*, **57**(5), pp. 289–326.

Ferrara, N. (2004) 'Vascular endothelial growth factor: basic science and clinical progress', *Endocr. Rev.*, **25**(4), pp. 581–611.

Ferrara, N., Gerber, H. P. and LeCouter, J. (2003) 'The biology of VEGF and its receptors', *Nat. Med.*, **9**(6), pp. 669–676.

Folkman, J. (1971) 'Tumour angiogenesis: therapeutic implications', *N. Engl. J. Med.*, **285**(21), pp. 1182–1186.

Folkman, J. (1995) 'Angiogenesis in cancer, vascular, rheumatoid and other disease', *Nat. Med.*, **1**(1), pp. 27–31.

Folkman, J., Haudenschild, C. C. and Zetter, B. R. (1979) 'Long-term culture of capillary endothelial cells', *Proc. Natl. Acad. Sci. USA*, **76**(10), pp. 5217–5221.

Folkman, J. and Shing, Y. (1992) 'Angiogenesis', *J. Biol. Chem.*, **267**(16), pp. 10931–10934.

Freedman, R. B., Hirst, T. R. and Tuite, M. F. (1994) 'Protein disulphide isomerase: building bridges in protein folding', *Trends Biochem. Sci.*, **19**(8), pp. 331–336.

Gallagher P. J. (2005) 'Blood vessels', in *Histology for Pathologists*, 2nd edn, Stephan S. and Sternberg M. D., eds., pp. 763–786.

Gallo, R., Kim, C., Kokenyesi, R., Adzick, N. S. and Bernfield, M. (1996) 'Syndecans-1 and -4 are induced during wound repair of neonatal but not fetal skin', *J. Invest. Dermatol.*, **107**(5), pp. 676–683.

Geng, J. G. (2003) 'Interaction of vascular endothelial cells with leukocytes, platelets and cancer cells in inflammation, thrombosis and cancer growth and metastasis', *Acta Pharmacol. Sin.*, **24**(12), pp. 1297–1300.

Gerhardt, H. and Betsholtz, C. (2003) 'Endothelial–pericyte interactions in angiogenesis', *Cell Tissue Res.*, **314**(1), pp. 15–23.

Gerritsen, M. E., Shen, C. P., McHugh, M. C., Atkinson, W. J. *et al.* (1995) 'Activation-dependent isolation and culture of murine pulmonary microvascular endothelium', *Microcirculation*, **2**(2), pp. 151–163.

Gerritsen, M. E., Soriano, R., Yang, S., Zlot, C. *et al.* (2003a) 'Branching out: a molecular fingerprint of endothelial differentiation into tube-like structures generated by Affymetrix oligonucleotide arrays', *Microcirculation*, **10**(1), pp. 63–81.

Gerritsen, M. E., Tomlinson, J. E., Zlot, C., Ziman, M. and Hwang, S. (2003b) 'Using gene expression profiling to identify the molecular basis of the synergistic actions of hepatocyte growth factor and vascular endothelial growth factor in human endothelial cells', *Br. J. Pharmacol.*, **140**(4), pp. 595–610.

Ghitescu, L. and Robert, M. (2002) 'Diversity in unity: the biochemical composition of the endothelial cell surface varies between the vascular beds', *Microsc. Res. Tech.*, **57**(5), pp. 381–389.

Giordano, F. and Mitola, S. (2000) 'Characterization of endothelial cells isolated by human meningiomas', *J. Neurosurg. Sci.*, **44**(4), pp. 177–185.

Glienke, J., Schmitt, A. O., Pilarsky, C., Hinzmann, B. *et al.* (2000) 'Differential gene expression by endothelial cells in distinct angiogenic states', *Eur. J. Biochem.*, **267**(9), pp. 2820–2830.

Gonlugur, U. and Efeoglu, T. (2004) 'Vascular adhesion and transendothelial migration of eosinophil leukocytes', *Cell Tissue Res.*, **318**(3), pp. 473–482.

Good, D. J., Polverini, P. J., Rastinejad, F., Le Beau, M. M. *et al.* (1990) 'A tumour suppressor-dependent inhibitor of angiogenesis is immunologically and functionally indistinguishable from a fragment of thrombospondin', *Proc. Natl. Acad. Sci. USA*, **87**(17), pp. 6624–6628.

Griffioen, A. W. (1997) 'Phenotype of the tumour vasculature; cell adhesion as a target for tumour therapy', *Cancer Journal*, **10**(5), pp. 249–255.

Grimwood, J., Bicknell, R. and Rees, M. C. (1995) 'The isolation, characterization and culture of human decidual endothelium', *Hum. Reprod.*, **10**(8), pp. 2142–2148.

Grossfeld, G. D., Ginsberg, D. A., Stein, J. P., Bochner, B. H. *et al.* (1997) 'Thrombospondin-1 expression in bladder cancer: association with p53 alterations, tumour angiogenesis, and tumour progression', *J. Natl. Cancer Inst.*, **89**(3), pp. 219–227.

Gupta, S. K., Hassel, T. and Singh, J. P. (1995) 'A potent inhibitor of endothelial cell proliferation is generated by proteolytic cleavage of the chemokine platelet factor 4', *Proc. Natl. Acad. Sci. USA*, **92**(17), pp. 7799–7803.

Gussin, H. A., Sharma, A. K. and Elias, S. (2004) 'Culture of endothelial cells isolated from maternal blood using anti-CD105 and CD133', *Prenat. Diagn.*, **24**(3), pp. 189–193.

Guyton, A. C. and Hall, J. E. (2005) 'Hemostasis and blood coagulation', in *Textbook of Medical Physiology*, W. B. Saunders Company, pp. 419–429.

Harvey, N. L. and Oliver, G. (2004) 'Choose your fate: artery, vein or lymphatic vessel?', *Curr. Opin. Genet. Dev.*, **14**(5), pp. 499–505.

Hewett, P. W. and Murray, J. C. (1993) 'Human lung microvessel endothelial cells: isolation, culture, and characterization', *Microvasc. Res.*, **46**(1), pp. 89–102.

Hirata, K., Ishida, T., Penta, K., Rezaee, M. *et al.* (2001) 'Cloning of an immunoglobulin family adhesion molecule selectively expressed by endothelial cells', *J. Biol. Chem.*, **276**(19), pp. 16223–16231.

Ho, M., Yang, E., Matcuk, G., Deng, D. *et al.* (2003) 'Identification of endothelial cell genes by combined database mining and microarray analysis', *Physiol Genomics*, **13**(3), pp. 249–262.

Hristov, M. and Weber, C. (2004) 'Endothelial progenitor cells: characterization, pathophysiology, and possible clinical relevance', *J. Cell Mol. Med.*, **8**(4), pp. 498–508.

Huminiecki, L. and Bicknell, R. (2000) 'In silico cloning of novel endothelial-specific genes', *Genome Res.*, **10**(11), pp. 1796–1806.

Huminiecki, L., Gorn, M., Suchting, S., Poulsom, R. and Bicknell, R. (2002) 'Magic roundabout is a new member of the roundabout receptor family that is endothelial specific and expressed at sites of active angiogenesis', *Genomics*, **79**(4), pp. 547–552.

Iivanainen, E., Kahari, V. M., Heino, J. and Elenius, K. (2003) 'Endothelial cell–matrix interactions', *Microsc. Res. Tech.*, **60**(1), pp. 13–22.

Jackson, D. E. (2003) 'The unfolding tale of PECAM-1', *FEBS Lett.*, **540**(1–3), pp. 7–14.

Jaffe, E. A., Nachman, R. L., Becker, C. G. and Minick, C. R. (1973) 'Culture of human endothelial cells derived from umbilical veins. Identification by morphologic and immunologic criteria', *J. Clin. Invest*, **52**(11), pp. 2745–2756.

Jeltsch, M., Kaipainen, A., Joukov, V., Meng, X. *et al.* (1997) 'Hyperplasia of lymphatic vessels in VEGF-C transgenic mice', *Science*, **276**(5317), pp. 1423–1425.

Jih, Y. J., Lien, W. H., Tsai, W. C., Yang, G. W. *et al.* (2001) 'Distinct regulation of genes by bFGF and VEGF-A in endothelial cells', *Angiogenesis*, **4**(4), pp. 313–321.

Jones, P. F. (2003) 'Not just angiogenesis–wider roles for the angiopoietins', *J. Pathol.*, **201**(4), pp. 515–527.

Jung, Y. D., Ahmad, S. A., Liu, W., Reinmuth, N. *et al.* (2002) 'The role of the microenvironment and intercellular cross-talk in tumour angiogenesis', *Semin. Cancer Biol.*, **12**(2), pp. 105–112.

Kahn, J., Mehraban, F., Ingle, G., Xin, X. *et al.* (2000) 'Gene expression profiling in an *in vitro* model of angiogenesis', *Am. J. Pathol.*, **156**(6), pp. 1887–1900.

Kaila, N. and Thomas, B. E. (2002) 'Design and synthesis of sialyl Lewis(x) mimics as E- and P-selectin inhibitors', *Med. Res. Rev.*, **22**(6), pp. 566–601.

Kakkar, A. K. and Lefer, D. J. (2004) 'Leukocyte and endothelial adhesion molecule studies in knockout mice', *Curr. Opin. Pharmacol.*, **4**(2), pp. 154–158.

Kallmann, B. A., Wagner, S., Hummel, V., Buttmann, M. *et al.* (2002) 'Characteristic gene expression profile of primary human cerebral endothelial cells', *FASEB J.*, **16**(6), pp. 589–591.

Kalluri, R. (2003) 'Basement membranes: structure, assembly and role in tumour angiogenesis', *Nat. Rev. Cancer*, **3**(6), pp. 422–433.

Kaneko, T., Fujii, S., Matsumoto, A., Goto, D. *et al.* (2003) 'Induction of tissue factor expression in endothelial cells by basic fibroblast growth factor and its modulation by fenofibric acid', *Thromb. J.*, **1**(1), p. 6.

Keiper, T., Santoso, S., Nawroth, P. P., Orlova, V. and Chavakis, T. (2005) 'The role of junctional adhesion molecules in cell–cell interactions', *Histol. Histopathol.*, **20**(1), pp. 197–203.

Kerbel, R. S. (2000) 'Tumour angiogenesis: past, present and the near future', *Carcinogenesis*, **21**(3), pp. 505–515.

Koch, A. E. (200) 'Angiogenesis as a target in rheumatoid arthritis', *Ann. Rheum. Dis.*, **62**(Suppl 2), pp. ii60–ii67.

Koivunen, E., Arap, W., Valtanen, H., Rainisalo, A. *et al.* (1999) 'Tumour targeting with a selective gelatinase inhibitor', *Nat. Biotechnol.*, **17**(8), pp. 768–774.

Kolber, D. L., Knisely, T. L. and Maione, T. E. (1995) 'Inhibition of development of murine melanoma lung metastases by systemic administration of recombinant platelet factor 4', *J. Natl. Cancer Inst.*, **87**(4), pp. 304–309.

Lai, L., Bohnsack, B. L., Niederreither, K. and Hirschi, K. K. (2003) 'Retinoic acid regulates endothelial cell proliferation during vasculogenesis', *Development*, **130**(26), pp. 6465–6474.

Lawson, N. D. and Weinstein, B. M. (2002) 'Arteries and veins: making a difference with zebrafish', *Nat. Rev. Genet.*, **3**(9), pp. 674–682.

Lawson, N. D., Scheer, N., Pham, V. N., Kim, C. H. *et al.* (2001) 'Notch signaling is required for arterial–venous differentiation during embryonic vascular development', *Development*, **128**(19), pp. 3675–3683.

Lawson, N. D., Vogel, A. M. and Weinstein, B. M. (2002) 'Sonic hedgehog and vascular endothelial growth factor act upstream of the Notch pathway during arterial endothelial differentiation', *Dev. Cell*, **3**(1), pp. 127–136.

LeCouter, J., Kowalski, J., Foster, J., Hass, P. *et al.* (2001) 'Identification of an angiogenic mitogen selective for endocrine gland endothelium', *Nature*, **412**(6850), pp. 877–884.

Li, J., Zhang, Y. P. and Kirsner, R. S. (2003) 'Angiogenesis in wound repair: angiogenic growth factors and the extracellular matrix', *Microsc. Res. Tech.*, **60**(1), pp. 107–114.

Lu, X., Le, N. F., Yuan, L., Jiang, Q. *et al.* (2004) 'The netrin receptor UNC5B mediates guidance events controlling morphogenesis of the vascular system', *Nature*, **432**(7014), pp. 179–186.

Luster, A. D., Greenberg, S. M. and Leder, P. (1995) 'The IP-10 chemokine binds to a specific cell surface heparan sulfate site shared with platelet factor 4 and inhibits endothelial cell proliferation', *J. Exp. Med.*, **182**(1), pp. 219–231.

Madden, S. L., Cook, B. P., Nacht, M., Weber, W. D. *et al.* (2004) 'Vascular gene expression in nonneoplastic and malignant brain', *Am. J. Pathol.*, **165**(2), pp. 601–608.

Makinen, T., Veikkola, T., Mustjoki, S., Karpanen, T. *et al.* (2001) 'Isolated lymphatic endothelial cells transduce growth, survival and migratory signals via the VEGF-C/D receptor VEGFR-3', *EMBO J.*, **20**(17), pp. 4762–4773.

Manconi, F., Markham, R. and Fraser, I. S. (2000) 'Culturing endothelial cells of microvascular origin', *Methods Cell Sci.*, **22**(2–3), pp. 89–99.

Maniotis, A. J., Folberg, R., Hess, A., Seftor, E. A. *et al.* (1999) 'Vascular channel formation by human melanoma cells *in vivo* and *in vitro*: vasculogenic mimicry', *Am. J. Pathol.*, **155**(3), pp. 739–752.

Marelli-Berg, F. M., Peek, E., Lidington, E. A., Stauss, H. J. and Lechler, R. I. (2000) 'Isolation of endothelial cells from murine tissue', *J. Immunol. Methods*, **244**(1–2), pp. 205–215.

Martin, A., Komada, M. R. and Sane, D. C. (2003) 'Abnormal angiogenesis in diabetes mellitus', *Med. Res. Rev.*, **23**(2), pp. 117–145.

McCarthy, S. A., Kuzu, I., Gatter, K. C. and Bicknell, R. (1991) 'Heterogeneity of the endothelial cell and its role in organ preference of tumour metastasis', *Trends Pharmacol. Sci.*, **12**(12), pp. 462–467.

McColl, B. K., Stacker, S. A. and Achen, M. G. (2004) 'Molecular regulation of the VEGF family – inducers of angiogenesis and lymphangiogenesis', *APMIS*, **112**(7–8), pp. 463–480.

McDonald, D. M. and Foss, A. J. (2000) 'Endothelial cells of tumour vessels: abnormal but not absent', *Cancer Metastasis Rev.*, **19**(1–2), pp. 109–120.

McDonald, D. M., Munn, L. and Jain, R. K. (2000) 'Vasculogenic mimicry: how convincing, how novel, and how significant?', *Am. J. Pathol.*, **156**(2), pp. 383–388.

McGinn, S., Poronnik, P., Gallery, E. D. and Pollock, C. A. (2004) 'A method for the isolation of glomerular and tubulointerstitial endothelial cells and a comparison of characteristics with the human umbilical vein endothelial cell model', *Nephrology (Carlton)*, **9**(4), pp. 229–237.

Meno, C., Shimono, A., Saijoh, Y., Yashiro, K. *et al.* (1998) 'lefty-1 is required for left-right determination as a regulator of lefty-2 and nodal', *Cell*, **94**(3), pp. 287–297.

Muller, A. M., Hermanns, M. I., Cronen, C. and Kirkpatrick, C. J. (2002) 'Comparative study of adhesion molecule expression in cultured human macro- and microvascular endothelial cells', *Exp. Mol. Pathol.*, **73**(3), pp. 171–180.

Muller, A. W. (2003) 'Finding extraterrestrial organisms living on thermosynthesis', *Astrobiology.*, **3**(3), pp. 555–564.

Mummert, M. E., Mohamadzadeh, M., Mummert, D. I., Mizumoto, N. and Takashima, A. (2000) 'Development of a peptide inhibitor of hyaluronan-mediated leukocyte trafficking', *J. Exp. Med.*, **192**(6), pp. 769–779.

Nilius, B. and Droogmans, G. (2001) 'Ion channels and their functional role in vascular endothelium', *Physiol Rev.*, **81**(4), pp. 1415–1459.

Nourshargh, S. and Marelli-Berg, F. M. (2005) 'Transmigration through venular walls: a key regulator of leukocyte phenotype and function', *Trends Immunol.*, **26**(3), pp. 157–165.

O'Reilly, M. S., Boehm, T., Shing, Y., Fukai, N. *et al.* (1997) 'Endostatin: an endogenous inhibitor of angiogenesis and tumour growth', *Cell*, **88**(2), pp. 277–285.

O'Reilly, M. S., Holmgren, L., Shing, Y., Chen, C. *et al.* (1994) 'Angiostatin: a novel angiogenesis inhibitor that mediates the suppression of metastases by a Lewis lung carcinoma', *Cell*, **79**(2), pp. 315–328.

Okada-Ban, M., Thiery, J. P. and Jouanneau, J. (2000) 'Fibroblast growth factor-2', *Int. J. Biochem. Cell Biol.*, **32**(3), pp. 263–267.

Parker, B. S., Argani, P., Cook, B. P., Liangfeng, H. *et al.* (2004) 'Alterations in vascular gene expression in invasive breast carcinoma', *Cancer Res.*, **64**(21), pp. 7857–7866.

Pasqualini, R., Koivunen, E. and Ruoslahti, E. (1997) 'Alpha v integrins as receptors for tumour targeting by circulating ligands', *Nat. Biotechnol.*, **15**(6), pp. 542–546.

Pasqualini, R., Koivunen, E., Kain, R., Lahdenranta, J. *et al.* (2000) 'Aminopeptidase N is a receptor for tumour-homing peptides and a target for inhibiting angiogenesis', *Cancer Res.*, **60**(3), pp. 722–727.

Pasqualini, R., Arap, W. and McDonald, D. M. (2002) 'Probing the structural and molecular diversity of tumour vasculature', *Trends Mol. Med.*, **8**(12), pp. 563–571.

Patan, S. (2000) 'Vasculogenesis and angiogenesis as mechanisms of vascular network formation, growth and remodeling 1', *J. Neurooncol.*, **50**(1–2), pp. 1–15.

Patel, V. A., Logan, A., Watkinson, J. C., Uz-Zaman, S. *et al.* (2003) 'Isolation and characterization of human thyroid endothelial cells', *Am. J. Physiol Endocrinol. Metab*, **284**(1), pp. E168–E176.

Price, G. C., Thompson, S. A. and Kam, P. C. (2004) 'Tissue factor and tissue factor pathway inhibitor', *Anaesthesia*, **59**(5), pp. 483–492.

Rajotte, D., Arap, W., Hagedorn, M., Koivunen, E. *et al.* (1998) 'Molecular heterogeneity of the vascular endothelium revealed by *in vivo* phage display', *J. Clin. Invest*, **102**(2), pp. 430–437.

Rastinejad, F., Polverini, P. J. and Bouck, N. P. (1989) 'Regulation of the activity of a new inhibitor of angiogenesis by a cancer suppressor gene', *Cell*, **56**(3), pp. 345–355.

Relou, I. A., Damen, C. A., van der Schaft, D. W., Groenewegen, G. and Griffioen, A. W. (1998) 'Effect of culture conditions on endothelial cell growth and responsiveness', *Tissue Cell*, **30**(5), pp. 525–530.

Renkin E. M. (1988) 'Transport pathways and processes', in *Endothelial Cell Biology in Health and Disease*, N. Simionescu and M. Simionescu, eds, Plenum Press, pp. 51–68.

Reuss, B. and von Bohlen und, H. O. (2003) 'Fibroblast growth factors and their receptors in the central nervous system', *Cell Tissue Res.*, **313**(2), pp. 139–157.

Rhee, J. S., Black, M., Schubert, U., Fischer, S. *et al.* (2004) 'The functional role of blood platelet components in angiogenesis', *Thromb. Haemost.*, **92**(2), pp. 394–402.

Ribatti, D. (2004) 'The involvement of endothelial progenitor cells in tumour angiogenesis', *J. Cell Mol. Med.*, **8**(3), pp. 294–300.

Ribatti, D., Nico, B., Vacca, A., Roncali, L. and Dammacco, F. (2002) 'Endothelial cell heterogeneity and organ specificity', *J. Hematother. Stem Cell Res.*, **11**(1), pp. 81–90.

Richard, L., Velasco, P. and Detmar, M. (1998) 'A simple immunomagnetic protocol for the selective isolation and long-term culture of human dermal microvascular endothelial cells', *Exp. Cell Res.*, **240**(1), pp. 1–6.

Sadler, J. E. (1997) 'Thrombomodulin structure and function', *Thromb. Haemost.*, vol. 78, no. 1, pp. 392–395.

Saharinen, P. and Petrova, T. V. (2004) 'Molecular regulation of lymphangiogenesis', *Ann. NY Acad. Sci.*, **1014**, pp. 76–87.

Saharinen, P., Tammela, T., Karkkainen, M. J. and Alitalo, K. (2004) 'Lymphatic vasculature: development, molecular regulation and role in tumour metastasis and inflammation', *Trends Immunol.*, **25**(7), pp. 387–395.

Sato, T. N. (2003) 'Emerging concept in angiogenesis: specification of arterial and venous endothelial cells', *Br. J. Pharmacol.*, **140**(4), pp. 611–613.

Schneeberger, E. E. and Lynch, R. D. (2004) 'The tight junction: a multifunctional complex', *Am. J. Physiol Cell Physiol*, **286**(6), pp. C1213–C1228.

Schneider, M., Othman-Hassan, K., Christ, B. and Wilting, J. (1999) 'Lymphangioblasts in the avian wing bud', *Dev. Dyn.*, **216**(4-5), pp. 311–319.

Scott, P. A. and Bicknell, R. (1993) 'The isolation and culture of microvascular endothelium', *J. Cell Sci.*, **105**(2), pp. 269–273.

Shubik, P. and Warren, B. A. (2000) 'Additional literature on "vasculogenic mimicry" not cited', *Am. J. Pathol.*, **156**(2), p. 736.

Simionescu, M., Gafencu, A. and Antohe, F. (2002) 'Transcytosis of plasma macromolecules in endothelial cells: a cell biological survey', *Microsc. Res. Tech.*, **57**(5), pp. 269–288.

Simionescu, M., Simionescu, N. and Palade, G. E. (1975) 'Segmental differentiations of cell junctions in the vascular endothelium. The microvasculature', *J. Cell Biol.*, **67**(3), pp. 863–885.

Sottile, J. (2004) 'Regulation of angiogenesis by extracellular matrix', *Biochim. Biophys. Acta*, **1654**(1), pp. 13–22.

St Croix, B., Rago, C., Velculescu, V., Traverso, G. *et al.* (2000) 'Genes expressed in human tumour endothelium', *Science*, **289**(5482), pp. 1197–1202.

Stan, R. V. (2002) 'Structure and function of endothelial caveolae', *Microsc. Res. Tech.*, **57**(5), pp. 350–364.

Stupack, D. G. and Cheresh, D. A. (2004) 'Integrins and angiogenesis', *Curr. Top. Dev. Biol.*, **64**, pp. 207–238.

Sullivan, D. C., Huminiecki, L., Moore, J. W., Boyle, J. J. *et al.* (2003) 'EndoPDI, a novel protein-disulfide isomerase-like protein that is preferentially expressed in endothelial cells acts as a stress survival factor', *J. Biol. Chem.*, **278**(47), pp. 47079–47088.

Sumpio, B. E., Riley, J. T. and Dardik, A. (2002) 'Cells in focus: endothelial cell', *Int. J. Biochem. Cell Biol.*, **34**(12), pp. 1508–1512.

Sun, L. (2004) 'Tumour-suppressive and promoting function of transforming growth factor beta', *Front Biosci.*, **9**, pp. 1925–1935.

Takai, Y. and Nakanishi, H. (2003) 'Nectin and afadin: novel organizers of intercellular junctions', *J. Cell Sci.*, **116**(1), pp. 17–27.

Terramani, T. T., Eton, D., Bui, P. A., Wang, Y. *et al.* (2000) 'Human macrovascular endothelial cells: optimization of culture conditions', *In Vitro Cell Dev. Biol. Anim*, **36**(2), pp. 125–132.

Thilo-Korner D. G. S and Heinrich D (1983) 'Historical development of endothelial cell research', in *The Endothelial Cell – a Pluripotent Control Cell of the Vessel Wall*, Thilo-Korner D. G. S and Freshney R. I, eds, Karger, pp. 1–12.

Thurston, G. (2003) 'Role of Angiopoietins and Tie receptor tyrosine kinases in angiogenesis and lymphangiogenesis 3', *Cell Tissue Res.*, **314**(1), pp. 61–68.

Torres-Vazquez, J., Gitler, A. D., Fraser, S. D., Berk, J. D. *et al.* (2004) 'Semaphorin-plexin signaling guides patterning of the developing vasculature', *Dev. Cell*, **7**(1), pp. 117–123.

van Beijnum, J. R. and Griffioen, A. W. (2005) 'In silico analysis of angiogenesis associated gene expression identifies angiogenic stage related profiles', *Biochim. Biophys. Acta*, Jul 25; **1755**(2), pp. 121–134.

van Buul, J. D. and Hordijk, P. L. (2004) 'Signaling in leukocyte transendothelial migration', *Arterioscler. Thromb. Vasc. Biol.*, **24**(5), pp. 824–833.

van der Schaft, D. W., Wagstaff, J., Mayo, K. H. and Griffioen, A. W. (2002) 'The antiangiogenic properties of bactericidal/permeability-increasing protein (BPI)', *Ann. Med.*, **34**(1), pp. 19–27.

Verheul, H. M., Jorna, A. S., Hoekman, K., Broxterman, H. J. *et al.* (2000) 'Vascular endothelial growth factor-stimulated endothelial cells promote adhesion and activation of platelets', *Blood*, **96**(13), pp. 4216–4221.

Vincent, P. A., Xiao, K., Buckley, K. M. and Kowalczyk, A. P. (2004) 'VE-cadherin: adhesion at arm's length', *Am. J. Physiol Cell Physiol*, **286**(5), pp. C987–C997.

Wang, H. U., Chen, Z. F. and Anderson, D. J. (1998) 'Molecular distinction and angiogenic interaction between embryonic arteries and veins revealed by ephrin-B2 and its receptor Eph-B4', *Cell*, **93**(5), pp. 741–753.

Wang, J. L., Liu, Y. H., Lee, M. C., Nguyen, T. M. *et al.* (2000) 'Identification of tumour angiogenesis-related genes by subtractive hybridization', *Microvasc. Res.*, **59**(3), pp. 394–397.

Wary, K. K., Thakker, G. D., Humtsoe, J. O. and Yang, J. (2003) 'Analysis of VEGF-responsive genes involved in the activation of endothelial cells', *Mol. Cancer*, **2**(1), p. 25.

Weinstein, B. M., Stemple, D. L., Driever, W. and Fishman, M. C. (1995) 'Gridlock, a localized heritable vascular patterning defect in the zebrafish', *Nat. Med.*, **1**(11), pp. 1143–1147.

Wigle, J. T., Harvey, N., Detmar, M., Lagutina, I. *et al.* (2002) 'An essential role for Prox1 in the induction of the lymphatic endothelial cell phenotype', *EMBO J.*, **21**(7), pp. 1505–1513.

Wilting, J., Papoutsi, M., Christ, B., Nicolaides, K. H., *et al.* (2002) 'The transcription factor Prox1 is a marker for lymphatic endothelial cells in normal and diseased human tissues', *FASEB J.*, **16**(10), pp. 1271–1273.

Xu, Q., Wang, Y., Dabdoub, A., Smallwood, P. M. *et al.* (2004) 'Vascular development in the retina and inner ear: control by Norrin and Frizzled-4, a high-affinity ligand-receptor pair 1', *Cell*, **116**(6), pp. 883–895.

Yadav, R., Larbi, K. Y., Young, R. E. and Nourshargh, S. (2003) 'Migration of leukocytes through the vessel wall and beyond', *Thromb. Haemost.*, **90**(4), pp. 598–606.

Yuan, L., Moyon, D., Pardanaud, L., Breant, C. *et al.* (2002) 'Abnormal lymphatic vessel development in neuropilin 2 mutant mice', *Development*, **129**(20), pp. 4797–4806.

Zetter B. R. (1988) 'Endothelial heterogeneity: Influence of vessel size, organ localizaton, and species specificity on the properties of cultured endothelial cells', in *Endothelial Cells*, vol. 2, Ryan U.S., Ed., CRC Press, pp. 64–79.

Zhang, H. T., Gorn, M., Smith, K., Graham, A. P. *et al.* (1999) 'Transcriptional profiling of human microvascular endothelial cells in the proliferative and quiescent state using cDNA arrays', *Angiogenesis*, **3**(3), pp. 211–219.

Zhang, Y., Deng, Y., Luther, T., Muller, M. *et al.* (1994) 'Tissue factor controls the balance of angiogenic and antiangiogenic properties of tumour cells in mice', *J. Clin. Invest*, **94**(3), pp. 1320–1327.

Zhong, T. P., Rosenberg, M., Mohideen, M. A., Weinstein, B. and Fishman, M. C. (2000) 'gridlock, an HLH gene required for assembly of the aorta in zebrafish', *Science*, **287**(5459), pp. 1820–1824.

Zhu, D. Z., Cheng, C. F. and Pauli, B. U. (1991) 'Mediation of lung metastasis of murine melanomas by a lung-specific endothelial cell adhesion molecule', *Proc. Natl. Acad. Sci. USA*, **88**(21), pp. 9568–9572.

2

Endothelial cell proliferation assays

Wen-Sen Lee

Abstract

Angiogenesis is the generation of new blood vessels which occurs when a vascular bed is extended, such as in tissue growth, wound healing and tumour formation. In experimental studies, a useful, indirect estimation of angiogenesis activity is obtained by assessing the intensity of proliferation of the endothelial cell component of vascular tissue. Such endothelial cells are often isolated and maintained using common cell culture procedures, so that the effects of various mitogens (growth promoters) on cellular proliferation may be observed *in vitro*. A number of techniques for assessing cell proliferation are now in common use. In this chapter, we describe the basic methods and relevant principles. The most frequently used *in vitro* model for studying endothelial cell proliferation employs cultured human umbilical vein endothelial cells (HUVEC), which can be easily obtained. Each of the methods entails some advantages and some intrinsic weaknesses. To achieve the most reliable results, the combined use of two or more methods is to be recommended.

Keywords

angiogenesis; thymidine incorporation; MTT assay; flow cytometry; BrdU

2.1 Introduction

Primary endothelial cell cultures

Endothelial cell proliferation is one of the major events thought to be essential for angiogenesis. To study cell proliferation, endothelial cells are first isolated from

Angiogenesis Assays   Edited by Carolyn A. Staton, Claire Lewis and Roy Bicknell
© 2006 John Wiley & Sons, Ltd

the vasculature and cultivated. Although the validity of the cultured cell as a model of physiological functions *in vivo* has been criticized as being too different from the natural cellular environment, it can become a very valuable tool as long as the culture conditions are adjusted to mimic closely the *in vivo* situation.

Vascular endothelial cells form a simple squamous epithelium that consists of a continuous, single cell layer, which lines the inner surface of blood vessels. Despite its apparent morphological simplicity, increasing evidence has indicated that vascular endothelial cells exhibit morphological and functional heterogeneity (Garlanda and Dejana, 1997). They vary not only in different organs but also in vessels of different calibres within an organ. For *in vitro* study of endothelial cell proliferation, both primary cells, which are derived from normal, healthy vasculature, and propagated cell lines, which can be induced by a variety of conditions, have been used in the laboratory. Although cell lines provide a relatively homogeneous source of material, and can maintain indefinite growth, they may express numerous altered characteristics. In contrast, primary cells provide the closest approximation to the characteristic cells *in vivo* (see Chapter 1).

Endothelial cells from the vasculature of different organs have been dissected and investigated in great detail for the purpose of experimentation. At present, cultured human umbilical vein endothelial cells (HUVEC) are the most frequently employed as an *in vitro* model to study endothelial proliferation (Figure 2.1). Although angiogenesis is ordinarily not observed in the human umbilical vein, this tissue itself is abundantly available from the obstetrical ward, and endothelial cells extracted from such vein specimens, and maintained in culture, can generally be induced to undergo proliferation.

A more appropriate cell type for studying angiogenesis activity is the microvascular endothelial cell, which can be isolated from human foreskin, another tissue often available from the obstetrician (see Chapter 1). To achieve purification of these cells, a cell sorting technique is required and this is a challenging and arduous procedure that could be discouraging to many researchers.

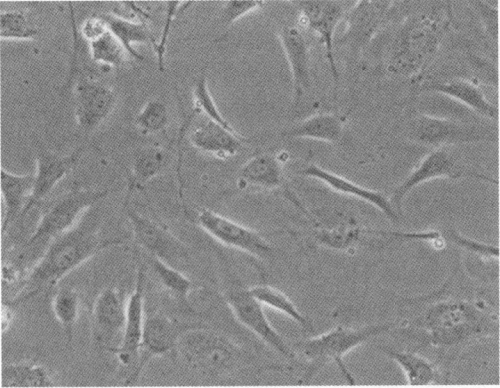

Figure 2.1 Phase-contrast photomicrograph of human umbilical vein endothelial cells (HUVEC) grown in culture medium supplemented with 10 per cent FBS.

Vascular endothelial cells isolated from other species including bovine, rabbit, sheep, pig, rat and mouse have been cultivated and used in biomedical research. One of the advantages in using animal tissue is the large amount of material available. Moreover, certain endothelial cell isolation techniques can be carried out on animals but not humans. However, the biochemical characteristics of some cells may be different in different species. Therefore, in research on the behaviour of human cells, human primary endothelial cell culture is more appropriate.

A major limitation that has to be overcome in order to study endothelial cell proliferation *in vitro* is the difficulty in establishing culture systems that allow the survival and proliferation of the small number of vascular endothelial cells that are isolated from vasculature by perfusion with enzymes such as collagenase or trypsin. Unlike the transformed cell lines, the cultured primary cells have only a limited lifespan. Eventually, the cells become senescent and the culture is lost. However, there is less concern about pathological changes present in the untransformed cells. In our studies of endothelial cell proliferation assays, the vascular endothelial cells from passages 3–10 are used.

Many cultured animal cells will only grow, proliferate and function when attached to surfaces. Most cells require attachment and spreading on a substrate in order to proliferate (anchorage dependent). Inadequate spreading due to poor attachment or confluence will usually lead to impaired cell proliferation. Glass or plastic are commonly used as culture substrates, however, these culture substrates might not be able to support the adhesion of some primary cells. Increasing evidence has indicated that treatment of the substrate with attachment factors, such as gelatin or collagen (purified extracellular matrix components), promotes cell adhesion to common culture substrates. For *in vitro* study of endothelial cell proliferation, the plates or flasks are often coated with sterilized 1 per cent gelatin in PBS before use (Folkman *et al.*, 1979).

An appropriate cell density is also very important for the study of proliferation *in vitro*. By nature, cells should grow and proliferate at low densities. For most primary cultured cells, including vascular endothelial cells, a marked reduction in proliferative activity occurs when they reach confluence, that is, cell–cell contact inhibition occurs when all available growth area becomes occupied. Density-dependent inhibition of proliferation reflects diminished nutrient supply and the accumulation of cell-derived factors (including waste products) into the medium. Since endothelial cell proliferation is strongly inhibited after confluence is reached, a cell density lower than 70–80 per cent confluence is recommended for evaluating any particular endothelial cell proliferation effect.

2.2 Cell proliferation

Normally, vascular endothelial cells are quiescent and proliferation occurs only during embryonic development, or in female reproductive organs during the ovarian cycle and for wound healing. In contrast, vascular endothelial cells in culture resume proliferative activity and enter into the cell cycle if the *in vitro* environment meets the fundamental requirements for this cellular function. The cell cycle consists of four phases including G_1, S, G_2 and M. DNA synthesis occurs only during the S-phase of the cell cycle. The progression of cell cycle activity is largely controlled by coordinated successive activation of certain cyclin-dependent kinases (CDK). The sequential activation of members of the CDK family of reactants and the consequent phosphorylation of certain substrates promotes the progression through the cell cycle.

Since investigations of endothelial cell proliferation assays are performed in HUVEC in our laboratory, techniques and conditions appropriate for cultivating HUVEC are described in this chapter. These techniques can, however, be easily adapted to culture other vascular endothelial cell types.

2.3   Cell proliferation assays

Cell counting assay

In cultured resting cells, the intracellular signalling proteins and genes that can be activated by growth factors are inactive and the actual number of cells in a culture will remain constant. When the cells are stimulated by growth factors or mitogens, these signalling proteins and genes become active and the cells proliferate. Cell proliferation can be measured by the number of cells that are dividing. One easy way of measuring this parameter is by cell-counting assays. In these assays, a defined number of cells are seeded into an appropriate culture chamber, and after a certain culture period (the length of a cell cycle for most primary culture endothelial cells falls in the range 18–26 h) to allow proliferation to occur, the increase in the number of cells is measured by a suitable cell counting device such as the Coulter counter (an electronic particle counter) or the haemocytometer (using a light microscope). The former is the more convenient, accurate and rapid and can be used to count low density of cells (Figure 2.2), but is expensive; the latter is often the simplest and cheapest method of counting cells, but requires higher density of cells and is more prone to sampling error (most of the errors occur by incorrect sampling and transfer of cells to the counting chamber). However, haemocytometer counting allows a visual estimation of the 'death' of the cells and, combined with a dye exclusion method such as trypan blue, can be used to estimate cell viability.

[3H]Thymidine incorporation assay (Combs *et al.*, 1965; Yu *et al.*, 2004)

One way that is commonly employed to gain an estimation of cell proliferation in a cell culture is to observe the intensity of DNA synthesis. This is often achieved by

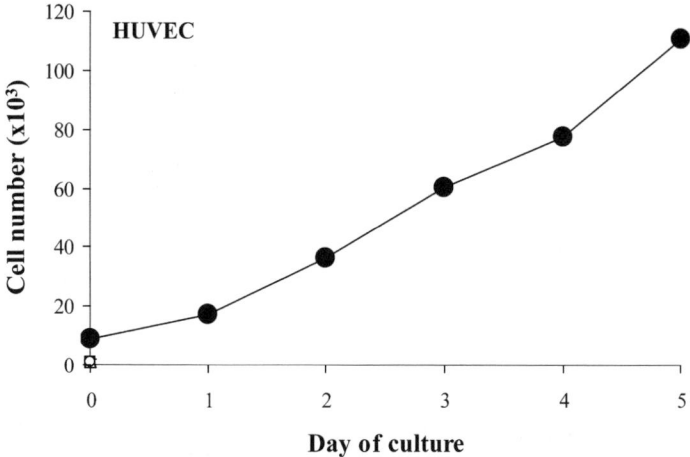

Figure 2.2   Cell growth curve HUVEC were grown in culture media containing 10 per cent FBS. Media are changed daily until cell counting. The number of cells is measured by the Coulter counter.

using a radiolabelled nucleotide (such as tritiated thymidine or tritiated deoxycytidine) as a precursor, and thus as a marker to follow the rate of incorporation of the nucleotide into newly synthesized DNA. During cell proliferation, cells undergoing mitosis for cell division must replicate DNA to produce daughter genes, which are segregated into the daughter cells. It is assumed that the rate of incorporation of [^{3}H]thymidine faithfully reflects the rate of nucleotide utilization for *de novo* synthesis of DNA in the process of the gene replication for cellular mitosis.

In actual practice, in order to minimize data scatter and to optimize reproducibility, cell cultures are usually pre-incubated in culture medium with reduced (0.5 per cent instead of 10 per cent) fetal bovine serum (FBS) content for 24 h in order to constrain mitotic activity and thereby synchronize the entire cell population into the G0 phase (Hsu *et al.*, 2003). Thereafter, to begin the assay, the cells are restored to 10 per cent FBS, and [^{3}H]thymidine marker is added during the S-phase (DNA synthesis phase) of the cell cycle (the S-phase of HUVEC begins at 18–20 h after challenge with 10 per cent FBS as shown in Figure 2.3) and [^{3}H]thymidine incorporation is allowed to proceed during the last 2–4 h of the incubation. Then the [3H]-labelled DNA is precipitated with trichloroacetic acid (TCA) and dissolved in 0.2N NaOH for radioactivity analysis by liquid scintillation counting. The radioactivity is proportional to the neosynthesis of DNA. Cells that incorporate radiolabelled thymidine can also be identified by autoradiography using light microscopy. The [^{3}H]thymidine labelling of DNA recorded in this manner represents the percentage of cells that are in S-phase during the time of exposure to [^{3}H]thymidine and certainly reflects neosynthesis of DNA. Labelling activity is commonly used to quantify the response of a cell population to a mitogenic or chemical stimulus. However, an increase of this labelling might in

(a)   (b)

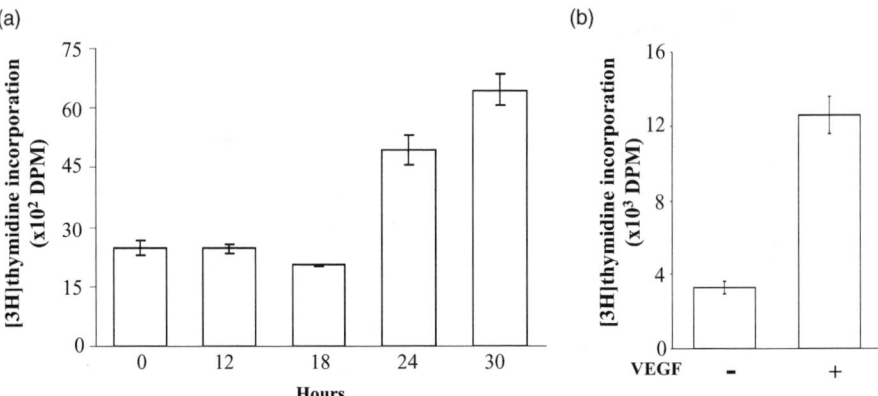

Figure 2.3   Effect of serum and growth factor on DNA synthesis (a) Time-dependent effect of FBS on the [³H]thymidine incorporation into HUVEC. [³H]thymidine incorporation was conducted after HUVEC release from quiescence by incubation in culture media supplemented with 10 per cent FBS. The increased [³H]thymidine incorporation was observed at 24 h after challenged with 10 per cent FBS. (b) VEGF induced an increase of [³H]thymidine incorporation into HUVEC.

fact be due to DNA repair activities not directly involved in cell mitosis and a decrease of this labelling might be due to cell death rather than cell growth retardation.

MTT cell proliferation assay (Lee *et al.*, 1997)

Although [³H]thymidine incorporation and cell number counting assays have been commonly used in the determination of cell proliferation, their results might not always reflect the status of cell proliferation. For instance, a decrease of [³H]thymidine and cell number might be due to cell death instead of a decrease of cell proliferation. MTT [3-(4,5-dimethylthiazol-2-yl)-2,5-diphenyltetrazolium bromide] assays have been applied to overcome this problem and are now recognized as a safe, accurate alternative to radiometric testing. The MTT assay, first described by Mosmann in 1983, has been used for the quantitative determination of cellular proliferation and viability (Mosmann, 1983). Recently, this principle has also been used successfully for susceptibility testing of yeasts and filamentous fungi and for assessment of chemosensitivity testing in established cell lines. This technique depends on the fact that viable cells contain a mitochondrial dehydrogenase enzyme which cleaves the yellow tetrazolium salt MTT taken up by cells to produce purple/blue formazan crystals which are insoluble, are impermeable to cell membranes, and accumulate within living cells. For the purpose of the assay, these formazan crystals may be dissolved by adding the solubilization reagent (often DMSO), and the solubilized formazan product can be quantitated using a multiwell scanning spectrophotometer (ELISA reader).

Since dehydrogenase content is relatively consistent among cells of a specific type, the number of living cells present in each well will be directly proportional to the colour intensity detected by spectrophotometry, which in turn is proportional to the cellular production of the formazan pigment. An increase in the number of living cells will accordingly be reflected by a stronger colour reading. On the other hand, the ability of cells to reduce MTT provides an indication of the mitochondrial integrity and activity, which, in turn, may be interpreted as a measure of cell number and/or viability. The advantage of this technique, as compared with the Coulter cell counting method, is that this MTT assay detects living cells so that dead cells or cell fragments will not give false-positive readings. Moreover, this method is simple to perform, is readily automated for data processing and is not expensive to set up. Although the MTT assay provides rapid detection of cell proliferation, there are several limitations of this technique. A major weak point of this assay is that contamination and some chemical agents might alter the activity of the dehydrogenase and therefore will alter the colorimetric reading. Certain compounds may selectively affect the mitochondria of the cells resulting in an overestimated or underestimated level of toxicity. Moreover, the MTT assay is not readily usable for cells with intrinsically low mitochondrial activity. Different cell lines may give different absorbance levels even at similar degrees of cultured cell confluence. A cautionary concern is that MTT is mutagenic and, therefore, constitutes a health hazard in the laboratory. Finally, since the coloured formazan product deteriorates over time, once the assay is initiated it must be completed without delay.

Identification of cell proliferation markers (Dolbeare *et al.*, 1983; Begg *et al.*, 1985; Lengronne *et al.*, 2001; Yoshizumi *et al.*, 1995; Lee *et al.*, 1990; Lin *et al.*, 2001)

Although measurement of [3H]thymidine incorporation into DNA and the MTT assay are the more commonly used method for observing cell proliferation, they cannot be used to analyse cell division at the level of the individual cell. Detection of proliferating cells in the individual cell can be achieved by a number of methods. Here, we introduce two approaches using the bromodeoxyuridine (BrdU) incorporation assay or PCNA (proliferating cell nuclear antigen) staining techniques. Both BrdU and PCNA have been used as markers of cell proliferation.

   The incorporation of BrdU is a convenient non-radioactive method for assessing proliferative activity of growing cells. Techniques using the nucleotide analogue BrdU, which competes with thymidine during the S-phase of the cell cycle, is firmly incorporated into newly synthesized DNA, and is detected by immunocytochemistry, to allow observation of cell division at the level of individual cells. Accordingly, detection of the BrdU incorporated into newly synthesized DNA allows direct identification of individual cells undergoing mitosis which have progressed through the S-phase of the cell cycle during the BrdU labelling period.

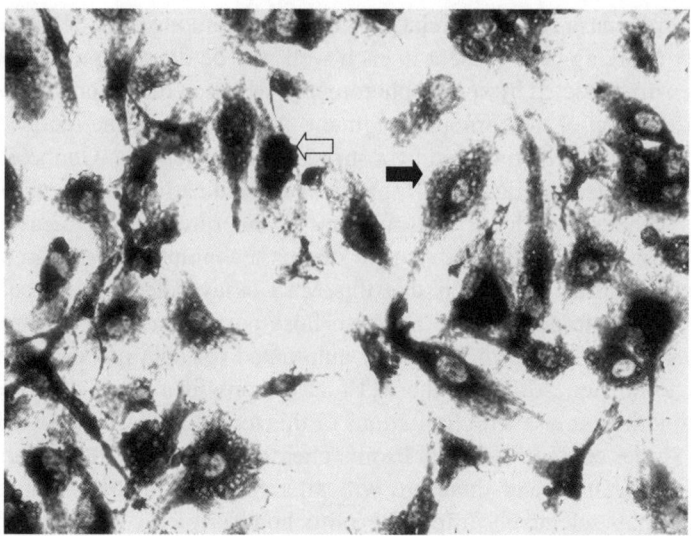

Figure 2.4 Expression of PCNA in HUVEC the cells were grown in culture media containing 10 per cent FBS. PCNA immunoreactivity was detected in cultured HUVEC by single immunoperoxidase methods. A cell that expressed PCNA is indicated by an open arrow. An endothelial cell that did not express PCNA is indicated by a filled arrow. The cells were counterstained with eosin to visualize the cytoplasm.

Although BrdU immunocytochemistry has proved useful for identifying S-phase cells, disadvantages include BrdU being incorporated into cell DNA during DNA repair and BrdU is a mutagen and is not suitable for genetic toxicology studies *in vivo*. For many years, the most common method used to identify proliferating cells without the use of BrdU has been immunocytochemical detection of PCNA. PCNA, a cofactor of cyclin D, is a much conserved and highly regulated protein necessary for cell cycle progression and has been identified as a polymerase-associated protein. During the cell cycle, PCNA translocates from the cytoplasm to the nucleus at the beginning of the S-phase, and forms multiple complexes with CDK2 at late G_1- or early S-phase. Expression of PCNA increases during the G_1-phase, peaks at the S-phase, and declines during G_2/M-phases of the cell cycle. Hence PCNA has been used to distinguish proliferating cells from resting cells. Immuncytochemical analysis examining the expression of PCNA in the cell may be employed to provide visual and quantitative evidence of cell proliferation (Figure 2.4).

BrdU and PCNA can be immunocytochemically detected *in vitro* as well as *in vivo* by means of anti-BrdU and PCNA antibodies respectively, and allow the specific identification of cells that are dividing during the period of BrdU uptake and PCNA expression. Such immunocytochemical staining techniques allow both the microscopic visualization of dividing cells and the detailed observation of tissue morphology by counterstaining (e.g. with haematoxylin/eosin or methyl green). Thus, it is possible to visualize cells, which have incorporated BrdU into DNA or express PCNA and hence are undergoing proliferation, and to observe the localization of these cells within a specific target tissue. The amount of BrdU incorporated into DNA or PCNA-expressing cells may be quantified either by microscopically identifying and counting the number of the labelled nuclei, or by chemical measurement of the total amount of BrdU or PCNA in a population of cells using Western blot analysis techniques. The intensity of incorporation of the BrdU into DNA or PCNA expression is directly proportional to the degree of ongoing cell division.

Although this methodology is a valuable tool for detecting proliferation at the level of the individual cell, we recognize the limitation of using any nucleotide analogue as a quantitative measure of proliferation. As described above, BrdU is also incorporated into cell DNA during DNA repair. PCNA is also involved in DNA repair, which suggests that PCNA may be expressed by cells that are not cycling. Accordingly, incorporation of thymidine analogues or expression of PCNA does not always correlate with true increases in DNA content and cell proliferation (Neckers *et al.*, 1995). A methodological drawback to the application of this technique is that over-fixation can reduce the signal and affect the accuracy of detection of labelled cells with immunocytochemical staining. Hence these techniques do require meticulous attention to the details of histochemical methodologies. Moreover, since this methodology is involved in the antibody/antigen interaction, the availability of antigen for antibody binding and cross-reactivity of antibody is always an issue to be considered.

Flow cytometry assay (Crissman and Tobey, 1974; Darzynkiewicz *et al.*, 1977; Ho *et al.*, 2004; Shih *et al.*, 2004)

The most significant cellular parameter used to monitor cell proliferation activity is DNA content per cell. Flow cytometry is a very powerful tool for rapid, reliable and sensitive measurement of the DNA content of cells using a Fluorescence Activated Cell Sorter (FACS), which counts only live cells and excludes dead and damaged cells. Like the Coulter Counter, the FACS also passes cells suspended in culture medium in single file through a detector which counts the total number of cells and segregates them according to cell size, large from small, into groups containing cells of the same size. In addition, FACS detects cells previously treated with a fluorescing dye, such as propidium iodide (PI), which links specifically to DNA, and segregates cells into groups containing cells in G0/G1 mitotic phase (containing 1X amount of DNA), in S-phase (containing mixed 1X to 2X DNA) and G2/M-phase (containing 2X DNA). By this means, proliferative activity in a culture of cells will be revealed by the increase of cell population in the S-phase of the cell cycle (Figure 2.5). The cell-cycle distribution and proliferative state of a population of cells are important parameters when studying many live-cell functions. Of course, as in the case of the MTT Cell Proliferation Assay, the

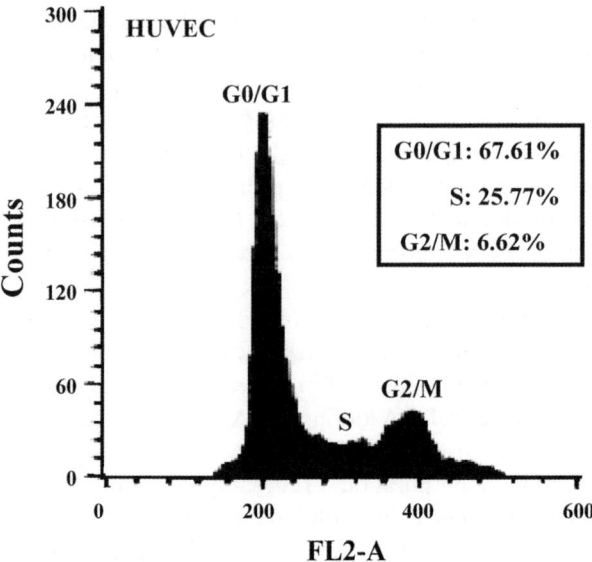

Figure 2.5 FACS analysis of DNA content was performed 24 h after release from quiescence by incubation in culture media supplemented with 10 per cent FBS. Percentage of cells at the G0/G1-, S-, or G2/M-phase of the cell cycle was determined using established CellFIT DNA analysis software.

final value for the total number of cells using flow cytometry analysis will include only living cells, and will not count dead or dying cells in the culture. One major limitation in using this technique is that cellular autofluorescence is a common obstacle in flow cytometry, and may cause difficulty in analysis when the overall level of fluorescence is low.

2.4   Conclusions

The measurement of cell proliferation is a valuable tool in a wide range of research areas including angiogenesis. Several approaches have been used in the past. In this chapter, we described the basic methods and relevant principles of endothelial cell proliferation assays. Cell counting is a very simple and direct method for measuring cell number changes. However, the results might not always reflect accurately the level of cell proliferation. Incorporation of thymidine analogues is a quick way commonly employed to gain an estimation of cell proliferation, but may not always correlate precisely with true increases in DNA content and cell proliferation. The advantage of MTT assay is that this method rapidly detects living cells so that dead cells or cell fragments will not give false-positive readings. A significant weak point of MTT assay is that some contamination or chemical agents might alter the activity of cellular dehydrogenases and alter the basic chromogenic reaction. Immunocytochemical detection of cell proliferation markers can be used to analyse cell division at the level of the individual cell. However, the accessibility of the intracellular proliferation marker antigens for binding to the antibodies which are applied to defect and identify them might be a limitation of this technique. Also, as with all immunological procedures, there could be questions regarding the specificity or cross-reactivity of the antibody probes which are employed. The key advantage of flow cytometry is that a very large number of cells can be evaluated in a very short time. A limitation of flow cytometry, however, is the concentration of cells present in the samples, and hence the large amount of cultured cells required.

Here we described several approaches commonly used for assessment of proliferative activity. Each test method has its merits and weak points. To achieve the most reliable results, the combined use of two or more methods is to be recommended.

Acknowledgements

I am grateful to Professor Winton Tong and Professor Ling-Ru Lee for helpful discussions during the preparation of the manuscript and the technical assistance of Ms Pei-Yin Ho.

References

Begg, A. C., McNally, N. J., Shrieve, D. C. and Karcher, H. (1985) 'A method to measure the duration of DNA synthesis and potential doubling time from a single sample', *Cytometry*, **6**, pp. 620–626.

Combs, J. W., Lagunoff, D. and Benditt, E. P. (1965) 'Differentiation and proliferation of embryonic mast cells of the rat', *J. Cell Biol.*, **25**, pp. 577–592.

Crissman, H. A. and Tobey, R. A. (1974) 'Cell-cycle analysis in 20 minutes', *Science*, **184**, pp. 1297–1298.

Dolbeare, F. A., Gratzner, H. G., Pallavicini, M. G. and Gray, J. W. (1983) 'Flow cytometric measurement of total DNA content and incorporated bromodeoxyuridine', *Proc. Natl. Acad. Sci. USA*, **80**, pp. 5573–5577.

Darzynkiewicz, Z., Traganos, F., Sharpless, T. K. and Melamed M. R. (1977) 'Cell cycle-related changes in nuclear chromatin of stimulated lymphocytes as measured by flow cytometry', *Cancer Res.*, **37**, pp. 4635–4540.

Folkman, J., Haudenschild, C. C. and Zatter, B. R. (1979) 'Long-term culture of capillary endothelial cells', *Proc. Natl. Acad. Sci. USA*, **76**, pp. 5217–5221.

Garlanda, C. and Dejana, E. (1997) 'Heterogeneity of endothelial cells: specific markers', *Arterioscler. Thromb. Vasc. Biol.*, **17**, pp. 1193–1202.

Ho, P. Y., Liang, Y. C., Ho, Y. S., Chen, C. T. and Lee W. S. (2004) 'Inhibition of human vascular endothelial cells proliferation by terbinafine', *Int. J. Cancer*, **111**, pp. 51–59.

Hsu, H. K., Juan, S. H., Ho, P. Y., Liang, Y. C. *et al.* (2003) 'YC-1 inhibits proliferation of human vascular endothelial cells through a cyclic GMP-independent pathway', *Biochem. Pharmacol.*, **66**, pp. 263–271.

Lee, W. S., Smith, M. S. and Hoffman G. E. (1990) 'Luteinizing hormone-releasing hormone neurons express Fos protein during the proestrous surge of luteinizing hormone', *Proc. Natl. Acad. Sci. USA*, **87**, pp. 5163–5167.

Lee, W. S., Harder, J. A., Yoshizumi, M., Lee, M. E. and Haber, E. (1997) 'Progesterone inhibits arterial smooth muscle cell proliferation', *Nat. Med.*, **3**, pp. 1005–1008.

Lengronne, A., Pasero, P., Bensimon, A. and Schwob, E. (2001) 'Monitoring S phase progression globally and locally using BrdU incorporation in TK$^+$ yeast strains', *Nucleic Acids Res.*, **29**, pp. 1433–1442.

Lin, S. Y., Chang, Y. T., Liu, J. D., Yu, C. H. *et al.* (2001) 'Molecular mechanisms of apoptosis induced by magnolol in colon and liver cancer cells', *Mol. Carcinogenesis*, **32**, pp. 73–83.

Mosmann, T. (1983) 'Rapid colorimetric assay for cellular growth and survival: application to proliferation and cytotoxicity assays', *J. Immunol. Methods*, **65**, pp. 55–63.

Neckers, L. M., Funkhouser, W. K., Trepel, J. B., Cossman, J. and Gratzner, H. G. (1995) 'Significant non-s-phase DNA synthesis visualized by flow cytometry in activated and in malignant human lymphoid cells', *Exp. Cell Res.*, **156**, pp. 429–438.

Shih, C. R., Wu, J., Liu, Y., Liang, Y. C. *et al.* (2004) 'Anti-proliferation effect of 5,5-diphenyl-2-thiohydantoin (DPTH) in human vascular endothelial cells', *Biochem. Pharmacol.*, **67**, pp. 67–75.

Yoshizumi, M., Lee, W. S., Hsieh, C. M., Tsai, J. C. *et al.* (1995) 'Disappearance of cyclin A correlates with permanent withdrawal of cardiomyocytes from the cell cycle in human and rat hearts', *J. Clin. Invest.*, **95**, pp. 2275–2280.

Yu, C. H., Wu, J., Su, Y. F., Ho, P. Y. *et al.* (2004) 'Anti-proliferation effect of 3-amino-2-imino-3,4-dihydro-2H-1,3-benzothiazin-4-one (BJ-601) on human vascular endothelial cells: G0/G1 p21-associated cell cycle arrest', *Biochem. Pharmacol.*, **67**, pp. 1907–1916.

3

Endothelial cell migration assays

Christos Polytarchou, Maria Hatziapostolou and **Evangelia Papadimitriou**

Abstract

Migration of endothelial cells plays a central role in many multi-step and complex physiological and pathological events, such as the repair of the blood vessel wall after tissue injury, in vasculogenesis during embryonic development and in angiogenesis during tumour growth. A number of *in vitro* endothelial cell migration assays have been developed, in an effort to isolate endothelial cell migration from this complexity. Among these are the transfilter assay, the wound healing assay, the Teflon fence assay and the phagokinetic track assay. There is also increasing interest and effort in developing assay(s) for assessing endothelial cell migration *in vivo*. Such approaches include the aortic ring assay, the Matrigel plug assay and, most recently, the zebrafish assay. A critical appraisal of the strengths and weaknesses, as well as recent applications and developments of these assays are discussed.

Keywords

migration; endothelial cells; angiogenesis; transfilter assay; wound healing assay; Teflon fence assay; phagokinetic track assay

3.1 Introduction

Angiogenesis, the formation of new blood vessels from pre-existing ones, is a complex, multi-step process which characterizes a variety of physiological and malignant conditions (Carmeliet and Jain, 2000). Endothelial cells are the main cellular component of the vessel wall and have important roles in the process of angiogenesis. During this active process, endothelial cells are stimulated to degrade the pre-existing basement membrane and to migrate into the perivascular stroma, where they proliferate and form new blood vessels. Inhibition of

Angiogenesis Assays   Edited by Carolyn A. Staton, Claire Lewis and Roy Bicknell
© 2006 John Wiley & Sons, Ltd

endothelial cell activation inhibits/retards angiogenesis (Jekunen and Kairemo, 2003) and the *in vitro* models for studying angiogenesis mainly focus on the effect of exogenous or endogenous agents on endothelial cell proliferation, migration and tube formation on extracellular matrices.

Several tests have been developed to estimate, in a qualitative and/or quantitative manner, the migratory response of endothelial cells to pro- or anti-angiogenic factors. The most commonly used assays are outlined in this chapter.

3.2   Transfilter assay

The transfilter assay, a modification of the Boyden chamber assay (Boyden, 1962), is the most frequently used system to quantitatively assess endothelial cell migration. It is a three-dimensional *in vitro* assay, performed in modified or unmodified chambers. It is used for evaluation of the involvement of adhesion molecules, extracellular matrix components and endogenously expressed or exogenously supplied molecules in endothelial cell migration. Due to its simplicity, it is also a useful assay for screening pharmacological agents.

This assay is based on the migration of cells to an attractant, through a porous, inert polycarbonate or polypropylene filter, which has pores (8.0 μm diameter) that allow only active passage of the cells. The endothelial cells are placed on top of the filter and migrate across the latter to a pro- or anti-angiogenic factor placed in the lower chamber, usually within 4 to 6 h. In order to assess the migratory effect, non-migratory cells remaining in the upper compartment are removed with a cotton swab. Cells that migrate through the filter and remain on the lower surface are fixed, stained (e.g. with haematoxylin and eosin, toluidine blue or other dyes) and quantified by counting either the entire area or different fields of each filter, using a grid and a light microscope (Figure 3.1) (Hagan *et al.*, 2003; Miao *et al.*, 2000; Staton *et al.*, 2004). The filter may be coated with single extracellular matrix (ECM) molecules, such as collagen or fibronectin, or complex matrices, such as Matrigel (Ohmori *et al.*, 2001; Albini *et al.*, 2004), in an effort to simulate the *in vivo* microenvironment.

The transfilter assay, modified or not, has several advantages. First, it is highly sensitive to small differences in concentration gradients and highly reproducible (Falk *et al.*, 1980). Its short duration in comparison to other assays (4–6 h) avoids the confounding effects of endothelial cell proliferation and/or matrix synthesis and in its simplest version there is no need of high tech imaging. The filter could also be examined for specific changes in protein expression and location.

The assay can be used to distinguish between chemotaxis (directional migration: biased and persistent movement of cells in response to a gradient of soluble stimuli, attractive or repulsive) and chemokinesis (random motility: the persistent random movement of cells that over a long period of time results in non-directional displacement). Positive, negative and null gradient conditions are created, by adding different amounts of the test substance to the lower and/or upper

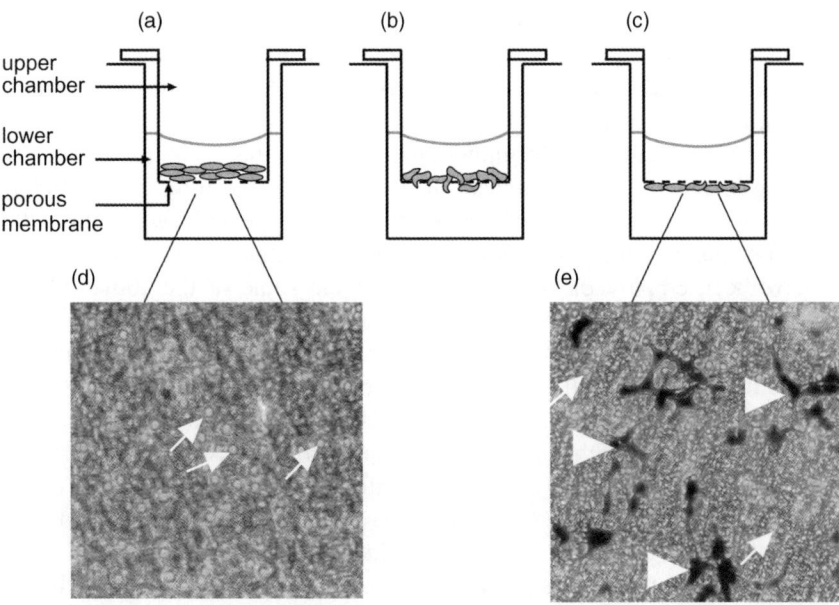

Figure 3.1 Schematic representation of the transfilter assay (a) The endothelial cells are plated on top of the filter inside the upper chamber of the transwell, while the test substance is placed in the lower chamber. (b) The pores of the filter are small enough to allow only the active passage of the cells. Chemotactic agents induce the migration of cells through the filter into the lower compartment. (c) Non-migratory cells remaining in the upper compartment are removed. Staining of the migrated cells and cell enumeration by eye provides the extent of migration. (d) and (e) Representative pictures showing a polycarbonate filter without cells (d) or with stained transmigrating cells (e). Small white arrows indicate pores; big white arrowheads indicate endothelial cells.

compartments (Taraboletti *et al.*, 1990). It is also possible to create a gradient of substrate-bound substances between the two sides of the filter, in order to study haptotactic motility (phenomenon similar to chemotaxis except in response to a surface-bound gradient) (Taraboletti *et al.*, 1993).

A major problem when testing exogenously supplied agents is to maintain transfilter gradients for prolonged periods of time, as the concentration equilibrates between compartments over time. Moreover, the lack of specific information about the gradient generated in the porous mesh and the inability to observe cell motion during experiments can be limiting factors of its utility (Smith *et al.*, 2004). It should also be kept in mind that cells migrate through an 'alien' matrix, a polycarbonate or polypropylene filter (Cary and Guan, 1999). When modified chambers are used, the assay time is increased due to the necessity of pre-coating the inserted filter with the desired ECM protein. Usually, this is achieved with overnight coating. Both the concentration of the ECM component and the coating procedure should be optimized so that the pores of the filter will not be occluded. However, the manual coating increases inter-assay variations, leading to the requirement of multiple samples and repeats. Today, there are several commercially available chemotaxis chambers with the bottom side of the filter pre-coated with ECM proteins.

The standard method of staining and counting the cells by eye is time consuming and when small numbers of cells traverse the filter, it is difficult to obtain accurate and statistically significant results, especially when the cellular distribution and/or staining are uneven. Alternative methods of counting migrating cells have been employed, in order to circumvent some of the above disadvantages. In all cases, the removal of non-migratory cells, using a cotton swab, is necessary. One possibility is to use computers equipped with digital cameras that scan the filter of each well systematically and software that recognizes stained cells. Nevertheless, the computer software cannot always distinguish between the pores of the filter and the stained cells (Fuller *et al.*, 2003; Debeir *et al.*, 2004).

Endothelial cells can also be metabolically labelled prior to use. After the removal of non-migratory cells, migration is quantified through radioactive counting (Dvorak *et al.*, 1977). This procedure requires proper equipment (such as scintillation counter and separate laboratory area) and generates radioactive waste. In order to avoid radioactive labelling, the number of viable, migrating cells can be quantified by measurement of the reduction of 3-(4,5-dimethylthiazol-2-yl)-2,5-diphenol tetrazolium bromide (Shi *et al.*, 1993). Another possibility is to stain migrating cells (e.g. with crystal violet) and then elute the stain with an extraction buffer, transfer to a microplate and measure spectrophotometrically. Optical density of the stained cells is correlated with cell migration (Santiago and Erickson, 2002).

Fluorescent labelling of the cells permits accurate measurement of the number of cells that transmigrate and provides time-related data. Migrating cells can be measured with a fluorescence plate reader, or visualized and counted under a fluorescence microscope (Gildea *et al.*, 2000; Jones *et al.*, 2001). The removal of non-migratory cells is still necessary, because of the interference of the emitted light from the upper to the bottom compartment and vice versa. The replacement of the transparent material of the filter with a light shielding material (such as polyethylene terephthalate) overcomes this disadvantage. Thus, fluorescently labelled non-migratory cells above the filter are invisible and the removal of non-migratory cells is not required. The major advantage of this approach is the capability of time-course analysis avoiding the destruction of the insert (Goukassian *et al.*, 2001).

Initiation of cell migration requires localized actin polymerization, which leads to formation of a dominant leading pseudopodium and a rear cellular compartment (Lauffenburger and Horwitz, 1996; Parent and Devreotes, 1999). A molecular understanding of this process has been limited due to the inability biochemically to separate the leading pseudopodium from the rear of the cell. A modification of the transfilter assay offers the opportunity to monitor pseudopodia extension. Cells are placed into the upper compartment of a chamber equipped with a $3.0\,\mu m$

porous polycarbonate filter, coated on both sides with an optimal amount of ECM protein. The chemoattractant is placed in the lower chamber to establish a gradient. Cells are allowed to extend pseudopodia through the pores toward the direction of the gradient for various time points. To initiate pseudopodia retraction, the chemoattractant is removed or an equivalent amount of chemoattractant is placed in the upper chamber to create a uniform concentration. The cell body on the upper surface is manually removed with a cotton swab and the total pseudopodia protein on the lower surface of the filter is determined. It is also possible to stain pseudopodia (e.g. with crystal violet) and then elute the dye with acetic acid, which can be measured in an ELISA plate reader (Klemke *et al.*, 1998; Cho and Klemke, 2002). Similar approaches could be applied to endothelial cells to study changes in the cytoskeleton that play a role on endothelial cell migration.

3.3 'Wound healing' assay

An alternative type of migration assay is based on the idea that endothelial cell migration into a denuded area is a pivotal event in wound healing *in vivo*. 'Wound healing' is a classic and commonly used method for studying endothelial cell migration and its underlying biology and is described as a two-dimensional (2D) *in vitro* assay (Lampugnani, 1999).

To perform a wound healing assay, endothelial cell monolayers are permitted to reach confluence. Using a scraping tool (a pipette tip or syringe needle or a micropipette guided by a micromanipulator), a portion of the monolayer is cleared of endothelial cells, providing a margin (Wong and Gotlieb, 1984). In order to fill the denuded area, endothelial cells migrate to the wound margin. The monolayers recover and heal the wound in a process that can be observed over a time course of 3–24 h (Figure 3.2). The rate and extent of endothelial cell migration is then monitored by manually imaging samples fixed at various time-points or by time-lapse microscopy (Pepper *et al.*, 1990).

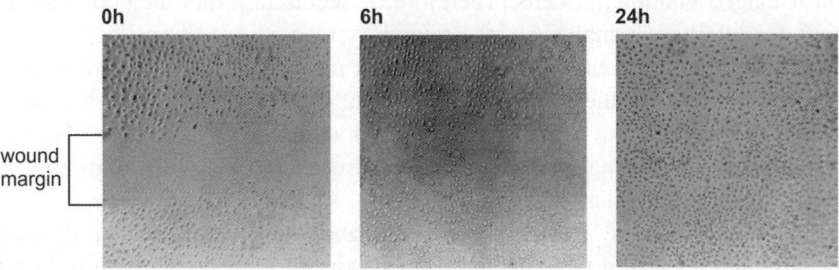

Figure 3.2 'Wound Healing' assay the pictures show human umbilical vein endothelial cells that migrate to heal a wound. The photographs were taken on an optical microscope 0, 6 and 24 h after wounding of the monolayer.

One major advantage of the wound healing assay is that it offers the opportunity to monitor cell migration over time, thus estimating the rate of the migratory response. As the confluent monolayer has been 'wounded' by scraping and the endothelial cells are migrating back to re-form the monolayer, this assay represents one significant aspect of wound healing and has been used for the study of several physiological and pathological processes that include inflammation (Philp et al., 2003; Saga et al., 2003). Furthermore, it has been used with multiple cell types for detailed biological studies, such as the investigation of the role of the Rho family GTPases, Rac, Rho, and Cdc 42 in the establishment of polarity and the regulation of actin cytoskeletal structures (Nobes and Hall, 1999; Etienne-Manneville and Hall, 2001; Fenteany et al., 2000), the investigation of the role of p53 in migration (Sablina et al., 2003), and the study of the orientation of the microtubule organization centre and the Golgi apparatus (Yvon et al., 2002; Magdalena et al., 2003a; Magdalena et al., 2003b). Finally, it has also been used for the discovery and validation of molecules that possibly affect cell migration (Mc Henry et al., 2002; Klein-Soyer et al., 1994; Fischer et al., 1990).

Different approaches have been developed in order to characterize the sequence of events occurring during wound repair in vitro. Fluorescence and immunofluorescence microscopy and time-lapse cinemicrophotography have been used in order to correlate the cytoskeletal events (organization of actin microfilaments, microtubules and centrosomes) with the motile activity of the cells occurring during the closure of the wound. Such approaches revealed that wound repair in an in vitro endothelial monolayer is a multi-step process involving spreading, proliferation and migration events (Wong and Gotlieb, 1988; Coomber and Gotlieb, 1990; Zahm et al., 1997). This makes the wound healing assay inaccurate for the determination of the net migration effect.

Another major disadvantage of this assay is the technical difficulty to run control and experimental groups under identical growth conditions of confluence and to introduce uniformly sized wounds in all samples of an experiment (Auerbach et al., 1991). Moreover, quantification is difficult and arbitrary, and results may vary depending on the size of the wound or the cell growth, which is limited to the wound area. The results may also be influenced by soluble components released from damaged endothelial cells. Therefore, it seems that this method is suitable mainly for qualitative analysis of the migratory response of the cells. Finally, the wound healing assay measures endothelial cell migration over a period of several hours, during which the cells synthesize an endogenous, endothelial-derived matrix, in addition to the matrix component on which they were originally plated. Thus, the wound healing assay cannot measure endothelial cell migration on a defined matrix component (Cai et al., 2000).

In order to overcome at least some of the above difficulties and to develop a high-throughput assay, a new method was introduced for creating uniformly sized wounds in the same position of each well. This method is an adaptation of the wound healing assay to a 384-well format, with the use of a 96-well floating-pin transfer device to transfer solutions between plates. The readouts are provided by four different imaging technologies: automated microscopy, fluorescence and transmitted-light scanners, and fluorescence microscopy. This version of the wound healing assay allows greater than 10 000 perturbations to be screened per day with a quantitative, high-content readout or can be used to characterize small numbers of perturbations in detail (Yarrow *et al.*, 2004).

3.4 Teflon fence assay

In the Teflon fence assay, confluent endothelial cells migrate as a sheet on a substratum of endogenous extracellular matrix or purified matrix component (Pratt *et al.*, 1984). Endothelial cells are initially confined to a region of a well or a glass slide, with a Teflon fence. After plating, they are allowed to proliferate within the confined area till confluency. Immediately following fence removal, surfaces are rinsed to remove any non-adherent cells, fixed and stained (e.g. with Wright Giemsa stain), in order to quantify changes in cell density within increasing distances from the site of the initial cell population, at several time points after fence removal (Figure 3.3). Several modifications of this technique have also been described (Cary and Guan, 1999; Cai *et al.*, 2000).

Using the Teflon fence assay, migration of endothelial cells on gradients of surface-bound ligands can be studied (Smith *et al.*, 2004). This method provides additional information concerning the effect of several surface-bound ligands on endothelial cell migration, in a concentration dependent manner. In contrast to the transfilter and wound healing assays, the cells migrate on a defined surface, allowing measurement of endothelial migration on any matrix component(s). Thus, specific mechanisms of cell–substrate interactions can be analysed.

The major disadvantage is the technical difficulty of setting up the assay. Producing increasing gradients is laborious and time-consuming work. Furthermore,

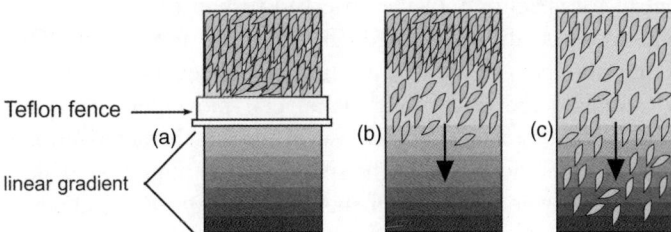

Figure 3.3 Schematic representation of the teflon fence assay (a) The figure illustrates a glass slide with a confluent monolayer of cells behind the Teflon fence. The Teflon fence is removed (b) and cells migrate (c) towards a linearly increasing gradient (indicated by the arrow) of a surface-bound chemoattractant.

the gradient generated should always be confirmed prior to performing this assay. Net migration is difficult to discriminate from cell proliferation, especially when experiments exceed 24 h. The confounding effects of endothelial cell–cell interactions and matrix synthesis can be restricted when the assay is performed immediately after cell attachment within the confined area. Thus, only the effect of the haptotactic gradient on endothelial cell migration can be determined (Smith *et al.*, 2004).

3.5   Phagokinetic track assay

Phagokinetic track methods are used for studying two-dimensional (2D) cell movement *in vitro*. In the first version of this assay, colloidal gold-plated coverslips were used as a substrate for the movement of cells. As cells migrate, they displace the colloidal gold and a track is generated that can be measured for directional properties and total distance (Zetter, 1987). An alternative method has been employed, in order to permit the rapid preparation and screening of multiple 96-well plates. A monolayer of 1 μm polystyrene beads is deposited on the bottom of 96-well plates. Endothelial cells are then plated on the well and migrate, leaving a track similar to that produced in colloidal gold. Cell movement is usually scored after 24 h (Obeso and Auerbach, 1984; Auerbach *et al.*, 1991).

A major disadvantage of these methods is that they measure endothelial cell migration on an 'alien' substrate and not one present *in vivo*. In comparison to the transfilter assay however, the major advantage of the phagokinetic track assays is the more accurate measurement of cell motility. In addition, one has the capability of studying directional effects on cell movement. This type of assay is equally useful for testing inhibitory factors, such as endostatin (Auerbach *et al.*, 2003), motility-enhancing factors, such as fibroblast growth factor-2 and vascular endothelial growth factor (Albrecht-Buehler, 1977) and ECM components (Jozaki *et al.*, 1990). Furthermore, because of its ease, the modified phagokinetic track assay permits utilization of several different endothelial and non-endothelial cell types to control for specificity of the observed response.

One problem is the difficulty in quantifying the movement of cells, which requires the measurement of the cell position across several frames. Mathematical models that simulate individual endothelial cell movement paths can facilitate quantification. Such approaches can be useful for the characterization of different parameters of cell migration, such as the speed and the migration angle (Schienbein and Gruler, 1993), velocity components (Distasi *et al.*, 2002), mean-square displacement (Stokes *et al.*, 1991; Stokes, 1992), persistence time and random motility coefficient (Dickinson and Tranquillo, 1993). Several computer-assisted methods have been developed to analyse the motion of living cells, through automatic tracking of the cell paths recorded using time-lapse microscopy (Solls and Wessels, 1998; Tvarusko *et al.*, 1999; Roy *et al.*, 2002; Ariano *et al.*, 2005).

3.6   *In/ex vivo* approaches for studying endothelial cell migration

A close approximation to *in vivo* endothelial cell migration could possibly be achieved by combining organ culture techniques with cell migration analysis. An example of such an approach is the aortic ring assay (Nicosia and Ottinetti, 1990), which forms a bridge between *in vitro* and *in vivo* studies as detailed in Chapter 6. Visual counting or computer-assisted image analysis of the outgrowths can achieve quantification (Nicosia, 1998) with the use of fluorescein-labelled antibodies to CD31 allowing quantification by pixel counts (Plendl *et al.*, 1993; Plendl *et al.*, 2002). The major advantage of this assay is that the cells grow and migrate out from the aorta in a three-dimensional (3D) environment. Furthermore, this model allows the study of migration and vessel formation in a complex system, because besides endothelial cells, other cell types (such as fibroblasts) are present.

Another method of studying migration in a 3D *in vivo* situation is the use of endothelial cells suspended in Matrigel and injected subcutaneously into mice (see Chapter 9 for further details). To associate the apparently migrating endothelial cell phenotype (i.e. pseudopodia extension) with a molecular marker of cell migration, sections of the Matrigel plugs can be stained with phalloidin to detect polymerized F-actin, as assembly of actin polymers appears to be an absolute requirement for pseudopodia extension and cell migration (Alberts *et al.*, 2002; Friedl and Brocker, 2000; Lauffenburger and Horwitz, 1996). Single endothelial cells in Matrigel plugs are of two distinct phenotypes: non-migratory that do not extend pseudopodia and are F-actin low, and migratory that strongly express F-actin and avidly extend pseudopodia (Skovseth *et al.*, 2005). The major problem in both the *ex/in vivo* assays described above is that they cannot distinguish migration from differentiation, as both occur simultaneously.

Recently the capability of studying endothelial cell migration in the *in vivo* zebrafish model (see Chapter 16 for further details of this model) has been described. A rhodamine-labelled monoclonal antibody, Phy-V002, was used, which specifically labels activated vascular endothelial cells (without staining mature vessels or other tissues). The antibody, solubilized in fish water, was injected into the circulation of zebrafish and 24 h later the effect on *in vivo* endothelial cell migration and vessel formation were visually assessed, by whole-mount immunochemical staining in the transparent embryo and examination by fluorescence microscopy. Treatment of zebrafish with inhibitors of endothelial cell migration *in vitro* and staining with Phy-V002 led to the conclusion that using this assay it is possible visually to assess drug effects on endothelial cell migration *in vivo* at the resolution of a single cell (Seng *et al.*, 2004). The zebrafish model has several advantages: in this animal, embryogenesis occurs outside the mother's body, the embryos are transparent, vascular development is very rapid and large number of embryos can be easily manipulated. Furthermore, the homology between the molecular and signalling pathways that drive vessel development in the zebrafish and mammals is remarkable (Fouquet *et al.*, 1997; Lyons *et al.*, 1998; Liang *et al.*, 1998). The major disadvantage of this assay is the high cost for maintaining the animals in breeding condition.

3.7   Conclusion

As stated above, all endothelial cell migration assays used today have limitations, although each one yields valuable information, which is usually complementary to information obtained from other angiogenesis assays. Given the complexity of the situations where endothelial cells are activated, a combination of more than one assay may be necessary in order to confirm an effect on endothelial cell migration.

References

Alberts, B., Johnsen, A., Lewis, J., Raff, M. *et al.* (2002) *Molecular Biology of the Cell*, 4th edn, Garland Science, New York.

Albini, A., Benelli, R., Noonan, D. M. and Brigati, C. (2004) 'The "chemoinvasion assay": a tool to study tumor and endothelial cell invasion of basement membranes', *Int. J. Dev. Biol.*, **48**(5–6), pp. 563–571.

Albrecht-Buehler, G. (1977) 'Phagokinetic tracks of 3T3 cells: parallels between the orientation of track segments and of cellular structures which contain actin or tubulin', *Cell*, **12**(2), pp. 333–339.

Ariano, P., Distasi, C., Gilardino, A., Zamburlin, P. and Ferraro, M. (2005) 'A simple method to study cellular migration', *J. Neurosc. Meth.*, **141**(2), pp. 271–276.

Auerbach, R., Auerbach, W. and Polakowski, I. (1991) 'Assays for angiogenesis: a review', *Pharmacol. Ther.*, **51**(1), pp. 1–11.

Auerbach, R., Lewis, R., Shinners, B., Kubai, L. and Akhtar, N. (2003) 'Angiogenesis assays: a critical overview', *Clin. Chem.*, **49**(1), pp. 32–40.

Boyden, S. (1962) 'The chemotactic effect of mixtures of antibody and antigen on polymorphonuclear leucocytes', *J. Exp. Med.*, **115**, pp. 453–466.

Cai, G., Lian, J., Shapiro, S. S. and Beacham, D. A. (2000) 'Evaluation of endothelial cell migration with a novel *in vitro* assay system', *Methods Cell Sci.*, **22**(2–3), pp. 107–114.

Carmeliet, P. and Jain, R. K. (2000) 'Angiogenesis in cancer and other diseases', *Nature*, **407**(6801), pp. 249–257.

Cary, L. A. and Guan, J. L. (1999) 'Focal adhesion kinase in integrin-mediated signalling', *Front. Biosci.*, **4**, pp D102–113.

Cho, S. Y. and Klemke, R. L. (2002) 'Purification of pseudopodia from polarized cells reveals redistribution and activation of Rac through assembly of a CAS/Crk scaffold', *J. Cell Biol.*, **156**(4), pp. 725–736.

Coomber, B. L. and Gotlieb, A. I. (1990) '*In vitro* endothelial wound repair. Interaction of cell migration and proliferation', *Arteriosclerosis*, **10**(2), pp. 215–222.

Debeir, O., Camby, I., Kiss, R., Van Ham, P. and Decaestecker, C. (2004) 'A model-based approach for automated *in vitro* cell tracking and chemotaxis analyses', *Cytometry A.*, **60**(1), pp. 29–40.

Dickinson, R. B. and Tranquillo, R. T. (1993) 'A stochastic model for adhesion-mediated cell random motility and haptotaxis', *J. Math. Biol.*, **31**(6), pp. 563–600.

Distasi, C., Ariano, P., Zamburlin, P. and Ferraro, M. (2002) '*In vitro* analysis of neuron–glial cell interactions during cellular migration', *Eur. Biophys. J.*, **31**(2), pp. 81–88.

Dvorak, K. and Feit, J. (1977) 'Migration of neuroblasts through partial necrosis of the cerebral cortex in newborn rats-contribution to the problems of morphological development and developmental period of cerebral microgyria. Histological and autoradiographical study', *Acta Neuropathol.*, **38**(3), pp. 203–212.

Etienne-Manneville, S. and Hall, A. (2001) 'Integrin-mediated activation of Cdc42 controls cell polarity in migrating astrocytes through PKCzeta', *Cell*, **106**(4), pp. 489–498.

Falk, W., Goodwin, R. H. and Leonard, E. J. (1980) 'A 48-well micro chemotaxis assembly for rapid and accurate measurement of leukocyte migration', *J. Immunol. Methods*, **33**(3), pp. 239–247.

Fenteany, G., Janmey, P. A. and Stossel, T. P. (2000) 'Signaling pathways and cell mechanics involved in wound closure by epithelial cell sheets', *Curr. Biol.*, **10**(14), pp. 831–838.

Fischer, E. G., Stingl, A. and Kirkpatrick, C. J. (1990) 'Migration assay for endothelial cells in multiwells. Application to studies on the effect of opioids', *J. Immunol. Methods*, **128**(2), pp. 235–239.

Fouquet, B., Weinstein, B. M., Serluca, F. C. and Fishman, M. C. (1997) 'Vessel patterning in the embryo of the zebrafish: guidance by notochord', *Dev. Biol.*, **183**(1), pp. 37–48.

Friedl, P. and Brocker, E. B. (2000) 'The biology of cell locomotion within three-dimensional extracellular matrix', *Cell Mol. Life Sci.*, **57**(1), pp. 41–64.

Fuller, T., Korff, T., Kilian, A., Dandekar, G. and Augustin, H. G. (2003) 'Forward EphB4 signaling in endothelial cells controls cellular repulsion and segregation from ephrinB2 positive cells', *J. Cell Sci.*, **116**(12), pp. 2461–2470.

Gildea, J. J., Harding, M. A., Gulding, K. M. and Theodorescu, D. (2000) 'Transmembrane motility assay of transiently transfected cells by fluorescent cell counting and luciferase measurement', *Biotechniques*, **29**(1), pp. 81–86.

Goukassian, D., Diez-Juan, A., Asahara, T., Schratzberger, P. *et al.* (2001) 'Overexpression of p27(Kip1) by doxycycline-regulated adenoviral vectors inhibits endothelial cell proliferation and migration and impairs angiogenesis', *FASEB J.*, **15**(11), pp. 1877–1885.

Hagan, S., Hiscott, P., Sheridan, C. M., Wong, D. *et al.* (2003) 'Effects of the matricellular protein SPARC on human retinal pigment epithelial cell behaviour', *Mol. Vis.*, **9**, pp. 87–92.

Jekunen, A. and Kairemo, K. (2003) 'Inhibition of angiogenesis at endothelial cell level', *Microsc. Res. Tech.*, **60**(1), pp. 85–97.

Jones, L. J., Gray, M., Yue, S. T., Haugland, R. P. and Singer, V. L. (2001) 'Sensitive determination of cell number using the CyQUANT cell proliferation assay', *J. Immunol. Methods*, **254**(1–2), pp. 85–98.

Jozaki, K., Marucha, P. T., Despins, A. W. and Kreutzer, D. L. (1990) 'An *in vitro* model of cell migration: evaluation of vascular endothelial cell migration', *Anal. Biochem.*, **190**(1), pp. 39–47.

Klein-Soyer, C., Cazenave, J. P., Herbert, J. M. and Maffrand, J. P. (1994) 'SR 25989 inhibits healing of a mechanical wound of confluent human saphenous vein endothelial cells which is modulated by standard heparin and growth factors', *J. Cell Physiol.*, **160**(2), pp. 316–322.

Klemke, R. L., Leng, J., Molander, R., Brooks, P. C. *et al.* (1998) 'CAS/Crk coupling serves as a "molecular switch" for induction of cell migration', *J. Cell Biol.*, **140**(4), pp. 961–972.

Lampugnani, M. G. (1999) 'Cell migration into a wounded area *in vitro*', *Methods Mol. Biol.*, **96**, pp. 177–182.

Lauffenburger, D. A. and Horwitz, A. F. (1996) 'Cell migration: a physically integrated molecular process', *Cell*, **84**(3), pp. 359–369.

Liang, D., Xu, X., Chin, A. J., Balasubramaniyan, N. V. *et al.* (1998) 'Cloning and characterization of vascular endothelial growth factor (VEGF) from zebrafish, Danio rerio', *Biochim. Biophys. Acta*, **1397**(1), pp. 14–20.

Liao, W., Bisgrove, B. W., Sawyer, H., Hug, B. *et al.* (1997) 'The zebrafish gene cloche acts upstream of a flk-1 homologue to regulate endothelial cell differentiation', *Development*, **124**(2), pp. 381–389.

Lyons, M. S., Bell, B., Stainier, D. and Peters, K. G. (1998) 'Isolation of the zebrafish homologues for the tie-1 and tie-2 endothelium-specific receptor tyrosine kinases', *Dev. Dyn.*, **212**(1), pp. 133–140.

Magdalena, J., Millard, T. H. and Machesky, L. M. (2003a) 'Microtubule involvement in NIH 3T3 Golgi and MTOC polarity establishment', *J. Cell Sci.*, **116**(4), pp. 743–756.

Magdalena, J., Millard, T. H., Etienne-Manneville, S., Launay, S. *et al.* (2003b) 'Involvement of the Arp2/3 complex and Scar2 in Golgi polarity in scratch wound models', *Mol. Biol. Cell*, **14**(2), pp. 670–684.

Mc Henry, K. T., Ankala, S. V., Ghosh, A. K. and Fenteany, G. (2002) 'A non-antibacterial oxazolidinone derivative that inhibits epithelial cell sheet migration', *Chem. Biochem.*, **3**(11), pp. 1105–1111.

Miao, R. Q., Murakami, H., Song, Q., Chao, L. and Chao, J. (2000) 'Kallistatin stimulates vascular smooth muscle cell proliferation and migration *in vitro* and neointima formation in balloon-injured rat artery', *Circ. Res.*, **86**(4), pp. 418–424.

Nicosia, R. F. and Ottinetti, A. (1990) 'Growth of microvessels in serum-free matrix culture of rat aorta. A quantitative assay of angiogenesis *in vitro*', *Lab. Invest.*, **63**(1), pp. 115–122.

Nicosia, R. F. (1998) 'The rat aorta model of angiogenesis and its applications', in *Vascular morphogenesis: In vivo, in vitro, in mente*, Moronov, C. D. L. D. and Sage, E. H., eds, Birkhäuser, Boston.

Nobes, C. D. and Hall, A. (1999) 'Rho GTPases control polarity, protrusion, and adhesion during cell movement', *J. Cell Biol.*, **144**(6), pp. 1235–1244.

Obeso, J. L. and Auerbach, R. (1984) 'A new microtechnique for quantitating cell movement *in vitro* using polystyrene bead monolayers', *J. Immunol. Methods*, **70**(2), pp. 141–152.

Ohmori, T., Yatomi, Y., Okamoto, H., Miura, Y. *et al.* (2001) 'G(i)-mediated Cas tyrosine phosphorylation in vascular endothelial cells stimulated with sphingosine 1-phosphate: possible involvement in cell motility enhancement in cooperation with Rho-mediated pathways', *J. Biol. Chem.*, **276**(7), pp. 5274–5280.

Parent, C. A. and Devreotes, P. N. (1999) 'A cell's sense of direction', *Science*, **284**(5415), pp. 765–770.

Pepper, M. S., Belin, D., Montesano, R., Orci, L. and Vassalli, J. D. (1990) 'Transforming growth factor-beta 1 modulates basic fibroblast growth factor-induced proteolytic and angiogenic properties of endothelial cells *in vitro*', *J. Cell Biol.*, **111**(2), pp. 743–755.

Philp, D., Huff, T., Gho, Y. S., Hannappel, E. and Kleinman, H. K. (2003) 'The actin binding site on thymosin beta4 promotes angiogenesis', *FASEB J.*, **17**(14), pp. 2103–2105.

Plendl, J., Gilligan, B. J., Wang, S. J., Lewis, R. *et al.* (2002) 'Primitive endothelial cell lines from the porcine embryonic yolk sac', *In Vitro Cell Dev. Biol. Anim.*, **38**(6), pp. 334–342.

Plendl, J., Hartwell, L. and Auerbach, R. (1993) 'Organ-specific change in Dolichos biflorus lectin binding by myocardial endothelial cells during *in vitro* cultivation', *In Vitro Cell Dev. Biol.*, **29A**(1), pp. 25–31.

Pratt, B. M., Harris, A. S., Morrow, J. S. and Madri, J. A. (1984) 'Mechanisms of cytoskeletal regulation. Modulation of aortic endothelial cell spectrin by the extracellular matrix', *Am. J. Pathol.*, **117**(3), pp. 349–354.

Roy, P., Rajfur, Z., Pomorski, P. and Jacobson, K. (2002) 'Microscope-based techniques to study cell adhesion and migration', *Nat. Cell Biol.*, **4**(4), pp. 91–96.

Sablina, A. A, Chumakov, P.M. and Kopnin, B. P. (2003) 'Tumor suppressor p53 and its homologue p73alpha affect cell migration', *J. Biol. Chem.*, **278**(30), pp. 27362–27771.

Saga, Y., Mizukami, H., Takei, Y., Ozawa, K. and Suzuki, M. (2003) 'Suppression of cell migration in ovarian cancer cells mediated by PTEN overexpression', *Int. J. Oncol.*, **23**(4), pp. 1109–1113.

Santiago, A. and Erickson, C. A. (2002) 'Ephrin-B ligands play a dual role in the control of neural crest cell migration', *Development*, **129**(15), pp. 3621–3632.

Schienbein, M. and Gruler, H. (1993) 'Langevin equation, Fokker-Planck equation and cell migration', *Bull. Math. Biol.*, **55**(3), pp. 585–608.

Seng, W. L., Eng, K., Lee, J. and McGrath, P. (2004) 'Use of a monoclonal antibody specific for activated endothelial cells to quantitate angiogenesis *in vivo* in zebrafish after drug treatment', *Angiogenesis*, **7**(3), pp. 243–253.

Shi, Y., Kornovski, B. S., Savani, R. and Turley, E. A. (1993) 'A rapid, multiwell colorimetric assay for chemotaxis', *J. Immunol. Methods*, **164**(2), pp. 149–154.

Skovseth, D. K., Veuger, M. J., Sorensen, D. R., De Angelis, P. M. and Haraldsen, G. (2005) 'Endostatin dramatically inhibits endothelial cell migration, vascular morphogenesis, and perivascular cell recruitment *in vivo*', *Blood*, **105**(3), pp. 1044–1051.

Smith, J. T., Tomfohr, J. K., Wells, M. C., Beebe, T. P. *et al.* (2004) 'Measurement of cell migration on surface-bound fibronectin gradients', *Langmuir*, **20**(19), pp. 8279– 8286.

Solls, D. R. and Wessels, D. (1998) *Motion Analysis of Living Cells*, John Wiley Inc., New York.

Staton, C. A., Stribbling, S. M., Tazzyman, S., Hughes, R. *et al.* (2004) 'Current methods for assaying angiogenesis *in vitro* and *in vivo*', *Int. J. Exp. Pathol.*, **85**(5), pp. 233–248.

Stokes, C. L. (1992) 'Endothelial cell migration and chemotaxis in angiogenesis', *EXS*, **61**, pp. 118–124.

Stokes, C. L., Lauffenburger, D. A. and Williams, S. K. (1991) 'Migration of individual microvessel endothelial cells: stochastic model and parameter measurement', *J. Cell Sci.*, **99**(2), pp. 419–430.

Taraboletti, G., Roberts, D., Liotta, L. A. and Giavazzi, R. (1990) 'Platelet thrombospondin modulates endothelial cell adhesion, motility, and growth: a potential angiogenesis regulatory factor', *J. Cell Biol.*, **111**(2), pp. 765–772.

Taraboletti, G., Belotti, D., Dejana, E., Mantovani, A. and Giavazzi, R. (1993) 'Endothelial cell migration and invasiveness are induced by a soluble factor produced by murine endothelioma cells transformed by polyoma virus middle T oncogene', *Cancer Res.*, **53**(16), pp. 3812–3816.

Tvarusko, W., Bentele, M., Misteli, T., Rudolf, R. *et al.* (1999) 'Time-resolved analysis and visualization of dynamic processes in living cells', *Proc. Natl. Acad. Sci. USA*, **96**(14), pp. 7950–7955.

Wong, M. K. and Gotlieb, A. I. (1984) '*In vitro* reendothelialization of a single-cell wound. Role of microfilament bundles in rapid lamellipodia-mediated wound closure', *Lab. Invest.*, **51**(1), pp. 75–81.

Wong, M.K. and Gotlieb, A. I. (1988) 'The reorganization of microfilaments, centrosomes, and microtubules during *in vitro* small wound reendothelialization', *J. Cell Biol.*, **107**(5), pp. 1777–1783.

Yarrow, J. C., Perlman, Z. E., Westwood, N. J. and Mitchison, T. J. (2004) 'A high-throughput cell migration assay using scratch wound healing, a comparison of image-based readout methods', *BMC Biotechnol.*, **4**(1), pp. 21–28.

Yvon, A. M., Walker, J. W., Danowski, B., Fagerstrom, C. *et al.* (2002) 'Centrosome reorientation in wound-edge cells is cell type specific', *Mol. Biol. Cell*, **13**(6), pp. 1871–1880.

Zahm, J. M., Kaplan, H., Herard, A. L., Doriot, F. *et al.* (1997) 'Cell migration and proliferation during the *in vitro* wound repair of the respiratory epithelium', *Cell. Motil. Cytoskeleton*, **37**(1), pp. 33–43.

Zetter, B. R. (1987) 'Assay of capillary endothelial cell migration', *Methods Enzymol.*, **147**, pp. 135–144.

4

Tubule formation assays

Ewen J. Smith and **Carolyn A. Staton**

Abstract

In vitro tubule assays are regarded as representative of the later stages of angiogenesis
(differentiation) and are used extensively to assay potential stimulators or inhibitors
of angiogenesis. Endothelial cells typically present a cobblestone morphology during
culture but during extended starvation or when plated onto or within Matrigel, fibrin
or some collagens, will differentiate into a network of tubules that mimic a capillary
network.

Most two-dimensional (2D) tubule assays are now performed on Matrigel while
fibrin and collagen are used to provide the matrices for three-dimensional (3D)
assays. Many types of endothelial cells have been used though human umbilical vein
endothelial cells are the most popular. Other cell types such as fibroblasts can be
included to provide an *in vitro* model of tubulogenesis which more closely mimics
the *in vivo* situation. Addition of pro- or anti-angiogenic agents alters the level of
tubule formation which is often assessed by measuring tubule length, area or number.
In this chapter we review the advantages and limitations of the principal tubule
formation assays in use, together with the most recent advances.

Keywords

tubule formation; Matrigel; collagen; fibrin; co-culture

4.1 Introduction

Angiogenesis involves the migration, proliferation and differentiation of endothelial cells. *In vitro* assays often attempt to mimic one or more parts of the
angiogenic process in order to test whether, and how, a compound can affect

Angiogenesis Assays   Edited by Carolyn A. Staton, Claire Lewis and Roy Bicknell
© 2006 John Wiley & Sons, Ltd

endothelial cell function. *In vitro* tubule assays are regarded as representative of the later stages of angiogenesis (differentiation) and are used extensively to assay novel compounds for anti-angiogenic effects. They are also considered a model of *in vivo* capillary development and used to further our understanding of the essential biology of the developing vasculature.

The typical tubule formation assay involves plating endothelial cells onto or into a layer of gel matrix (which could be collagen, fibrin or Matrigel) in the presence of potential stimulators or inhibitors of angiogenesis. Tubule development is observed over a 4–24 h time period and recorded using a camera. The extent of tubule formation is measured qualitatively or quantitatively sometimes using computer-based image analysis programs.

4.2   Endothelial cell sources

Endothelial cells from macrovascular and microvascular, primary and immortalized, and animal and human origin have been used in tubule formation assays. This can make comparison between different sets of results difficult.

Cell types include primary and immortalized bovine aortic endothelial cells (BAEC), primary human umbilical vein endothelial cells (HUVECs), primary human microvascular endothelial cells (HMECs), primary human dermal microvascular endothelial cells (HDMECs) and immortalized microvascular endothelial cells (HMEC-1).

HUVECs are popular as they are primary cells that are reasonably easy to isolate in a normal laboratory (Bachetti and Morbidelli, 2000). However, results obtained using primary cells can be difficult to compare with each other as the cells are derived from a variety of donors using various cell extraction methods (Bouis *et al.*, 2001). Primary cell sources also senesce after a period of time, usually less than 10 passages, making use of these cells from a commercial source expensive.

Immortalized cell lines are useful because they allow standardized experimental conditions and reproducibility but immortalizing will, by definition, alter cellular characteristics (Bouis *et al.*, 2001). Other factors that affect endothelial cell behaviour include culture conditions, species of origin, organ of origin, age of the organism, macrovascular vs. microvascular and arterial vs. venous (Bouis *et al.*, 2001). These can all cause variations in many cell attributes including antigen and cytokine expression (Auerbach *et al.*, 2000), junctions (Craig *et al.*, 1998), anchorage dependency, angiotensin converting enzyme activity, vWf expression, possession of Weibel-palade bodies (Bouis *et al.*, 2001).

Importantly to angiogenesis research, macrovascular and microvascular endothelial cells are different. Recorded differences between microvascular and macrovascular cells include MMP secretion (Jackson and Nguyen, 1997), ultrastructure, antigen expression, lectin binding, VCAM and MHC expression (Craig *et al.*, 1998). Angiogenesis only occurs in microvascular cells, however, historically endothelial cells from macrovascular sources (HUVECs, BAEC) have been used. Reasons for this have included availability of macrovascular sources (umbilical veins), relative ease of isolation and the general purity of cells obtained (Bachetti and Morbidelli, 2000).

All the factors described above must be considered when choosing which cell type to use for a particular assay and when translating the result to an *in vivo* situation. Of the four cell types listed, primary HMEC are likely to be the most appropriate for angiogenesis research as they are capillary endothelial cells and of human origin. HDMEC (Promocell or TCS) are a good choice for microvascular endothelium (Bouis *et al.*, 2001) because they are well characterized. However, this has to be weighed against other factors such as cell maintenance costs and availability which suggests that HUVECs will continue to be used regularly. For a more detailed comparison of endothelial cell types see Chapter 1.

4.3   Endothelial cell morphology and tubule formation

Under normal *in vitro* culture conditions endothelial cells observe typical 'cobblestone' morphology (Figure 4.1). However, endothelial cells of all origins appear able to form tubules spontaneously *in vitro*, given sufficient time to lay down appropriate extracellular matrix components (Folkman and Haudenschild, 1980). Maciag and colleagues (1982) showed that HUVECs could be induced to form tubules in cell culture if deprived of endothelial cell growth factor (ECGF) for 4 to 6 weeks. The rate of tubule formation could be increased to 2–3 weeks if the cells were grown, in the same media, on a fibronectin matrix. If the fibronectin matrix

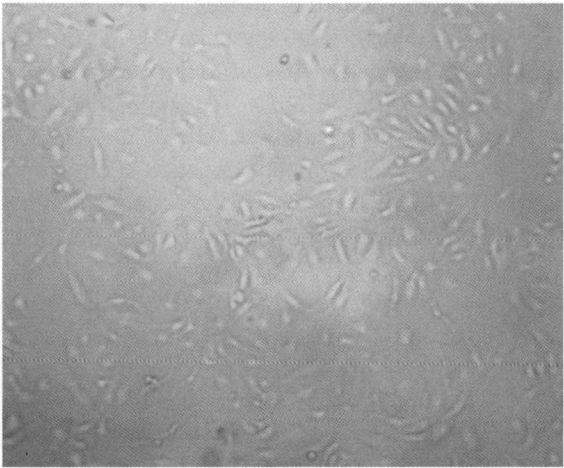

Figure 4.1   Human dermal microvascular endothelial cells under normal culture conditions showing cobblestone morphology.

was pre-treated with proteases including trypsin and plasmin then the rate of tubule formation increased up to fourfold. This resembles the situation *in vivo* where protease activity is one of the earliest stages of angiogenesis. Proteolysis alters the conformational form of some extracellular matrix proteins revealing cryptic sites, which can impart distinct functional influences on the vascular endothelial cells including proliferation, migration and differentiation signals (Kalluri, 2003). It was also observed that differentiation into tubules was not a terminal response as the cells could return to a proliferative state following administration of ECGF and serum (Maciag *et al.*, 1982).

4.4   Tubule assay matrices

Plating endothelial cells on different types of extracellular matrix alters the rate at which tubules develop. For example, culturing endothelial cells on collagens I and III (interstitial collagens) leads to cell proliferation and occasional tubule formation whilst plating on collagens IV and V (basement membrane collagens) leads to extensive tubule formation with only minimal proliferation (Madri and Williams, 1983) demonstrating that extracellular matrix components can modulate endothelial cell activity. Tubule formation assays are usually performed on matrices consisting of fibrin, collagen or Matrigel (the basement membrane extract of the EHS sarcoma), which stimulate the attachment, migration and differentiation of endothelial cells into tubules in a manner that mirrors the *in vivo* situation (Lawley and Kubota, 1989).

Lawley and Kubota (1989) demonstrated that the rate of differentiation of endothelial cells was massively increased by plating onto Matrigel, with tubules beginning to form in 1 h and completed in 8–12 h. Matrigel contains large amounts of basement proteins including collagen IV and laminin and is a liquid at 4°C but gels irreversibly above room temperature. It is now used extensively in endothelial cell tubule formation assays (Hernandez *et al.*, 2004; Osusky *et al.*, 2004; Zhong *et al.*, 2005; Matou *et al.*, 2005; Zaccagnini *et al.*, 2005; Labrecque *et al.*, 2005). However, cultured cells of non-endothelial origins including primary human fibroblasts, human prostate carcinoma cells (PC3) and glioblastoma cells (U87-MG) have also been shown to form 'tubules' on Matrigel (Donovan *et al.*, 2001). This limitation means that some care must be taken in interpreting results.

Following plating onto Matrigel endothelial cells elaborate dynamic cellular protrusions and then form capillary like aggregates or cords. The formation of tight junctions between the endothelial cells has been confirmed by electron microscopy (Auerbach *et al.*, 2003) although these early structures lack lumen. Microtubule dynamics are required for this initial shape change, however, actin polymerization and myosin dependent contractility are not. The later two processes are required for the second stage of differentiation into tubules where cells migrate to form lumen containing tubule-like structures (Connolly *et al.*, 2002).

There is some dispute over how much the tubules formed on Matrigel resemble capillaries. Lawley and Kubota state that tubules possess lumen and that in some instance lumen formation appeared to take place within the cells (Lawley and Kubota, 1989). Macaig and colleagues (1982) described the simplest tubules as being formed by one cell folding over to form a junction with itself (Macaig *et al.*, 1982). The presence of lumen in tubules on Matrigel was confirmed by Grant and colleagues (1991) and Connolly and colleagues (2002) using electron and light microscopy respectively, although it was disputed by Bikfalvi and colleagues (1991) who could not observe lumen (Grant *et al.*, 1991; Bikfalvi *et al.*, 1991; Connolly *et al.*, 2002).

HUVEC lumen formation in 3D collagen gels and 2D Matrigel gels has been compared (Montanez *et al.*, 2002). Lumen in 3D collagen gels were formed by HUVECs aligning in solid tubular structures and the inner cells dying. The degeneration led to formation of an intercellular space resembling a lumen. On Matrigel, lumen were formed by a dynamic folding and a cellular fusion mechanism where elongated cord like cells lined up either side of a central space wrapped round the space to form a lumen. In some cases this wrapping was incomplete and a cleft was formed. It has been demonstrated that apoptosis is an essential part of tubule development on Matrigel or collagen (Segura *et al.*, 2002). Tubule formation by HUVECs on Matrigel or in 2D or 3D collagen assays is inhibited by caspase inhibitors. Apoptosis occurs both outside and inside the lumen of the tubules.

Zimrin *et al.* (1995) investigated how endothelial cells respond when plated onto different extracellular matrices. They showed that differentiation of tubules on Matrigel was not inhibited by actinomycin D, cyclohexamide or puromycin, implying that neither transcriptional or translational events were involved, however a protein kinase C inhibitor (H7), a microtubule inhibitor (nocodazole) and a general growth factor receptor inhibitor (suramin) all inhibited differentiation suggesting that multiple post-translational mechanisms may be important for tubule formation on Matrigel. However, it has also been shown that tubule formation was dependent on gene expression in the first few hours of plating on Matrigel (Cockerill *et al.*, 1998). In contrast, differentiation of HUVECs plated onto fibrin was inhibited by transcriptional and translational inhibitors (Zimrin *et al.*, 1995). These results suggest the mechanism of action of endothelial cell tubule formation is at least partly dependent on the matrix.

4.5   2D assay protocols

Three-dimensional fibrin and collagen gels have been used since the 1980s and as such there has been a gradual waning in the use of 2D fibrin gels and collagen gels (Montesano *et al.*, 1983; Madri *et al.*, 1988). Instead, 2D tubule formation assays are now conducted using Matrigel.

A typical protocol for 2D fibrin based assays involved dissolving bovine fibrinogen in media or PBS and inducing polymerization with thrombin (Zimrin *et al.*, 1995; Nehls *et al.*, 1994). Endothelial cells were plated in media with/ without test agents. The cells were allowed to attach and then non-adherents washed off the next day. Aprotinin (200 U/ml) was often included in the gel matrix and media to prevent excess fibrinolysis (Nehls *et al.*, 1994). A similar protocol was employed by Vailhe and colleagues (1998), though the cells were suspended in low serum media. Tubule growth was assessed by photography and measurement of tubule lengths (Vailhe *et al.*, 1998).

The typical protocol for a 2D collagen based assay was to mix rat tail collagen type-I with media, pipette this into the wells of a tissue culture plate and allow to gelate at 37°C. Confluent endothelial cells were seeded on top of the collagen matrix, cultured for up to 5 days and formation of capillary-like structures was analysed by phase contrast microscopy (Plaisier *et al.*, 2004; Sato *et al.*, 1993).

To investigate Matrigel based tubule assay designs 31 recent papers were reviewed for tubule assay protocols (Table 4.1). The volume of Matrigel used per well varied considerably between research groups. The groups used 6, 12, 24, 48 and 96-well plates, the lowest volume of Matrigel used was 0.17 μl/mm^2 (Staton *et al.*, 2004) and the greatest 3.53 μl/mm^2 (Zhong *et al.*, 2005; Chen *et al.*, 2005). Average volume was 1.7 μl/mm^2 (95% CI = 1.34–2.06 μl). In practice the surface tension of Matrigel means that using too low a volume creates a well effect allowing cells to pool in the centre of the well. We have found that using less than 34 μl Matrigel per well of a 96-well plate (1.2 μl/mm^2) can lead to this effect. Only 15 per cent of protocols reviewed used less than 1.2 μl/mm^2. It is probable that using extremely low levels of Matrigel (Staton *et al.*, 2004) also avoids this effect, although using so little Matrigel requires manual spreading to distribute the Matrigel over the well which may lead to uneven Matrigel distribution.

Recently, a second form of Matrigel has been developed, growth factor-reduced (GFR-) Matrigel, in which the levels of stimulatory cytokines and growth factors have been markedly reduced. Endothelial cells do not require growth factors to induce differentiation on Matrigel or GFR-Matrigel and extensive tubule networks can be formed without any exogenous growth factor being added. However, using GFR-Matrigel allows for more selective determination of the efficacy of pro-angiogenic agents. BD biosciences (manufacturers of GFR-Matrigel) recommend GFR-Matrigel be used as a layer at least 0.5 mm thick (0.5 μl/mm^2).

A wide variety of endothelial cells have been used in Matrigel based tubule formation assays. These include HMEC (Narmoneva *et al.*, 2005), HMEC-1 (Hernandez *et al.*, 2004), HDMEC (Staton *et al.*, 2004) and bovine pulmonary microvascular endothelial cells (BPMEC) (Ashino *et al.*, 2003) though HUVECs (Hernandez *et al.*, 2004; Osusky *et al.*, 2004; Zhong *et al.*, 2005; Matou *et al.*, 2005; Zaccagnini *et al.*, 2005; Labrecque *et al.*, 2005) are the most popular cell type. Cell plating density varied from 56.6–2829.7 cells/mm^2. Average cell density was 563 cells/mm^2 (95% CI = 346–780 cells/mm^2). Plating and/or culturing density may affect the tubule formation response to growth factors (Liu *et al.*, 1999; Liu *et al.*, 2002). Too high plating densities can result in large areas of clustered,

Table 4.1 Matrigel and cell plating density in 2D tubule formation assays n/a = not available, n/s = not stated

Ref	Matrix	Cell type	Plate	Well area (mm²)	Matrigel (μl)	Matrigel (μl/mm²)	Cells/well	Cell density/mm²
Hernandez et al., 2004	GFR-Matrigel	HUVEC, HUAEC, HMEC-1	24	176.7	n/s	n/a	50000	283.0
Osusky et al., 2004	GFR-Matrigel	HUVEC	24	176.7	300	1.70	48000	271.6
Li et al., 2004	Matrigel	HUVEC	96	28.3	50	1.77	10000	353.4
Jee et al., 2004	Matrigel	HUVEC	24	176.7	300	1.70	200000	1131.9
Skurk et al., 2005	GFR-Matrigel	HUVEC	12	346.4	n/s	n/a	30000	86.6
Chang et al., 2005	Matrigel	HUVEC	12	346.4	n/s	n/a	60000	173.2
Labrecque et al., 2005	Matrigel	HUVEC	96	28.9	50	1.73	25000	865.1
Lee et al., 2005	GFR-Matrigel	HUVEC	24	176.7	200	1.13	100000	565.9
Zhong et al., 2005	Matrigel	HUVEC	96	28.3	100	3.53	50000	1766.8
Matou et al., 2005	Matrigel	HUVEC	48	95	n/s	n/a	30000	315.8
Zaccagnini et al., 2005	Matrigel	HUVEC	24	176.7	400	2.26	50000	283.0
Chen et al., 2005	GFR-Matrigel	HUVEC	96	28.3	100	3.53	15000	530.0
Natori et al., 2005	GFR-Matrigel	HUVEC	24	176.7	n/s	n/a	20000	113.2
Lee et al., 2005	Matrigel	HUVEC	24	176.7	320	1.81	30000	169.8
Personal	GFR-Matrigel	HDMEC	96	28.3	34	1.20	15000	530.0
Staton et al., 2004	GFR-Matrigel	HDMEC	24	176.7	30	0.17	40000	226.4
Galvez et al., 2001	Matrigel	HUVEC	96	28.3	80	2.83	40000	1413.4
Aozuka et al., 2004	Matrigel	HUVEC	48	95	100	1.05	30000	315.8
Unger et al., 2002	Matrigel	HDMEC	24	176.7	100	0.57	100000	565.9
Yahata et al., 2003	GFR-Matrigel	HDMEC	24	176.7	n/s	n/a	40000	226.4

(continued)

Table 4.1 (Continued)

Ref	Matrix	Cell type	Plate	Well area (mm^2)	Matrigel (μl)	Matrigel $(\mu l/mm^2)$	Cells/well	Cell density/mm^2
Bischof et al., 2004	Matrigel	HDMEC and HUVEC	24	176.7	300	1.70	48000	271.6
Sakurai et al., 2004	Matrigel	HUVEC	6	962.1	1000	1.04	50000	52.0
Dasgupta et al., 2004	Matrigel	HAEC	96	28.3	100	3.53	15000	530.0
Qian et al., 2004	GFR-Matrigel	HUVEC	24	176.7	300	1.70	50000	283.0
Hahm et al., 2004	Matrigel	HUVEC	24	176.7	n/s	n/a	500000	2829.7
Chen et al., 2004	Matrigel	HUVEC	24	176.7	250	1.41	40000	226.4
Mousa and Mohamed, 2004	GFR-Matrigel	HUVEC	24	176.7	250	1.41	40000	226.4
Ashino et al., 2003	GFR-Matrigel	BPMEC	24	176.7	250	1.41	200000	1131.9
Busby et al., 2003	GFR-Matrigel	HUVEC	12	346.4	250	0.72	10000	28.9
Golubkov et al., 2003	Matrigel (dilute)	HUVEC	24	176.7	300	1.70	200000	1131.9
Vucenik et al., 2004	Matrigel	HUVEC	24	176.7	200	1.13	100000	565.9

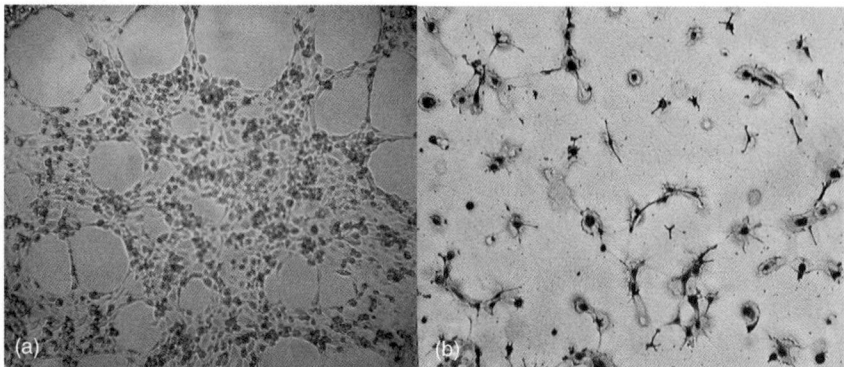

Figure 4.2 Suboptimal plating density can inhibit tubule formation. (a) Too high plating densities can result in large areas of clustered, undifferentiated cells, (b) too low plating densities can prevent tubule formation as cells do not make contact.

undifferentiated cells (Figure 4.2a), too low plating densities can prevent tubule formation (Figure 4.2b). Incubation time on Matrigel prior to analysis varied from 4–24 h.

When setting up a tubule formation assay all of the above issues need to be considered. We suggest doing the following. Use GFR-Matrigel if possible because the reduced levels of cytokines and growth factors should enable you to achieve a more consistent response. Both forms of Matrigel are expensive so you want to use the minimum possible by using 24 or even better 96-well plates. It is probably best not to reduce the amount of Matrigel per well to below 1.2 µl/mm^2 or the cells may pool in the centre of the well. As shown above many types of endothelial cells have been used in these assays. Using a microvascular cell is likely to be most representative of the *in vivo* situation, however HUVECs are a commonly used alternative. Finally, these assays can have fairly high intra-assay variability so keep n $\geq$ 3.

4.6   Analysis of tubule formation in 2D assays

Tubule formation in most 2D assays is observable using a standard inverted microscope from as little as 4 h post-plating depending on the matrix. Tubule formation is typically recorded using a camera attached to an inverted microscope or phase contrast microscope. It is possible clearly to make out the tubules against the background without any staining, although some groups do use H & E staining to improve contrast (Staton *et al.*, 2004; Osusky *et al.*, 2004). It is also possible to use CD31 and vWf staining to discriminate tubule formation from background (Sengupta *et al.*, 2003). Recording tubule formation can involve some degree of selection of areas of tubule formation to record and can thus lead to unconscious bias within an experiment. A standard digital camera microscope combination cannot record tubule formation in the whole well of any plate. In a 96-well plate light refraction within the well creates a shadowed area around the edge of the well where tubule formation is unrecordable. In larger wells multiple areas within the well can be photographed. Steps must be taken to prevent subjective bias in selecting the images for each well. We have found that using 96-well plates will substantially reduce subjective bias as one photo can record all visible tubule formation per well.

There are four primary variables that are commonly used to determine the extent of tubule formation. They are total tubule length, number of tubules, average tubule length and tubule area. Their advantages and limitations are detailed below.

Total tubule length has been used to assess tubule formation in many papers (Isaji *et al.*, 1997; Hernandez *et al.*, 2004). In many ways it would seem the most useful variable as it directly measures the formation of tubules. Perhaps the major limitation is accurately measuring total tubule length. This can be performed by hand but takes a considerable amount of time. It can also be done using image analysis software as long as the software can reliably determine tubules from background and undifferentiated cells (see below). Counting the number of tubules is also widely used to assess tubule formation (Smith and Hoffman, 2005; Wu *et al.*, 1997) probably because it is relatively easy to assess with or without image analysis software. However, it is also a time-consuming process if done by hand. Average tubule length is determined by dividing total tubule length by the number of tubules. As it is based on two other variables it cannot strictly be seen as a separate independent variable.

Measuring tubule area is a commonly used technique to assess tubule development (Leopold *et al.*, 2003; Staton *et al.*, 2004). It is easy to perform with an image analysis program and many images can be processed in a short space of time. However, there are two significant problems with using tubule area. The first is that image analysis programs cannot always differentiate between tubules and undifferentiated cells (see below) thus the variable could be more accurately named 'cell and tubule area'. The second is that under certain circumstances, particularly at high cell density, 'cell and tubule area' actually decreases under angiogenic conditions. This was neatly illustrated by Liu *et al.* (2002) where tubule formation endothelial cells plated at high and low density on Matrigel was assessed. Using a similar experimental design to Liu and colleagues (2002), we found that when HDMEC are plated at high density (1400 cells/mm^2), combined 'cell and tubule area' is maximal under low angiogenic (control) conditions (Figure 4.3b). Where tubule formation has occurred (Figure 4.3d), due to VEGF stimulation, 'cell and tubule area' has decreased. This problem can also occur to a lesser degree when HDMEC are plated at low density (530 cells/mm^2) (Figure 4.3a and 4.3c). So caution must be taken when using tubule area to assess tubule development.

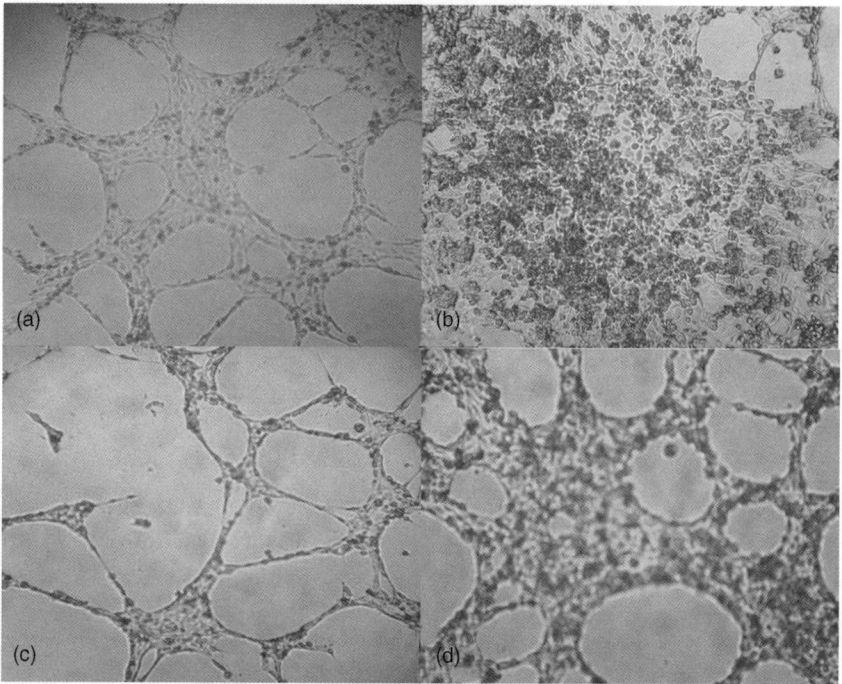

Figure 4.3 Reduction in 'Cell and Tubule Area' of VEGF-stimulated HDMEC plated at low and high density in HDMEC plated at a high density on Matrigel, VEGF stimulation has caused a decrease in 'cell and tubule area'. This has also occurred to a lesser degree when HDMEC are plated at low density. (a) HDMEC plated at 1.5×10^4 cells/well (low density) tubule formation at 10 h post-plating, no exogenous growth factor added. (b) Tubule formation 10 h post-plating at high density, no exogenous growth factor added. (c) Tubule formation 10 h post-plating at low density, VEGF 30 ng/ml added. (d) Tubule formation 10 h post-plating at high density, VEGF 30 ng/ml added.

Each of these variables can, to varying degrees, be measured by hand or by using image analysis programs. The use of image analysis programs can considerably speed up measurements but care has to be taken to ensure that a true assessment is being performed. For example, image analysis programs cannot always fairly discriminate between tubules, individual cells and background. This has been our experience when using Scion (an effective, simple to use, freely available image analysis software). When using this software to determine tubule area the program uses thresholding to differentiate between cells/tubules and background. Thresholding takes a greyscale image and converts all pixels into black or white pixels. The choice of whether a grey pixel is to be white or black is determined by the level of greyness to which the threshold is set. The program can then count the number of black pixels and this can be related to tubule development. It is highly difficult to differentiate between narrow tubules and background as they can be a

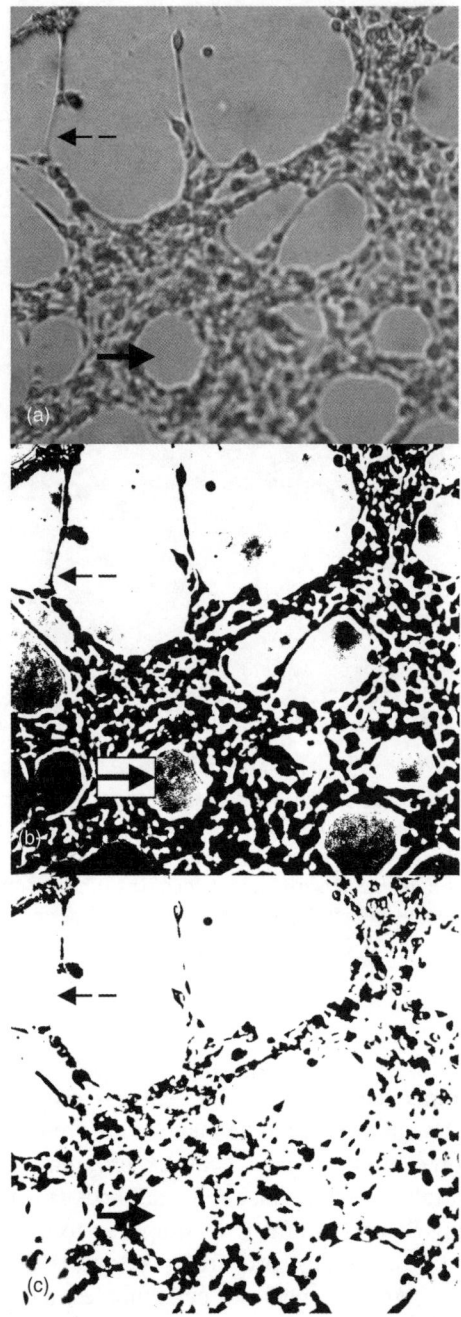

Figure 4.4 The effect of thresholding on discriminating between tubules, undifferentiated cells, and background (a) original image, (b) high threshold level enabling all tubules to be measured, (c) low threshold level to prevent measuring background. The dashed arrow highlights a tubule visible in the original image but invisible at low thresholding, the solid black arrow highlights an area of background in the original image that is measured as 'cell and tubule area' at high thresholding.

very similar degree of grey. This can result in some tubules being classified as 'below threshold' and not being counted, which will result in lower tubule area than is true.

Thresholding can also result in individual non-tubule forming cells, actual tubules and background being counted as 'tubule area' – thus inaccurately increasing the recorded value for the variable. Examples of these problems are demonstrated in Figure 4.4. The original image (Figure 4.4a) shows that tubule development has occurred, however there are still extensive numbers of undifferentiated cells present. When thresholding is set high enough to record all tubules then background and undifferentiated cells are also recorded (Figure 4.4b), whilst reducing the thresholding levels to eliminate background results in the loss of some narrower tubules (Figure 4.4c).

Problems in truly identifying tubules were also evident in a demo disc version of Angiosys. This program attempted to measure total tubule length by reducing all thresholded areas to 1 pixel wide and counting the total number of pixels present. This worked fairly well in assessing the length of clearly defined tubules, but less well when it had to deal with an area of undifferentiated cells. The program would typically reduce the area into many short 1 pixel wide lengths of black pixels and thus measure tubule length in an area where no tubules were present.

A variety of other image analysis programs have been used by groups including Quantimet (Montanez *et al.*, 2002), Image-Pro Plus (Yahata *et al.*, 2003; Hernandez *et al.*, 2004) and Image J (Osusky *et al.*, 2004), all of which have their own advantages and limitations.

4.7 Recent developments in 2D assays

Recently Sanz and colleagues (2002) attempted to adapt the Matrigel technique, to circumvent some of the problems associated with the assay, namely the large quantity of expensive Matrigel used per test, the inability to quantify a whole well with ease, and the subjective choice of field of view, with the associated bias and time-consuming nature of analysis. This was achieved by scaling down the assay for use in 384-well and 1536-well high-density formats where the endothelial cells yielded highly reproducible tubule formation per well. A computer-assisted integrated platform for capturing and processing images of complete wells was then used. This system evaluated the total number of nodes, connected and unconnected tubules as well as their lengths, and was tested in a double-blind experiment, which validated the results (Sanz *et al.*, 2002). This procedure has eliminated some of the problems of the Matrigel assays with respect to reproducibility and to the difficulties in analysing the tubule formation seen in a total well accurately and has resulted in an easy computerized system for analysing tubule formation. However, according to the research paper it has only been assessed with anti-angiogenic agents. It would be interesting to see the response to pro-angiogenic agents in this assay.

4.8   3D assays

Three-dimensional tubule formation assays have been developed in an attempt to provide an *in vitro* model of tubulogenesis which more closely mimics the *in vivo* situation. The matrix used in 3D assays can significantly affect tubule formation. The addition of collagen I to a fibrin matrix inhibited tubule formation by immortalized HUVECs by inhibiting fibrinolysis (Kroon *et al.*, 2002). Tubule formation of HUVECs into a fibrin matrices involves intracellular vacuolation and subsequent coalescing into lumens (Bayless *et al.*, 2000). Tubule formation could be inhibited by anti-$\alpha_v\beta_3$ and anti-$\alpha5$ integrins. Cellular morphology was also studied by Dye and colleagues (2004). They found that HUVECs underwent elongation, lattice formation and vacuolation on collagen-1 and Matrigel but showed little response on fibrin. There are various different versions of the 3D tubule formation assay as detailed below.

Microcarriers in a fibrin gel

Endothelial cells have been attached to microcarriers and then these used to disperse the cells throughout a 3D gel (Figure 4.5). Endothelial cells were allowed to adhere to gelatine coated cytodex-3 microcarriers. In a 24-well plate the microcarriers were evenly distributed in a fibrinogen solution (1 mg/ml in EBM) which was polymerized with thrombin (Nehls and Drenckhahn, 1995; Sun *et al.*, 2004). Following polymerization media was laid on top of the matrix and growth factors were administered both into the media and into the fibrinogen solution pre-polymerization. HMVECs form a capillary like network in 5–7 days under these conditions (Sun *et al.*, 2004). Capillary sprout formation was assessed using a digital camera

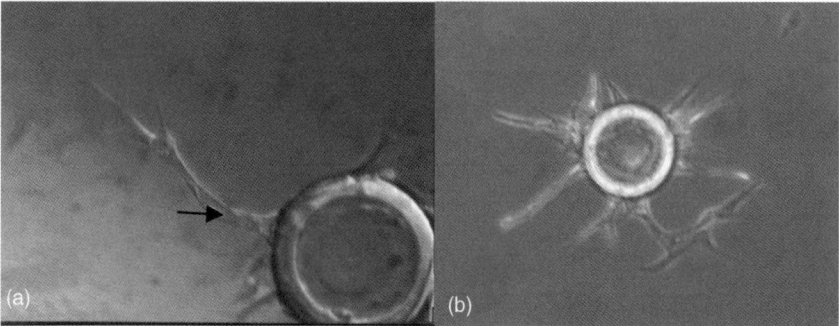

Figure 4.5   Tubule formation of endothelial cells in 3D culture endothelial cells were plated onto microcarriers before dispersion into the fibrin matrix. (a) Close up on a single tubule. The arrow highlights endothelial cells wrapping around and creating a lumen. (b) Multiple tubules originating from endothelial cells on a microcarrier.

and inverted microscope. A grid was applied to each image and sprout length measured by comparison to the grid (Sun *et al.*, 2004). Tubule formation could also be assessed by counting the number of tubules visible that were greater in length than the diameter of a microcarrier and were made up of at least three endothelial cells (Nehls and Drenckhahn, 1995). In practice, one limitation of this assay has been that the microcarriers sink to the bottom of the gel.

Without microcarriers in a fibrin gel

In a different experimental design endothelial cells were plated as a 2D layer and then allowed to grow up into a fibrin gel, forming a 3D tubular network. The endothelial cells were plated in a confluent monolayer in gelatin coated wells and then covered with a fibrinogen solution containing further endothelial cells. Thrombin was used to polymerize the fibrinogen and the matrix was then overlaid with medium. Primary endothelial cells formed tubules in the horizontal plane by day 7, by day 15 the endothelial tubules branched upwards and penetrated the gel to form a 3D network of tubules (Gagnon *et al.*, 2002). Both tubule formation at the base of the matrix (net and cord like structures) and within the fibrin gel (tubule-like, vesicle and stellate shaped structures) were measured. A common limitation with this assay is that the gel has to be fairly thin to allow oxygen and nutrients to diffuse through. If the gel is too thick considerable cell death can be observed.

Without microcarriers in a collagen gel

Collagen can also be used to construct the matrix for 3D assays. An ice-cold solution of collagen type I in DMEM was mixed with HUVECs and added to a filter chamber set into the well of a 24-well tissue culture plate and allowed to set at room temperature. One ml medium was added per well which completely covered the collagen gel. Test compounds were added to the medium surrounding the gel. After 7 days tubule formation was assessed. The gels were fixed using formalin or paraformaldehyde, dehydrated, sliced into 5 µm sections and stained with haematoxylin-eosin (Ilan *et al.*, 1998). Again a common limitation with this assay is that the gel has to be fairly thin to allow oxygen and nutrients to diffuse through. In addition the exposure of endothelial cells to ice-cold collagen solution can shock them and lead to cell death.

4.9   Analysis of 3D assays

The greatest advantage of the 3D assays over the 2D assays is that they more closely mimic the *in vivo* situation. However, quantification of cell behaviour in three dimensions, which is notoriously difficult to analyse, remains a challenge. The analysis typically involves taking pictures at different heights (e.g. every 50–100 μm) in the gel from the bottom to the top. The length (both width, for those in the horizontal plane, and height, for those in the vertical plane) and largest diameter of each vessel are measured (Gagnon et al., 2002). The microvessel density can also be calculated using the Chalkley grid method where an eyepiece-mounted, 25-point Chalkley array graticule is used to assess the vascular density by counting the number of points on the Chalkley grid coinciding with tubules at different heights through the gel. The limitations are obvious with only a proportion of tubules, area or volume of the gel being analysed. The assays also take considerably longer to run (5–15 days versus <24 h).

4.10   Co-culture assays

Another form of tubule assay involves the co-culture of endothelial cells with stromal cells, with or without the provision of an extracellular matrix (Montesano et al., 1993; Nehls and Drenckhahn, 1995; Bishop et al., 1999). Again this was developed in an attempt to provide an in vitro model of tubulogenesis which more closely mimics the in vivo situation. The stromal cells used may be fibroblasts, smooth muscle cells or blood vessel explants (like the aortic ring assay, see Chapter 6) containing numerous cell types (Nicosia and Ottinetti 1990; Sakuda et al., 1992). The following are a variety of co-culture systems described in the literature.

When human fibroblasts are co-cultured with endothelial cells the fibroblasts secrete the necessary matrix components that act as a scaffold for tubule formation (Bishop et al., 1999). This assay has been shown to produce tubules that contain lumen, and a more heterogeneous pattern of tubule lengths with some longer tubules than in the Matrigel assays, which more closely resemble the capillary bed in vivo (Bishop et al., 1999; Donovan et al., 2001). However, this assay takes 12–14 days and is therefore more time consuming than the Matrigel assay (typically 4–24 h), and is a less well-characterized assay as the matrix components secreted by the fibroblasts have not been defined. Also, undefined interactions between endothelial cells and the fibroblasts may occur, making the effect(s) of drugs (or other compounds) specifically on endothelial cells hard to identify.

Montesano and colleagues (1993) describe two co-culture systems. The 'collagen bilayer' system has a 1–1.5 ml layer of collagen containing fibroblasts or epithelial cells applied to the base of a 35 mm plastic dish. After gelling, a cell-free collagen layer is applied on top and then, finally, after the second layer has gelled, endothelial cells are seeded on top in media composed of a 1:1 mix of endothelial and the other cell media and incubated for 10–20 days with medium changes every 2–3 days. Tubule development was in three dimensions, however a simpler assay kept tubule formation to essentially two dimensions. The 'collagen disc' system consists of two concentric cell containing collagen gels (the outer gel contains fibroblasts, the innermost gel contains endothelial cells.) separated by a torus shaped cell free collagen gel. Endothelial cell tubules develop horizontally into the cell free gel. The cells are incubated at 37°C for up to 15 days. In both cases tubule formation was recorded using a phase contrast microscope and camera (Montesano *et al.*, 1993).

A co-culture system has also been used to show how tumour cells induce an activated phenotype in HUVECs (Khodarev *et al.*, 2003). This assay was set up using transwell inserts. U87 glioblastoma cells transfected with green fluorescent protein (GFP-U87) were plated in the chambers and HUVECs were plated in the inserts. Tubule formation was assessed using the removed, stained inserts.

A comparison between co-culture, Matrigel and GFR-Matrigel assays was performed by Donovan *et al.* (2001). They studied the comparative morphology and response to VEGF. Total tubule length, tubule area, branch points and average tubule length were measured using a combination of manual measurement and the image analysis program 'Scion'. Tubules formed in the co-culture system were typically twice as long as those on Matrigel though the number of tubules and number of branch points was reduced. Total cell and tubule area was similar for each assay. The response to VEGF varied between the assays. VEGF only significantly increased total tubule length, area and average length in the GFR-Matrigel assay. It did not significantly increase any variable in the Matrigel assay but it did significantly increase total tubule length and area in the co-culture system.

TCS Cellworks markets a complete 'Human angiogenesis model for analysing the pro-angiogenic or anti-angiogenic properties of test compounds'. This kit is based on the assay developed by Bishop and colleagues (1999) and contains human endothelial cells mixed with human fibroblasts in a special medium. Over a period of 12–14 days the endothelial cells proliferate, migrate and form tubules. The tubules can be stained for vWf, CD31 or ICAM-1 and angiogenesis is assessed by scoring tubule development by eye or with the 'Angiosys' image analysis program. This has been used to investigate the angiogenic effects of VEGF and HGF (Beilmann *et al.*, 2004).

TCS Cellworks has recently introduced an ELISA kit for determining tubule formation. The kit 'has been designed to give a rapid colorimetric output with the TCS CellWorks Angiogenesis Kit using CD31 as a marker for visualizing tubule formation at the final stage of the assay. It uses an alkaline phosphatase-coupled anti-CD31 antibody linked to a soluble chromogenic substrate. This provides colour development in the AngioKit wells proportional to the degree of tubule development'. We do not know of any published literature using this kit yet, however a similar assay has been described by Friis and colleagues (2003) which uses a co-culture of human dermal fibroblasts and HUVECs and generates results in 72 h. Endothelial proliferation and tubule formation is quantified by ELISA using antibodies to CD31, vWf and/or collagen IV. However, it should be noted that the quantitative expression of CD31 and vWf should only be varied by changes in cell number not by differentiation into tubules.

4.11   Conclusions

Tubule formation assays are a useful *in vitro* model of the later stages of angiogenesis. However, tubule formation is only one step in the angiogenic pathway and should not be considered in isolation, indeed tubule formation is dependent on endothelial cell migration for alignment and apoptosis for lumen formation.

The 2D Matrigel assay is useful for quickly assessing the angiogenic modulatory activity of novel agents and will generate results within 24 h. As many groups use this assay, results can easily be compared against the literature. The newly developed GFR-Matrigel with reduced levels of tubule promoters may be particularly useful in testing pro-angiogenic agents. The longer running co-culture assays may more closely mimic the *in vivo* situation where other cells are present, but are more time consuming.

As tubule development *in vivo* occurs in three dimensions the *in vitro* 3D tubule formation assays are likely to be more representative than the 2D assays in developing our understanding of physiological tubule development. These assays are not as convenient as 2D Matrigel assays for screening the activity of angiogenic compounds as they take longer to run and are more difficult to analyse. With all tubule formation assays analysing and measuring tubule formation remains a substantial challenge, with a definite need for the further development of image analysis programs that can reliably measure the parameters of tubule development.

References

Aozuka, Y., Koizumi, K., Saitoh, Y., Ueda, Y. *et al.* (2004) 'Anti-tumor angiogenesis effect of aminopeptidase inhibitor bestatin against B16-BL6 melanoma cells orthotopically implanted into syngeneic mice', *Cancer Lett.*, **216**(1), pp. 35–42.

Ashino, H., Shimamura, M., Nakajima, H., Dombou, M. *et al.* (2003) 'Novel function of ascorbic acid as an angiostatic factor', *Angiogenesis*, **6**(4), pp. 259–269.

Auerbach, R., Akhtar, N., Lewis, R. L. and Shinners, B. L. (2000) 'Angiogenesis assays: problems and pitfalls', *Cancer Metastasis Rev.*, **19**, pp. 167–172.

Auerbach, R., Lewis, R., Shinners, B., Kubai, L. and Akhtar, N. (2003) 'Angiogenesis assays: a critical overview', *Clin. Chem.*, **49**(1), pp. 32–40.

Bachetti, T. and Morbidelli, L. (2000) 'Endothelial cells in culture: a model for studying vascular functions', *Pharmacol. Res.*, **42**, pp. 9–19.

Bayless, K. J., Salazar, R. and Davis, G. E. (2000) 'RGD-dependent vacuolation and lumen formation observed during endothelial cell morphogenesis in three-dimensional fibrin matrices involves the alpha(v)beta(3) and alpha(5)beta(1) integrins', *Am. J. Pathol.*, **156**(1), pp. 1673–1683.

Beilmann, M., Birk, G. and Lenter, M. C. (2004) 'Human primary co-culture angiogenesis assay reveals additive stimulation and different angiogenic properties of VEGF and HGF', *Cytokine*, **26**(4), pp. 178–185.

Bikfalvi, A., Cramer, E. M., Tenza, D. and Tobelem, G. (1991) 'Phenotypic modulations of human umbilical vein endothelial cells and human dermal fibroblasts using two angiogenic assays', *Biol. Cell*, **72**(3), pp. 275–278.

Bischof, M., Abdollahi, A., Gong, P., Stoffregen, C. *et al.* (2004) 'Triple combination of irradiation, chemotherapy (pemetrexed), and VEGFR inhibition (SU5416) in human endothelial and tumor cells', *Int. J. Radiat. Oncol. Biol. Phys.*, **60**(4), pp. 1220–1232.

Bishop, E. T., Bell, G. T., Bloor, S., Broom, I. J. *et al.* (1999) 'An *in vitro* model of angiogenesis: basic features', *Angiogenesis*, **3**(4), pp. 335–344.

Bouis, D., Hospers, G. A. P., Meijer, C., Molema, G. and Mulder, N. H. (2001) 'Endothelium *in vitro*: A review of human vascular endothelial cell lines for blood vessel-related research', *Angiogenesis*, **4**, pp. 91–102.

Busby, J. E., Shih, S. J., Yang, J. C., Kung, H. J. and Evans, C. P. (2003) 'Angiogenesis is not mediated by prostate cancer neuropeptides', *Angiogenesis*, **6**(4), pp. 289–293.

Chang, X., Tou, J. C., Hong, C., Kim, H. A., *et al.* (2005) '3,3'-Diindolylmethane inhibits angiogenesis and the growth of transplantable human breast carcinoma in athymic mice', *Carcinogenesis*, **26**(4), pp. 771–778.

Chen, Y., Zhang, Y. X., Li, M. H., Zhao, W. M. *et al.* (2005) 'Antiangiogenic activity of 11,11'-dideoxyverticillin, a natural product isolated from the fungus *Shiraia bambusicola*', *Biochem. Biophys. Res. Commun.*, **329**(4), pp. 1334–1342.

Chen, Z., Zhang, X., Li, M., Wang, Z. *et al.* (2004) 'Simultaneously targeting epidermal growth factor receptor tyrosine kinase and cyclooxygenase-2, an efficient approach to inhibition of squamous cell carcinoma of the head and neck', *Clin. Cancer Res.*, **10**(17), pp. 5930–5939.

Cockerill, G. W., Varcoe, L., Meyer, G. T., Vadas, M. A. and Gamble J. R. (1998) 'Early events in angiogenesis: cloning an alpha-prolyl 4-hydroxylase-like gene', *Int. J. Oncol.*, **13**(3), pp. 595–600.

Connolly, J. O., Simpson, N., Hewlett, L. and Hall, A. (2002) 'Rac regulates endothelial morphogenesis and capillary assembly', *Mol. Biol. Cell*, **13**(7), pp. 2474–2485.

Craig, L. E., Spelman, J. P., Stradberg, J. D. and Zink, M. C. (1998) 'Endothelial cells from diverse tissues exhibit differences in growth and morphology', *Microvascular Res.*, **55**, pp. 65–76.

Dasgupta, P., Sun, J., Wang, S., Fusaro, G. *et al.* (2004) 'Disruption of the Rb–Raf-1 interaction inhibits tumor growth and angiogenesis', *Mol. Cell Biol.*, **24**(21), pp. 9527–9541.

Donovan, D., Brown, N. J., Bishop, E. T. and Lewis, C. E. (2001) 'Comparison of three *in vitro* human 'angiogenesis' assays with capillaries formed *in vivo*', *Angiogenesis*, **4**(2), pp. 113–121.

Dye, J., Lawrence, L., Linge, C., Leach, L. *et al.* (2004) 'Distinct patterns of microvascular endothelial cell morphology are determined by extracellular matrix composition', *Endothelium*, **11**(3–4), pp. 151–167.

Folkman, J. and Haudenschild, C. (1980) 'Angiogenesis *in vitro*', *Nature*, **288**(5791), pp. 551–556.

Friis, T., Sorensen, B., Engel, A. M., Rygaard, J. and Houen, G. (2003) 'A quantitative ELISA-based co-culture angiogenesis and cell proliferation assay', *APMIS*, **111**(6), pp. 658–668.

Gagnon, E., Cattaruzzi, P., Griffith, M., Muzakare, L. *et al.* (2002) 'Human vascular endothelial cells with extended life spans: *in vitro* cell response, protein expression, and angiogenesis', *Angiogenesis*, **5**(1–2), pp. 21–33.

Galvez, B. G., Matias-Roman, S., Albar, J. P., Sanchez-Madrid, F. and Arroyo, A. G. (2001) 'Membrane type 1-matrix metalloproteinase is activated during migration of human endothelial cells and modulates endothelial motility and matrix remodeling', *J. Biol. Chem.*, **276**(40), pp. 37491–37500.

Golubkov, V., Hawes, D. and Markland, F. S. (2003) 'Anti-angiogenic activity of contortrostatin, a disintegrin from Agkistrodon contortrix contortrix snake venom', *Angiogenesis*, **6**(3), pp. 213–224.

Grant, D. S., Lelkes, P. I., Fukuda, K. and Kleinman, H. K. (1991) 'Intracellular mechanisms involved in basement membrane induced blood vessel differentiation *in vitro*', *In Vitro Cell Dev. Biol.*, **27A**(4), pp. 327–336.

Hahm, E. R., Gho, Y. S., Park, S., Park, C. *et al.* (2004) 'Synthetic curcumin analogs inhibit activator protein-1 transcription and tumor-induced angiogenesis' *Biochem. Biophys. Res. Commun.*, **321**(2), pp. 337–344.

Hernandez, J. L., Coll, T. and Ciudad C. J. (2004) 'A highly efficient electroporation method for the transfection of endothelial cells', *Angiogenesis*, **7**(3), pp. 235–241.

Ilan, N., Mahooti, S. and Madri, J. A. (1998) 'Distinct signal transduction pathways are utilized during the tube formation and survival phases of *in vitro* angiogenesis', *J. Cell Sci.*, **111**(24), pp. 3621–3631.

Isaji, M., Miyata, H., Ajisawa, Y., Takehana, Y. and Yoshimura, N. (1997) 'Tranilast inhibits the proliferation, chemotaxis and tube formation of human microvascular endothelial cells *in vitro* and angiogenesis *in vivo*', *Br. J. Pharmacol.*, **122**(6), pp. 1061–1066.

Jackson, C. J. and Nguyen, M. (1997) 'Human microvascular endothelial cells differ from macrovascular endothelial cells in their expression of matrix metalloproteases', *Int. J. Biochem. Cell Biol.*, **29**, pp. 1167–1177.

Jee, S. H., Chu, C. Y., Chiu, H. C., Huang, Y. L. *et al.* (2004) 'Interleukin-6 induced basic fibroblast growth factor-dependent angiogenesis in basal cell carcinoma cell line via JAK/STAT3 and PI3-kinase/Akt pathways', *J. Invest. Dermatol.*, **123**(6), pp. 1169–1175.

Kalluri, R. (2003) 'Basement membranes: structure, assembly and role in tumour angiogenesis', *Nat. Rev. Cancer*, **3**(6), pp. 422–433.

Khodarev, N. N., Yu, J., Labay, E., Darga, T. *et al.* (2003) 'Tumour-endothelium interactions in co-culture: coordinated changes of gene expression profiles and phenotypic properties of endothelial cells', *J. Cell Sci.*, **116**(6), pp. 1013–1022.

Kroon, M. E., van Schie, M. L., van der Vecht, B., van Hinsbergh, V. W. and Koolwijk, P. (2002) 'Collagen type 1 retards tube formation by human microvascular endothelial cells in a fibrin matrix', *Angiogenesis*, **5**(4), pp. 257–265.

Labrecque, L., Lamy, S., Chapus, A., Mihoubi, S. *et al.* (2005) 'Combined inhibition of PDGF and VEGF receptors by ellagic acid, a dietary-derived phenolic compound', *Carcinogenesis*, **26**(4), pp. 821–826.

Lawley, T. J. and Kubota, Y. (1989) 'Induction of morphologic differentiation of endothelial cells in culture', *J. Invest. Dermatol.*, **93**(2S), pp. 59S–61S.

Lee, K. H., Choi, H. R. and Kim, C. H. (2005) 'Anti-angiogenic effect of the seed extract of Benincasa hispida Cogniaux', *J. Ethnopharmacol.*, **97**(3), pp. 509–513.

Leopold, J. A., Walker, J., Scribner, A. W., Voetsch, B. *et al.* (2003) 'Glucose-6-phosphate dehydrogenase modulates vascular endothelial growth factor-mediated angiogenesis' *J. Biol. Chem.*, **278**(34), pp. 32100–32106.

Li M.-H., Miao Z.-H., Tan W.-F., Yue J.-M., Zhang C., Lin L.-P., Zhang X.-W., Ding J. (2004) 'Pseudolaric Acid B inhibits angiogenesis and reduces hypoxia-inducible factor 1α by promoting proteasome-mediated degradation', *Clinical Cancer Research*, **10**, 8266–8274.

Liu, J., Razani, B., Tang, S., Terman, B. I. *et al.* (1999) 'Angiogenesis activators and inhibitors differentially regulate caveolin-1 expression and caveolae formation in vascular endothelial cells. Angiogenesis inhibitors block vascular endothelial growth factor-induced down-regulation of caveolin-1', *J. Biol. Chem.*, **274**(22), pp. 15781–15785.

Liu, J., Wang, X. B., Park, D. S. and Lisanti, M. P. (2002) 'Caveolin-1 expression enhances endothelial capillary tubule formation', *J. Biol. Chem.*, **277**(12), pp. 10661–10668.

Macaig, T., Kadish, J., Wilkins, L., Stemerman, M. B. and Weinstein, R. (1982) 'Organizational behavior of human umbilical vein endothelial cells', *J. Cell Biol.*, **94**(3), pp. 511–520.

Madri, J. A. and Williams, S. K. (1983) 'Capillary endothelial cell cultures: phenotypic modulation by matrix components', *J. Cell Biol.*, **97**(1), pp. 153–165.

Madri, J. A., Pratt B. M. and Tucker A. M. (1988) 'Phenotypic modulation of endothelial cells by transforming growth factor-beta depends upon the composition and organization of the extracellular matrix', *J. Cell Biol.*, **106**(4): 1375–1384.

Matou, S., Colliec-Jouault, S., Galy-Fauroux, I., Ratiskol, J. *et al.* (2005) 'Effect of an oversulfated exopolysaccharide on angiogenesis induced by fibroblast growth factor-2 or vascular endothelial growth factor *in vitro*', *Biochem. Pharmacol.*, **69**(5), pp. 751–759.

Montanez, E., Casaroli-Marano, R. P., Vilaro, S. and Pagan, R. (2002) 'Comparative study of tube assembly in three-dimensional collagen matrix and on Matrigel coats', *Angiogenesis*, **5**(3), pp. 167–172.

Montesano, R., Pepper, M. S. and Orci, L. (1993) 'Paracrine induction of angiogenesis *in vitro* by Swiss 3T3 fibroblasts', *J. Cell Sci.*, **105**(4), pp. 1013–1024.

Mousa, S. A. and Mohamed, S. (2004) 'Inhibition of endothelial cell tube formation by the low molecular weight heparin, tinzaparin, is mediated by tissue factor pathway inhibitor', *Thromb. Haemost.*, **92**(3), pp. 627–633.

Narmoneva, D. A., Oni, O., Sieminski, A. L., Zhang, S. *et al.* (2005) 'Self-assembling short oligopeptides and the promotion of angiogenesis', *Biomaterials*, **26**(23), pp. 4837–4846.

Natori, T., Sata, M., Nagai, R. and Makuuchi, M. (2005) 'Cimetidine inhibits angiogenesis and suppresses tumor growth', *Biomed. Pharmacother.*, **59**(1–2), pp. 56–60.

Nehls, V., Schuchardt, E. and Drenckhahn, D. (1994) 'The effect of fibroblasts, vascular smooth muscle cells, and pericytes on sprout formation of endothelial cells in a fibrin gel angiogenesis system', *Microvasc. Res.*, **48**(3), pp. 349–363.

Nehls, V. and Drenckhahn, D. (1995) 'A novel, microcarrier-based *in vitro* assay for rapid and reliable quantification of three-dimensional cell migration and angiogenesis', *Microvasc. Res.*, **50**(3), pp. 311–322.

Nicosia, R. F. and Ottinetti, A. (1990) 'Modulation of microvascular growth and morphogenesis by reconstituted basement membrane gel in three-dimensional cultures of rat aorta: a comparative study of angiogenesis in matrigel, collagen, fibrin, and plasma clot', *In Vitro Cell Dev. Biol.*, **26**(2), pp. 119–128.

Osusky, K. L., Hallahan, D. E., Fu, A., Ye, F. *et al.* (2004) 'The receptor tyrosine kinase inhibitor SU11248 impedes endothelial cell migration, tubule formation, and blood vessel formation *in vivo*, but has little effect on existing tumor vessels', *Angiogenesis*, **7**(3), pp. 225–233.

Plaisier, M., Kapiteijn, K., Koolwijk, P., Fijten, C. *et al.* (2004) 'Involvement of membrane-type matrix metalloproteinases (MT-MMPs) in capillary tube formation by human endometrial microvascular endothelial cells: role of MT3-MMP', *J. Clin. Endocrinol. Metab.*, **89**(11), pp. 5828–5836.

Qian, D. Z., Wang, X., Kachhap, S. K., Kato, Y. *et al.* (2004) 'The histone deacetylase inhibitor NVP-LAQ824 inhibits angiogenesis and has a greater antitumor effect in combination with the vascular endothelial growth factor receptor tyrosine kinase inhibitor PTK787/ZK222584', *Cancer Res.*, **64**(18), pp. 6626–6634.

Sakuda, H., Nakashima, Y., Kuriyama, S. and Sueishi, K. (1992) 'Media conditioned by smooth muscle cells cultured in a variety of hypoxic environments stimulates *in vitro* angiogenesis. A relationship to transforming growth factor-beta 1', *Am. J. Pathol.*, **141**(6), pp. 1507–1516.

Sakurai, D., Tsuchiya, N., Yamaguchi, A., Okaji, Y. *et al.* (2004) 'Crucial role of inhibitor of DNA binding/differentiation in the vascular endothelial growth factor-induced activation and angiogenic processes of human endothelial cells', *J. Immunol.*, **173**(9), pp. 5801–5809.

Sanz, L., Pascual, M., Munoz, A., Gonzalez, M. A. *et al.* (2002) 'Development of a computer-assisted high-throughput screening platform for anti-angiogenic testing', *Microvasc. Res.*, **63**(3), pp. 335–339.

Sato, Y., Okamura, K., Morimoto, A., Hamanaka, R. *et al.* (1993) 'Indispensable role of tissue-type plasminogen activator in growth factor-dependent tube formation of human microvascular endothelial cells *in vitro*', *Exp. Cell Res.*, **204**(2), pp. 223–229.

Segura, I., Serrano, A., De Buitrago, G. G., Gonzalez, M. A. *et al.* (2002) 'Inhibition of programmed cell death impairs *in vitro* vascular-like structure formation and reduces *in vivo* angiogenesis', *FASEB J.*, **16**(8), pp. 833–841.

Sengupta, S., Sellers, L. A., Li, R. C., Gherardi, E. *et al.* (2003) 'Targeting of mitogen-activated protein kinases and phosphatidylinositol 3 kinase inhibits hepatocyte growth factor/scatter factor-induced angiogenesis', *Circulation*, **107**(23), pp. 2955–2961.

Skurk, C., Maatz, H., Rocnik, E., Bialik, A. *et al.* (2005) 'Glycogen-Synthase Kinase3beta/beta-catenin axis promotes angiogenesis through activation of vascular endothelial growth factor signaling in endothelial cells', *Circ. Res.*, **96**(3), pp. 308–318.

Smith, E. and Hoffman, R. (2005) 'Multiple fragments related to angiostatin and endostatin in fluid from venous leg ulcers', *Wound Repair Regen.*, **13**(2), pp. 148–157.

Staton, C. A., Brown, N. J., Rodgers, G. R., Corke, K. P. *et al.* (2004) 'Alphastatin, a 24-amino acid fragment of human fibrinogen, is a potent new inhibitor of activated endothelial cells *in vitro* and *in vivo*', *Blood*, **103**(2), pp. 601–606.

Sun, X. T., Ding, Y. T., Yan, X. G., Wu, L. Y. *et al.* (2004) 'Angiogenic synergistic effect of basic fibroblast growth factor and vascular endothelial growth factor in an *in vitro* quantitative microcarrier-based three-dimensional fibrin angiogenesis system', *World J. Gastroenterol.*, **10**(17), pp. 2524–2528.

Unger, R. E., Krump-Konvalinkova, V., Peters, K. and Kirkpatrick, C. J. (2002) '*In vitro* expression of the endothelial phenotype: comparative study of primary isolated cells and cell lines, including the novel cell line HPMEC-ST1.6R', *Microvasc. Res.*, **64**(3), pp. 384–397.

Vailhe, B., Lecomte, M., Wiernsperger, N. and Tranqui, L. (1998) 'The formation of tubular structures by endothelial cells is under the control of fibrinolysis and mechanical factors', *Angiogenesis*, **2**(4), pp. 331–344.

Vucenik, I., Passaniti, A., Vitolo, M. I., Tantivejkul, K. *et al.* (2004) 'Anti-angiogenic activity of inositol hexaphosphate (IP6)', *Carcinogenesis*, **25**(11), pp. 2115–2123.

Wu, Z., O'Reilly, M. S., Folkman, J. and Shing, Y. (1997) 'Suppression of tumor growth with recombinant murine angiostatin', *Biochem. Biophys. Res. Commun.*, **236**(3), pp. 651–654.

Yahata, Y., Shirakata, Y., Tokumaru, S., Yamasaki, K. *et al.* (2003) 'Nuclear translocation of phosphorylated STAT3 is essential for vascular endothelial growth factor-induced human dermal microvascular endothelial cell migration and tube formation', *J. Biol. Chem.*, **278**(41), pp. 40026–40031.

Zaccagnini, G., Gaetano, C., Della Pietra, L., Nanni, S. *et al.* (2005) 'Telomerase mediates vascular endothelial growth factor-dependent responsiveness in a rat model of hind limb ischemia', *J. Biol. Chem.*, **280**(15), pp. 14790–14798.

Zhong, L., Guo, X. N., Zhang, X. H., Wu, Z. X. *et al.* (2005) 'Expression and purification of the catalytic domain of human vascular endothelial growth factor receptor 2 for inhibitor screening', *Biochim. Biophys. Acta*, **1722**(3), pp. 254–261.

Zimrin, A. B., Villeponteau, B. and Maciag, T. (1995) 'Models of *in vitro* angiogenesis: endothelial cell differentiation on fibrin but not matrigel is transcriptionally dependent', *Biochem. Biophys. Res. Commun.*, **213**(2), pp. 630–638.

5

Modelling the effects of the haemodynamic environment on endothelial cell responses relevant to angiogenesis

Gerard B. Nash and **Stuart Egginton**

Abstract

There is much evidence showing that the phenotype and functional responses of endothelial cells are conditioned by local environmental factors, including the haemodynamic forces that act upon them. Haemodynamic conditioning may thus be predicted to modify responses of endothelial cells relevant to angiogenesis, including cell migration and proliferation, as well as production of cytokines, growth factors and matrix re-modelling enzymes. For example, *in vivo* studies show that angiogeneis can be induced in response to increase in fluid shear stress applied by chronically elevated blood flow. It is therefore important to study 'angiogenic' responses of endothelial cells to flow forces imposed on them, either acting alone or in concert with soluble pro- and anti-angiogenic agents. Here we outline methods by which endothelial cells can be exposed to relevant flow forces *in vitro*, and how information relevant to angiogenesis can be extracted from such studies.

Keywords

haemodynamics; shear stress; flow models

5.1 Introduction

Previous chapters have shown how key angiogenic responses of endothelial cells (EC), particularly proliferation and migration, can be investigated using

---

in vitro assays. However, for a number of reasons, angiogenic responses should ideally be studied under conditions of flow. First, *in vivo*, EC continually experience forces applied by the haemodynamic environment: shear stress applied by flowing blood, compressive pressure and circumferential tension generated by the cardiac cycle, and a pressure wave travelling along the wall. In capillaries where angiogenesis originates, the pressure wave has dissipated and circumferential strain is not obviously evident, as diameter is constant throughout the cardiac cycle. However, capillary pressure and shear stress vary on a longer time scale, depending on changes in the upstream arteriolar tone. In addition, flow is typically intermittent in individual capillaries of most capillary beds. Thus the natural environment of the EC is one in which it constantly experiences shear force, albeit of a variable magnitude. It has become increasingly evident in recent years that EC are sensitive to the shear stress applied to them (Barakat and Lieu, 2003; Davies *et al.*, 2002). Responses to acute changes in flow may be rapid, based for example, on changes in production of reactive oxygen species and/or nitric oxide. Slower, prolonged adaptations to shear conditions depend on changes in gene expression. As a consequence, the physical environment is recognized as an important modifier of a range of endothelial functions including proliferation, apoptosis, motility, adhesion, and matrix deposition or degradation (Lelkes, 1999). Moreover, since the responses of EC to inflammatory cytokines have been shown to be conditioned by exposure to shear stress (Sheikh *et al.*, 2003), it seems likely that the effects of pro- or anti-angiogenic growth factors or cytokines on EC will also vary depending on whether investigations are made under static or flow conditions.

Secondly, changes in the local physical environment may directly induce angiogenesis, for example, in normal mature cardiac and skeletal muscle. In animal models, capillaries exposed to hyperaemia and hence increased wall shear stress, or to a sustained strain, both showed angiogenic responses (Milkiewicz *et al.*, 2001; Zhou *et al.*, 1998) (see Figure 5.1). The responses happened in parallel with and appeared reliant on the regulation of key chemical mediators of angiogenesis, but with different time courses. For instance, during shear-induced capillary growth there was a rapid increase in capillary-located vascular endothelial growth factor-A (VEGF-A) expression that matched cell proliferation at these sites and capillary density correlated with increased VEGF expression (Milkiewicz *et al.*, 2001), whereas fibroblast growth factor (FGF) did not appear to be important (Brown *et al.*, 1998). Under conditions of low shear the angiogenic effect of VEGF was critically dependent on the relative expression of its two cognate receptors, VEGF-R1 and R2 (Milkiewicz *et al.*, 2003). Levels of matrix metalloproteinases (MMPs) were upregulated in EC in parallel with increased shear stress *in vivo* and *in vitro* and inhibition of their activity *in vivo* prevented angiogenesis (Haas *et al.*, 2000). These studies again point to the relevance of studies under flow for understanding endothelial responses relevant to angiogenesis, and to the desirability of combining studies of shear stress with classical soluble angiogenic factors.

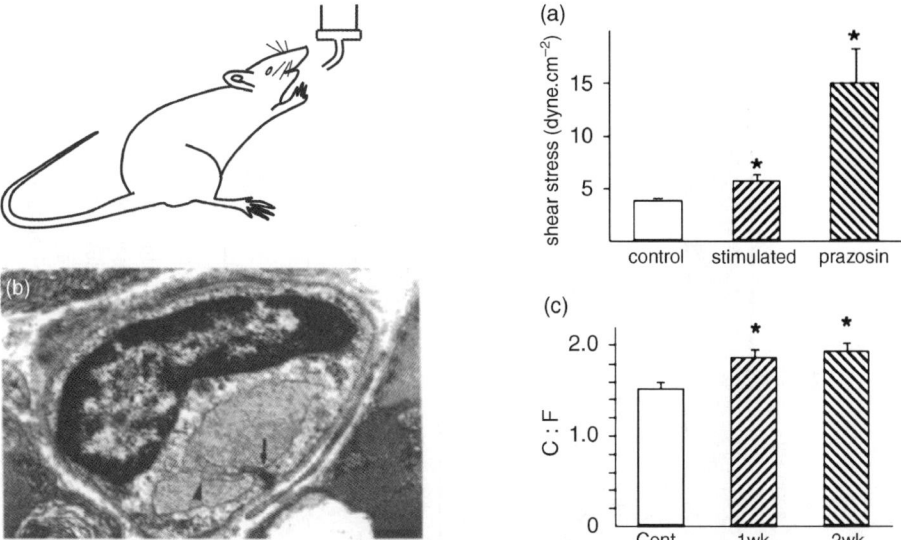

Figure 5.1 Chronic increases in muscle blood flow, induced by pharmacological blockade of peripheral vascular tone by the α-1 adrenoreceptor antagonist prazosin given in drinking water, caused: (a) elevated shear stress; (b) angiogenesis by intraluminal splitting of capillaries (arrowed); (c) increased capillary to muscle fibre ratio (C:F). Data from Milkiewicz *et al.* (2001) and Zhou *et al.* (1998).

In this chapter we review the patterns of flow that may be relevant to studies of angiogenesis and the methods for culturing endothelial cells under such conditions. We also give an overview of the relevant readouts that might be obtained in such studies, and comment on desirable developments of these approaches.

5.2 Definitions

Laminar flow describes the situation where streamlines run in parallel and mixing does not occur, and can exist in curved and branching vessels, depending on the flow rate. In curved vessels, and at bifurcation and junctions, complex patterns of flow can occur at higher flow rates, with flow separation and vortices established. Even in regions where vortices occur, the flow can be considered as laminar if the recirculation is stable, as there are effectively two separate regions of non-mixing laminar flow. The local conditions of flow at a surface can be defined by the wall shear rate (γ_w; velocity gradient perpendicular to the wall) acting at the surface of the endothelial monolayer, and this will be determined by the volumetric flow rate and the geometry of the 'vessel' (Goldsmith and Turitto, 1986). The wall shear stress (τ_w) is the product of the shear rate and the fluid viscosity, and defines the force per unit area acting on the surface (measured in Nm^{-2} or Pascals, Pa, in SI units, or dyn/cm^2 in cgs units; $1\,Pa = 10\,dyn/cm^2$). *In vitro*, endothelial cells are

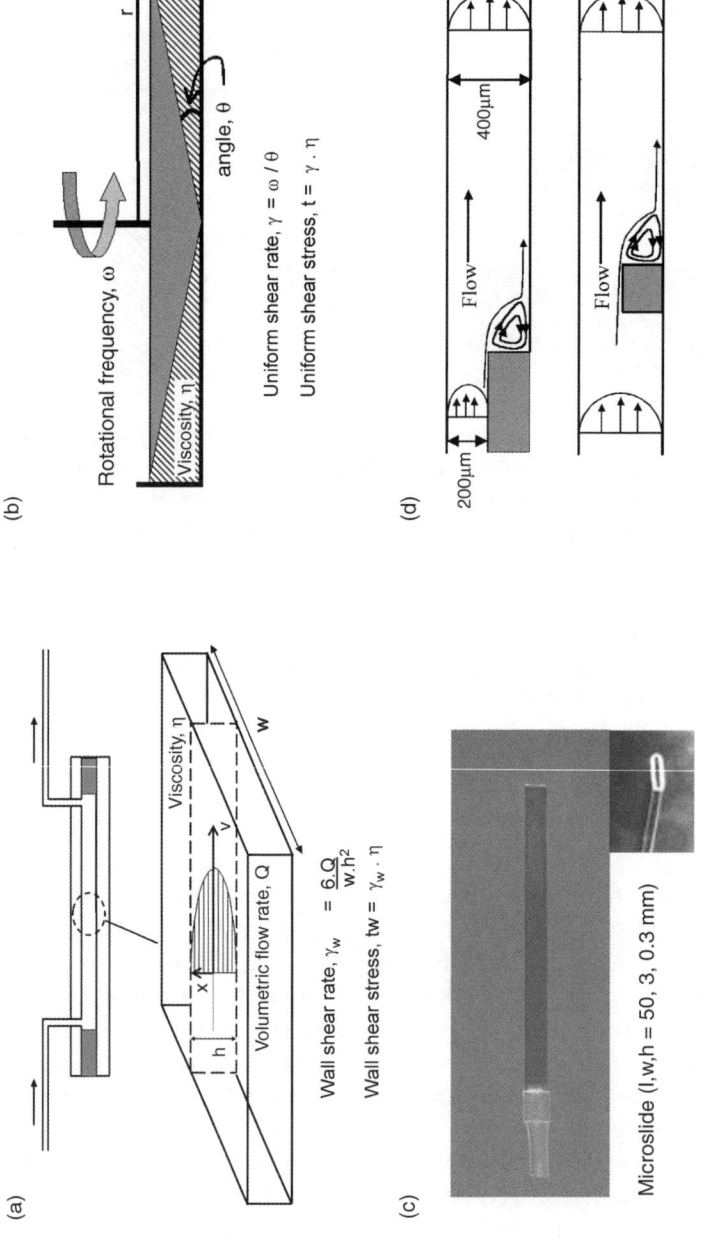

(a)

Viscosity, η

w

Volumetric flow rate, Q

Wall shear rate, $\gamma_w = \dfrac{6.Q}{w.h^2}$

Wall shear stress, $tw = \gamma_w \cdot \eta$

(b)

Rotational frequency, ω

Viscosity, η

angle, θ

r

Uniform shear rate, $\gamma = \omega / \theta$

Uniform shear stress, $t = \gamma \cdot \eta$

(c)

Microslide (l,w,h = 50, 3, 0.3 mm)

(d)

200µm

Flow

400µm

Flow

Figure 5.2 Chambers or devices used for exposing EC to shear stress: (a) parallel-plate flow chamber, with equations relating wall shear rate (γ_w) and wall shear stress (τ_w) to geometry and flow parameters; (b) cone-plate flow device, with equations relating the uniform shear rate (γ) and shear stress (τ) to geometry and rotational velocity; (c) prefabricated glass capillary (microslide (25); Camlab Ltd., Cambridge, UK); (d) flow chambers with backward facing step or indentation which induce a region of recirculating flow with spatial variation in wall shear stress. In a, b and d, EC are cultured on a glass slide or coverslip which is incorporated as the lower plate of the assembly. In c, EC are seeded and grown directly in the capillary.

most commonly exposed experimentally to laminar flow with uniform shear rate and shear stress over a surface. Such patterns are illustrated in Figure 5.2, along with formulae for predicting γ_w and τ_w. Spatial gradients in shear stress (i.e. changes in the level of shear stress along the wall), or oscillating or pulsatile (time-varying) stress may also be applied. *In vivo*, patterns of shear rate and stress are made more complex because blood has a particulate nature and non-newtonian viscosity, which varies depending on the shear rate. However, experimentally, blood is rarely used for studies of endothelial responses to shear stress.

5.3 Experimental patterns of shear exposure and relevance to angiogenesis assays

Endothelial cells are typically cultured and passaged *in vitro* on the surfaces of stationary dishes or flasks. Early studies took such 'static' cultures, passaged them onto convenient surfaces, incorporated them in a perfusion system (Levesque and Nerem, 1989; Nollert *et al.*, 1989) or in the base of a shearing device (Franke *et al.*, 1984; Lan *et al.*, 1994), and then exposed the EC to flow which provided a chosen steady level of shear stress. Responses to shear stress *per se* were evaluated over short or prolonged periods, by comparing characteristics of the EC which were exposed to flow with those of EC which were maintained under static conditions. The wall shear stress experienced by the EC might vary between quite low levels found in the venous circulation ($\sim$0.1–1 Pa) and higher levels relevant to the arterial circulation ($>$1 Pa). To model more complex situations, a pulsatile element was added, such as a sinusoidal variation in flow rate through a chamber (Chappell *et al.*, 1998; Helmlinger *et al.*, 1991). The flow might nevertheless be continually in one forward direction (with a fluctuation around the mean), or might reverse completely. Spatial (as opposed to temporal) variations in shear stress have also been imposed by placing a discontinuity or step in the wall of a flow chamber so that a vortex was set up downstream (Chiu *et al.*, 1998; Phelps and DePaola, 2000). Under these conditions, EC downstream of the 'step' were exposed to a flow which varied strongly over a short distance and which included a component perpendicular to the wall as well as parallel to it. In the extreme, turbulent flow was introduced, imposing chaotic and unpredictable patterns of shear on the EC (Davies *et al.*, 1986).

Recent studies have taken more account of the fact that EC are typically continually exposed to shear *in vivo*. Rather than comparing responses to 'static' cells after a limited period, studies have compared different types of flow (steady vs. pulsatile) (Brooks *et al.*, 2002), or altered the level of shear after a period of flow conditioning, e.g. by stopping flow to mimic ischaemia (Wei *et al.*, 1999). In addition, studies have increasingly combined shear exposure with addition of bioactive agents. In this way studies can be made, not only of effects of shear itself, but also of the effects shear has on responses to other agents (i.e. shear conditioning of responses to growth factors or cytokines) (Sheikh *et al.*, 2003, 2005; Surapisitchat *et al.*, 2001).

Not all of the above systems are obviously relevant if one wishes to model the *in vivo* processes of angiogenesis. In the microvasculature, turbulence is not expected, and pulsatility is largely damped out. Simple comparison of static and flow cultures is worthwhile, if combined with angiogenic factors, as this may reveal information about variable sensitivity of EC to disturbance of their quiescent state. In terms of effects of flow *per se*, then the most obvious patterns of interest, based on studies of ischaemia or of flow augmentation, would seem to be to increase or decrease the fluid shear stress from some pre-imposed steady state. The behaviour of EC is clearly not the same *in vitro* when flow-conditioning is followed by stasis, as when cultures are static throughout. There may be little spatial variation in shear along uniform capillary segments, but shear gradients will be greater near the entrances into and exits from capillaries. In addition, at sites of outgrowth or division of a capillary, and in developing buds, local shear patterns will be complex and non-uniform. Thus, broadly, it may be interesting to vary the levels of shear to which EC are exposed (e.g. from a static or initial conditioning level), to examine responses to cessation of flow, or to examine how spatial variation in shear influence relevant responses. In all cases (not only in simple static vs. flow comparisons) it will be relevant to add soluble agonists or antagonists of angiogenesis, as the shear environment is unlikely to act in isolation in influencing endothelial responses.

5.4   Methods for studying responses of endothelial cells exposed to shear stress

Culture surfaces and chambers

Typical flow devices are illustrated in Figure 5.2. To expose EC to shear, one either perfuses a flow channel (usually with rectangular cross-section) with one surface coated with EC (Figure 5.2a), or uses a rotating cone-plate device (based on a standard viscometer design) to expose a lower, coated plate to uniform shear (Figure 5.2b). Provided any rectangular flow chamber has a width $\gg$ depth, fully developed laminar flow with a parabolic velocity profile should be established a short distance downstream from the entrance, and the wall shear rate and stress can be calculated from the volumetric flow rate, the chamber dimensions and the viscosity of the perfused fluid (Figure 5.2a). The wall shear rate and stress will decrease near the side walls, and measurements are ideally made near to the centre line of the chamber. In the rotating cone-plate device, the shear rate is uniform both across the gap in the vertical direction and at all radial positions, being determined by the rotational frequency and the angle of the cone (Figure 5.2b).

Most groups have first cultured the EC on a plate (usually a glass slide or coverslip), and then assembled it into the flow device and exposed the cells to shear. We have seeded EC directly inside pre-formed glass capillaries with rectangular cross-section (microslides; Figure 5.2c), and either exposed them immediately to flow or allowed the monolayers a period to stabilize before perfusing the capillaries (Sheikh *et al.*, 2004). Multiple, parallel, porous cylinders in cartridges have been used to obtain a large surface area (e.g. to collect released substances) but these are not suitable for direct visual observation or studies of functional responses such as migration (Wei *et al.*, 1999). To obtain shear gradients, EC are again typically cultured on a glass plate, which is then put into the base of a flow chamber whose depth increases halfway along (i.e. flow over a backward facing step) (Chiu *et al.*, 1998) or a rectangular obstruction if placed on the culture which fill some portion of the height of the chamber (Phelps and DePaola, 2000) (Figure 5.2d).

The intended culture surface can itself be pre-coated with proteins. Fibronectin, for instance, has commonly been used as a substrate for cell growth. Typically glass is first acid-washed, and we have used a method where it is then coated with a protein-binding agent, aminopropyltriethoxy-silane (APES), to improve attachment (Cooke *et al.*, 1993). While there is good reason to believe that functional responses of EC can be modified by their substrate (Aird, 2003), there have been few studies of functional responses to flow on different surfaces. Cultured EC deposit basement membrane constituents, and although the major proteins can be detected within days, it may take up to 20 days to obtain a structure resembling that observed *in vivo* (Huber and Weiss, 1989). Not surprisingly, most studies of flow-induced responses by endothelial cells have used purified matrix constituents (Jalali *et al.*, 2001; Orr *et al.*, 2005). It should also be borne in mind that responses of EC may depend critically on the confluence of monolayers, and constituents of the culture medium. While endothelium is confluent in intact vessels, responses of subconfluent monolayers may be relevant to the situation of newly developing vessels. These factors are not specific to flow-based culture, or indeed studies of angiogenesis. Nevertheless, there has been surprisingly little systematic study of the effects of culture conditions *per se*, and they should be borne in mind when comparing studies between different laboratories using quite widely different culture conditions.

Flow systems

There are several ways to generate flow and hence shear stress in flow chambers or capillaries (Figure 5.3). Typically a flow loop is used, which circulates medium through the chamber. The medium may run through the chamber under gravity from an upper to a lower reservoir giving steady flow determined by the pressure head. The fluid is then pumped back to the upper reservoir where it is equilibrated with 5 per cent CO_2 and warmed (Figure 5.3a). If flow is circulated through the

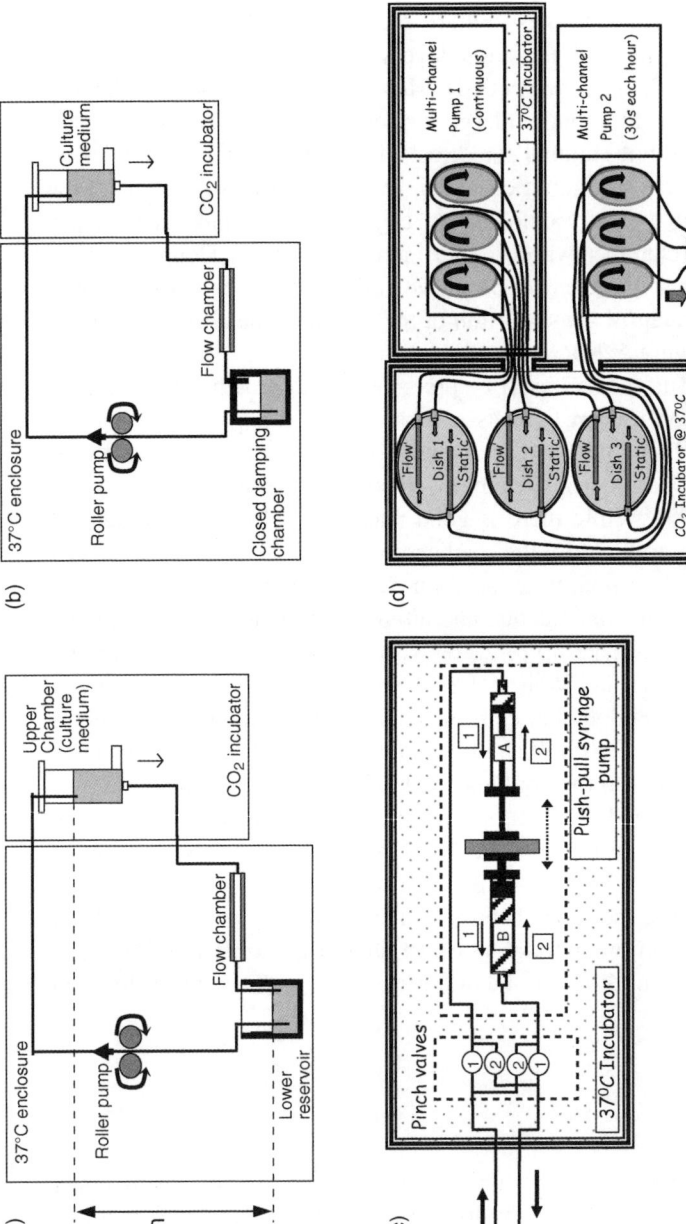

Figure 5.3 Flow systems for perfusing EC grown in chambers or capillaries in (a) flow is driven from the upper to lower reservoir, via the flow chamber, by the hydrostatic pressure head h. The roller pump returns flow to the upper chamber. In (b), flow is driven around a closed loop by the roller pump. The damping chamber serves to reduce pulsatility induced by the pump. In (c), steady flow is delivered by a 'push–pull' syringe pump (two independent pumps could also be used). At the start of phase 1 of perfusion, pinch valves 1 opens (and valves 2 closes) and syringe A (which is empty) withdraws medium from a reservoir via an attached chamber. Syringe B (which is full) replaces medium in the reservoir. At the start of phase 2 of perfusion the pump reverses, pinch valves 2 open (and 1 closes) and syringe B withdraws medium via the chamber, and syringe A replaces medium in the reservoir. In (d) continuous flow is delivered to three microslide capillaries in separate dishes by a roller pump and a return circuit. Static control cultures are exposed to identical medium in the same three dishes, with perfusion supplied for 30 s in each hour by a separate roller pump which pumps to waste.

chamber from and then to a single reservoir, then a damping chamber may be incorporated in the flow line to reduce pulsatility typically induced by roller or peristaltic pumps (Figure 5.3b). Steady flow can also be generated by syringe pumps (two pumps or in a back-to-back arrangement), working in opposite directions so that one fills as the other empties (Shiu *et al.*, 2000). Periodic switching of direction occurs, linked with cross-over valves, so that flow is always in one direction through the chamber (Figure 5.3c). In a variant on such arrangements, we immerse coated microslides in the medium reservoir (a specially made glass dish with tubes fused into the walls) and pump directly out of the reservoir through the capillaries and back by another line (Figure 5.3d) (Sheikh *et al.*, 2004). This obviates any need for bubble traps in the system, which is to some extent self-priming. As described, these systems deliver steady flow, or flow with a fluctuation about the mean.

When a shear gradient is desired, then this is delivered by the design of the step chamber noted above, and the flow system will generally be the same as described above. Use of peristaltic or roller pumps imparts a somewhat unpredictable (although measurable) fluctuation in flow rate. If a controlled degree of pulsatility or oscillation is required, then an oscillating syringe pump can be incorporated into the perfusion system, e.g. between the feed reservoir and the flow chamber. The oscillation can be imposed on a very low forward flow so that flow actually reverses periodically, or it can be imposed on a high forward flow so that one obtains an oscillation about forward flow that does not reverse (DeKeulenaer *et al.*, 1998; Helmlinger *et al.*, 1991).

In cone-plate devices, the rotating shallow-angled cone generates a uniform, linear shear field on the coated plate without the need for a perfusion system. In general, the surface area to volume ratio of these devices is great and, unless modified, there is no fluid throughput. Thus released substances will accumulate. However, it may be possible to engineer an inlet and outlet to allow change in medium during shearing for prolonged periods.

Practical problems and design features

In all the systems described, substances will be released from the endothelial cells that could feed back into the response to be observed. This may not be desirable, and to mimic the situation *in vivo*, the surface area to volume of the perfusion system should be large. In our own initial studies, we found that specific responses of endothelial cells were highly sensitive to agents leached from tubing and to slight cooling ($\sim2°C$) because of failure to control the temperature of the circulating system (Sheikh *et al.*, 2004). These problems were solved by testing various tubing and modifying apparatus design, but they illustrate an important point about controls. In our system (Figure 5.3d), the control cells experienced the same media and conditions as the flow-conditioned cells. Using totally independent cultures in tubes without flow in separate dishes, the effects in the flow 'controls' became obvious. Controls in other systems are not usually so closely matched, and under these conditions it is important to prove that factors related to medium, temperature, surface area/volume etc. are not influencing comparative results.

A general problem of flow cultures is low throughput associated with the complexity of the apparatus. While short-term responses to flow exposure may be studied quite quickly, if cells are conditioned for periods of one or more days, then a series of parallel flow systems is needed if variations in conditions and treatments are to be compared. Our capillary based system was originally designed for three dishes holding each flow and static cultures exposed to different conditions. We have gone on to use greater numbers of capillaries (as many as 21 including static controls) but this requires quite complex experimental and apparatus design. In addition, the quantity of cells may not be great, e.g. for mRNA or protein extraction. Although chambers can be engineered to have different surface areas, cell numbers will rarely be $>10^6$ cells, and in our capillary system, we have had to use multiple capillaries, e.g. for Western blotting. Multi-fibre cartridge systems can expose larger numbers of cells to flow (Wei et al., 1999), and although not useful for studies requiring visualization of cells, they can yield more protein and mRNA.

5.5 Readouts in flow cultures relevant to angiogenesis

In principle, the models outlined above can be used to measure directly responses relevant to angiogenesis such as cell migration, proliferation and apoptosis (although the latter should occur at very low levels in healthy cultures), to evaluate expression of target genes or large panels of genes, and to measure intracellular, surface presentation or secretion of proteins such as adhesion receptors, cytokines and proteinases. These readouts are no different from those that might be used in any culture system, and so the problem comes down to adapting the measurement systems for the perfusion system or to the relatively low numbers of cells commonly used in flow models.

Microscopic studies have shown that velocity and direction of migration of endothelial cells are modified by exposure to flow (Hsu et al., 2001). We have examined endothelial motion in confluent and sub-confluent monolayers over 24 h using time-lapse video microscopy, but this is necessarily a very low throughput system. Proliferation of endothelial cells exposed to flow can also be studied in situ in the cultures by staining for proliferating cell nuclear antigen (PCNA) or BrdU incorporation. Turbulent flow or high levels of laminar shear stress have been found to increase rates of division (Ando et al., 1987), while low levels of shear inhibited proliferation (Lin et al., 2000). DNA synthesis was increased in regions of disturbed flow (Chiu et al., 1998) and when flow conditioned cells were exposed to stasis (Milovanova et al., 2004). In addition, apoptosis is typically suppressed by exposure to laminar flow compared to static conditions (Bartling *et al.*, 2000).

Flow-based studies are amenable to analysis of released substances in the media, although there may be great dilution in circulated systems. We have, for example, measured release of interleukin-8 in static and flow cultures, with and without addition of the cytokine tumour necrosis factor-α (Sheikh *et al.*, 2003). Surface proteins can be measured on intact monolayers using enzyme-linked immunoassay, or on a cell-by-cell basis using immunofluorescence and flow cytometry, upon dissociation of the monolayer. The cells from the culture systems can be retrieved, and protein or mRNA extracted for standard molecular biological analyses.

The most popular indirect readout in flow-based endothelial studies has been gene expression. Studies may be based on analysis of relatively small numbers of genes relevant to a process of interest. Adhesion molecules and chemokines have been examined most commonly in the past, because of wide interest in the role of flow conditioning in regulating inflammatory responses and especially development of atheroma. Large screens of genes using microarray technology have been carried out, e.g. comparing static and flow-exposed cells or cells exposed to different flow regimens (laminar, turbulent or oscillatory flow) (Brooks *et al.*, 2002; McCormick *et al.*, 2003; Ohura *et al.*, 2003). The results are a matter of record, and although, again, studies were mainly aimed at atherosclerotic and inflammatory responses, the data are without bias. Thus they can be mined for results relevant to angiogenesis. Future studies might usefully extend these approaches to combination of angiogenic factors and flow, or specific patterns or changes in flow relevant to angiogenic responses.

The other most widely evaluated responses relate to signal transduction pathways and transcription factors upstream of gene expression, which are affected by or induced by shear stress (Lan *et al.*, 1994; Orr *et al.*, 2005; Surapisitchat *et al.*, 2001). The underlying mechanisms of mechanotransduction are relevant to all functional responses, as potentially are signalling and gene regulatory pathways. Specificity will again arise from studies of responses induced by specific patterns of flow exposure or of modifications in responses to specific angiogenic factors. Definition of a functional modification by flow conditioning may be relatively straightforward given an adequate experimental set-up. However, definition of the mechanism of modification will generally be more difficult based on detailed analyses of the associated signalling and genetic responses.

5.6 Critical considerations and conclusions

Studies of endothelial responses to flow or to changes in flow conditions have tended to be piecemeal – at signalling, genomic, protein or functional levels – and rarely integrated over this range. While it is technically difficult to cover such a range in a single study, if it is found that a key response is modified by haemodynamic factors, it is clearly desirable to follow the response back to its origin if one is going to be in a position to usefully modify it. From a purely practical point of view, this requires development of systems that give adequate throughput, material and availability of functional readouts. Moreover, the systems need to generate flow conditions relevant to the situation *in vivo*, with adequate variability in level or pattern of shear stress, and ability to study prolonged rather than short-term exposures. If this were not enough, it must also be borne in mind that in the long term, effects of flow cannot be considered in isolation, and need to be combined with humoral and structural factors relevant to the microcirculation.

The complexity of *in vivo* control of angiogenesis demands a reductionist approach (epitomized by the methods outlined above) to dissect the potential mechanisms involved, but future developments will need to incorporate multiple influences on cellular responses, e.g. flow, cytokine and stromal interactions. Proliferation and migration of EC, and expression of key genes have been followed *in vitro*, but responses to different levels of shear have usually been studied without added growth factors known to be altered *in vivo*. Thus, at present we do not know how exposure to haemodynamic forces generated by blood flow modifies the responses of endothelial cells (EC) to pro-angiogenic growth factors. Moreover, the endothelial glycocalyx and the underlying matrix are both likely to influence endothelial shear sensing, and mural cells are clearly important in modulating angiogenic signals. Thus, while flow-based studies are a step towards physiologically-relevant modelling of angiogenesis, there is a long way to go before *in vitro* studies can reproduce the situation *in vivo*. While there is an understandable temptation to equate the observed processes *in vitro* with those occurring in true angiogenesis *in vivo*, it is important to recognize them as putative mechanisms which require validation in an integrative system. For example, use of a substrate such as Matrigel may offer a model that appears closer to the real world than flat monocultures. However, the subsequent tube formation is not a unique property of endothelium, being observed with epithelial cells and fibroblasts, and the mechanism of lumen formation is different from that found in tissue (Egginton and Gerritsen, 2003). In addition, each step of the angiogenic cascade described in reviews based primarily on tissue culture experiments is known to have exceptions in capillary growth (Egginton *et al.*, 2001). It is important therefore that in developing 'realistic' manipulable models *in vitro*, which incorporate effects of haemodynamic forces, phenomena observed are linked to those observed in relevant animal models.

Acknowledgements

Work on effects of shear stress on EC in the laboratory of GN is supported by a programme grant from the British Heart Foundation. Work on angiogenesis in the laboratory of SE is supported by grants from the British Heart Foundation and MRC.

References

Aird, W. C. (2003) 'Endothelial cell heterogeneity', *Crit. Care Med.*, **31**(4) Suppl, pp. S221–S230.

Ando, J., Nomura, H. and Kamiya, A. (1987) 'The effect of fluid shear stress on the migration and proliferation of cultured endothelial cells', *Microvascular Res.*, **33**(1), pp. 62–70.

Barakat, A. and Lieu, D. (2003) 'Differential responsiveness of vascular endothelial cells to different types of fluid mechanical shear stress', *Cell Biochem. Biophys.*, **38**(3), pp. 323–343.

Bartling, B., Tostlebe, H., Darmer, D., Holtz, J. *et al.* (2000) 'Shear stress-dependent expression of apoptosis-regulating genes in endothelial cells', *Biochem. Biophys. Res. Commun.*, **278**, pp. 740–746.

Brooks, A. R., Lelkes, P. I. and Rubanyi, G. M. (2002) 'Gene expression profiling of human aortic endothelial cells exposed to disturbed flow and steady laminar flow', *Physiol. Gen.*, **9**(27), pp. 27–41.

Brown, M. D., Walter, H., Hansen-Smith, F. M., Hudlicka, O. and Egginton, S. (1998) 'Lack of involvement of basic fibroblast growth factor (FGF-2) in capillary growth in skeletal muscles exposed to long-term contractile activity', *Angiogenesis*, **2**(1), pp. 81–91.

Chappell, D. C., Varner, S. E., Nerem, R. M., Medford, R. M. and Alexander, R. W. (1998) 'Oscillatory shear stress stimulates adhesion molecule expression in cultured human endothelium', *Circ. Res.*, **82**, pp. 532–539.

Chiu, J. J., Wang, D. L., Chien, S., Skalak, R. and Usami, S. (1998) 'Effects of disturbed flow on endothelial cells', *J. Biomech. Eng.*, **120**, pp. 2–8.

Cooke, B. M., Usami, S., Perry, I. and Nash, G. B. (1993) 'A simplified method for culture of endothelial cells and analysis of adhesion of blood cells under conditions of flow', *Microvascular Res.*, **45**(1), pp. 33–45.

Davies, P. F., Polacek, D. C., Shi, Y. F. and Helmke, B. P. (2002) 'The convergence of haemodynamics, genomics, and endothelial structure in studies of the focal origin of atherosclerosis', *Biorheology*, **39**, pp. 299–306.

Davies, P. F., Remuzzi, A., Gordon, E. J., Forbes Dewey, C. J. and Gimbrone, M. A., Jr (1986) 'Turbulent fluid shear stress induces vascular endothelial cell turnover *in vitro*', *Proc. Natl. Acad. Sci. USA.*, **83**, pp. 2114–2117.

DeKeulenaer, G. W., Chappell, D. C., Ishizaka, N., Nerem, R. M. *et al.* (1998) 'Oscillatory and steady laminar shear stress differentially affect human endothelial redox state. Role of superoxide-producing NADH oxidase', *Circ. Res.*, **82**, pp. 1094–1101.

Egginton, S. and Gerritsen, M. (2003) 'Lumen formation: *in vivo* versus *in vitro* observations', *Microcirculation*, **10**(1), pp. 45–61.

Egginton, S., Zhou, A. L., Brown, M. D. and Hudlicka, O. (2001) 'Unorthodox angiogenesis in skeletal muscle', *Cardiovasc. Res.*, **49**(3), pp. 634–646.

Franke, R. P., Gräfe, M., Schnittler, H., Seiffge, D. and Mittermayer, C. (1984) 'Induction of human vascular endothelial stress fibres by fluid shear stress', *Nature*, **307**, pp. 648–649.

Goldsmith, H. L. and Turitto, V. T. (1986) 'Rheological aspects of thrombosis and haemostasis: basic principles and applications', *Thromb. Haemo.*, **55**(3), pp. 415–35.

Haas, T. L., Milkiewicz, M., Davis, S. J., Zhou, A. L. *et al.* (2000) 'Matrix metalloproteinase activity is required for activity-induced angiogenesis in rat skeletal muscle', *Am. J. Physiol.*, **279**(4), pp. H1540–H1547.

Helmlinger, G., Geiger, R. V., Schreck, S. and Nerem, R. M. (1991) 'Effects of pulsatile flow on cultured vascular endothelial cell morphology', *J. Biomech. Eng.*, **113**, pp. 123–131.

Hsu, P.-P., Li, S., Li, Y.-S., Usami, S. *et al.* (2001) 'Effects of flow patterns on endothelial cell migration into a zone of mechanical denudation', *Biochem. Biophys. Res. Commun.*, **285**, pp. 751–759.

Huber, A. R. and Weiss, S. J. (1989) 'Disruption of the subendothelial basement membrane during neutrophil diapedesis in an *in vitro* construct of a blood vessel wall', *J. Clin. Invest.*, **83**, pp. 1122–1136.

Jalali, S., del Pozo, M. A., Chen, K. D., Miao, H. *et al.* (2001), 'Intergrin-mediated mechanotransduction requires its dynamic interaction with specific extracellular matrix (ECM) ligands', *Proc. Natl. Acad. Sci. USA*, **98**(3), pp. 1042–1046.

Lan, Q., Mercurius, K. O. and Davies, P. F. (1994) 'Stimulation of transcription factors NF kappa B and AP1 in endothelial cells subjected to shear stress', *Biochem. Biophys. Res. Commun.*, **201**(2), pp. 950–956.

Lelkes, P. I. (1999) *Mechanical Forces and the Endothelium*, Harwood Academic Press, Amsterdam.

Levesque, M. J. and Nerem, R. M. (1989) 'The study of rheological effects on vascular endothelial cells in culture', *Biorheology*, **26**, pp. 345–357.

Lin, K., Hsu, P.-P., Chen, B. P., Yuan, S. *et al.* (2000) 'Molecular mechanism of endothelial growth arrest by laminar shear stress', *Proc. Natl. Acad. Sci. USA*, **97**(17), pp. 9385–9389.

McCormick, S. M., Frye, S. R., Eskin, S. G., Teng, C. L. *et al.* (2003) 'Microarray analysis of shear stressed endothelial cells', *Biorheology*, **40**(1–3), pp. 5–11.

Milkiewicz, M., Brown, M. D., Egginton, S. and Hudlicka, O. (2001) 'Association between shear stress, angiogenesis, and VEGF in skeletal muscles *in vivo*', *Microcirculation*, **8**(4), pp. 229–241.

Milkiewicz, M., Hudlicka, O., Verhaeg, J., Egginton, S. and Brown, M. D. (2003) 'Differential expression of Flk-1 and Flt-1 in rat skeletal muscle in response to chronic ischaemia: favourable effect of muscle activity', *Clin. Sci. (Lond)*, **105**(4), pp. 473–482.

Milovanova, T., Manevich, Y., Haddad, A., Chatterjee, S. *et al.* (2004) 'Endothelial cell proliferation associated with abrupt reduction in shear stress is dependent on reactive oxygen species', *Antioxidants and Redox Signaling*, **6**(2), pp. 245–258.

Nollert, M. U., Hall, E. R., Eskin, S. G. and McIntire, L. V. (1989) 'The effect of shear stress on the uptake and metabolism of arachidonic acid by human endothelial cells', *Biochim. Biophys. Acta*, **1005**, pp. 72–78.

Ohura, N., Yamamoto, K., Ichioka, S., Sokabe, T. *et al.* (2003) 'Global analysis of shear stress-responsive genes in vascular endothelial cells', *J. Atherosclerosis Thromb.*, **10**(5), pp. 304–313.

Orr, A. W., Sanders, J. M., Bevard, M., Coleman, E. *et al.* (2005) 'The subendothelial extracellular matrix modulates NF-kappaB activation by flow: a potential role in athero-sclerosis', *J. Cell Biol.*, **169**(1), pp. 191–202.

Phelps, J. E. and DePaola, N. (2000) 'Spatial variations in endothelial barrier function in disturbed flows *in vitro*', *Am. J. Physiol.*, **278**, pp. H469–H476.

Sheikh, S., Rainger, G. E., Gale, Z., Rahman, M. and Nash, G. B. (2003) 'Exposure to fluid shear stress modulates the ability of endothelial cells to recruit neutrophils in response to tumor necrosis factor-alpha: a basis for local variations in vascular sensitivity to inflammation', *Blood*, **102**(8), pp. 2828–2834.

Sheikh, S., Gale, Z., Rainger, G. E. and Nash, G. B. (2004) 'Methods for exposing multiple cultures of endothelial cells to different fluid shear stresses and to cytokines, for subsequent analysis of inflammatory function', *J. Immunol. Methods*, **288**(1–2), pp. 35–46.

Sheikh, S., Rainger, G. E., Gale, Z., Luu, N.-T. *et al.* (2005) 'Differing mechanisms of leukocyte recruitment and sensitivity to conditioning by shear stress for endothelial cells treated with tumour necrosis factor-α or interleukin-1', *Br. J. Pharmacol.*, **145**, pp. 1052–1061.

Shiu, Y. T., Udden, M. M. and McIntire, L. V. (2000) 'Perfusion with sickle erythrocytes up-regulates ICAM-1 and VCAM-1 gene expression in cultured human endothelial cells', *Blood*, **95**(10), pp. 3232–3241.

Surapisitchat, J., Hoefen, R. J., Pi, X., Yoshizumi, M. *et al.* (2001) 'Fluid shear stress inhibits TNF-α activation of JNK but not ERK1/2 or p38 in human umbilical vein endothelial cells: Inhibitory crosstalk among MAPK family members', *Proc. Natl. Acad. Sci. USA.*, **98**(11), pp. 6476–6481.

Wei, Z., Costa, K., Al-Mehdi, A. B., Dodia, C. *et al.* (1999) 'Simulated ischemia in flow-adapted endothelial cells leads to generation of reactive oxygen species and cell signaling', *Circ. Res.*, **85**, pp. 682–689.

Zhou, A. L., Egginton, S., Brown, M. D. and Hudlicka, O. (1998) 'Capillary growth in overloaded, hypertrophic adult rat skeletal muscle: an ultrastructural study', *Anatomical Rec.*, **252**(1) pp. 49–63.

6

Whole or partial vessel outgrowth assays

Cindy H. Chau and **William D. Figg**

Abstract

In recent years, the assessment of angiogenesis has often involved the use of whole or partial vessel outgrowth assays such as that of the rat aortic ring, chick aortic arch, porcine carotid artery, human saphenous vein, placental vein disc, foetal mouse bone explant and rat vena cava model. The development of this *ex vivo* organ culture method closely recapitulates the complexities of angiogenesis, forming a bridge between *in vitro* and *in vivo* studies. Herein we will provide detailed descriptions of the individual assays followed by a discussion of their unique strengths and weaknesses. We will also highlight the emergence of recent developments and applications of this methodology that have been incorporated into common practice.

Keywords

aortic ring assay; vessel outgrowth; ex vivo organ culture

6.1 Introduction

Studies of angiogenesis almost always rely on *in vitro* or *in vivo* models to assess various aspects of the angiogenic process. The recognition that angiogenesis *in vivo* involves a network of endothelial cells and their surrounding cells in constant communication with the microenvironment has led to the development of organ culture methods in an attempt closely to evaluate this complex process. These *ex vivo* assays represent various aspects of angiogenesis from endothelial cell proliferation, migration and microvessel formation to invasion through extracellular matrices. In this approach, segments of the specific tissue type

Angiogenesis Assays   Edited by Carolyn A. Staton, Claire Lewis and Roy Bicknell
© 2006 John Wiley & Sons, Ltd

Table 6.1 Strengths and weakness of species-specific organ culture assays

Species	Organ type	Advantages	Disadvantages
Rat	Aorta	Easy to manipulate	Intra-species statistical variation Large vessel
	Inferior vena cava	Greater angiogenic response Mimics *in vivo* angiogenesis	Long preparation time Technical difficulty Vessel integrity
Mouse	Aorta	Study of gene function in transgenics	Large vessel
	Foetal bone metatarsal	Microvascular tissue	Derived from growing embryos
Chick	Embryonic aorta arch	Microvascular tissue Rapid assay time	Derived from growing embryos
Porcine	Carotid artery	Unlimited tissue supply Large assay number High throughput screening	Single donor selection bias Large vessel
Human	Placental blood vessels	Eliminates species-specificity Clinical relevance	Limited tissue supply Large vessel
	Saphenous veins	Eliminates species-specificity Clinical relevance	Limited tissue supply Large vessel Lack of control of donor characteristics

are cultured in a three-dimensional matrix and monitored for microvessel outgrowths. By using intact vascular explants, this model reproduces more accurately the environment in which angiogenesis takes place than existing *in vitro* assays.

Each *ex vivo* model varies in the origin and type of tissue selected as well as the culture conditions established. This chapter will review the different methods currently being used to study angiogenesis, address the strengths and weaknesses of each approach (Table 6.1), and provide an overview of the recent developments and applications of this methodology.

6.2   Rat aortic ring assay

One of the most widely used organ-culture assays in angiogenesis research is the rat aortic ring assay, an *ex vivo* model of rat aortic explants cultured in a

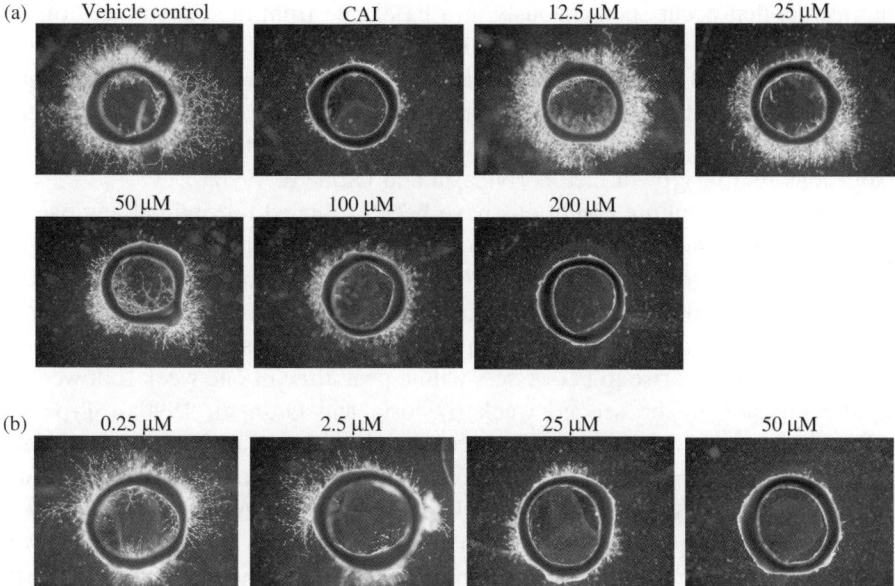

Figure 6.1 Representative images of rat aortic rings treated with the thalidomide analogue, CPS11 (a), or celecoxib (b) at the indicated concentrations. Aortic rings were cultured on Matrigel coated wells, treated daily with the vehicle (0.5 per cent DMSO), CAI (12 μg/ml), CPS11 (12.5–200 μM), or celecoxib (0.25–50 μM) for 4 days, and photographed on the 5th day using a ×2.5 objective. Carboxyamidotriazole (CAI), a known anti-angiogenic agent, was used at a higher than clinically achievable concentration to serve as the positive control.

three-dimensional matrix (Auerbach *et al.*, 2003). Originally developed by Nicosia and Ottinetti (1990a) this assay consists of an isolated rat aorta cut into segments, embedded in a matrix-containing environment, and cultured in a basal nutrient growth medium optimized for microvascular endothelial cells. A comprehensive detailed protocol of the assay can be found elsewhere (Nicosia and Zhu, 2004; Burbridge and West, 2001). Briefly, thoracic aortas are excised from 8- to 12-week-old rats followed by removal of the fibroadipose tissue (Figure 6.1). The areas between the arterial branches of the thoracic rat aorta are the best for harvesting ring cultures with optimal outgrowth (Kruger *et al.*, 2001) . The aortas are cut into the optimal length of 1 mm cross-sections generating approximately 30 rings per aorta (Burbridge and West, 2001). While shorter lengths are more difficult to handle and position correctly and give rise to a lower level of outgrowth, longer lengths result in a lower yield of rings per aorta. In general, greater uniformity in fragment size results in better reproducibility of the outgrowths (Stiffey-Wilusz *et al.*, 2001). Rings are then placed on the matrix-coated wells and covered with an additional layer of the matrix to allow embedding of the aortic rings. Over the next 7–14 days explants are monitored for microvessel outgrowths that occur spontaneously at a basal rate from the cut surfaces of the aortic rings in the unsupplemented minimal medium. Although the angiogenic process is triggered by injury caused during the dissection procedure (Villaschi and Nicosia, 1993) and mediated by endogenous growth factors produced from the aorta, microvessel proliferation is minimal and self-limited in the absence of exogenous sera or growth factors (Nicosia and Ottinetti, 1990a).

Various matrix culture conditions have been evaluated for optimal angiogenic responses involving embedding the aortic sections in a plasma clot, Matrigel, fibrin or collagen gels (Nicosia and Ottinetti, 1990a; Nicosia and Ottinetti, 1990b; Bonanno et al., 2000). The microvascular growth curves generated from these assays were characteristic for each type of gel. In these serum-free conditions, type I collagen gave rise to neovessels with a peak time of one week followed by rapid regression in the second week (Nicosia and Ottinetti, 1990a). Type IV collagen promoted neovessel elongation and survival in a dose-dependent manner, stabilizing outgrowths and preventing regression at high concentrations (300 µg/ml) and with significant stimulatory effects observed only at the intermediate concentration (30 µg/ml) (Bonanno et al., 2000). Fibrin gels, however, stimulated angiogenesis by 170 per cent, protected the neovessels from early regression, and supported growth through the second week (Nicosia and Ottinetti, 1990a). Moreover, recent modifications in culture conditions have predominantly used Matrigel, a basement membrane substance, and this resulted in shorter culture times (Kruger et al., 1998; Kruger et al., 2000; Bauer et al., 1998; Bauer et al., 2000). Microvessel formation was evident after 5 days of culture showing robust sprouting, Factor VIII positive outgrowths, and ultrastructural changes consistent with lumen formation. The use of Matrigel as the matrix provided a better substratum for attachment than collagen or fibrin. From a technical perspective, the Matrigel experimental protocol was efficient and easy to manipulate since it did not solidify as rapidly as fibrin or collagen (Kruger et al., 2001; Stiffey-Wilusz et al., 2001). However, because growth factors are present in Matrigel, a formulation of reduced-growth factor Matrigel is recommended for these assays. Nevertheless, the rat aortic ring assay is the most likely form of angiogenesis assay to evolve sufficiently to meet the expanding needs of angiogenesis researchers.

6.3   Mouse aorta models

The development of genetic manipulation technology to generate transgenic animals has allowed the study of gene function in various (patho)physiological conditions. Specifically, transgenic mice are used to model human genetic diseases by evaluating the impact of gain or loss of function of a targeted protein. The use of transgenics has proven to be a powerful tool in the generation of mouse models of vascular development and tumour angiogenesis (Hanahan et al., 1996; Carmeliet et al., 1998; Sato, 1999). Moreover, the efficacy of anti-angiogenic drugs is often evaluated in mice prior to human testing in clinical trials. As such, the aortic ring assay has been adapted to the mouse, including the use of murine aortas from transgenic models to study the molecular pharmacology of agents whose targets are deleted or altered (McDonald *et al.*, 2004).

Initial mouse aortic explants revealed the inefficiency of this system to produce spontaneously an angiogenic response under serum-free conditions as compared with the rat aortic rings. Studies using mouse aorta derived from 14-day-old embryos of C57BL/6 mice embedded in type I collagen resulted in a low microvessel yield after 8 days of culture in 5 per cent serum (Akimoto *et al.*, 2002). A more robust outgrowth of cells was observed only when the explants were cultured in reduced oxygen (5%) and upon the addition of exogenous VEGF. Moreover, thoracic aorta isolated from 8- to 12-week-old mice with a mixed genetic background of C57BL/6 and 129 SV/SL strains cultured in the same matrix showed that the addition of 2.5 per cent serum to the medium was an absolute requisite for sprouting outgrowths from the mouse aorta rings (Masson *et al.*, 2002).

Subsequent studies by Zhu *et al.* (2003) demonstrated the significant influence of ageing and genetic background in determining the angiogenic response of the mouse aorta. Unlike rat aortic explants, cultures of mouse aortic rings in collagen gels required supplementation with basic fibroblast growth factor (bFGF) or vascular endothelial growth factor (VEGF) to stimulate angiogenesis with the speed and extent of the sprouting being dependent on the genetic background of the animal (Zhu *et al.*, 2003). In the presence of bFGF, aortic rings from 129/SVJ mice formed much stronger and more stable vessels than those of the C57BL/6 or BALB/c strains, which, in turn, were more angiogenic than the FVB aortic explants. The addition of VEGF to culture medium resulted in a greater angiogenic response observed for aortas of C57BL/6 mice compared with those of 129/SVJ, BALB/c and FVB mice. These findings were consistent with a previous report that found different strains of mice eliciting variable angiogenic responses (Rohan *et al.*, 2000). Furthermore, the angiogenic response was also affected by age. Angiogenic sprouting in response to both VEGF and bFGF was delayed in 6- to 10-month-old mice as compared with 1- to 2-month-old mice (Zhu *et al.*, 2003). Given the dramatic variations in aortic microvessel outgrowth dependent on age and genetic background of the animals, the inclusion of proper internal controls in these assays is essential in the interpretation of the results. Nevertheless, the mouse aortic ring assay can be a useful tool to evaluate gene function in genetically modified mice; however, variables including type of angiogenic stimulation, genetic background of the animal, and ageing may influence the angiogenic response of these aortic explants.

6.4 Chick aortic arch assay

Adapted from the rat aortic ring assay, the chick aortic arch assay was originally developed for the specific purpose of testing thalidomide which had previously been shown to have limited effects in rodents but strong effects in chick embryos (Auerbach *et al.*, 2003). In this model, the aortic arches were dissected from 12- to 14-day-old chick embryos, cut into rings similar to those of the rat aorta, and placed on Matrigel (Muthukkaruppan *et al.*, 2000). Substantial outgrowth of cells occurred within 48 h. If the aortic arch was everted before explanting, the time was further reduced to 24 h.

Overcoming some of the limitations of the original rat organ-culture assay, this chick explant assay avoids the use of laboratory animals, is rapid, with an assay time of 1–3 days, and can be carried out in serum-free conditions (Muthukkaruppan *et al.*, 2000). Embryonic aortic arch cultures grow readily in the absence of serum, and the phenotype of the embryonic endothelial cells more closely resembles microvascular endothelial cells than that of adult aortic endothelial cells (Auerbach *et al.*, 2000). However, more experimentation with this model is needed before its validity can be fully evaluated (Auerbach *et al.*, 2000).

6.5   Porcine carotid artery assay

Studies by Stiffey-Wilusz *et al.* (2001) described a modification of the rat aortic ring assay utilizing commercial porcine carotid artery. Given the close approximation of porcine to human vasculature (Jarmolych *et al.*, 1968), carotid arteries are ideal for establishing *ex vivo* angiogenesis inhibitor screens due to their physiological relevance and are easily purchased from an abattoir. Carotid arteries were selected as a tissue source over aortas or venular tissues because they provided the best combination of size, robustness of growth, reproducibility and ease of handling. The carotid artery is splayed flat open and cut into 1.5 mm sections. These fragments are then placed into a reduced-growth factor Matrigel matrix and cultured in 2 per cent serum. Explants are monitored for microvessel outgrowths over the next 3 weeks with optimal sprouting observed at 12–14 days. Immunohistochemistry revealed positive Factor VIII staining of vessels consistent with endothelial cell morphology. Treatment with angiogenesis inhibitors suramin, 2-methoxyestradiol, and the matrix metalloprotease inhibitor Batimastat caused microvessel inhibition. The feasibility of the model as an angiogenesis inhibitor drug-screening assay and the accessibility of tissues are attractive features of this porcine assay.

6.6   Human explant cultures

The need to establish human models of angiogenesis resulted in the development of a humanized form of ring assays using tissues from placental blood vessels and saphenous veins. This allows for the screening of enhancers and inhibitors of human angiogenesis and the assessment of targeted therapies specific for human receptors (McDonald *et al.*, 2004). One of the original humanized organ-culture assays used fragments of human placental arterial vessels cultured in a fibrin gel without the addition of exogenous growth factors. The explants were then monitored for microvessel outgrowths over a period of 7 to 21 days (Brown *et al.*, 1996). The angiogenic response was detected in serum-free conditions and was further enhanced by stimulation with growth factors. Moreover, a stimulatory effect was observed when Matrigel was incorporated into the fibrin gel, indicating that components of the extracellular matrix play an important role in governing the strength of the angiogenic response. This system proved useful for screening potential inhibitors of angiogenesis as explants treated with the angiogenesis antagonists, hydrocortisone or suramin, significantly inhibited the angiogenic response.

Subsequent studies by Jung *et al.* (2001) using human placental vein disc cultures further confirmed the feasibility of a humanized assay. In this model, blood vessels on the surface of the placenta were harvested within 3 h of delivery to optimize endothelial cell viability. The adventitial tissue was dissected from the outside of the vessel and the vein segment was opened longitudinally to produce a flat film of full-thickness venous tissue. Vein discs (2 mm in diameter) were then embedded in fibrin–thrombin clots cultured in 20 per cent serum. This model was used to demonstrate that treatment with high-dose steroid or heparin/steroid combination for 15 days substantially decreased the angiogenic response observed in these humanized explants (Jung *et al.*, 2001).

Another humanized *ex vivo* model described by Kruger *et al.* (2000) is the human saphenous vein assay. This method involved specimens of human saphenous veins, which were acquired through a routine surgical procedure and immediately transferred to cell culture media for sectioning. Veins were cut into 2-mm long cross-sections, rinsed in endothelial cell media, embedded into Matrigel, and cultured under the same conditions as described for the rat aortic ring assay (Kruger *et al.*, 2000; Kruger *et al.*, 2001). Control experiments with sections of human adrenal veins indicated the optimal incubation time would be approximately 10–14 days; thus, explants are monitored for 14 days after which photographs are taken for image analysis (Kruger *et al.*, 2000). Immunohistochemistry demonstrated CD34 and Factor VIII positive staining on segments of the rings confirming the presence of endothelial cells in the microvessel outgrowths as well as structures consistent with vascular luminal formations (Kruger *et al.*, 2000; Kruger *et al.*, 2001). This assay has proved to be useful for the screening and testing of anti-angiogenic compounds such as endostatin, carboxyamidotriazole and thalidomide metabolites (Kruger *et al.*, 2000; Macpherson *et al.*, 2003; Kruger *et al.*, 2001; Price *et al.*, 2002). The main disadvantages of this model are the limited availability of tissues and the lack of control of donor characteristics (Kruger *et al.*, 2001). However, the existence of species differences in the response to anti-angiogenic drugs mandates the essential need of humanized assays to substantiate the effectiveness of these drugs for clinical use.

6.7   Recent developments

The foetal mouse bone explant assay described by Deckers *et al.* (2001) involves the culture of 17-day-old foetal mouse metatarsals. As these tissues are derived from growing embryos, they contain endothelial cells from the perichondrium. The isolated metatarsals were cultured in medium containing 10 per cent foetal calf serum over a 2-week period. Within 72 h of culture, the metatarsals become attached to the culture plastic and exhibit outgrowths by day 7. These tube-like structures display endothelial-like morphology and stain positively for PECAM-1, an endothelial cell marker. Thus, spontaneous outgrowths are observed in the absence of a coating of extracellular matrix and exogenous growth factor supplementation. Upon stimulation with VEGF-A, tube-like structural formation is enhanced. Moreover, when metatarsals are cultured on type I collagen or fibrin gels, the rate and area of outgrowth are no different from culture plastic. This assay is suitable for the testing of angiogenesis inhibitors since treatment of the explants with endostatin resulted in inhibition of the tube-like structures (Deckers *et al.*, 2001).

A very recent development by Nicosia *et al.* (2005) involves a new *ex vivo* model to study venous angiogenesis using rat inferior vena cava explants. Although *in vivo* angiogenesis primarily initiates from the venous side of the vascular bed, most organ culture models are based on arterial vessels (Folkman, 1982) with the exception of the human saphenous vein and placental vein disc models. This vena cava model can therefore provide further understanding of the vasoformative properties of veins and their regulation of the angiogenic process (Nicosia *et al.*, 2005). Preparation of the inferior vena cava explants can be challenging and involves the isolation of vessels from 1- to 2-month old rats, removal of fibroadipose tissue surrounding the vena cava, and cross-sectioning it into 1- to 2-mm long segments. Explants are cultured in collagen gel under serum-free conditions and microvascular sprouts composed of endothelial cells and pericytes are observed within 2 days and peaked by day 6–7. The explants produce angiogenic outgrowths that can be dose-dependently stimulated with exogenous bFGF and VEGF. Addition of these growth factors results in the formation of prominently elongated microvessels and extensive vascular networks. A blocking antibody against VEGF significantly reduced the spontaneous angiogenic response indicating the usefulness of this assay to study the effects of angiogenic inhibitors on venous angiogenesis.

Nicosia *et al.* (2005) also evaluated co-culture models of vena cava explants with rat aortic rings (Figure 6.2). The two explants were embedded in the same collagen gel set 1–2 mm apart from each other. Co-culturing of the vena cava with aortic rings dramatically enhanced venous angiogenic outgrowths comparable to levels observed for explants stimulated with growth factors (Nicosia *et al.*, 2005; Nicosia *et al.*, 1997; Villaschi and Nicosia, 1993). Venous explants in turn also stimulated sprouting from the aorta. The greater angiogenic response of the vena cava over the aorta explants suggested that venous endothelial cells have a greater

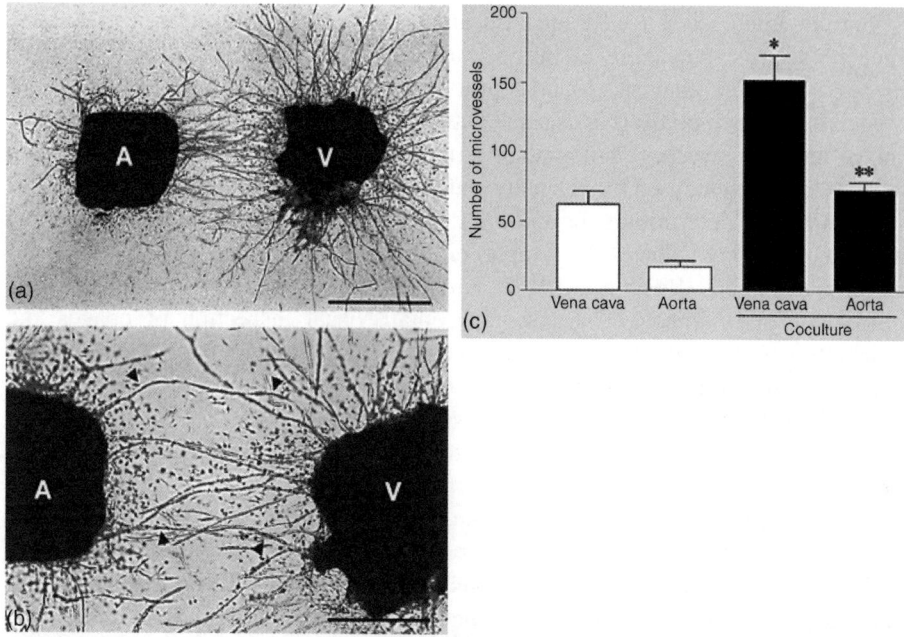

Figure 6.2   Angiogenic co-cultures of rat abdominal aorta and inferior vena cava (a) Neovessels of venous (V) and arterial (A) origin sprout towards each other forming (b) anastomoses which connect the two explants. Scale bars – 1,000 μm (c) the two explants reciprocally stimulate their angiogenic response (*P < 0.001; **P < 0.05). (Copyright with permission – Figure 8 from J. Vasc. Res. 2005; 42:111–119).

capacity to form neovessels than arterial endothelial cells. The vena cava-aorta co-cultures demonstrate the cross-talk between arterial and venous vessels, suggesting that endothelial cells from the two sides of the circulation communicate to influence each other's directional sprouting. This model represents an advancement over existing venous endothelial cell assays because it reproduces *ex vivo* the angiogenic response of a native venous endothelium that has not been modified by repeated passages in culture (Nicosia *et al.*, 2005).

6.8   Quantification

The quantification of angiogenesis in organ culture assays is challenging and needs to be standardized to allow for interpretation of results. In general, quantification is achieved by measuring the number and length of microvessel outgrowths from the primary explant, but since individual outgrowths are clustered together, the area covered by outgrowth is more commonly measured (Staton *et al.*, 2004). Proper quantification requires determining the number and length of branching microvessels, the size and form of aortic rings, and the number and spatial distribution of fibroblast-like cells (Blacher *et al.*, 2001).

Current approaches to the quantification of microvessels include qualitative methods such as manual processing and measurement or newer quantitative methods employing image analysis. Manual processing involves counting the number of vessels and branch points (Nicosia and Ottinetti, 1990a; Brown *et al.*, 1996) whereas image analysis involves photographing ring cultures to generate a digital image. Images are then analysed by manually encircling the area of outgrowth and the area is quantified by determining the sprouting index (the ratio between the cellular area and the mean grey value of the sample) or the mean area density by computing the mean square pixels (Bocci *et al.*, 1999; Kruger *et al.*, 2000). These methods are time consuming and subjective because they entail visual inspection of images and manually counting the newly formed microvessels that extend from the cultured aortic ring. In addition, manual counting does not take into consideration the length or width of the microvessels (Go and Owen, 2003).

Another method uses an automated image analysis-based procedure for binary processing of grey level image and determination of the number and total area of microvessels at a fixed distance from the aortic ring (Nissanov *et al.*, 1995). This approach offers the attractive feature of being automatic and suitable for screening as it runs approximately 2.5 times faster than manual quantification. Studies by Blacher *et al.* (2001) describe an improved technique using computer-assisted image analysis as a tool to allow quantification of the number, branching, length and distribution of microvessels as well as quantification, positioning and distribution of fibroblast-like cells. The use of computer-image analysis requires maintaining the same microscopic settings throughout the entire experiment, particularly in standardizing the amount of illumination to visualize the microvessels (Go and Owen, 2003). With newer computer software becoming available, automated image analysis methods have been modified to enable compatibility with different program tools. Current image processing softwares include the Image-Pro Plus 4.1 (Blatt *et al.*, 2004) and Adobe Photoshop (Berger *et al.*, 2004) (Figure 6.3).

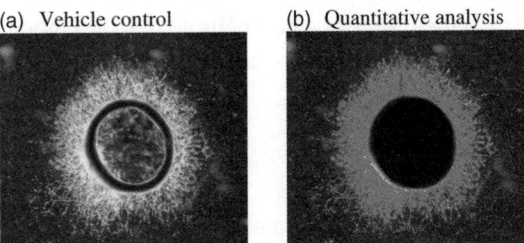

(a) Vehicle control   (b) Quantitative analysis

Figure 6.3 Image analysis of microvessel outgrowth of aortic rings. (a) Aortic ring treated with vehicle control showing microvessel formation. (b) The area of angiogenic sprouting was quantified using Adobe PhotoShop. The lumen of the aorta is first blacked out then pixels of the outgrowth are highlighted based on their intensity. The *Similar* and *Grow* tools are used together to fully select the area of outgrowth, followed by pixel quantification using the *Histogram* feature. Percentage inhibition is defined as: (growth in pixels of treated ring/growth in pixels of control ring).

Furthermore, Wang *et al.* (2004) developed a simple quantitative method for evaluating these aortic ring models by using a colorimetric assay to obtain an index for assessing the outgrowth of the capillary-like endothelial cells. The colorimetric assay measures a volumetric density of the endothelial growing activity in the three-dimensional aorta culture system, providing a low cost and efficient protocol that is feasible for performing massive screenings. Owing to the existence of many experimental variables, it is critical to define standardized criteria for the assessment of quantification of these microvessel outgrowth assays to enable valid interpretation of the angiogenic response.

6.9 Strengths and weaknesses of organ culture assays

The aortic ring organ-culture system is considered by many to come closest to simulating the *in vivo* angiogenesis environment because the system includes the surrounding non-endothelial cells (such as smooth muscle cells and pericytes) and a supporting matrix. In addition, the endothelial cells have not been pre-selected or modified by repeated passages and thus are not proliferating at the time of explantation. This system is hence more representative of the situation found *in vivo* where angiogenesis is triggered and quiescent endothelial cells respond by becoming proliferative, migrating out from the existing vessels and differentiating into tubules (Staton *et al.*, 2004). However, the model is not entirely representative of the microvascular environment encountered during angiogenic reactions induced by tumours or inflammatory mediators as the large number of different factors released by tumour cells and the tumour cells themselves are not present (Auerbach *et al.*, 2000). Nonetheless, this assay offers the advantage of low cost, easy manipulation of treatment conditions, lack of inflammatory complications seen with other *in vivo* models, and the possibility of generating many assays from one animal since approximately 30 rings may be obtained from each isolated aorta (Kruger *et al.*, 2001; Burbridge and West, 2001).

Modification of the rat aortic ring assay with the use of aortic fragments from mice has several advantages. Mice are common hosts for orthotopic and heterotopic tumour models. As such, this murine assay can be used to show differences in aortic disc angiogenesis in tumour bearing versus non-tumour-bearing mice, as well as differences in the assay between mice treated with an oral anti-angiogenic agent (Berger *et al.*, 2004). Additionally, the mouse aortic assay can be used to study differences in gene function in transgenic and/or knockout mice.

Although the aortic ring assay has gained wide acceptance as a tool to study angiogenesis, there are limitations associated with its use. The assay is time consuming to perform, quantification is challenging, and growth requirements differ between the explant and the cell outgrowth. There have been reports of significant variability in vessel outgrowth between explants from different animals, making the use of a large number of donors necessary. The porcine carotid artery explant assay is considered an improvement on the rat aortic ring assay, in that the larger blood vessel derived from the pig enables the running of a number of experimental assays from a single carotid artery, thus eliminating statistical variation between animals (Stiffey-Wilusz *et al.*, 2001). This model also reduces the number of animals used per assay and eliminates the need to house research animals, providing a substantial amount of social and economic savings. However, drawbacks to the porcine explant method include the use of a single donor, which may not be representative of the true population, and the use of a large rather than small vessel (Staton *et al.*, 2004).

Since angiogenesis is normally a microvascular event, the use of large vessels in the aortic ring assay is far from ideal. The use of 17-day-old foetal mouse metatarsal explants has the advantage of being microvascular in origin, thus more closely representing real angiogenesis (Deckers *et al.*, 2001). However, since the tissues used are obtained from growing embryos, they will be undergoing proliferation before explantation and therefore are not truly representative of the stimulation of non-proliferative endothelial cells that occurs *in vivo*. A modification of the rat aortic ring assay is the chick aortic arch model (Muthukkaruppan *et al.*, 2000), which has the advantage of being a rapid experiment that can be carried out in serum-free conditions. Embryonic arch endothelial cells share many properties with microvascular endothelial cells. However, like the foetal mouse metatarsal tissues, these aortic arches are derived from growing embryos, which undergo rapid cell division before explantation and exposure to angiogenic mediators (Auerbach *et al.*, 2003).

A criticism of using non-human tissues in these organ culture assays is their applicability as preclinical screening assays, whereby the responses seen with various drugs/test substances may be species-specific. In an effort to overcome this species-specificity factor, human models of angiogenesis were developed with the use of human placental blood vessels and saphenous veins (Brown *et al.*, 1996; Jung *et al.*, 2001; Kruger *et al.*, 2000). Again since outgrowth of endothelial cells from these explants are derived from large vessels, whether this type of assay is truly representative of *in vivo* angiogenesis remains an issue.

The most recent development of an *ex vivo* model has focused on venous angiogenesis and the relevancy of using vena cava explants because angiogenesis typically stems from the venous side of the vascular bed (Folkman, 1982). The vascular context in which angiogenesis occurs *in vivo* may be more closely reproduced with this model than the aortic ring assay (Nicosia *et al.*, 2005). Although the vena cava-aorta co-culture model adds an additional layer of complexity to the original simplified organ culture method, this new model closely mimics the environment *in vivo*, as our vasculature is comprised of both venous and arterial vessels. The challenge in performing this assay lies in the technical difficulty in acquiring the inferior vena cava tissue. In addition to the long preparation time in vessel isolation, the venous explants have a tendency to collapse and may vary somewhat in size as compared with the aortic rings which maintain their shape and can easily be cut into uniform length (Nicosia *et al.*,

2005). It remains to be seen whether this co-culture system is a more sensitive assay and better screening tool for evaluating potential anti-angiogenic inhibitors.

Vascular organ culture models of angiogenesis may be influenced by a number of variables including the species-specificity of the tissue source and the type of blood vessels used for the assay. It is also important to bear in mind that while these organ culture assays prove useful as a screening assay, some anti-angiogenic compounds will not demonstrate activity if they require metabolic activation (Bauer *et al.*, 1998; Kruger *et al.*, 2000).

6.10 Applications

Organ culture models have been designed to integrate *in vitro* and *in vivo* models of angiogenesis. Specifically, the aortic ring assay has gained wide acceptance as a practical tool to study angiogenesis and its surrounding environment as well as the molecular mechanisms of its regulation. The model provides a unique microenvironment to measure the interaction of various cell types and biological factors that may influence the angiogenic process. This assay has been used to evaluate the ability of growth factors to induce angiogenesis from quiescent vascular explants and how modifications to the extracellular matrix affect the angiogenic process (Villaschi and Nicosia, 1993; Nicosia *et al.*, 1994; Zhu *et al.*, 2000). While several other groups have used variations of the aortic ring assay for screening purposes to study anti-angiogenic compounds, a human *ex vivo* angiogenesis assay similar to that described by Brown *et al.* (1996) has also been applied as a screening for natural compounds and herbal drug preparations (Baronikova *et al.*, 2004).

In developing these assays to screen and monitor the efficacy of angiogenesis inhibitors, an important aspect to consider that is largely underestimated by current angiogenesis assays is *in vivo* protein binding effects. Studies by Kruger and Figg (2001) investigated a modification of the rat aortic ring angiogenesis assay to account for serum protein–drug interactions, thus providing a bioassay for anti-angiogenic activity. This human serum rat aortic ring bioassay was shown to predict the concentrations of protein-bound anti-angiogenic agents required for free fraction biological activity and compared those doses with clinically achievable concentrations and therapeutic outcome (Kruger and Figg, 2001). In addition, the aortic ring assay was used to detect angiogenic activity in serum and plasma and to test for potential angiogenic factors or inhibitors in biological fluids (Go *et al.*, 2003). It was demonstrated to be a promising system to support the isolation of angiogenic factors from blood and other physiological fluids. Finally, a recent human tissue-based angiogenesis assay has applied the concept of the original organ culture method to the study of tumour angiogenesis (Woltering *et al.*, 2003; Gulec and Woltering, 2004). This assay used fragments of tumour tissue discs embedded in fibrin gels and cultured in 20 per cent foetal bovine serum. Angiogenic sprouting was observed depicting a vessel compartment that was separate from the tumour fragment. This dual-compartmental structure allowed for the functional assessment of the angiogenic potential of human tumours and the simultaneous evaluation of the anti-tumour and anti-angiogenic effects of a therapeutic agent.

Other modifications of the aortic ring model have facilitated its use in combination with other techniques to study angiogenesis. A thin prep modification of the assay has significantly simplified the procedure and allows the staining of aortic outgrowths as whole mounts (Zhu and Nicosia, 2002). Studies using microarray technology were also performed to evaluate the modulation of gene expression patterns in angiogenesis that occur when rat aortic rings undergo treatment with anti-angiogenic agents (Zogakis *et al.*, 2002). Expanding on this idea, isolation of RNA from the sub-culturing of endothelial cells from the murine aortic disc model will enable researchers to determine differences on the genomic level of cells from discs treated with various anti-angiogenic agents (Berger *et al.*, 2004). These mouse aorta models have also allowed the study of gene function through transgenic animals, particularly the knockout mouse (Devy *et al.*, 2002). Furthermore, the possibility to transduce anti-angiogenic genes into aortic rings by adenovirus-mediated gene transfer demonstrated the feasibility of this technology to assess the potential therapeutic use of gene products for gene therapy (Masson *et al.*, 2002).

6.11 Conclusions

Organ culture assays have gained wide acceptance as one of the most efficient and effective methods to study angiogenesis. As outlined in this chapter, each experimental approach offers specific advantages and limitations. Although the models may vary in technical detail, tissue source and culture conditions, making results difficult to compare, standardization of certain procedures and inclusion of proper controls should help minimize these statistical variations. Selection of the most relevant and appropriate model to use depends on the specific purpose of the assay. Overall they are assays suitable for screening new anti-angiogenic agents. As drug development progresses, so will the demand for quantitative angiogenic assays. With advancements in technology on the horizon, these whole or partial vessel outgrowth assays will continue to evolve to adapt to current practical needs.

References

Akimoto, T., Liapis, H. and Hammerman, M. R. (2002) 'Microvessel formation from mouse embryonic aortic explants is oxygen and VEGF dependent', *Am. J. Physiol. Regul. Integr. Comp. Physiol.*, **283**(2), pp. R487–R495.

Auerbach, R., Akhtar, N., Lewis, R. L. and Shinners, B. L. (2000) 'Angiogenesis assays: problems and pitfalls', *Cancer Metastasis Rev.*, **19**(1–2), pp. 167–172.

Auerbach, R., Lewis, R., Shinners, B., Kubai, L. and Akhtar, N. (2003) 'Angiogenesis assays: a critical overview', *Clin. Chem.*, **49**(1), pp. 32–40.

Baronikova, S., Apers, S., Vanden Berghe, D., Cos, P. *et al.* (2004) 'An ex-vivo angiogenesis assay as a screening method for natural compounds and herbal drug preparations', *Planta Med.*, **70**(10), pp. 887–892.

Bauer, K. S., Cude, K. J., Dixon, S. C., Kruger, E. A. and Figg, W. D. (2000) 'Carboxyamido-triazole inhibits angiogenesis by blocking the calcium-mediated nitric-oxide synthase-vascular endothelial growth factor pathway', *J. Pharmacol. Exp. Ther.*, **292**(1), pp. 31–37.

Bauer, K. S., Dixon, S. C. and Figg, W. D. (1998) 'Inhibition of angiogenesis by thalidomide requires metabolic activation, which is species-dependent', *Biochem. Pharmacol.*, **55**(11), pp. 1827–1834.

Berger, A. C., Wang, X. Q., Zalatoris, A., Cenna, J. and Watson, J. C. (2004) 'A murine model of ex vivo angiogenesis using aortic disks grown in fibrin clot', *Microvasc. Res.*, **68**(3), pp. 179–187.

Blacher, S., Devy, L., Burbridge, M. F., Roland, G. *et al.* (2001) 'Improved quantification of angiogenesis in the rat aortic ring assay', *Angiogenesis*, **4**(2), pp. 133–142.

Blatt, R. J., Clark, A. N., Courtney, J., Tully, C. and Tucker, A. L. (2004) 'Automated quantitative analysis of angiogenesis in the rat aorta model using Image-Pro Plus 4.1', *Comput. Methods Programs Biomed.*, **75**(1), pp. 75–79.

Bocci, G., Danesi, R., Benelli, U., Innocenti, F. *et al.* (1999) 'Inhibitory effect of suramin in rat models of angiogenesis in vitro and in vivo', *Cancer Chemother. Pharmacol.*, **43**(3), pp. 205–212.

Bonanno, E., Iurlaro, M., Madri, J. A. and Nicosia, R. F. (2000) 'Type IV collagen modulates angiogenesis and neovessel survival in the rat aorta model', *In Vitro Cell Dev. Biol. Anim.*, **36**(5), pp. 336–340.

Brown, K. J., Maynes, S. F., Bezos, A., Maguire, D. J. *et al.* (1996) 'A novel in vitro assay for human angiogenesis', *Lab. Invest.*, **75**(4), pp. 539–555.

Burbridge, M. F. and West, D. C. (2001) 'Rat aorta ring, 3D model of angiogenesis in vitro', in *Methods in Molecular Medicine: Angiogenesis Protocols*, Vol. 46, Murray, J. C., Ed., Humana Press, Totowa, pp. 185–204.

Carmeliet, P., Moons, L. and Collen, D. (1998) 'Mouse models of angiogenesis, arterial stenosis, atherosclerosis and hemostasis', *Cardiovasc. Res.*, **39**(1), pp. 8–33.

Deckers, M., van der Pluijm, G., Dooijewaard, S., Kroon, M. *et al.* (2001) 'Effect of angiogenic and antiangiogenic compounds on the outgrowth of capillary structures from fetal mouse bone explants', *Lab. Invest.*, **81**(1), pp. 5–15.

Devy, L., Blacher, S., Grignet-Debrus, C., Bajou, K. *et al.* (2002) 'The pro- or antiangiogenic effect of plasminogen activator inhibitor 1 is dose dependent', *Faseb J.*, **16**(2), pp. 147–154.

Folkman, J. (1982) 'Angiogenesis: initiation and control', *Ann. NY Acad. Sci.*, **401**, pp. 212–227.

Go, R. S. and Owen, W. G. (2003) 'The rat aortic ring assay for in vitro study of angiogenesis', *Methods Mol. Med.*, **85**, pp. 59–64.

Go, R. S., Ritman, E. L. and Owen, W. G. (2003) 'Angiogenesis in rat aortic rings stimulated by very low concentrations of serum and plasma', *Angiogenesis*, **6**(1), pp. 25–29.

Gulec, S. A. and Woltering, E. A. (2004) 'A new in vitro assay for human tumor angiogenesis: three-dimensional human tumor angiogenesis assay', *Ann. Surg. Oncol.*, **11**(1), pp. 99–104.

Hanahan, D., Christofori, G., Naik, P. and Arbeit, J. (1996) 'Transgenic mouse models of tumour angiogenesis: the angiogenic switch, its molecular controls, and prospects for preclinical therapeutic models', *Eur. J. Cancer*, **32A**(14), pp. 2386–2393.

Jarmolych, J., Daoud, A. S., Landau, J., Fritz, K. E. and McElvene, E. (1968) 'Aortic media explants. Cell proliferation and production of mucopolysaccharides, collagen, and elastic tissue', *Exp. Mol. Pathol.*, **9**(2), pp. 171–188.

Jung, S. P., Siegrist, B., Wade, M. R., Anthony, C. T. and Woltering, E. A. (2001) 'Inhibition of human angiogenesis with heparin and hydrocortisone', *Angiogenesis*, **4**(3), pp. 175–186.

Kruger, E. A. and Figg, W. D. (2001) 'Protein binding alters the activity of suramin, carboxyamidotriazole, and UCN-01 in an ex vivo rat aortic ring angiogenesis assay', *Clin. Cancer Res.*, **7**(7), pp. 1867–1872.

Kruger, E. A., Blagosklonny, M. V., Dixon, S. C. and Figg, W. D. (1998) 'UCN-01, a protein kinase C inhibitor, inhibits endothelial cell proliferation and angiogenic hypoxic response', *Invasion Metastasis*, **18**(4), pp. 209–218.

Kruger, E. A., Duray, P. H., Price, D. K., Pluda, J. M. and Figg, W. D. (2001) 'Approaches to preclinical screening of antiangiogenic agents', *Semin. Oncol.*, **28**(6), pp. 570–576.

Kruger, E. A., Duray, P. H., Tsokos, M. G., Venzon, D. J. *et al.* (2000) 'Endostatin inhibits microvessel formation in the ex vivo rat aortic ring angiogenesis assay', *Biochem. Biophys. Res. Commun.*, **268**(1), pp. 183–191.

Macpherson, G. R., Ng, S. S., Forbes, S. L., Melillo, G. *et al.* (2003) 'Anti-angiogenic activity of human endostatin is HIF-1-independent in vitro and sensitive to timing of treatment in a human saphenous vein assay', *Mol. Cancer Ther.*, **2**(9), pp. 845–854.

Masson, V. V., Devy, L., Grignet-Debrus, C., Bernt, S. *et al.* (2002) 'Mouse Aortic Ring Assay: a new approach of the molecular genetics of angiogenesis', *Biol. Proced. Online*, **4**, pp. 24–31.

McDonald, D. M., Teicher, B. A., Stetler-Stevenson, W., Ng, S. S. *et al.* (2004) 'Report from the society for biological therapy and vascular biology faculty of the NCI workshop on angiogenesis monitoring', *J. Immunother.*, **27**(2), pp. 161–175.

Muthukkaruppan, V. R., Shinneers, B. L., Lewis, R., Park, S. J. *et al.* (2000) 'The chick embryo aortic arch assay: a new, rapid, quantifiable in vitro method for testing the efficacy of angiogenic and anti-angiogenic factors in a three-dimensional, serum-free organ culture system', *Proc. Am. Assoc. Cancer Res.*, **41**, p. 65.

Nicosia, R. F. and Ottinetti, A. (1990a) 'Growth of microvessels in serum-free matrix culture of rat aorta. A quantitative assay of angiogenesis in vitro', *Lab. Invest.*, **63**(1), pp. 115–122.

Nicosia, R. F. and Ottinetti, A. (1990b) 'Modulation of microvascular growth and morphogenesis by reconstituted basement membrane gel in three-dimensional cultures of rat aorta: a comparative study of angiogenesis in matrigel, collagen, fibrin, and plasma clot', *In Vitro Cell Dev. Biol.*, **26**(2), pp. 119–128.

Nicosia, R. F. and Zhu, W. H. (2004) 'Rat aortic ring assay of angiogenesis', in *Methods in Endothelial Cell Biology*, Augustin, H., Ed., Springer, Berlin, pp. 125–144.

Nicosia, R. F., Nicosia, S. V. and Smith, M. (1994) 'Vascular endothelial growth factor, platelet-derived growth factor, and insulin-like growth factor-1 promote rat aortic angiogenesis in vitro', *Am. J. Pathol.*, **145**(5), pp. 1023–1029.

Nicosia, R. F., Lin, Y. J., Hazelton, D. and Qian, X. (1997) 'Endogenous regulation of angiogenesis in the rat aorta model. Role of vascular endothelial growth factor', *Am. J. Pathol.*, **151**(5), pp. 1379–1386.

Nicosia, R. F., Zhu, W. H., Fogel, E., Howson, K. M. and Aplin, A. C. (2005) 'A new ex vivo model to study venous angiogenesis and arterio-venous anastomosis formation', *J. Vasc. Res.*, **42**(2), pp. 111–119.

Nissanov, J., Tuman, R. W., Gruver, L. M. and Fortunato, J. M. (1995) 'Automatic vessel segmentation and quantification of the rat aortic ring assay of angiogenesis', *Lab. Invest.*, **73**(5), pp. 734–739.

Price, D. K., Ando, Y., Kruger, E. A., Weiss, M. and Figg, W. D. (2002) '5′-OH-thalidomide, a metabolite of thalidomide, inhibits angiogenesis', *Ther. Drug Monit.*, **24**(1), pp. 104–110.

Rohan, R. M., Fernandez, A., Udagawa, T., Yuan, J. and D'Amato, R. J. (2000) 'Genetic heterogeneity of angiogenesis in mice', *Faseb J.*, **14**(7), pp. 871–876.

Sato, T. N. (1999) 'Gene trap, gene knockout, gene knock-in, and transgenics in vascular development', *Thromb. Haemost.*, **82**(2), pp. 865–869.

Staton, C. A., Stribbling, S. M., Tazzyman, S., Hughes, R. *et al.* (2004) 'Current methods for assaying angiogenesis in vitro and in vivo', *Int. J. Exp. Pathol.*, **85**(5), pp. 233–248.

Stiffey-Wilusz, J., Boice, J. A., Ronan, J., Fletcher, A. M. and Anderson, M. S. (2001) 'An ex vivo angiogenesis assay utilizing commercial porcine carotid artery: modification of the rat aortic ring assay', *Angiogenesis*, **4**(1), pp. 3–9.

Villaschi, S. and Nicosia, R. F. (1993) 'Angiogenic role of endogenous basic fibroblast growth factor released by rat aorta after injury', *Am. J. Pathol.*, **143**(1), pp. 181–190.

Wang, H. S., Hwang, L. L., Sue, H. F., Lee, K. M. and Chen, C. T. (2004) 'A simple quantitative method for evaluation of angiogenesis activity', *Assay Drug Dev. Technol.*, **2**(1), pp. 31–38.

Woltering, E. A., Lewis, J. M., Maxwell, P. J. T., Frey, D. J. *et al.* (2003) 'Development of a novel in vitro human tissue-based angiogenesis assay to evaluate the effect of antiangiogenic drugs', *Ann. Surg.*, **237**(6), pp. 790–798; discussion 798–800.

Zhu, W. H. and Nicosia, R. F. (2002) 'The thin prep rat aortic ring assay: a modified method for the characterization of angiogenesis in whole mounts', *Angiogenesis*, **5**(1–2), pp. 81–86.

Zhu, W. H., Guo, X., Villaschi, S. and Francesco Nicosia, R. (2000) 'Regulation of vascular growth and regression by matrix metalloproteinases in the rat aorta model of angiogenesis', *Lab. Invest.*, **80**(4), pp. 545–555.

Zhu, W. H., Iurlaro, M., MacIntyre, A., Fogel, E. and Nicosia, R. F. (2003) 'The mouse aorta model: influence of genetic background and aging on bFGF- and VEGF-induced angiogenic sprouting', *Angiogenesis*, **6**(3), pp. 193–199.

Zogakis, T. G., Costouros, N. G., Kruger, E. A., Forbes, S. *et al.* (2002) 'Microarray gene expression profiling of angiogenesis inhibitors using the rat aortic ring assay', *Biotechniques*, **33**(3), pp. 664–670.

7
Assaying endothelial–mural cell interactions

Melissa K. Nix and Karen K. Hirschi

Abstract

Recent studies in genetically malleable embryonic model systems have enabled the identification of factors required for blood vessel formation. However, it is not possible in most *in vivo* systems to dissect carefully the exact cellular behaviours, as well as cell–cell and cell–matrix interactions that such factors modulate. Thus, it has been imperative to develop *in vitro* systems to analyse further the cellular role of vascular-associated factors, and elucidate their intracellular signalling pathways. Such systems are also helpful for identifying new factors that regulate vascular cell growth, migration and differentiation. Although much information has been gained from investigating cellular behaviours in solo cultures of endothelial or mural cells, factors involved in heterocellular interactions such as junctional proteins, adhesion molecules and soluble paracrine effectors can only be studied in co-culture assay systems. Herein, we will discuss two-dimensional (2D) and three-dimensional (3D) co-culture systems that have been developed to address these issues.

Keywords

2D co-culture; 3D co-culture; endothelial cells; mural cells; cell–cell interaction

7.1 2D models to study endothelial–mural cell interactions

There are generally two types of co-culture systems to study heterocellular interactions: contacting co-cultures that enable direct, physical cell–cell interaction; and non-contacting co-cultures that enable heterocellular interactions via diffusible paracrine factors. We will discuss examples and the utility of each, as well as their advantages and disadvantages.

Angiogenesis Assays   Edited by Carolyn A. Staton, Claire Lewis and Roy Bicknell
© 2006 John Wiley & Sons, Ltd

Contacting co-cultures

Direct contact between different cell types is most commonly achieved via co-plating distinct cell populations, such as endothelial and mural cells, within the same culture well. Cells are either co-plated simultaneously or one cell type is plated first, allowed to adhere, and the second plated on top. In such systems, the influence of one cell type on the other's behaviour (i.e. growth, differentiation) can be directly monitored if one population is labelled prior to co-culture, so that the cell types can be clearly distinguished after time in culture. This is useful, even if the cell types are morphologically distinct initially, because direct heterocellular contact can alter the phenotype of one or both cell types over time in co-culture (Hirschi *et al.*, 1998).

Cell labelling

There are various means of labelling vascular cells or progenitors for co-culture, including isolation of one of the populations from a genetically altered animal that expresses a reporter gene in all cells (i.e. eGFP mice (Okabe *et al.*, 1997) that ubiquitously express Green Fluorescent Protein, GFP), or in the cell type of interest (i.e. Tie-2-LacZ mice (Sato *et al.*, 1993) that express LacZ in endothelial cells but not mural cells). Fluorescent reporters are particularly useful for monitoring cell behaviour (migration, proliferation) in real time over the course of the experiment. Vascular cells and progenitors isolated from wild-type animals or human tissue can also be labelled via transient or stable transfection with an expression plasmid or infection with a viral vector, expressing a reporter gene, or via permanent membrane labelling with a fluorescent dye, such as PKH26, which can be retained in cells up to 100 doublings (Horan and Slezak, 1989). The effects of co-culture on cell behaviour can be monitored by various means depending on the behaviour of interest.

Measurement of proliferation

To measure the influence of direct co-culture on cell growth, one can trypsinize the cells in culture into a single cell suspension, then count (Coulter Counter, hemocytometer) the total number of cells after specific incubation times, knowing the number plated. Cell cycle analysis via fluorescence activated cell sorting (FACS) can also be performed on similarly prepared cells that were fixed and stained with the DNA binding dye propidium iodide (Lai *et al.*, 2003). However, this approach would not reveal which cell population(s) was dividing, unless the cells were separated prior to counting using cell type-specific fluorescent labels and FACS. Another approach is to label the cells *in situ* in the co-culture well (Hirschi *et al.*, 1999). For example, live cell cultures can be incubated with bromo-deoxyuridine (BrdU), which is taken up by all actively dividing cells, then fixed and immunostained with antibodies against BrdU to identify proliferating cells. Co-staining for cell type-specific markers or reporter genes would enable quantification of growth of each population.

Assessment of differentiation

Direct co-culture systems have also been useful for studying the effect of one cell type on the differentiation of another, such as the effects of endothelial cells on the differentiation of mural cell precursors, also referred to as mesenchymal cells. During vessel formation, endothelial tubes form first and then recruit mesenchymal cells, via secretion of platelet-derived growth factor-BB (Hirschi *et al.*, 1998; Lindhal *et al.*, 1997). It was found in contacting co-cultures, that the latent form of transforming growth factor-beta (TGF-β), which is produced by both endothelial and mesenchymal cells, is activated upon direct cell–cell interaction (Antonelli-Orlidge *et al.*, 1989) and promotes the transcription of cytoskeletal and contractile proteins in the mesenchymal cells, inducing mural cell differentiation (Hirschi *et al.*, 1998; Hirschi *et al.*, 2001). In these experiments, it was necessary to pre-label the endothelial cells in order clearly to distinguish the cell types and co-stain for the upregulation of mural cell specific protein expression in the mesenchymal cells that were in direct contact with endothelial cells.

Analysis of gap junction communication

In subsequent studies, also using direct co-culture with pre-labelled endothelial cells, it was determined that in order for TGF-β to be activated upon endothelial–mesenchymal cell contact, the two cell types must form functional gap junction channels that enable heterocellular communication (Hirschi *et al.*, 2003). In these co-cultures, in which endothelial cells were pre-labelled with the fluorescent dye PKH26, it was possible to distinguish them in live cultures using epifluorescence microscopy, then using dual whole-cell voltage clamp, determine to what extent they were communicating specifically via gap junction channels, and what impact heterocellular communication had on endothelial-induced mural cell differentiation. In order to demonstrate a role for specific gap junction proteins in mediating endothelial-induced mural cell differentiation, mesenchymal cells were then isolated from genetically modified mice lacking gap junction protein Cx43 (Cx43−/−). In co-cultures between endothelial and wild-type (Cx43+/+) mesenchymal cells, gap junction communication was established and mural cell differentiation was induced. However, in co-cultures of endothelial cells and Cx43−/− mesenchymal cells, heterocellular communication via gap junctions was not established, TGF-β was not activated, and mural cell differentiation was not induced in response to endothelial cell contact. Re-expressing Cx43 in the

Cx43–/– cells enabled mesenchymal cells to re-establish communication with endothelial cells, activate TGF-β and undergo endothelial-induced mural cell differentiation (Hirschi *et al.*, 2003), demonstrating the need for Cx43-mediated gap junction communication in this process. Whether other Cx proteins can also enable heterocellular communication between endothelial and mesenchymal cells and support TGF-β activation and endothelial-induced mural cell differentiation is under investigation using similar methods (Hirschi Lab., unpublished).

Advantages/disadvantages

The major advantages of a contacting co-culture system are that it is not technically challenging, all adherent vascular cell types and progenitors can be used, and the density of each cell type can be manipulated to modulate the degree of heterocellular contact achieved within the culture. When used in combination with cell labelling techniques, one can easily assess the effects of co-culture on the growth and differentiation of either or both cell types, as well as assess degree of communication and adhesion established between them. One disadvantage of the direct co-culture is the inability to assess the extent of cell–cell interactions mediated via diffusible paracrine factors made by one cell type. It would not be possible to detect a gradient of soluble factor established by one cell type to influence the other in the co-culture because both cell types would be evenly dispersed throughout the culture dish to optimize one-on-one interactions. For similar reasons, it would also not be possible to monitor the effects of one cell type on the migration of the other. However, both of these processes, establishment of a gradient of diffusible effectors and monitoring of directed migration, can be easily assessed and measured in a non-contacting assay system known as the Under Agarose Assay, as discussed below.

Non-contacting co-cultures

In contrast to the co-cultures previously described, non-contacting co-cultures do not typically allow two cell types, such as endothelial and mural cells, to achieve direct cell–cell interaction, so the establishment and function of heterocellular junctions and adhesion complexes cannot be studied in these systems.

Transwell membrane assay

The most common non-contacting co-culture is established using a commercially available transwell membrane apparatus (i.e. Millipore). In such systems, there are culture dishes into which one cell type can be plated in a well, and then another cell type added onto a matrix-coated membrane chamber that fits into the culture well, thus enabling suspension of the second cell type above the first. This type of co-culture allows for the exchange of soluble, diffusible signals that can alter growth and migration, both of which can be measured in this system. The effects of co-culture on cell growth are typically measured by trypsinizing cells and counting the total number within the well or chamber region after co-incubation. Pre-labelling of the cells is not necessary, as they never intermingle in the culture dish. An index of directed migration of the cells in the upper chamber can also be obtained via counting the number of cells that migrate through the upper membrane and remain on the bottom of the chamber, in response to signals from the cell type in the bottom well. If observed cellular effects are suspected to be mediated via specific soluble effectors, neutralizing reagents against the candidate effector can be added into the bottom well or chamber to determine whether they suppress the cellular effects. Conversely, the soluble candidate effector can be added to the well or chamber instead of the cells to determine if it recapitulates the effects of the cells.

Under Agarose Assay

Another co-culture method that can be used to achieve a non-contacting co-culture is the Under Agarose Assay (Hirschi and D'Amore, 1998). In this system, culture wells are filled with 1 per cent agarose dissolved in DMEM. Once solidified (at 4°C for 30+ min), 5 mm holes that are ~2 mm apart are punched into the agarose using a template and cutter. The agarose is removed from one hole and the first cell type plated into the agarose well in the same medium. Once adhered (4–6 h), the second cell type is similarly plated in the other well. The cells in both wells are attached to the bottom of the same culture well, and once in the incubator, the agarose surrounding the cells becomes semi-solid, enabling the cells to migrate toward each other, moving under the agarose. Soluble diffusible factors can also be exchanged between the cell types, and their effects specifically tested using neutralizing reagents, as described above, however, they must be dissolved in the agarose during its preparation prior to plating of the cells (Hirschi *et al.*, 1998).

Measurement of migration

In the Under Agarose Assay, since both cell types are plated into the same well and able to exchange diffusible factors, the effects of a specific factor(s) produced by one cell type on the migration and proliferation of the other cell type can be readily measured (Hirschi *et al.*, 1999). To assess directed cell migration in response to co-culture, the two cell types are plated and co-incubated for 24–48 h, then fixed with paraformaldehyde and stained with Coomassie Blue. One can then measure the distance migrated by the cells at the front, directly adjacent to the other cell type, relative to the cells at the back of the well, facing the plastic wall of the well. In these assays, one must subtract the background random migration of cells from similarly prepared agarose wells in response to control media, instead of cells (Hirschi *et al.*, 1999).

Measurement of proliferation

To assess the effects of co-culture on cell proliferation, the live cell cultures can be incubated with bromo-deoxyuridine (BrdU), as described for contacting cultures (Hirschi *et al.*, 1999), then fixed and immunostained with antibodies against BrdU to identify which cells are proliferating, and count the proportion relative to the entire population. It is likely in this assay system that cellular behaviours such as growth will be affected in a gradient-like manner. That is, the cells closest to the other well should have the highest mitotic index in response to adjacent cells.

Assessment of differentiation

Since both cell types are plated in the same well, and able migrate toward each other, if they are left in culture for 5–7 days, the cell types can come into contact in the junctional region between the two agarose wells, and the effects of cell–cell interactions on cell phenotype can be monitored. If one cell population is labelled prior to co-culture, the two cell types can be distinguished upon fixing and immunostaining for markers of cell differentiation. It was using this model system that determined that direct contact with endothelial cells promotes the differentiation of mesenchymal cells toward a mural cell phenotype (Hirschi *et al.*, 1998).

Advantages/disadvantages

The major advantages of a non-contacting co-culture system are that it can be used to assess the effects of co-culture on the growth and migration of either or both cell types without pre-labelling the cell populations. Using the Under Agarose Assay, in combination with pre-labelling, one can also determine effects of co-culture on cellular phenotype, since this non-contacting culture system becomes a contacting system over time, and cell–cell contact is known to modulate vascular cell phenotype (Hirschi *et al.*, 1998).

7.2   3D models to study blood vessel assembly

Although there are distinct advantages to studying the effects of one cell type on another in the contacting and non-contacting co-culture systems described above, there is one disadvantage common to them all. That is, the cellular behaviours are studied in two dimensions, but cells *in vivo* exist in 3D structures, and their behaviours are modulated therein. To circumvent this problem, researchers have developed 3D co-culture systems that can be used to assess endothelial cell tube formation and the role of mural cells in vessel stabilization.

When comparing gene expression profiles and phenotypes of endothelial cells cultured in 2D to *in vivo* endothelial cells and those cultured in 3D, it is apparent that the 3D culture systems better represent *in vivo* endothelial cells than the 2D systems. For example, endothelial cells grown in 3D culture conditions undergo morphogenesis and differentiation (vacuole formation and elongation) as compared with endothelial cells grown in a monolayer, which retain a more undifferentiated phenotype (Matsumoto *et al.*, 2002; Yang *et al.*, 2004). In the adult system, endothelial cells are maintained in a quiescent state (Engerman *et al.*, 1967; Hobson and Denekamp, 1984), however, once endothelial cells are cultured in a monolayer, they are no longer quiescent and begin to synthesize growth factors such as platelet-derived growth factor-B (PDGF-B) (Barrett *et al.*, 1984; Collins *et al.*, 1987; DiCorleto and Bowen-Pope, 1983) which is expressed in only the most immature endothelial cells *in vivo* (Hellstrom *et al.*, 1999). These *in vivo* expression patterns of PDGF-B are recapitulated in the 3D assay of endothelial cells co-cultured with smooth muscle cells (Korff *et al.*, 2001). The remainder of this chapter will give an overview of the various 3D systems designed to investigate endothelial cell tube formation and cell–cell interactions between endothelial cells and mural cells and how this promotes blood vessel maturation.

Endothelial tube formation

The molecular regulation of endothelial cell tube formation has been the subject of much research. Due to the complexity of this process, culturing endothelial cells in a 2D system does not allow for a comprehensive investigation (see Chapter 4). As previously mentioned, endothelial cells cultured in a 3D system undergo vacuolization, elongation and coalescence into tube-like structures, thereby recapitulating the events that take place *in vivo* (Yang *et al.*, 2004). Thus far, there are two main types of 3D matrices upon which endothelial cells are cultured, collagen and fibrin.

The concept of growing endothelial cells on a collagen matrix came from the observation that, *in vivo*, endothelial cells are constantly in contact with components of the extracellular matrix (ECM) and that in order for endothelial cells to take on a more *in vivo* phenotype they need to be surrounded by ECM components (Delvos *et al.*, 1982). Since collagen is the major component of ECM, researchers decided to plate endothelial cells within collagen matrices and noted a drastic change in their morphology so that they resembled the microvasculature *in vivo* (Delvos *et al.*, 1982; Montesano *et al.*, 1983). In these initial experiments, the 3D matrices were produced by collagen coating flasks and plating bovine aortic endothelial cells (BAEC) on the surface of these gels. After the 3 days of culture, the growth medium was removed and a second layer of collagen was poured on top of the endothelial cells. More recently, there are two ways in which researchers typically create their 3D collagen matrix. The first of which, developed by Davis and Camarillo (1996), involves mixing collagen, 10X Medium 199 and NaOH at 0°C. After mixing, endothelial cells are added and this mixture is plated into microwells and allowed to gel at 37°C. After gelatinization is complete culture media is then applied (Davis and Camarillo, 1996). Another method for culturing on a 3D collagen matrix has been developed by D'Amore's group (Beck *et al.*, 2004) wherein endothelial cells are plated onto collagen gel matrices and after 2 days of culture, the gels are dislodged, thereby creating a floating collagen gel. Upon removal of the collagen gels off the plate, the endothelial cells begin to take on a more *in vivo* phenotype and genotype so that their mitotic index returns to *in vivo* levels as well as their expression levels of PDGF-B.

While collagen matrices are the most widely used out of the 3D matrices, there is yet another 3D matrix that is commonly utilized, called the fibrin matrix. Fibrin is an elastic protein which forms a fibrous network upon conversion from fibrinogen, a protein found in blood plasma. This process occurs during blood coagulation, therefore the fibrin matrix is frequently used when recreating an injury response model. The process of making a fibrin gel involves suspending endothelial cells in a mixture of plasminogen-and uPA-free human fibrinogen (in serum-free medium). Thrombin is added and the mixture is aliquoted into the desired culturing container and allowed to clot. Serum containing endothelial cell media plus angiogenic factors is added and then cultured for at least 3 days whereupon endothelial tube-like structures are then observed (Lafleur *et al.*, 2002; Bayless and Davis, 2003). Collagen and fibrin matrices have been very informative in regard to the molecular regulation of endothelial cell tube formation. Table 7.1 depicts a variety of genes that have been shown to be required for this process via investigations using these 3D gel assays.

Endothelial–mural cell interactions

After endothelial cells have coalesced to form a tube-like structure, mural cells are recruited to maintain quiescence and structural stability. Therefore, creating a 3D co-culture system of endothelial cells with mural cells is of great importance to determine how the interaction between these two cell types regulates blood vessel formation and stability. Two 3D model systems have been devised to allow for investigation of this cell–cell interaction, the spheroidal model and the co-culture model, with 10T1/2 cells as mural cell precursors.

Culturing cells in a spheroid was originally devised to investigate differentiation and proliferation rates of tumour cells (Barrett *et al.*, 1984; Lincz *et al.*, 1997) and embryonic stem cells (Vittet *et al.*, 1996; Itskovitz-Eldor *et al.*, 2000). Researchers first adapted the spheroidal model to investigate endothelial cell maturation in this system to determine if they could recapitulate the *in vivo* endothelial cell differentiation and apoptosis rates (Korff and Augustin, 1998). Upon confirmation

Table 7.1 Genes needed for endothelial cell tube formation demonstrates which gene families and members are required for the various aspects of tube formation

Family of genes	Genes Shown to be Required for Lumen Formation	Function
Integrins	$\alpha2\beta1,\alpha1\beta1$-collagen matrices (Delvos *et al.*, 1982; Senger *et al.*, 1997) $\alpha5\beta1$, $\alpha v\beta3$-fibrin matrices (Bayless *et al.*, 2000)	Vacuolization and Lumen Formation
Membrane Type Matrix Metalloproteinases (MT-MMPs)	MT1-MMP (LaFleur *et al.*, 2002; Galvez *et al.*, 2001)	Tube Network Formation
Phosphorylated Lipids	Sphingosine 1-phosphate (S1P) (Bayless and Davis, 2003)	Induces Integrin and MT-MMP Regulated EC Morphogenesis
Rho GTPases and cytoskeleton	RhoA (Davis *et al.*, 2002) Rac1 (Bayless and Davis, 2002) Cdc42 (Bayless and Davis, 2002)	Endothelial Cell Sprouting Vacuole and Lumen Formation Vacuole and Lumen Formation and EC Branching and Sprouting

that endothelial cells could mimic the *in vivo* rates, the model was adjusted to include smooth muscle cells in the system (Korff *et al.*, 2001). The process of generating a co-culture spheroid consists of suspending the endothelial cells and mural cells separately in their corresponding media with carboxymethylcellulose. The suspensions are mixed together and then seeded in non-adherent round-bottom 96-well plates. Per well, only one spheroid will form. The spheroids are cultured anywhere from 1 to 4 days (Korff *et al.*, 2001). It is to be noted that unlike the *in vivo* situation, the mural cells are in the centre of the spheroid and the endothelial cells are on the outside (Figure 7.1).

The second model system used to investigate these cell–cell interactions is another co-culture system. This method is more widely employed due to its ease of use. To create the 3D system, Matrigel™ is allowed to polymerize on either a coverslip or on the bottom of a 24-well plate. Endothelial and 10T1/2 mesenchymal cells are plated on top of the Matrigel™ in media consisting of growth medium with 2 per cent fetal calf serum, and invade down into the Matrigel. Morphogenesis can begin to be seen after 6 h of culture (Darland *et al.*, 2003; Ding *et al.*, 2004; Akimoto *et al.*, 2000). This co-culture model resembles more of the *in vivo* phenotype wherein the mesenchymal cells are recruited by the endothelial cells to form a surrounding mural cell layer. As with the endothelial tube formation systems, the 3D co-culture models have provided invaluable data as to what regulates endothelial cell and smooth muscle cell interactions, as shown in Table 7.2.

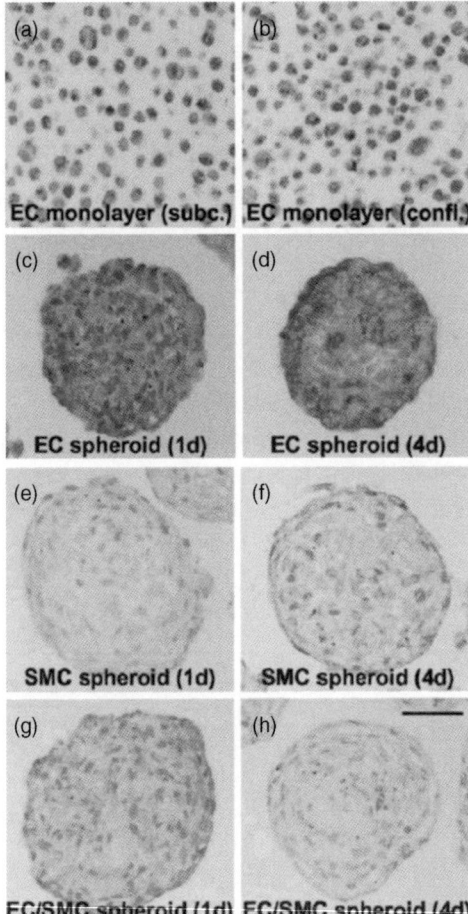

Figure 7.1   The spheroid 3D culture system (a) and (b) represent sub-confluent and confluent monolayer endothelial cells (EC), respectively. (c) and (d) illustrate EC spheroids, (e) and (f) demonstrate smooth muscle cell (SMC) spheroids, and (g) and (h) show EC/SMC spheroids. Adapted from Korff *et al.* (2001).

Table 7.2   Genes needed to regulate cell–cell interactions between endothelial and smooth muscle cells demonstrates which genes are needed to mediate the cell–cell interactions between endothelial and smooth muscle cells

Gene	Function
Transforming Growth Factor Beta (TGFβ) (Darland and D'Amore, 2001)	Regulates Cord Formation and Promotes Smooth Muscle Cell Differentiation
Vascular Endothelial Growth Factor (VEGF) (Darland *et al.*, 2003; Kitahara *et al.*, 2005	Endothelial Cell Survival and Differentiation
VEGF and Ang2 (Korff *et al.*, 2001)	Endothelial Cell Sprouting

Monitoring of flow

The 3D *in vitro* systems all have one major limitation in that they cannot recreate plasma and/or blood flow as seen *in vivo*. Laminar flow and oscillatory flow forces of blood circulation are of great significance in that they can change the behaviour of endothelial cells (see Chapter 5). For example, upon exposure of endothelial cells to laminar shear stress, their proliferation decreases (Akimoto *et al.*, 2000), apoptosis is inhibited (Dimmeler *et al.*, 1997; Dimmeler *et al.*, 1998) and migration increases (Sprague *et al.*, 1997). In order to observe these changes *in vivo*, multiphoton confocal imaging has risen to the forefront of the field.

The mechanism by which multiphoton laser-scanning microscopy (MPLSM) works *in vivo* is depicted in Figure 7.2. This system was initially created to monitor tumour angiogenesis (Brown *et al.*, 2001). In brief, dorsal skinfold chambers (Chapter 13) are implanted into severe combined immunodeficient (SCID) mice and adenocarcinomas are implanted into the centre of the viewing window. These tumours are allowed to grow within the mice for 2 weeks. After 2 weeks, the vasculature is visualized by MPLSM. Using this technique, researchers are able to

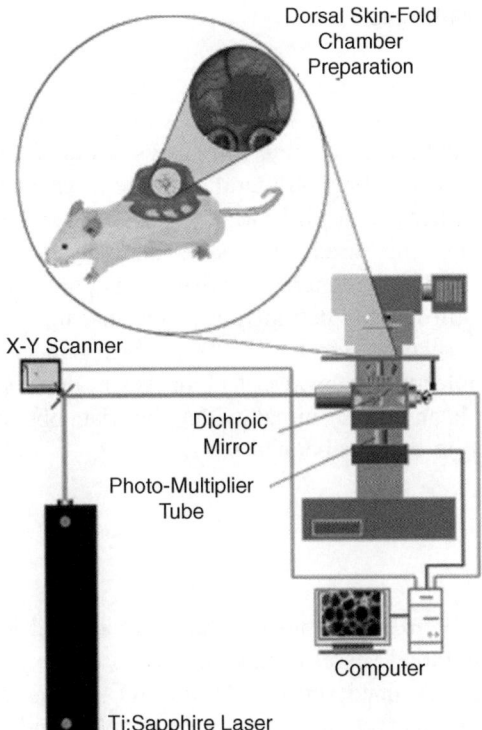

Figure 7.2 The experimental design of the multiphoton laser-scanning microscopy adapted from Brown *et al.*, 2001.

image gene expression, vessel permeability, endothelial/leukocyte interactions as well as red blood cell velocity over endothelial cells (Brown *et al.*, 2001). Researchers have now made it so that they can capture a time series of events spanning 7 days, observing angiogenic changes as a result of tumour formation (Abdul-Karim *et al.*, 2003).

In an effort to combine 3D *in vitro* culture systems with *in vivo* imaging, Jain's group has recently engineered long-lasting blood vessels (Koike *et al.*, 2004). These have been implanted into mice and appear to be functional and stable for at least one year *in vivo*. The method by which they accomplished this was by co-culturing human endothelial cells with 10T1/2 mesenchymal cells in a 3D collagen assay. These 3D constructs were then implanted into SCID mice containing transparent viewing windows. Within the first 2 weeks of transplantation, numerous amounts of perfused blood vessels formed that were connected to the mouse's circulatory system. Immunohistochemical analysis revealed that the lumens of the engineered vessels were surrounded by endothelial cells and the 10T1/2 mesenchymal cells efficiently differentiated into mural cells (Koike *et al.*, 2004).

Advantages/disadvantages

The major advantages of the 3D systems are that the cell types retain a more *in vivo* phenotype and genotype as compared with the 2D systems, leading to a better understanding of how the complex processes of tube formation and cell–cell interactions between endothelial and mural cells are molecularly regulated. Also, these model systems, used with MPLSM, can account for the effects of flow on blood vessel stability. The disadvantages of these systems are that they are typically more technically challenging relative to 2D cultures and, even with the 3D model, there is still no way definitively to say that the tubes formed and the cell–cell interactions seen are identical to that which happens *in vivo*. The only way in which the molecular pathways of blood vessel development will be fully accepted is by proof in an *in vivo* animal model, thus data obtained in 2D or 3D co-culture systems need to be corroborated *in vivo*.

7.3   Summary

In summary, both *in vitro* and *in vivo* systems are needed to discover and test factors that modulate blood vessel formation, and to define their cellular roles in this process. We have outlined various 2D and 3D systems used specifically to analyse heterocellular interactions between endothelial and mural cells during vessel assembly. Importantly, such systems can be adapted to study heterocellular interactions that are required for the formation of many other multicellular tissues.

References

Abdul-Karim, M. A., Al-Kofahi, K., Brown, E. B., Jain, R. K. and Roysam, B. (2003) 'Automated tracing and change analysis of angiogenic vasculature from *in vivo* multi-photon confocal image time series', *Microvasc. Res.*, **66**, pp. 113–125.

Akimoto, S., Mitsumata, M., Sasaguri, T. and Yoshida, Y. (2000) 'Laminar shear stress inhibits vascular endothelial cell proliferation by inducing cyclin-dependent kinase inhibitor p21(Sdi1/Cip1/Waf1)', *Circ. Res.*, **86**, pp. 185–190.

Antonelli-Orlidge, A., Saunders, K. B., Smith, S. R. and D'Amore, P. A. (1989) 'An activated form of transforming growth factor ß is produced by cocultures of endothelial cells and pericytes', *Proc. Natl. Acad. Sci. USA*, **86**, pp. 4544–4548.

Barrett, T. B., Gajdusek, C. M., Schwartz, S. M., McDougall, J. K. and Benditt, E. P. (1984) 'Expression of the sis gene by endothelial cells in culture and *in vivo*', *Proc. Natl. Acad. Sci.*, **81**, pp. 6772–6774.

Bayless, K. J. and Davis, G. E. (2002) 'The Cdc42 and Rac1 GTP ases are required for capillary lumen formation in three-dimensional extracellular matrices', *J. Cell Sci.*, **115**, pp. 1123–1136.

Bayless, K. J. and Davis, G. E. (2003) 'Sphingosine-1-phosphate markedly induces matrix metalloproteinase and integrin-dependent human endothelial cell invasion and lumen formation in three-dimensional collagen and fibrin matrices', *Biochem. Biophys. Res. Commun.*, **312**, pp. 903–913.

Bayless, K. J., Salazar, R. and Davis, G. E. (2000) 'RGD-dependent vacuolation and lumen formation observed during endothelial cell morphogenesis in three-dimensional fibrin matrices involves the αvβ3 and α5β1 integrins', *Am. J. Pathol.*, **156**, pp. 1673–1683.

Beck, L. H. Jr., Goodwin, A. M. and D'Amore, P. A. (2004) 'Culture of large vessel endothelial cells on floating collagen gels promotes a phenotype characteristic of endothelium *in vivo*', *Differentiation*, **72**, pp. 162–170.

Brown, E. B., Campbell, R.B., Tsuzuki, Y., Xu, L. *et al.* (2001) '*In vivo* measurement of gene expression, angiogenesis and physiological function in tumours using multiphoton laser scanning microscopy', *Nature Med.*,**7**, pp. 864–868.

Collins, T., Pober, J. S., Gimbrone, M. A. Jr., Hammacher, A. *et al.* (1987) 'Cultured human endothelial cells express platelet-derived growth factor A chain', *Am. J. Pathol.*, **126**, pp. 7–12.

Darland, D. C. and D'Amore, P. A. (2001) 'TGFβ is required for the formation of capillary-like structures in three-dimensional cocultures of 10T1/2 and endothelial cells', *Angiogenesis*, **4**, pp. 11–20.

Darland, D. C., Massingham, L. J., Smith, S. R., Piek, E. *et al.* (2003) 'Pericyte production of cell-associated VEGF is differentiation-dependent and is associated with endothelial survival', *Dev. Biol.*, **264**, pp. 275–288.

Davis, G. E. and Camarillo, C. W. (1996) 'An α2β1 integrin-dependent pinocytic mechanism involving intracellular vacuole formation and coalescence regulates capillary lumen and tube formation in three-dimensional collagen matrix', *Exp. Cell Res.*, **224**, pp. 39–51.

Davis, G. E., Bayless, K. J. and Mavila, A. (2002) 'Molecular basis of endothelial cell morphogenesis in three-dimensional extracellular matrices', *Anat. Rec.*, **268**, pp. 252–275.

Delvos, U., Gajdusek, C., Sage, H., Harker, L. A. and Schwartz, S. M. (1982) 'Interactions of vascular wall cells with collagen gels', *Lab. Invest.*, **46**, pp. 61–72.

DiCorleto, P. E. and Bowen-Pope D. F. (1983) 'Cultured endothelial cells produce a platelet-derived growth factor-like protein', *Proc. Natl. Acad. Sci.*, **80**, pp. 1919–1923.

Dimmeler, S., Haendeler, J., Nehls, M. and Zeiher, A. M. (1997) 'Suppression of apoptosis by nitric oxide via inhibition of interleukin-1beta-converting enzyme (ICE)-like and cysteine protease protein (CPP)-32-like proteases', *J. Exp. Med.*, **185**, pp. 601–607.

Dimmeler, S., Assmus, B., Hermann, C., Haendeler, J. and Zeiher, A. M. (1998) 'Fluid shear stress stimulates phosphorylation of Akt in human endothelial cells: involvement in suppression of apoptosis', *Circ. Res.*, **83**, pp. 334–341.

Ding, R., Darland, D. C., Parmacek, M. S. and D'Amore, P. A. (2004) 'Endothelial–mesenchymal interactions *in vitro* reveal molecular mechanisms of smooth muscle/pericyte differentiation', *Stem Cells Dev.*, **13**, pp. 509–520.

Engerman, R. L., Pfaffenbach, D. and Davis, M. D. (1967) 'Cell turnover of the capillaries', *Lab. Invest.*, **17**, pp. 738–743.

Galvez, B. G., Matias-Roman, S., Albar, J. P., Sanchez-Madrid, F. and Arroyo, A. G. (2001) 'Membrane type 1-matrix metalloproteinase is activated during migration of human endothelial cells and modulates endothelial motility and matrix remodeling', *J. Biol. Chem.*, **276**, pp. 37491–37500.

Hellstrom, M., Kalen, M., Lindahl, P., Abramsson, A. and Betsholtz, C. (1999) 'Role of PDGF-B and PDGFR-beta in recruitment of vascular smooth muscle cells and pericytes during embryonic blood vessel formation in the mouse', *Development*, **126**, pp. 3047–3055.

Hirschi, K. K. and D'Amore, P. A. (1998) '*In vitro* coculture models of vessel formation and function', in *Vascular Morphogenesis: In vivo, In vitro, In Mente*, Little, C., Mironov, V. and Sage, H., eds, Birkhauser Publications, pp. 132–140.

Hirschi, K. K., Rohovsky, S. A. and D'Amore, P. A. (1998) 'PDGF, TGF-β and heterotypic cell–cell interactions mediate the recruitment and differentiation of 10T1/2 cells to a smooth muscle fate'. *J. Cell Biol.*, **141**, pp. 805–814.

Hirschi, K. K., Rohovsky, S. A., Beck, L. H., Smith, S. and D'Amore, P. A. (1999) 'Endothelial cells modulate the proliferation of mural cell precursors via PDGF-BB and heterotypic contact', *Circ. Res.*, **84**, pp. 298–305.

Hirschi, K. K., Lai, L., Belaguli, N. S., Dean, D. *et al.* (2001) 'TGF-β induction of a smooth muscle cell phenotype requires transcriptional and post-transcriptional control of serum response factor', *J. Biol. Chem.*, **277**, pp. 6287–6295.

Hirschi, K. K., Burt, J. M., Hirschi, K. D. and Dai, C. (2003) 'Gap junction communication mediates TGF-β activation and endothelial-induced mural cell differentiation', *Circ. Res.*, **93**, pp. 429–437.

Hobson, B. and Denekamp, J. (1984) 'Endothelial proliferation in tumours and normal tissues', *Br. J. Cancer*, **49**, pp. 405–413.

Horan P. K. and Slezak, S. E. (1989) 'Stable cell membrane labeling', *Nature*, **340**, pp. 167–168.

Itskovitz-Eldor, J., Schuldiner, M., Karsenti, D., Eden, A. *et al.* (2000) 'Differentiation of human embryonic stem cells into embryoid bodies compromising the three embryonic germ layers', *Molec. Med.*, 88–95.

Kitahara, T., Hiromura, K., Ikeuchi, H., Yamashita, S. *et al.* (2005) 'Mesangial cells stimulate differentiation of endothelial cells to form capillary-like networks in a three-dimensional culture system', *Nephrol., Dialysis, Transplantation*, **20**, pp. 42–49.

Koike, N., Fukumura, D., Gralla, O., Au, P. et al. (2004) 'Creation of long-lasting blood vessels', *Nature*, **428**, pp. 138–139.

Korff, T. and Augustin, H. G. (1998) 'integration of endothelial cells in multicellular spheroids prevents apoptosis and induces differentiation', *J. Cell Biol.*, **143**, pp. 1341–1352.

Korff, T., Kimmina, S., Martiny-Baron, G. and Augustin, H. G. (2001) 'Blood vessel maturation in a 3-dimensional spheroidal coculture model: direct contact with smooth muscle cells regulates endothelial cell quiescence and abrogates VEGF responsiveness', *FASEB*, **15**, pp. 447–457.

Lafleur, M. A., Handsley, M. M., Knauper, V., Murphy, G. and Edwards, D. (2002) 'Endothelial tubulogenesis within fibrin gels specifically requires the activity of membrane-type-matrix metalloproteinases (MT-MMPs)', *J. Cell Sci.*, **115**, pp. 3427–3438.

Lai, L., Bohnsack, B. L., Niederreither, K. and Hirschi, K. K. (2003) 'Retinoic acid signaling regulates endothelial cell proliferation during vasculogenesis', *Development*, **130**, pp. 6465–6474.

Lincz, L. F., Buret, A. and Burns, G. F. (1997) 'Formation of spheroid structures in a human colon carcinoma cell line involves a complex series of intercellular rearrangements', *Differentiation*, **61**, pp. 261–274.

Lindahl, P., Johansson, B., Leveen, P. and Betsholtz, C. (1997) 'Pericyte loss and micro-aneurysm formation in platelet-derived growth factor B-chain-deficient mice', *Science*, **277**, pp. 242–245.

Matsumoto, T., Turesson, I., Book, M., Gerwins, P. and Claesson-Welsh, L. (2002) 'p38 MAP kinase negatively regulates endothelial cell survival, proliferation, and differentiation in FGF-2-stimulated angiogenesis', *J. Cell Biol.*, **156**, pp. 149–160.

Montesano, R., Orci, L. and Vassalli, P. (1983) '*In vitro* rapid organization of endothelial cells into capillary-like networks is promoted by collagen matrices', *J. Cell Biol.*, **97**, pp. 1648–1652.

Okabe M., Ikawa M., Kominami K., Nakanishi T. and Nishimune Y. (1997) '"Green mice" as a source of ubiquitous green cells', *FEBS Lett.*, **407**, pp. 313–319.

Sato, T. N., Qin, Y., Kozak, C. and Audus, K. L. (1993) '*tie-1* and *tie-2* define another class of putative receptor tyrosine kinase genes expressed in early embryonic vascular system', *Proc. Natl. Acad. Sci. USA*, **90**, pp. 9355–9358.

Senger, D. R., Claffey, K. P., Benes, J. E., Perruzzi, C. A. et al. (1997) 'Angiogenesis promoted by vascular endothelial growth factor: regulation through alpha1beta1 and alpha2beta1 integrins', *Proc. Natl. Acad. Sci.*, **94**, pp. 13612–13617.

Sprague, E. A., Luo, J. and Palmaz, J. C. (1997) 'Human aortic endothelial cell migration onto stent surfaces under static and flow conditions', *J. Vasc. Interventional Radiol.*, **8**, pp. 83–92.

Vittet, D., Prandini, M. H., Berthier, R., Schweitzer, A. et al. (1996) 'Embryonic stem cells differentiate *in vitro* to endothelial cells through successive maturation steps', *Blood*, **88**, pp. 3424–3431.

Yang, B., Cao, D. J., Sainz, I., Colman, R. W. and Guo, Y. L. (2004) 'Different roles of ERK and p38 MAP kinases during tube formation from endothelial cells cultured in 3-dimensional collagen matrices', *J. Cell. Physiol.*, **200**, pp. 360–369.

8

Assays for membrane and intracellular signalling events

Vittorio Tomasi, Cristiana Griffoni, Spartaco Santi, Patrizia Lenaz, Rosa Anna Iorio, Antonio Strillacci and **Enzo Spisni**

Abstract

The concept that cell signalling is topologically restricted and vectorially orientated has stimulated much work. Restriction has been explained by the identification of signalling protein complexes in membrane microdomains (rafts and caveolae). The participation of endothelial cell caveolae to angiogenesis has been well reported and a distinction between normal and tumour-induced angiogenesis has been clearly documented. These advancements in the comprehension of signal transduction events preceding or concomitant to the angiogenic process required dramatic advancements in the methodology employed. We will describe the main techniques employed to evaluate the connection of endothelial cell signal transduction to angiogenesis, focusing on their strength and limitations.

Keywords

angiogenesis, endothelial cells, caveolae, RNA interference, protein–protein interaction

8.1 Introduction

Angiogenesis, the development of new capillaries from the pre-existing vascular bed, is a multi-step process that involves a large number of regulatory molecules. Some of these steps are extracellular events (such as basement membrane degradation), while other steps are cellular events, localized at the membrane or inside the endothelial cells (EC). For example, most pro-angiogenic signalling events start at the plasma membrane of EC, where angiogenic factor receptors

Angiogenesis Assays   Edited by Carolyn A. Staton, Claire Lewis and Roy Bicknell
© 2006 John Wiley & Sons, Ltd

(for example, vascular endothelial growth factor receptors) are localized. Signalling cascades proceed from the plasma membrane to the nucleus and involve many
cytoplasmic amplification and modulation steps. The dissection of the molecular
events involved in angiogenesis requires the use of many different techniques
aimed at identifying the molecules involved in the different steps as well as their
physiological and physical interaction with other molecular partners of their
physiome. In this chapter we utilized caveolae, specialized plasmalemmal vesicles,
as paradigmatic organelles suitable for dissection of molecular angiogenic events
occurring both at the plasma membrane level and in the cytosol of EC. In fact,
Caveolin-1 (Cav-1), the main structural protein of EC caveolae represents a
physical border between the membrane and the cytosolic environment. Cav-1 is
a transmembrane protein with a unique hairpin structure within the membrane,
with a central hydrophobic segment inserted into the membrane that splits the
molecule into two cytoplasmic domains called NH2 and COOH-terminal domains.
While the transmembrane domain of Cav-1 may interact with other membrane-
linked proteins, the two terminal tails of caveolin-1 remain entirely cytoplasmic
and are therefore accessible for cytoplasmic protein interactions. It has recently
been argued that signalling molecules required for angiogenesis exist in the
quiescent vasculature as modular complexes held together by Cav-1 (Hoffman,
2004). These signalosomes should disassemble to enable angiogenic factors to
promote angiogenesis. In the following sections, we will analyse biochemical,
biophysical and molecular biology techniques aimed at dissecting some of the
caveolae molecular events involved in angiogenesis.

8.2   The endothelial caveolae as the organizers of efficient spatial signal transduction mechanisms

First described by George Palade and co-workers as plasmalemmal vesicles,
caveolae are flask-shaped plasma membrane invaginations, which were initially
studied as effectors of transcytosis, mainly in endothelial cells (Anderson and
Jacobson, 2002; Stahlhut *et al.*, 2000). A 22 KDa integral membrane protein
termed Caveolin-1 which functions as scaffolding protein is the main structural
constituent and organizer of caveolae. Three different isoforms of caveolin exist,
with cell-specific expression and endothelial cells contain isoform 1 and 2, while
muscle cells contain only isoform 3 (Anderson and Jacobson, 2002; Stahlhut *et al.*,
2000; Cohen *et al.*, 2004).

Caveolae are known to be internalized and recycled in functional EC (Anderson,
1998; Minshall *et al.*, 2003) and this internalization has been linked to Cav-1
phosphorylation (Aoki *et al.*, 1999). Caveolae, derived from membranes of EC and
other cell types, were found to be a rich source of signalling proteins. This led
Lisanti and co-workers to hypothesize a signalling role for this organelle (Lisanti
et al., 1995). Many signalling molecules within caveolae directly interact with
Cav-1, through a region of this protein called caveolin scaffolding domain (CSD).

Table 8.1 Signalling molecules that regulate angiogenesis and associate with Cav-1

Signalling molecule	Stage of angiogenesis	Associates with caveolae	Immunopre-cipitates with Cav-1	Activity inhibited by Cav-1	References
Caspase 3	apoptosis	+			Oxhorn and Buxton, 2003
Endostatin	Apoptosis	+	+		Wickstrom *et al.*, 2003
eNOS	permeability	+	+	Yes	Garcia-Cardena *et al.*, 1996
ERα	proliferation	+	+	No	Acconcia *et al.*, 2005
Glypican-1	proliferation	+	+		Wickstrom *et al.*, 2003
Integrin α5β1	migration		+		Wickstrom *et al.*, 2003
PGI2S	permeability	+	+	No	Spisni *et al.*, 2001a and b
VEGFR2 (KDR)	proliferation	+	+	Yes	Labrecque *et al.*, 2003

Since Cav-1 forms homo-oligomers of 14–16 molecules, it may promote the formation of multiple signalling complexes. Many of the signalling proteins involved in angiogenesis interact, directly or indirectly, with Cav-1 (see Table 8.1). It is very likely that signalling molecules involved in angiogenesis form modular complexes with Cav-1 as the scaffolding protein. Since the majority of angiogenesis molecules are inactive when bound to Cav-1, it is also likely that caveolae represent inactive membrane-localized signalling modules. The strategic location of Cav-1 is such that it communicates with extracellular proteins involved in angiogenesis such as glypican-1 and integrin $\alpha_5\beta_1$ (Wickstrom *et al.*, 2003), and directly interact with cytosolic angiogenic modulators, such as endothelial nitric oxide synthase (eNOS) and prostacyclin synthase (PGIs).

8.3 The prostaglandins forming system: regulation of signal transduction and angiogenesis

The physiological role of prostaglandins in angiogenesis

The discovery of an endothelial specific growth factor, vascular endothelium growth factor (VEGF), was a milestone in the field of angiogenesis regulation. VEGF acts through, at least, two tyrosine kinases receptors (flt-1 and KDR) which, when activated, preferentially use Src and MAP kinase (ERK 1/2), signalling pathways. VEGF-KDR receptor uses both eNOS and PGIS for downstream signalling (Fidler and Ellis, 2004; Yancopoulos *et al.*, 2000; He *et al.*, 1999).

While the involvement of eNOS in angiogenesis has received more attention, less consideration has been dedicated to the role of PGI_2 in angiogenesis (Cooke and Losordo, 2002). PGI_2 is generated by an enzyme called prostacyclin synthase (PGIs), belonging to the cytochrome P450 family, when the substrate endoperoxide (PGH_2) is delivered by cyclooxygenases. The coupling of COX-2/PGIs suggested in several papers (Smith *et al.*, 2000; Simmons *et al.*, 2004) has been demonstrated by localizing this complex to caveolae (Liou *et al.*, 2001; Spisni *et al.*, 2001a and b; Spisni *et al.*, 2003; Massimino *et al.*, 2002).

PGIs is a constitutively expressed enzyme, while COX-2 is almost absent in resting endothelial cells and has to be induced (Spisni *et al.*, 1995; Spisni *et al.*, 2001a and b; Liou *et al.*, 2001). Two receptors have been detected for PGI_2: a classical plasma membrane receptor coupled to adenylate cyclase (IP receptor) and a more recently discovered nuclear receptor (PPAR-δ), whose role in angiogenesis is still unclear. Dey and co-workers (Lim and Dey, 2002) have observed that PGI_2 signalling depends mainly on its nuclear PPAR-δ receptor. The most studied PPAR-δ role during angiogenesis is the induction of implantation after fertilization, a step requiring a strong angiogenic response. However, in PPAR-δ knockout mice implantation does not seem to be compromised (Lim and Dey, 2002; Smyth and Fitzgerald, 2002; Barak *et al.*, 2002; Peters *et al.*, 2000).

Reviewing the literature it could be assumed that the generation of PGI_2 represents a signal used by EC to initiate a commitment towards angiogenesis (Leahy *et al.*, 2000). PGI_2 has also been described as an apoptotic signal, and is therefore useful for discriminating between cells committed or not committed to angiogenesis (Dimmeler and Zeiher, 2000; Hatae *et al.*, 2001).

Alteration of signal transduction during tumour-induced angiogenesis: the case of prostaglandins

The role of prostaglandins in tumour-induced angiogenesis was proposed in studies carried out several years ago (Spisni and Tomasi, 1997) using mainly prostaglandin E_2 (PGE_2). Many independent studies unambiguously demonstrated that while COX-2 is induced only during inflammation in endothelial cells (thus in normal endothelial cells only COX-1 is fully active), during malignant transformation COX-2 may be constitutively overexpressed not only by tumour cells, but also in tumour vasculature. In these conditions EC receive signals that are not normally generated for a long period of time, which may be responsible for apoptosis inhibition of EC and therefore an increase in tumour vascularization (Fidler and Ellis, 2004; Yancopoulos *et al.*, 2000). Many angiogenic factors seem to be able to act on COX-2 gene, even if controversial results are frequent. Transformed cells have constitutively elevated levels of COX-2 and this has been connected to tumour-induced angiogenesis and to the relevant role of COX-2 specific inhibitors in the treatment of certain cancers such as, for example, colon cancer. It is widely accepted that the disregulation of COX-2 pathway favours clonal expansion of mutant colonocytes (Cutler *et al.*, 2003).

Chang and co-workers (2004) believe that PGE_2 may represent an angiogenic switch in COX-2 induced breast cancer progression. Wang and DuBois (2004) speculated that PGE_2 generation represents an inducer of angiogenesis at the earliest stage of tumour development. However, we believe that PGE_2 generation by tumour cells (but not by endothelial cells) represents an important modulation of the immune system as we have previously demonstrated (Tomasi et al., 1984) that cytotoxic T-lymphocytes become anergic and quiescent when exposed to elevated levels of PGE2 produced enzymatically or by spontaneous conversion of PGH_2 to PGE_2 by tumour cells.

Smith et al. (2000) carefully reviewed this field thereby generating a list of several growth factors and tumour promoters involved in the regulation of COX-2 expression. Many of these inducers converge in the activation of NFkB signalling pathway and in the connection with MAP kinases signalling pathway. Nevertheless, it is still not clear how COX-2 gene is upregulated in cancer (Jiang et al., 2004).

8.4 Use of antisense oligo and siRNA to evaluate the function of single signalling molecules

Nucleic acid-based inhibition of gene expression: from antisense oligonucleotides and ribozymes to RNA interference

Exploring the complexity of cells at the molecular level requires new powerful tools which allow perturbation of normal functions of the cellular machinery thereby identifying the role of single molecules in complex biological processes. Over the last two decades the accumulation of a large amount of information regarding the human genome triggered several attempts to realize innovative experimental approaches and novel therapeutic strategies based on the use of exogenous nucleic acids. During this period there have been thousands of published reports describing the application of antisense nucleic acids derivatives for selective inhibition of gene function by a sequence-specific mRNA knockdown.

The notion that small oligodeoxynucleotides (ODNs) could be used specifically to inhibit gene expression was first introduced by Zamecnik and Stephenson (Stephenson and Zamecnik, 1978; Zamecnik and Stephenson, 1978). Some years after these elegant experiments investigators realized that ODNs had potential for gene inhibition via antisense base pairing. The mechanism of action of ODNs is via RNAse-H mediated cleavage of target mRNAs or by blocking splicing, translation or nuclear-cytoplasmic transport, due to a steric hindrance. The introduction of backbone modifications to enhance ODNs stability and biological activity and improvements in cell delivery methods allowed large applications of antisense as gene expression inhibitors and triggered the development of clinical trials for a variety of cancer and inflammatory diseases. However, only one antisense drug has been approved by USA Food and Drug Administration for the treatment of cytomegalovirus-associated retinitis and in general the interest in antisense oligos declined because their use as therapeutic agents is limited by practical constraints (Scherer and Rossi, 2003).

The second wave of interest in the use of nucleic acids as inhibitors of gene expression followed the discovery of catalytic RNAs (ribozymes) in the early 1980s. Ribozymes are RNA molecules that act as enzymes: they have the catalytic activity of breaking or forming covalent bonds with excellent specificity and increase the spontaneous rates of targeted reactions by many orders of magnitude. Ribozymes occur naturally, but can also be artificially engineered for specific targeting and consequent cleavage of any mRNA sequences. The therapeutic use of ribozymes also depends on the introduction of stabilizing backbone modifications. Concerning delivery, ribozymes have an advantage over ODNs in that they can be delivered to cells with plasmid or viral vectors and there have been numerous ribozymes studied for downregulating specific cellular and viral targets (Breaker, 2004).

Another interesting category of site-specific cleaving nucleic acids is that of catalytic DNAs. Small DNAs capable of site specifically cleaving RNA targets have been developed *in vitro*, because no known DNA enzymes occur in nature. Cleavage of RNA targets results in their destruction and then DNAzymes recycle and cleave multiple substrates. Catalytic DNAs are relatively inexpensive to synthesize and have good catalytic properties, making them useful substitutes for either antisense DNA or ribozymes. Several applications of DNAzymes in cell culture have been published. It is also possible to perform backbone modifications of DNAzymes which allow them to be systematically delivered in the absence of a carrier agent (Santoro and Joyce, 1997).

The most recent exciting advances in the antisense world followed the discoveries of Mello and co-workers in *Caenorhabditis elegans* (Fire *et al.*, 1998), that double-stranded RNAs (dsRNAs) elicit potent and specific degradation of complementary RNA sequences. This mechanism, named RNA interference (RNAi), occurs in a wide variety of eukaryotic organisms. It is triggered by dsRNA precursors different in length and origin, which are processed into short interfering RNA (siRNAs) duplexes of 21–28 nucleotides and then incorporated into a silencing complex called RISC (RNA-induced silencing complex), which guides the recognition and the consequent cleavage or translational repression of complementary single-stranded RNAs. RNAi silencing mechanisms were first recognized as a biological response to protect cells from viral double-stranded RNAs or prevent the random integration of transposable elements. However, the general role of silencing in the regulation of gene expression became evident following the observation that specific genes in plants and animals encode short forms of fold-back dsRNAs which are processed to form microRNAs (miRNAs) and trigger the cleavage or the inhibition of translation of sequence-complementary mRNAs (Meister and Tuschl, 2004).

When Elbashir and co-workers (2001) reported that gene expression in mammalian cell lines can be efficiently downregulated by transfecting cells with synthetic siRNAs approximately 21 to 23 nucleotides in length, RNAi appeared a powerful genetic tool for understanding the function of a gene by silencing its expression. RNAi appears to be more potent than antisense ODNs, ribozymes or DNAzymes for targeted mRNA destruction, probably because this mechanism exploits cellular machinery that efficiently guides the antisense component of siRNA molecules to the target mRNA for site-directed cleavage. At the same time, the prospect of using siRNAs as potent and specific inhibitors to any target gene raised a great interest in therapeutic applications.

The relative strengths and weaknesses of different antisense technologies are reported in Table 8.2.

Table 8.2 Relative strengths and weaknesses of antisense technologies (modified from Scherer and Rossi, 2003)

Approach	Advantages	Disadvantages
Antisense ODNs	Can be designed also to target introns or nuclear RNAs Can be modified to improve selectivity and efficacy Easy to make	Side effects due to induction of interferon system (if long and rich of CpG motifs) Can bind proteins (aptamer activity) Only exogenous delivery possible (synthetic ODNs) Off-target effects
Ribozymes	Can target introns and particular subcellular compartments Can discriminate single base polymorphisms Can be used to correct defects Inducible or tissue-specific expression possible	Require GUC triplet at the cleavage site–limitation for the choice of target Bind proteins (aptamer activity)
DNAzymes	Inexpensive to make Good catalytic properties Can be modified for systemic delivery	Only exogenous delivery possible
RNAi	Effective at lower concentrations with respect to antisense, ribozymes and DNAzymes Can be delivered by multiple pathways (chemically synthesized siRNAs or vector-based shRNAs) Inducible or tissue-specific expression possible Nontoxic? Lasts longer?	It is not clear if they can also target nuclear RNAs or introns No option for improving if target is refractive (a perfect antisense/target base pairing is necessary) Some reports of off-target effects Some reports of interferon system activation

Application of antisense and RNA interference mechanisms to the study of angiogenesis-related signalling events

Antisense technology is a versatile, efficient and specific method for inhibition of gene expression and identification of genes that are important in controlling angiogenic signalling events. An antisense oligo can be designed for any gene for which the sequence is available and the types of genes that can be targeted are not limited unlike traditional pharmacological inhibitors. Another technology that focuses on inhibition of gene expression is the development of mice with single-gene knockouts. With respect to antisense approaches, gene knockout has the advantage that the gene is absent in all tissues, while ODNs are not equally distributed in all tissues. However, knockout technology is time-consuming and expensive. Moreover, it is sometimes limited to the study of genes that are not fundamental for development and the target gene cannot be manipulated to determine how much inhibition is necessary to provide a particular phenotype (Zambrowicz and Sands, 2003).

Antisense approaches are powerful tools for target validation, the process that scientists follow to assign a functional role to genes in selected diseases. The application of ODNs for target validation in the area of angiogenesis and the process of design, synthesis and testing of ODNs as inhibitors for the study of single gene functions have recently been reviewed (Henry *et al.*, 2004). Starting from a known gene target sequence, the identification of optimal ODNs inhibitors requires an empirical process in which different ODNs are designed to hybridize with many regions of the target mRNA. All of these are transfected into cells, using commercially available transfection reagents following standard protocols, and their effects are analysed by evaluating the reduction of target mRNA and protein levels using RT-PCR and Western blotting respectively. The most effective ODN for each gene is subsequently screened for functional activity, comparing it with that of positive and negative controls.

To identify genes involved in angiogenesis, assays based on the use of human vein endothelial cells (HUVEC) have been developed. Primary screening includes measuring cell function and gene expression of marker genes that are considered important in angiogenesis. ODNs which show inhibition in primary screening assays are further studied in more complex secondary assays, including apoptosis, migration, 3D tube formation and gene arrays. Inhibitors blocking the expression of the angiogenic phenotype in secondary assays then need to be fully validated in animal models.

Even if the objective is the identification of genes which are crucial for a physiological or pathological process, negative data are also interesting because they exclude the involvement of some genes in these processes. In the case of angiogenesis, inhibitors of around 650 genes demonstrated activity in primary screening tests, but only a small fraction of them produced the desired phenotypic changes in secondary assays. In this fraction, enzymes of the Rho-family of

GTPases and mitogen-activated protein (MAP) kinase cascade are highly represented (Henry *et al.*, 2004).

Vascular endothelial growth factor (VEGF), VEGF receptors and some integrins are considered validated targets for angiogenesis (Yu and Sato, 1999; Kroll and Waltenberger, 1997; Eliceiri and Cheresh, 1998). In particular, several studies have shown that the VEGF signalling pathway is important in the initiation and progression of angiogenesis as utilization of ODNs directed against VEGF showed inhibition in angiogenesis animal models and tumour experimental models (Robinson *et al.*, 1996; Glade-Bender *et al.*, 2003).

However, targeting a single growth factor or its receptor inhibits only one of the several potential molecules which are involved in angiogenesis and acts only on early events of the process. Thus, it is important to identify cell signalling targets which might affect many aspects of the angiogenic process and complement anti-VEGF strategies. For example, both VEGF and $\alpha_v\beta_3$-integrin trigger the MAP kinase signalling pathway and the inhibition of a common signal cascade should be beneficial, considering that VEGF provides a signal that initiates angiogenesis, while $\alpha_v\beta_3$-integrin provides the signal necessary for terminal differentiation. ODNs have been used to inhibit this signalling pathway, for example antisense targeting protein kinase C (Wang *et al.*, 2002) and, in particular, antisense targeting C-Raf kinase (Monia *et al.*, 1996; Vinores *et al.*, 2000). We recently demonstrated that an antisense ODN directed against caveolin-1 mRNA impaired angiogenesis *in vitro* and *in vivo* (Griffoni *et al.*, 2000) and this observation has been confirmed by Lisanti and co-workers (Liu *et al.*, 2002), suggesting an important role of caveolar membrane microdomains as assembling sites for signalling molecules involved in the angiogenic process (Massimino *et al.*, 2002).

The recent emergence of RNAi as a silencing mechanism offered a more powerful approach to inhibit genes *in vitro* and *in vivo* and improved studies on the mechanism of action for many disease genes, including those involved in controlling the angiogenic process (Lu *et al.*, 2003). With the introduction of a clinically viable delivery vehicle, anti-angiogenesis RNAi agents appear to have a promising and novel role for the treatment of many serious diseases that result from excessive angiogenesis, such as cancer, ocular neovascularization and rheumatoid arthritis. The application of siRNAs to knockdown a specific gene and study its function follows a process similar to that described for ODNs: in particular, target mRNA accessibility, effective intracellular delivery and potent mRNA inhibition are required. RNAi can be mediated by chemically synthesized siRNAs or by gene expression vectors harbouring short hairpin RNAs (shRNAs), which allow a stable and long-lasting knockdown of target genes.

The functional validation of angiogenic factors for their specific role has been greatly facilitated by the use of RNA inhibitors, revealing a network involving the early activation of the VEGF pathway and interactions among matrix metalloproteinases (MMPs) and adhesion molecules, leading to the regulation of cellular signal transduction pathways.

An overview of RNAi agents used to investigate the functions of cytokines and receptors, matrix proteins and adhesion molecules, and cellular signalling factors in the angiogenic process has been recently published (Lu *et al.*, 2005) and data reviewed in that paper are summarized in Table 8.3.

Our group recently designed siRNAs molecules targeting cyclooxygenase-2 (COX-2), the inducible isoform of COX which is involved in several pathological processes. We tested these siRNAs in an angiogenesis model (HUVEC grown on 3D collagen gel) and in a colorectal cancer model (HT29 cells), in order to evaluate their effects as anti-angiogenic agents and anti-cancer therapeutics (Strillacci *et al.*, 2006).

The therapeutic potential of anti-angiogenesic siRNAs will only be realized when their *in vivo* activity can be achieved using clinically feasible delivery systems. Until now, anti-VEGF siRNA agents have been used in *in vivo* models for the treatment of cancer and ocular neovascularization diseases. Using intra-tumoural delivery of VEGF-targeted siRNA, the inhibition of tumour growth and angiogenesis was observed in human xenograft cancer models, raising a growing interest in siRNA-based therapeutics for the treatment of cancer (Xie, 2004; Takei *et al.*, 2004). Considering that the local delivery of anti-tumour agents is limited to only a few tumour types, systemic delivery of siRNAs will provide greater clinical benefits, especially for disseminated metastatic cancer. Thus, the *in vivo* activity of VEGF-siRNA agents has been further evaluated by systemic delivery to mice bearing neuroblastoma tumours using a ligand-directed nanoparticle carrier. In that model siRNAs showed inhibition of target expression, angiogenesis and tumour growth after repeated dosing, with a potency that suggests they may have clinical potential as anti-tumoural agents (Schiffelers *et al.*, 2004). Moreover, anti-angiogenic siRNAs have been used as therapeutic agents for the treatment of ocular neovascularization diseases in mouse models. In particular, potent siRNAs targeting VEGF, VEGF-R1 and VEGF-R2 were first validated *in vitro* and systematically delivered in a mouse model of herpetic stromal keratitis, by mixing the siRNAs cocktail with a polymer conjugate to form nanoparticles. This polymer self-assembles with negatively charged siRNAs into a particle with an RGD peptide exposed on its surface. The RGD peptide allows the specific binding to integrins $\alpha_v\beta_3$ and $\alpha_5\beta_1$ of activated endothelial cells in the neovasculature, while a polyethylene glycol (PEG) component prevents non-specific binding to other tissues. In this way, anti-VEGF siRNAs can be delivered by ligand-directed endocytosis and have a stronger effect than siRNAs delivered by subconjunctival administration. A siRNAs cocktail targeting multiple genes appears to be very useful in treating angiogenesis-related diseases which are caused by the abnormal overexpression of multiple genes, such as many human diseases (Kim *et al.*, 2004; Schiffelers *et al.*, 2004).

In conclusion, RNAi appears as powerful tool to validate the function of genes and factors involved in angiogenesis *in vitro* and *in vivo* and shows unique advantages over previous methods for gene function studies, including nucleic acids inhibitors as ODNs, ribozymes and DNAzymes (Table 8.2).

Table 8.3   RNAi-mediated studies of the functions of genes involved in angiogenesis

Molecular target	RNAi agent	Model of study and RNAi phenotype
a) Cytokines and receptors		
VEGF	siRNA and vector-based siRNA	VEGF-siRNA intratumoural administration suppressed tumour angiogenesis and growth in xenografts models (Xie, 2004; Takei *et al.*, 2004)
		VEGF-siRNA inhibited choroidal neovascularization in a murine retina model (Reich *et al.*, 2003)
		Systemic delivery of VEGF-siRNA mediated by ligand-directed nanoparticles suppressed neovascularization induced by herpes simplex virus infection in mice (Kim *et al.*, 2004)
VEGF receptors R1/R2	siRNA	Inhibition of ocular neovascularization in murine models and reduction of the growth of syngenic tumours (Kim *et al.*, 2004; Schiffelers *et al.*, 2004)
EGF receptor (erbB1)	siRNA	Inhibition of EGF-induced phosphorylation and induction of cell apoptosis in A431 human epidermal carcinoma cells (Nagy *et al.*, 2003)
Her-2/neu	Retroviral vector based siRNA	Reduction of proliferation, increase of apoptosis, increase of G0/G1 arrest and decrease of tumour growth in breast and ovarian tumour cells (Yang *et al.*, 2004)
S100A10 (plasminogen receptor)	Retroviral vector based siRNA	Abrogation of plasminogen-dependent cellular invasiveness in Colo 222 colorectal cancer cells (Zhang *et al.*, 2004)
CXCR4	siRNA	Inhibition of breast cancer cell migration *in vitro* (Chen *et al.*, 2003)
Amphiregulin (AR)	siRNA	Inhibition of cell proliferation and migration and inhibition of the survival mediator Akt/PKB activation (Gschwind *et al.*, 2003)
Mcl-1	siRNA	Reduction of proliferation induced by serum, VEGF and IL-6, induction of apoptosis (Le Gouill *et al.*, 2004)
SPK-1/SPK-2	siRNA	SPK-1 siRNA, but not SPK-2 siRNA, blocked VEGF-induced accumulation of Ras-GTP and phospho-ERK in T24 bladder tumour cells (Shu *et al.*, 2002)
PRB (progesterone receptor B)	siRNA	Abrogation of oestradiol-induced VEGF expression (Wu *et al.*, 2004)
b) Matrix proteins and adhesion molecules		
MMP-9	siRNA	Increase of surface E-cadherin levels and redistribution of β-catenin at the plasma membrane (Sanceau *et al.*, 2003)

(continued)

Table 8.3   *(Continued)*

Molecular target	RNAi agent	Model of study and RNAi phenotype
MMP-2	siRNA	Reversion of the inhibitory effect of conditioned medium from bovine aortic endothelial cells (BAEC) on the aortic smooth muscle cells migratory activity (von Offenberg Sweeney *et al.*, 2004)
Smad2	siRNA	Abrogation of the enhanced activation of ERK1/2 induced by TGF-β1 in gastric carcinoma cells (Lee *et al.*, 2004)
PINCH-1 (focal adhesion protein)	siRNA	PINCH-1 appears to be essential for cell spreading and motility, and it is crucial for cell survival (Fukuda *et al.*, 2003)
Integrin $\alpha_v\beta_3$	siRNA	Inhibition of proliferation and increase of apoptosis of cultured hepatic stellate cells (Zhou *et al.*, 2004)
Vimentin	siRNA	Endothelial cells assembled smaller than normal focal contacts and showed decreased adhesion to the substratum (Tsuruta and Jones, 2003)
Integrin $\alpha_4\beta_4$	siRNA	Decrease of invasion of tumour cells and decrease of migration on non-laminin substrates (Lipscomb *et al.*, 2003)
c) Cellular signalling factors		
Gas 1 (growth arrest specific 1)	SiRNA	Reduction of the anti-apoptotic protective effect of VEGF (Spagnuolo *et al.*, 2004)
cRAF-1	siRNA	Lowering of the levels of steady-state phosphorylated MEK and phosphorylated MAPK, inhibition of melanoma cells proliferation (Calipel *et al.*, 2003)
FAK (Focal Adhesion Kinase)	siRNA	Prevention of pressure-stimulate adhesion in human colon cancer cells (Thamilselvan and Basson, 2004)
ILK (integrin-linked kinase)	siRNA	Inhibition of HIF-1 (hypoxia inducible factor 1) and VEGF expression and reduction of angiogenesis *in vitro* and *in vivo* (Tan *et al.*, 2004)
PDK1 (3-phosphoinositide-dependent protein kinase-1)	siRNA	Reduction of the levels of steady-state phosphorylated MEK and MAPK, inhibition of tumour cells growth (Sato *et al.*, 2004)
Lyn (Src kinase)	siRNA	Inhibition of cell migration mediated by $\alpha_v\beta_3$ integrin in PDGF stimulated cells (Ding *et al.*, 2003)
TRIP6 (thyroid receptor interacting protein-6)	Vector based siRNA	Reduction of cell migration induced by lysophosphatidic acid (LPA) in ovarian cancer cells (Xu *et al.*, 2004)

8.5   Knocking down and knocking out Caveolin-1 gene

In endothelial cells, enzymes with fundamental vascular functions, such as eNOS and PGIs, have been found within caveolae (Razani and Lisanti, 2001; Spisni *et al.*, 2001a and b; Razani *et al.*, 2002). Caveolin-1, a 21–24 kDa protein, is the principal integral membrane component of caveolae membranes (Rothberg *et al.*, 1992; Glenney and Soppet, 1992), well conserved by evolution (Tang *et al.*, 1997). An interesting structural feature of Cav-1 is that it can form homotypic high molecular mass oligomers of 350 kDa, containing 15 individual molecules (Sargiacomo *et al.*, 1995; Fernandez *et al.*, 2002). It is believed that Cav-1 is instrumental in the caveolae-modulation of signalling systems. Emerging evidence suggests that most of the caveolae functions are directly regulated by Cav-1 (Liu *et al.*, 2002). In cultured cell models, Cav-1 is known to function as a negative regulator of eNOS, Ras-ERK-1/2 kinase cascade and as a transcriptional repressor of cyclin D1 gene expression. The fact that Cav-1 is directly involved in angiogenesis is demonstrated by the changes in Cav-1 levels observed during angiogenesis (Liu *et al.*, 2002). Moreover, we and others have shown that downregulation of Cav-1 in vascular cells impairs angiogenesis *in vitro* and *in vivo*, demonstrating that Cav-1 expression is necessary for capillary tubule formation (Griffoni *et al.*, 2000; Liu *et al.*, 2002). Nevertheless, the exact functional role of Cav-1 in angiogenesis still remains controversial.

In vitro Cav-1 transfection resulted in the inhibition of NO-dependent and VEGF-dependent tube formation on Matrigel (Brouet *et al.*, 2001), and angiogenesis activators have been shown to downregulate Cav-1 expression in EC (Liu *et al.*, 1999). Moreover, the anti-tumour effects observed *in vivo*, after Cav-1 gene delivery, have been mainly associated with the inhibition of tumour angiogenesis (Brouet *et al.*, 2005). These contrasting results may be explained by the hypothesis that Cav-1 may act at different levels during the different steps of the angiogenesic process. Accordingly, Cav-1 is downregulated during EC proliferation and upregulated during EC differentiation (Liu *et al.*, 2002).

Thus, it is interesting to review recent observations carried out on Cav-1 knockout mice, which show a relatively mild phenotype (Razani and Lisanti, 2001; Razani *et al.*, 2002; Parton, 2001) with Cav-1 gene knockout impairing the function of those cells where caveolae are most abundant and functionally well characterized. Thus, vascular dysfunctions and pulmonary defects are clearly related to the functions of caveolae in endothelial cells and in type-1 pneumocytes, while the fact that knockout mice are lean, resistant to diet-induced obesity and show hypertryglyceridaemia reflects the important function of caveolae in adipocytes (Ranzani *et al.*, 2002; Cohen *et al.*, 2003). In particular, it has been recently shown that in Cav-1 knockout mice endothelial cells have impaired signalling pathways, leading to endothelium hyperplasia, associated with an impaired NO synthesis (Razani *et al.*, 2001; Drab *et al.*, 2001).

Cav-1 is also involved in the control of tumour cell signalling events, with contrasting results. While in some tumour cells Cav-1 may function as a potent tumour suppressor protein (Williams *et al.*, 2004), in prostate cancer cells Cav-1 seems to function as a tumour promoter protein (Williams and Lisanti, 2005). Moreover, an augmented Cav-1 expression seems to promote metastases growth. It has been also collected results demonstrating that Cav-1 may function as a positive or negative regulator of cell transformation (Manara *et al.*, 2006; Ayala *et al.*, 2006). All together these data suggest that Cav-1 constitutes a key switch signalling protein through its function as an angiogenesis promoter and an angiogenesis inhibitor. While Cav-1 regulation of eNOS in EC has been extensively characterized (Bernatchez *et al.*, 2005), many more studies are still necessary to clarify the effect of Cav-1 on the modulation of other EC signalling molecules and surface receptors involved in angiogenesis.

8.6   Analysing protein–protein interactions in signalling molecules involved in angiogenesis

Immunoprecipitation experiments in the study of protein–protein interactions

One way to underscore the function of a protein is to identify its interacting partner. When caveolae were first isolated from tissue and cell cultures, it was evident that caveolae are rich in cell signal transduction molecules. Entire signalling modules (such as PDGFR-Ras-ERK) have been localized to caveolae, and they are fully functional even after caveolae isolation (Liu *et al.*, 1997). In the light of these discoveries it was obvious to think that Cav-1 could modulate signal transduction by attracting signalling molecules, as well as having a scaffolding function due to its ability to oligomerize (Lisanti *et al.*, 1994). Subsequently Couet and collaborators used synthetic scaffolding domains of Cav-1 (aa 61-101) to isolate two Cav-1 binding motifs, by phage display library (Couet *et al.*, 1997). Many co-immunoprecipitation studies have demonstrated the interaction between Cav-1 and a variety of signalling molecules (see Table 8.4). These interactions often suppressed the signalling activity of the molecules, but in some cases had no effect or even stimulated signalling through the bound molecule (Czarny *et al.*, 1999; Spisni *et al.*, 2001a and b). Despite the general agreement for the scaffolding hypothesis of Cav-1, other considerations need to be made. The wide variety of Cav-1 interacting molecules, including lipids and lipid anchors, demonstrate that Cav-1 is a promiscuous protein and that it is difficult to know the true extent of its promiscuity. In our hands, the scaffolding domain is a very hydrophobic sticky region that may non-specifically bind many proteins, at least *in vitro*. It also has a tendency to interact with membranes in a non-specific manner (Schlegel *et al.*, 1999). Methodologically, it is now evident that co-immunoprecipitation experiments are not sufficient to prove a direct protein–protein interaction, in particular for hydrophobic or sticky proteins. Thus, other

Table 8.4 Molecules that interact with Cav-1, as detected by immunoprecipitation experiments

Molecules	Binding region of human Cav-1	References
190-KDa pY	Whole protein	Liu *et al.*, 1996
30-KDa pY	Whole protein	Mastick and Saltiel, 1997
Adenyl cyclase/PLCβ2	82–101	Schreiber *et al.*, 2000
Connexin	82–101 and 135–178	Schubert *et al.*, 2002
COX-2	Whole protein	Liou *et al.*, 2001
Csk	PY14 Cav-1	Cao *et al.*, 2002
c-Src	61–101	Li *et al.*, 1996
EGFR	61–101	Couet *et al.*, 1997
eNOS	81–101	Garcia-Cardena *et al.*, 1996
Oestrogen receptor	82–101	Schlegel *et al.*, 2001
Fatty acids	Whole protein	Trigatti *et al.*, 1999
GD3	Whole protein	Kasahara *et al.*, 1997
GM1	Whole protein	Fra *et al.*, 1995
GRK1, 2 and 5	61–101	Carman *et al.*, 1999
HSP 56	Whole protein	Uittenbogaard *et al.*, 1998
Insulin receptor	81–101	Yamamoto *et al.*, 1998
Integrins	Whole protein	Wei *et al.*, 1999
		Wary *et al.*, 1998
Lyn	81–101	Muller *et al.*, 2001
PDGFR α and β	82–101	Yamamoto *et al.*, 1999
PGIS	61–101	Spisni *et al.*, 2001a and b
PKA	81–101	Razani *et al.*, 1999
PKCε	Whole protein	Wu *et al.*, 2002
PLD/PKCα	81–101	Kim *et al.*, 1999
Striatin	Whole protein	Gaillard *et al.*, 2001
TGFβ RI	61–101	Razani *et al.*, 2001
TrkA/p75NTR	Whole protein	Bilderback *et al.*, 1997
Trp1	Whole protein	Lockwich *et al.*, 2000
uPAR	Whole protein	Wei *et al.*, 1996

methods, such as confocal microscopy or fluorescence resonance energy transfer, that allow *in situ* analysis of the protein–protein interactions occurring in a single cellular compartment, constitutively or following activation-dependent stimuli, are utilized.

Confocal microscopy in the study of the co-localization of signalling molecules

To demonstrate the co-localization Cav-1/PGIS in HUVEC, we firstly adopted a classical confocal microscopic approach (Spisni *et al.*, 2001a and b). More recently

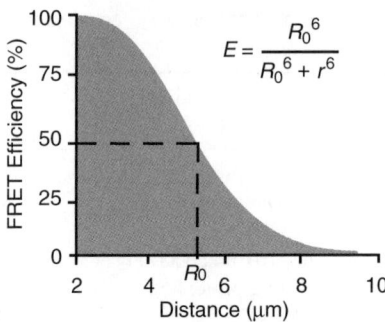

Figure 8.1 Fluorescence energy transfer (FRET). The curve represents the relationship between the efficiency of the fluorescence resonance energy transfer and the distance separating the donor and the acceptor. The efficiency of the transfer depends on the inverse sixth power of the distance between donor and acceptor. At the distance of the Föster radius R_0 between the molecules, the FRET efficiency is 50 per cent.

in our laboratory, we have established fluorescence resonance energy transfer (FRET), a technique for measuring spectroscopic distance and interaction between two fluorochrome-labelled molecules. Regular fluorescence occurs when a fluorophore absorbs electromagnetic energy of one wavelength (the excitation spectrum) and re-emits that energy at a different wavelength (the emission spectrum). In FRET, the emission peak of the donor fluorophore must overlap with the excitation peak of the acceptor fluorophore. Some of the electronic excitation energy can be efficiently transferred through dipole–dipole interactions (according to the theories of T. Förster, Ann. Phys. 1948, 2: 55) from the donor to the acceptor fluorophore that re-emits at its own emission wavelength. Because the efficiency of the transfer depends on the inverse sixth power of the distance between donor and acceptor (2–10 nm) (Figure 8.1), a FRET signal, corresponding to a particular localization within a specific cellular domain, provides an additional technique for investigating co-localization of biological molecules with the spatial resolution beyond the limits of conventional light microscopy (∼200 nm) and confocal laser scanning microscopy (∼160 nm). If the fluorophore distance is under 2 nm, such as within the protein complexes, no FRET signal will be observed. Ro is the distance between the donor and acceptor probe at which the energy transfer is (on average) 50% efficient. The Ro value is specific to a given FRET pair. There are number of combinations of FRET pairs that can be used depending on the biological applications (see Table 8.5). Recently, FRET has been used to study Ca^{2+}-regulated CREB activation in vascular smooth muscle cells, an event clearly involved in arterial responses to tumour-stimulated angiogenesis (Pulver *et al.*, 2004). In our laboratory, the molecular interaction between Cav-1 and PGIS in HUVE cells has been recently confirmed by using FRET.

FRET efficiency (E) can be obtained measuring the mean fluorescence intensities of the donor with acceptor (F_{DA}), the donor only (F_D) and the background

Table 8.5 Donor–acceptor pairs with emission and excitation peaks

Donor (Emission)	Awcceptor (Excitation)
CFP (477)	YFP (514)
BFP (450)	GFP (500)
BFP (450)	YFP (514)
CFP (477)	dsRED (568)
GFP (508)	Rhodamine (550)
GFP (508)	dsRED (568)
YFP (514)	dsRED (568)
YFP (514)	TRITC (550)
YFP (514)	Cy3 (550)
Alexa488 (515)	Alexa555 (555)
Alexa488 (515)	Cy3 (550)
FITC (520)	Cy3 (550)
FITC (520)	Rhodamine (550)
FITC (520)	TRITC (550)
Cy3 (566)	Cy5 (649)

(**B**) that was determined by using samples incubated without primary antibodies, according to the Förster equation (Table 8.6):

$$E = 1 - \frac{F_{DA} - B}{F_D - B}$$

More comments on the strengths and weaknesses of the FRET method is available in Riccio *et al.* (2004).

8.7 Final remarks

In order to assess the possible fate of the protein into microdomains by inspecting primary structure, Couet *et al.* (1997) have identified motifs in proteins allowing them to interact with the scaffolding domains of Cav-1 and Cav-3. These motifs are clusters of aromatic amino acids interacting with the scaffolding domain, but variations in these motifs have been detected studying PGI_2 synthase (Spisni *et al.*, 2001a and b; Tomasi *et al.*, 2000). The interaction between caveolar protein and Cav-1 (or Cav-3) may result in inhibition of enzymes as eNOS and adenylate cyclase, but this is not a general rule since PGI_2 synthase remains active. The outcome of such interactions may depend on the nature of the enzyme binding site (Tomasi *et al.*, 2000).

The main initiator for bringing a protein to microdomains is the palmitoylation of cysteine residues and the best way to demonstrate this is by using point mutated proteins transfected into cells. The methodology for this is described in several

Table 8.6   Schematic representation of FRET signal. The distance
requirement is described by the Förster equation for the efficiency of transfer

Channels	FITC		Cy3
Excitation	488		568
Emission	515		580
Distance		$>7\,\mu m$	
Excitation	488		–
Emission	–		580
Distance		$<7\,\mu m$	

papers from Lisanti's group (Lisanti *et al.*, 1994 for eNOS and Acconcia *et al.*, 2005 for oestrogen receptor alpha). Combination of this approach with confocal microscopy has a synergistic effect, especially if softwares to evaluate quantitatively co-localization are available (Spisni *et al.*, 2001a and b). In order to precisely evaluate protein–Cav-1 interactions, we recommend the use of glutathione transferase fusion proteins, which Spisni *et al.* (2001a and b) used to demonstrate that PGI$_2$, following the interaction PGI$_2$ synthase-Cav-1, in the presence of COX-2, is an angiogenic signalling molecule. Co-immunoprecipitation is a widely used technique in the field of membrane microdomains (Spisni *et al.*, 2001a and b; Griffoni *et al.*, 2003).

As a first approach to establish the participation of microdomains in angiogenesis, we recommend the use of antisense oligonucleotides against Cav-1 as described in Griffoni *et al.* (2000). Targeting Cav-1 destroys caveolae and completely blocks the formation of tube-like structures in collagen gel and/or the formation of chorioallantoic membrane (CAM) *in vivo*. Phosphorothioate antisenses are sufficiently stable to be used *in vivo*. We used them successfully in CAM experiments, to block angiogenesis and metastasis in mice. Using a tumour cell line (TS/A), originated from a mammary adenocarcinoma and capable of metastasizing in lungs (Nanni *et al.*, 1983), we detected a significant decrease in the number of metastasis in antisense-treated animals, with respect to control injected with scrambled antisense (Griffoni *et al.*, unpublished experiments).

A more difficult approach is the identification of the role played by any single caveolar protein in the recruitment of signalling complexes. This does not apply to enzymes as eNOS, PGI$_2$ synthase, adenylate cyclase and other enzymes, which when activated (or when the constraints are removed) release signals well identified as NO, PGI$_2$ or cyclic AMP.

Problems arise when membrane receptors are involved and we will focus on VEGF-R (KDR) and oestrogen receptor alpha (ERalpha). The reason why downregulation of endothelial caveolae results in a block of angiogenesis has to be connected mainly to the failure of caveolar signalling and proper activation of angiogenic molecules. For example, VEGF receptor connected to angiogenesis (KDR) is a caveolar receptor and its role is not vicariated by VEGF receptor flt-1 which is mainly involved in endothelial cell proliferation (Bauer *et al.*, 2005). To sustain angiogenesis, the role played by small molecules as NO, PGI_2, PGE_2 and cyclic AMP is important but not yet carefully evaluated.

In the field of microdomains the challenge is to evaluate the proteomic profile of rafts and caveolae. The methodology applied to soluble complexes may not be adequate to handle membrane protein composition. Moreover, as microdomains are dynamic structures, it has to be expected that changes in composition may involve different signal transduction pathways activation. This problem is being approached by two different methods, namely 1) isolation of membrane micro-domains by gradient centrifugation (flotation) in order to isolate and concentrate microdomain proteins; and 2) use of two-dimensional separation in the presence of detergents to ensure a better solubility of membrane proteins.

A particular attention has to be devoted to the mechanism of recruitment of cytosolic proteins into microdomains following phosphorylation. A well studied example is the recruitment of Fyn kynase in rafts and caveolae after activation. It appears that prion protein is the preferred partner of Fyn kynase and that Fyn may play the role of an adaptor protein, favouring perhaps the recruitment of other proteins (Toni *et al.*, in press).

In conclusion, in order to study signal transduction in angiogenesis, a number of approaches need to be employed, with analysis of initial over/under expression of genes using RT-PCR for measuring mRNA and Western blotting for analysis of protein synthesis forming the basis to any work. More sophisticated technologies are then employed depending on the questions to be answered, each with their different strengths and weaknesses as described in this chapter. These technologies include antisense and siRNA, to establish the requirement of a particular protein for signal transduction, immunoprecipitation to establish binding between pro-teins, and FRET to establish co-localization of signalling molecules. For a complete analysis of any particular signal transduction pathway it is anticipated that the majority of these techniques will be required.

Acknowledgement

The experiments reported were supported by grants from MIUR (PRIN 2004), FIRB (2004), ASI (2003) and Istituto Superiore di Sanità (1 per cent, 2003).

References

Acconcia, F., Ascenzi, P., Bocedi, A., Spisni, E. *et al.* (2005) 'Palmitoylation-dependent estrogen receptor alpha membrane localization: regulation by 17beta-estradiol', *Molec. Biol. Cell*, **16**(1), pp. 231–237.

Anderson, R. G. (1998) 'The caveolae membrane system', *Ann. Rev. Biochem.*, **67**, pp. 199–225.

Anderson, R. G. and Jacobson, K. (2002) 'A role for lipid shells in targeting proteins to caveolae, rafts, and other lipid domains', *Science*, **296**(5574), pp. 1821–1825.

Aoki, T., Nomura, R. and Fujimoto, T. (1999) 'Tyrosine phosphorylation of caveolin-1 in the endothelium', *Exp. Cell Res.*, **253**(2), pp. 629–636.

Ayala, G. E., Dai, H., Tahir, S. A., Li, R., Timme, T., Ittmann, M., Frolov, A., Wheeler, T. M., Rowley, D., Thompson, T. C. (2006) "Stromal antiapoptotic paracriue loop in perineural invasion of prostatic carcinoma", *Cancer Res.*, **66**(10), pp. 5159–5164.

Barak, Y., Liao, D., He, W., Ong, E. S. *et al.* (2002) 'Effects of peroxisome proliferator-activated receptor-δ on plancentation, adiposity and colorectal cancer', *Proc. Natl. Acad. Sci. USA*, **99**(1), pp. 303–308.

Bauer, P. M., Yu, J., Chen, Y., Hickey, R. *et al.* (2005) 'Endothelial-specific expression of caveolin-1 impairs microvascular permeability and angiogenesis', *Proc. Natl. Acad. Sci. USA*, **102**(1), pp. 204–209.

Bernatchez, P. N., Bauer, P. M., Yu, J., Prendergast, J. S. *et al.* (2005) 'Dissecting the molecular control of endothelial NO synthase by caveolin-1 using cell-permeable peptides', *Proc. Natl. Acad. Sci. USA*, **102**(3), pp. 761–766.

Bilderback, T. R., Grigsby, R. J. and Dobrowsky, R. T. (1997) 'Association of p75(NTR) with caveolin and localization of neurotrophin-induced sphingomyelin hydrolysis to caveolae', *J. Biol. Chem.*, **272**(16), pp. 10922–10927.

Breaker, R. R. (2004) 'Natural and engineered nucleic acids as tools to explore biology', *Nature*, **432**(7019), pp. 838–845.

Brouet, A., Sonveaux, P., Dessy, C., Moniotte, S. *et al.* (2001) 'Hsp90 and caveolin are key targets for the proangiogenic nitric oxide-mediated effects of statins', *Circ. Res.*, **89**(10), pp. 866–873.

Brouet, A., DeWever, J., Martinive, P., Havaux, X. *et al.* (2005) 'Antitumor effects of *in vivo* caveolin gene delivery are associated with the inhibition of the proangiogenic and vasodilatory effects of nitric oxide', *FASEB J.*, **19**(6), pp. 602–604.

Calipel, A., Lefevre, G., Pouponnot, C., Mouriaux, F. *et al.* (2003) 'Mutation of B-Raf in human choroidal melanoma cells mediates cell proliferation and transformation through the MEK/ERK pathway', *J. Biol. Chem.*, **278**(43), pp. 42409–42418.

Cao, H., Courchesne, W. E. and Mastick, C. C. (2002) 'A phosphotyrosine-dependent protein interaction screen reveals a role for phosphorylation of caveolin-1 on tyrosine 14: recruitment of C-terminal Src kinase', *J. Biol. Chem.*, **277**(11), pp. 8771–8774.

Carman, C. V., Lisanti, M. P. and Benovic, J. L. (1999) 'Regulation of G protein-coupled receptor kinases by caveolin', , *J. Biol. Chem.*, **274**(13), pp. 8858–8864.

Chang, S. H., Liu, C. H., Conway, R., Han, D. K., Nithipatikom, K., Trifan, O. C., Lane, T. F., Hla, T. (2004) "Role of prostaglandin E2-Dependent angiogenic switch in cyclooxygenase 2-induced breast cancer progression", *Proc. Natl. Acad. Sci. USA*. **101**(2), pp. 591–596.

Chen, Y., Stamatoyannopoulos, G. and Song, C. Z. (2003) 'Down-regulation of CXCR4 by inducible small interfering RNA inhibits breast cancer cell invasion *in vitro*', *Cancer Res.*, **63**(16), pp. 4801–4804.

Cohen, A. W., Razani, B., Wang, X. B., Combs, T. P. *et al.* (2003) 'Caveolin-1-deficient mice show insulin resistance and defective insulin receptor protein expression in adipose tissue', *Am. J. Physiol. Cell Physiol.*, **285**(1), pp. C222–235.

Cohen, A. W., Hnasko, R., Schubert, W. and Lisanti, M. P. (2004) 'Role of caveolae and caveolins in health and disease', *Physiol. Rev.*, **84**(4), pp. 1341–1379.

Cooke, J. P. and Losordo, D. W. (2002) 'Nitric oxide and angiogenesis', *Circulation*, **105**(18), pp. 2133–2135.

Couet, J., Li, S., Okamoto, T., Ikezu, T. and Lisanti, M. P. (1997) 'Identification of peptide and protein ligands for the caveolin-scaffolding domain. Implications for the interaction of caveolin with caveolae-associated proteins', *J. Biol. Chem.*, **272**(10), pp. 6525–6533.

Couet, J., Sargiacomo, M. and Lisanti, M. P. (1997) 'Interaction of a receptor tyrosine kinase, EGF-R, with caveolins. Caveolin binding negatively regulates tyrosine and serine/threonine kinase activities', *J. Biol. Chem.*, **272**(48), pp. 30429–30438.

Cutler, N. S., Graves-Deal, R., LaFleur, B., Gao, Z. *et al.* (2003) 'Stromal production of prostacyclin confers an antiapoptoic effect to colonic epithelial cells', *Cancer Res.*, **63**(8), pp. 1748–1751.

Czarny, M., Lavie, Y., Fiucci, G. and Liscovitch, M. (1999) 'Localization of phospholipase D in detergent-insoluble, caveolin-rich membrane domains. Modulation by caveolin-1 expression and caveolin-182-101', *J. Biol. Chem.*, **274**(5), pp. 2717–2724.

Dimmeler, S. and Zeiher, A. M. (2000) 'Endothelial cell apoptosis in angiogenesis and vessel regression', *Circ. Res.*, **87**(6), pp. 434–439.

Ding, Q., Stewart, J. Jr., Olman, M. A., Klobe, M. R. and Gladson, C. L. (2003) 'The pattern of enhancement of Src kinase activity on platelet-derived growth factor stimulation of glioblastoma cells is affected by the integrin engaged', *J. Biol. Chem.*, **278**(41), pp. 39882–39891.

Drab, M., Verkade, P., Elger, M., Kasper, M. *et al.* (2001) 'Loss of caveolae, vascular dysfunction, and pulmonary defects in caveolin-1 gene-disrupted mice', *Science*, **293**(5539), pp. 2449–2452.

Elbashir, S. M., Harborth, J., Lendeckel, W., Yalcin, A. *et al.* (2001) 'Duplexes of 21-nucleotide RNAs mediate RNA interference in cultured mammalian cells', *Nature*, **411**(6836), pp. 494–498.

Eliceiri, B. P. and Cheresh, D. A. (1998) 'The role of alphav integrins during angiogenesis', *Molec. Med.*, **4**(12), pp. 741–750.

Fernandez, I., Ying, Y., Albanesi, J. and Anderson, R. G. (2002) 'Mechanism of caveolin filament assembly', *Proc. Natl. Acad. Sci. USA*, **99**(17), pp. 11193–11198.

Fidler, I. J. and Ellis, L. M. (2004) 'Neoplastic angiogenesis – not all blood vessels are created equal', *New Eng. J. Med.*, **351**(3), pp. 215–216.

Fire, A., Xu, S., Montgomery, M. K., Kostas, S. A. *et al.* (1998) 'Potent and specific genetic interference by double-stranded RNA in C. elegans', *Nature*, **391**(66669), pp. 806–811.

Fra, A. M., Masserini, M., Palestini, P., Sonnino, S. and Simons, K. (1995) 'A photo-reactive derivative of ganglioside GM1 specifically cross-links VIP21-caveolin on the cell surface', *FEBS Letters*, **375**(1–2), pp. 11–14.

Fukuda, T., Chen, K., Shi, X. and Wu, C. (2003) 'PINCH-1 is an obligate partner of integrin-linked kinase (ILK) functioning in cell shape modulation, motility, and survival', *J. Biol. Chem.*, **278**(51), pp. 51324–51333.

Gaillard, S., Bartoli, M., Castets, F. and Monneron, A. (2001) 'Striatin, a calmodulin-dependent scaffolding protein, directly binds caveolin-1', *FEBS Letters*, **508**(1), pp. 49–52.

Garcia-Cardena, G., Fan, R., Stern, D. F., Liu, J. and Sessa, W. C. (1996) 'Endothelial nitric oxide synthase is regulated by tyrosine phosphorylation and interacts with caveolin-1', *J. Biol. Chem.*, **271**(44), pp. 27237–27240.

Glade-Bender, J., Kandel, J. J. and Yamashiro, D. J. (2003) 'VEGF blocking therapy in the treatment of cancer', *Expert Opin. Biol. Therapy*, **3**(2), pp. 263–276.

Glenney, J. R. Jr. and Soppet, D. (1992) 'Sequence and expression of caveolin, a protein component of caveolae plasma membrane domains phosphorylated on tyrosine in Rous sarcoma virus-transformed fibroblasts', *Proc. Natl. Acad. Sci. USA*, **89**(21), pp. 10517–10521.

Griffoni, C., Spisni, E., Santi, S., Riccio, M. *et al.* (2000) 'Knockdown of caveolin-1 by antisense oligonucleotides impairs angiogenesis *in vitro* and *in vivo*', *Biochem. Biophys. Res. Commun.*, **276**(2), pp. 756–761.

Griffoni, C., Toni, M., Spisni, E., Bianco, M. C. *et al.* (2003) 'The cellular prion protein: biochemistry, topology, and physiologic functions', *Cell Biochem. Biophys.*, **38**(3), pp. 287–304.

Gschwind, A., Hart, S., Fischer, O. M. and Ullrich, A. (2003) 'TACE cleavage of proamphiregulin regulates GPCR-induced proliferation and motility of cancer cells', *EMBO J.*, **22**(10), pp. 2411–2421.

Hatae, T., Wada, M., Yokoyama, C., Shimonishi, M. and Tanabe, T. (2001) 'Prostacyclin dependent apoptosis mediated by PPAR delta', *J. Biol. Chem.*, **276**(49), pp. 46260–46267.

He, H., Venema, V. J., Gu, X., Venema, R. C. *et al.* (1999) 'Vascular endothelial growth factor signals endothelial cell production of nitric oxide and prostacyclin through flk-1/KDR activation of c-Src', *J. Biol. Chem.*, **274**(35), pp. 25130–25135.

Henry, S. P., Marcusson, E. G., Vincent, T. M. and Dean, N. M. (2004) 'Setting sights on the treatment of ocular angiogenesis using antisense oligonucleotides', *Trends Pharmacol. Sci.*, **25**(10), pp. 523–527.

Hoffman, R. (2004) 'Do the signalling proteins for angiogenesis exist as a modular complexes?', *Med. Hypotheses*, **63**(4), pp. 675–680.

Jiang, H., Weyrich, A. S., Zimmerman, G. A. and McIntyre, T. M. (2004) 'endothelial cell confluence regulates cyclooxygenase-2 and prostaglandin E_2 production that modulate motility', *J. Biol. Chem.*, **279**(53), pp. 55905–55913.

Kasahara, K., Watanabe, Y., Yamamoto, T. and Sanai, Y. (1997) 'Association of Src family tyrosine kinase Lyn with ganglioside GD3 in rat brain. Possible regulation of Lyn by glycosphingolipid in caveolae-like domains', *J. Biol. Chem.*, **272**(47), pp. 29947–29953.

Kim, B., Tang, Q., Biswas, P. S., Xu, J. *et al.* (2004) 'Inhibition of ocular angiogenesis by siRNA targeting vascular endothelial growth factor-pathway genes; therapeutic strategy for herpetic stromal keratitis', *Am. J. Pathol.*, **165**(6), pp. 2177–2185.

Kim, J. H., Han, J. M., Lee, S., Kim, Y. *et al.* (1999) 'Phospholipase D1 in caveolae: regulation by protein kinase C alpha and caveolin-1', *Biochemistry*, **38**(12), pp. 3763–3769.

Kroll, J. and Waltenberger, J. (1997) 'The vascular endothelial growth factor receptor KDR activates multiple signal transduction pathways in porcine aortic endothelial cells', *J. Biol. Chem.*, **272**(51), pp. 32521–32527.

Labrecque, L., Royal, I., Surprenant, D. S., Patterson, C. *et al.* (2003) 'Regulation of vascular endothelial growth factor receptor-2 activity by caveolin-1 and plasma membrane cholesterol', *Molec. Biol. Cell*, **14**(1), pp. 334–347.

Leahy, K. M., Koki, A. T. and Masferrer, J. L. (2000) 'Role of cyclooxygenases in angiogenesis', *Curr. Med. Chem.*, **7**(11), pp. 1163–1170.

Le Gouill, S., Podar, K., Amiot, M., Hideshima, T. *et al.* (2004) 'VEGF induces MCL-1 upregulation and protects multiple myeloma cells against apoptosis', *Blood*, **104**(9), pp. 2886–2892.

Lee, M. S., Ko, S. G., Kim, H. P., Kim, Y. B. *et al.* (2004) 'Smad2 mediates Erk1/2 activation by TGF-β1 in suspended, but not in adherent, gastric carcinoma cells', *Int. J. Oncol.*, **24**(5), pp. 1229–1234.

Li, S., Couet, J. and Lisanti, M. P. (1996) 'Src tyrosine kinases, Galpha subunits, and H-Ras share a common membrane-anchored scaffolding protein, caveolin. Caveolin binding negatively regulates the auto-activation of Src tyrosine kinases', *J. Biol. Chem.*, **271**(46), pp. 29182–29190.

Lim, H. and Dey, S. K. (2002) Minirewiev: 'A novel pathway of prostacyclin signalling-hanging out with nuclear receptors', *Endocrinology*, **143**(9), pp. 3207–3210.

Lipscomb, E. A., Dugan, A. S., Rabinovitz, I. and Mercurio, A. M. (2003) 'Use of RNA interference to inhibit integrin (α6β4)-mediated invasion and migration of breast carcinoma cells', *Clinic Exp. Metastasis*, **20**(6), pp. 569–576.

Liou, J. Y., Deng, W. G., Gilroy, D. W., Shyue, S. K. and Wu, K. K. (2001) 'Co-localization and interaction cyclooxygenase-2 with caveolin-1 in human fibroblasts', *J. Biol. Chem.*, **276**(37), pp. 34975–34982.

Lisanti, M. P., Scherer, P. E., Tang, Z. and Sargiacomo, M. (1994) 'Caveolae, caveolin and caveolin-rich membrane domains: a signalling hypothesis', *Trends Cell Biol.*, **4**(7), pp. 231–235.

Lisanti, M. P., Tang, Z., Scherer, P. E., Kubler, E. *et al.* (1995) 'Caveolae, transmembrane signalling and cellular transformation', *Molec. Membrane Biol.*, **12**(1), pp. 121–124.

Liu, P., Ying, Y., Ko, Y. G. and Anderson, R. G. (1996) 'Localization of platelet-derived growth factor-stimulated phosphorylation cascade to caveolae', *J. Biol. Chem.*, **271**(17), pp. 10299–10303.

Liu, P., Ying, Y. and Anderson, R. G. (1997) 'Platelet-derived growth factor activates mitogen-activated protein kinase in isolated caveolae', *Proc. Natl. Acad. Sci. USA*, **94**(25), pp. 13666–13670.

Liu, J., Razani, B., Tang, S., Terman, B. I. *et al.* (1999) 'Angiogenesis activators and inhibitors differentially regulate caveolin-1 expression and caveolae formation in vascular endothelial cells. Angiogenesis inhibitors block vascular endothelial growth factor-induced down-regulation of caveolin-1' *J. Biol. Chem.*, **274**(22), pp. 15781–15785.

Liu, J., Wang, X. B., Park, D. S. and Lisanti, M. P. (2002) 'Caveolin-1 expression enhances endothelial capillary tubule formation', *J. Biol. Chem.*, **277**(12), pp. 10661–10668.

Lockwich, T. P., Liu, X., Singh, B. B., Jadlowiec, J. *et al.* (2000) 'Assembly of Trp1 in a signaling complex associated with caveolin-scaffolding lipid raft domains', *J. Biol. Chem.*, **275**(16), pp. 11934–11942.

Lu, P. Y., Xie, F. Y. and Woodle, M. C. (2003) 'siRNA-mediated antitumorigenesis for drug target validation and therapeutics', *Curr. Opin. Molec. Thera.*, **5**(3), pp. 225–234.

Lu, P. Y., Xie, F. Y. and Woodle, M. C. (2005) 'Modulation of angiogenesis with siRNA inhibitors for novel therapeutic', *Trends Molec. Med.*, **11**(3), pp. 104–113.

Massimino, M. L., Griffoni, C., Spisni, E., Toni, M. and Tomasi, V. (2002) 'Involvement of caveolae and caveolae-like domains in signalling, cell survival and angiogenesis', *Cell. Signalling*, **14**(2), pp. 93–98.

Mastick, C. C. and Saltiel, A. R. (1997) 'Insulin-stimulated tyrosine phosphorylation of caveolin is specific for the differentiated adipocyte phenotype in 3T3-L1 cells', *J. Biol. Chem.*, **272**(33), pp. 20706–20714.

Manara, M. C., Bernard, G., Lollioni, P. L., Nanni, P., Puntini, M., Landuzzi, L., Benini, S., Lattanti, G., Sciandra, M., Serra, M., Colombo, M. P., Bernard, A., Picci, P., Scotlandi, K.

(2006) "CD99 acts as an oncosuppressor in osteosarcoma", *Molec. Biol. Cell*, **17**(4), pp. 1910–1921.

Meister, G. and Tuschl, T. (2004) 'Mechanisms of gene silencing by double-stranded RNA', *Nature*, **431**(7006), pp. 343–349.

Minshall, R. D., Sessa, W. C., Stan, R. V., Anderson, R. G. and Malik, A. B. (2003) 'Caveolin regulation of endothelial function', *Am. J. Physiol. Lung Cellular Molec. Physiol.*, **285**(6), pp. 179–183.

Monia, B. P., Sasmor, H., Johnston, J. F., Freier, S. M. *et al.* (1996) 'Sequence-specific antitumor activity of a phosphorothioate oligodeoxyribonucleotide targeted to human c-raf kinase supports an antisense mechanism of action *in vivo*' *Proc. Natl. Acad. Sci. USA*, **93**(26), pp. 15481–15484.

Muller, G., Jung, C., Wied, S., Welte, S. *et al.* (2001) 'Redistribution of glycolipid raft domain components induces insulin-mimetic signaling in rat adipocytes', *Molec. Cellular Biol.*, **21**(14), pp. 4553–4567.

Nagy, P., Arndt-Jovin, D. J. and Jovin, T. M. (2003) 'Small interfering RNAs suppress the expression of endogenous and GFP-fused epidermal growth factor receptor (erbB1) and induce apoptosis in erbB1-overexpressing cells', *Exp. Cell Res.*, **285**(1), pp. 39–49.

Nanni, P., De Giovanni, C., Lollini, P. L., Nicoletti, G. and Prodi, G. (1983) 'TS/A: a new metastasizing cell line from a BALB/c spontenous mammary adenocarcinoma', *Clin. Exp Metastasis*, **1**(4), pp. 373–380.

Oxhorn, B. C. and Buxton, I. L. (2003) 'Caveolar compartmentation of caspase-3 in cardiac endothelial cells', *Cellular Signalling*, **15**(5), pp. 489–496.

Parton, R. G. (2001) 'Cell biology. Life without caveolae', *Science*, **293**(5539), pp. 2404–2405.

Peters, J. M., Lee, S. S., Li, W., Ward, J. M. *et al.* (2000) 'Growth, adipose, brain and skin alterations resulting from targeted disruption of the mouse peroxisome proliferator-activated receptor delta', *Molec. Cellular Biol.*, **20**(14), pp. 5119–5128.

Pulver, R. A., Rose-Curtis, P., Roe, M. W., Wellman, G. C. and Lounsbury, K. M. (2004) 'Store-operated Ca2+ entry activates the CREB transcription factor in vascular smooth muscle', *Circ. Res.*, **94**(10), pp. 1351–1358.

Razani, B. and Lisanti, M. P. (2001) 'Caveolin-deficient mice: insights into caveolar function human disease', *J. Clin. Invest.*, **108**(11), pp. 1553–1561.

Razani, B., Rubin, C. S. and Lisanti, M. P. (1999) 'Regulation of cAMP-mediated signal transduction via interaction of caveolins with the catalytic subunit of protein kinase A', *J. Biol. Chem.*, **274**(37), pp. 26353–26360.

Razani, B., Zhang, X. L., Bitzer, M., von Gersdorff, G. *et al.* (2001) 'Caveolin-1 regulates transforming growth factor (TGF)-beta/SMAD signaling through an interaction with the TGF-beta type I receptor' *J. Biol. Chem.*, **276**(9), pp. 6727–6738.

Razani, B., Combs, T. P., Wang, X. B., Frank, P. G. *et al.* (2002) 'Caveolin-1-deficient mice are lean, resistant to diet-induced obesity, and show hypertriglyceridemia with adipocyte abnormalities' *J. Biol. Chem.*, **277**(10), pp. 8635–8647.

Reich, S. J., Fosnot, J., Kuroki, A., Tang, W. *et al.* (2003) 'Small interfering RNA (siRNA) targeting VEGF effectively inhibits ocular neovascularization in a mouse model', *Molec. Vision*, **9**, pp. 210–216.

Riccio, M., Dembic, M., Cinti, C. and Santi, S. (2004) 'Multifluorescence labeling and colocalization analyses', *Methods Molec. Biol.*, **285**, pp. 171–177.

Robinson, G. S., Pierce, E. A., Rook, S. L., Foley, E. *et al.* (1996) 'Oligodeoxynucleotides inhibit retinal neovascularization in a murine model of proliferative retinopath', *Proc. Natl. Acad. Sci. USA*, **93**(10), pp. 4851–4856.

Rothberg, K. G., Heuser, J. E., Donzell, W. C., Ying, Y. S. *et al.* (1992) 'Caveolin, a protein component of caveolae membrane coats', *Cell*, **68**(4), pp. 673–682.

Sanceau, J., Truchet, S. and Bauvois, B. (2003) 'Matrix metalloproteinase-9 silencing by RNA interference triggers the migratory-adhesive switch in Ewing's sarcoma cells', *J. Biol. Chem.*, **278**(38), pp. 36537–36546.

Santoro, S. W. and Joyce, G. F. (1997) 'A general purpose RNA-cleaving DNA enzyme', *Proc. Natl. Acad. Sci. USA*, **94**(9), pp. 4262–4266.

Sargiacomo, M., Scherer, P. E., Tang, Z., Kubler, E. *et al.* (1995) 'Oligomeric structure of caveolin: implication for caveolae membrane organization', *Proc. Natl. Acad. Sci. USA*, **92**(20), pp. 9407–9411.

Sato, S., Fujita, N. and Tsuruo, T. (2004) 'Involvement of 3-phosphoinositide-dependent protein kinases-1 in the MEK/MAPK signal-transduction pathway', *J. Biol. Chem.*, **279**(32), pp. 33759–33767.

Scherer, L. J. and Rossi, J. J. (2003) 'Approaches for the sequence-specific knockdown of mRNA', *Nature Biotechnol.*, **21**(12), pp. 1457–1465.

Schlegel, A., Schwab, R. B., Scherer, P. E. and Lisanti, M. P. (1999) 'A role for the caveolin scaffolding domain in mediating the membrane attachment of caveolin-1. The caveolin scaffolding domain is both necessary and sufficient for membrane binding *in vitro*', *J. Biol. Chem.*, **274**(32), pp. 22660–22667.

Schlegel, A., Wang, C., Pestell, R. G. and Lisanti, M. P. (2001) 'Ligand-independent activation of oestrogen receptor alpha by caveolin-1', *Biochem. J.*, **359**(1), pp. 203–210.

Schreiber, S., Fleischer, J., Breer, H. and Boekhoff, I. (2000) 'A possible role for caveolin as a signaling organizer in olfactory sensory membranes', *J. Biol. Chem.*, **275**(31), pp. 24115–24123.

Schiffelers, R. M., Ansari, A., Xu, J., Zhou, Q. *et al.* (2004) 'Cancer siRNA therapy by tumor selective delivery with ligand-targeted sterically-stabilized nanoparticle', *Nucleic Acids Res.*, **32**(19), pp. e149.

Schubert, A. L., Schubert, W., Spray, D. C. and Lisanti, M. P. (2002) 'Connexin family members target to lipid raft domains and interact with caveolin-1', *Biochemistry*, **41**(18), pp. 5754–5764.

Shu, X., Wu, W., Mosteller, R. D. and Broek D. (2002) 'Sphingosine kinase mediates vascular endothelial growth factor-induced activation of ras and mitogen-activated protein kinases', *Molec. Cellular Biol.*, **22**(22), pp. 7758–7768.

Simmons, D. L., Botting, R. M., and Hla, T. (2004) 'Cyclooxygenase isozymes: the biology of prostaglandin synthesis and inhibition', *Pharmacol. Reviews*, **56**(3), pp. 387–437.

Smith, W. L., DeWitt, D. L. and Garavito, K. M. (2000) 'Cyclooxygenases: structural, cellular and molecular biology', *Ann. Rev. Biochem.*, **69**, pp. 145–182.

Smyth, E. M. and Fitzgerald, G. A. (2002) 'Human prostacyclin receptor', *Vitamins Hormones*, **65**, pp. 149–165.

Spagnuolo, R., Corada, M., Orsenigo, F., Zanetta, L. *et al.* (2004) 'Gas1 is induced by VE-cadherin and vascular endothelial growth factor and inhibits endothelial cell apoptosis', *Blood*, **103**(8), pp. 3005–3012.

Spisni, E. and Tomasi, V. (1997) 'Involvement of prostanoids in angiogenesis', in *Tumor Angiogenesis*, Bicknell, R., Lewis, C. E. and Ferrara, N, eds, pp. 291–300.

Spisni, E., Bartolini, G., Orlandi, M., Belletti, B. *et al.* (1995) 'Prostacyclin (PGI2) synthase is a constitutively expressed enzyme in human endothelial cells', *Exp. Cell Res.*, **219**(2), pp. 507–513.

Spisni, E., Griffoni, C., Santi, S., Riccio, M. *et al.* (2001a) 'Colocalization prostacyclin (PGI2) synthase-caveolin-1 in endothelial cells and new roles for PGI2 in angiogenesis', *Exp. Cell Res.*, **266**(1), pp. 31–43.

Spisni, E., Griffoni, C., Santi, S., Riccio, M. *et al.* (2001b) 'Co-localization of PGI_2 synthase and caveolin-1 in endothelial cells underscores, new roles of PGI_2 in angiogenesis', in *Advances in Prostaglandin and Leukotrienes Research, Basic Science and New Clinical Applications*, Samuelsson, B, Paoletti, R, Folco, G. C., Granstrom, E. and Nicosia, S., eds, Kluwer Academic Publisher, pp. 139–142.

Spisni, E., Bianco, M. C., Griffoni, C., Toni, M. *et al.* (2003) 'Mechanosensing role of caveolae and caveolar constituents in human endothelial cells', *J. Cell. Physiol.*, **197**(2), pp. 198–204.

Stahlhut, M., Sandvig, K. and Van Deurs, B. (2000) 'Caveolae: uniform structures with multiple functions in signaling, cell growth, and cancer', *Exp. Cell Res.*, **261**(1), pp. 111–118.

Strillacci, A., Griffoni, C., Spisni, E., Manara, M. C., Tomasi, V. (2006) "RNA interference as a key to knockdown overexpressed cyclooxygenase-2 gene in tumour cells", *British J Cancer*, **94**(9), pp. 1300–1310.

Stephenson, M. L. and Zamecnik, P. C. (1978) 'Inhibition of Rous sarcoma viral RNA translation by a specific oligodeoxyribonucleotide' *Proc. Natl. Acad. Sci. USA*, **75**(1), pp. 285–288.

Takei, Y., Kadomatsu, K., Yuzawa, Y., Matsuo, S. and Muramatsu, T. (2004) 'A small interfering RNA targeting vascular endothelial growth factor as cancer therapeutics', *Cancer Res.*, **64**(10), pp. 3365–3370.

Tan, C., Cruet-Hennequart, S., Troussard, A., Fazli, L. *et al.* (2004) 'Regulation of tumor angiogenesis by integrin-linked kinase (ILK)', *Cancer Cell*, **5**(1), pp. 79–90.

Tang, Z., Okamoto, T., Boontrakulpoontawee, P., Katada, T. *et al.* (1997) 'Identification, sequenc, and expression of an invertrebate caveolin gene family from the nematode Caenarhabditis elegans. Implication for the molecular avolution of mammalian caveolin genes', *J. Biol. Chem.*, **272**(4), pp. 2437–2445.

Thamilselvan, V. and Basson, M. D. (2004) 'Pressure activates colon cancer cell adhesion by inside-out focal adhesion complex and actin cytoskeletal signaling', *Gastroenterology*, **126**(1), pp. 8–18.

Tomasi, V., Mastacchi, R., Bartolini, G., Fadda, S. *et al.* (1984) 'The relationships between the high production of prostaglandins by tumors and their action on lymphocytes as suppressive agents', in: *Genetic and Phenotypic Markers of Tumors*, Aaronson, S. A., Frati, L. and Verona, R., eds, Plenum Press, pp. 235–260.

Tomasi, V., Spisni, E., Griffoni, C. and Guarnieri, T. (2000) 'Caveolae, caveolar enzymes and angiogenesis', *Curr. Topics Biochem. Res.*, **3**, pp. 81–90.

Toni, M., Spisni, E., Griffoni, C., Santi, S., Riccio, M., Lenaz, P., and Tomasi, V. (2006) "Cellular prion protein and caveolin-1 interaction in a neuronal cell line precedes Fyn/Erk 1/2 signal transduction", *J Biomed Biotech*, in press.

Trigatti, B. L., Anderson, R. G. and Gerber, G. E. (1999) 'Identification of caveolin-1 as a fatty acid binding protein', *Biochem. Biophys. Res. Commun.*, **255**(1), pp. 34–39.

Tsuruta, D. and Jones, J. C. (2003) 'The vimentin cytoskeleton regulates focal contact size and adhesion of endothelial cells subjected to shear stress', *J. Cell Sci.*, **116**(24), pp. 4977–4984.

Uittenbogaard, A., Ying, Y. and Smart, E. J. (1998) 'Characterization of a cytosolic heat-shock protein-caveolin chaperone complex. Involvement in cholesterol trafficking', *J. Biol. Chem.*, **273**(11), pp. 6525–6532.

Vinores, S. A., Seo, M. S., Okamoto, N., Ash, J. D. *et al.* (2000) 'Experimental models of growth factor-mediated angiogenesis and blood–retinal barrier breakdown', *General Pharmacol.*, **35**(5), pp. 233–239.

Von Offenberg Sweeney, N., Cummins, P. M., Cotter, E. J., Fitzpatrick, P. A. *et al.* (2004) 'Cyclic strain-induced endothelial MMP-2: role in vascular smooth muscle cell migration', *Biochem. Biophys. Res. Commun.*, **329**(2), pp. 325–333.

Wang, A., Nomura, M., Patan, S. and Ware, J. A. (2002) 'Inhibition of protein kinase Calpha prevents endothelial cell migration and vascular tube formation *in vitro* and myocardial neovascularization *in vivo*', *Circulation Res.*, **90**(5), pp. 609–616.

Wang, D. and DuBois, R. (2004) 'Cyclooxygenase-2 derived prostaglandin E_2 regulates the angiogenic switch', *Proc. Natl. Acad. Sci. USA*, **101**(2), pp. 415–416.

Wary, K. K., Mariotti, A., Zurzolo, C. and Giancotti, F. G. (1998) 'A requirement for caveolin-1 and associated kinase Fyn in integrin signaling and anchorage-dependent cell growth', *Cell*, **94**(5), pp. 625–634.

Wei, Y., Lukashev, M., Simon, D. I., Bodary, S. C. *et al.* (1996) 'Regulation of integrin function by the urokinase receptor', *Science*, **273**(5281), pp. 1551–1555.

Wei, Y., Yang, X., Liu, Q., Wilkins, J. A. and Chapman, H. A. (1999) 'A role for caveolin and the urokinase receptor in integrin-mediated adhesion and signaling', *J. Biol. Chem.*, **144**(6), pp. 1285–1294.

Wickstrom, S. A., Alitalo, K. and Keski-Oja, J. (2003) 'Endostatin associates with lipid rafts and induces reorganization of the actin cytoskeleton via down-regulation of RhoA activit', *J. Biol. Chem.*, **278**(39), pp. 37895–37901.

Williams, T. M. and Lisanti, M. P. (2005) 'Caveolin-1 in oncogenic transformation, cancer, and metastasis', *Am. J. Physiol. Cell Physiol.*, **288**(3), pp. C494–506.

Williams, T. M., Medine, F., Badaw, I., Hazan, R. B. *et al.* (2004) 'Caveolin-1 gene disruption promotes mammary tumorigenesis and dramatically enhances lung metastasis *in vivo*. Role of cav-1 in cell invasiveness and matrix metalloproteinase (MMP-2/9) secretion', *J. Biol. Chem.*, **279**(49), pp. 51630–51646.

Wu, D., Foreman, T. L., Gregory, C. W., McJilton, M. A. *et al.* (2002) 'Protein kinase cepsilon has the potential to advance the recurrence of human prostate cancer', *Cancer Res.*, **62**(8), pp. 2423–2429.

Wu, J., Richer, J., Horwitz, K. B. and Hyder, S. M. (2004) 'Progestin-dependent induction of vascular endothelial growth factor in human breast cancer cells: preferential regulation by progesterone receptor B', *Cancer Res.*, **64**(6), pp. 2238–2244.

Xie, F. Y. (2004) 'Delivering siRNA to animal disease models for validation of novel drug targets *in vivo*', *PharmaGenomics*, July/August, pp. 28–31.

Xu, J., Lai, Y. J., Lin, W. C. and Lin, F. T. (2004) 'TRIP6 enhances lysophosphatidic acid-induced cell migration by interacting with the lysophosphatidic acid 2 receptor', *J. Biol. Chem.*, **279**(11), pp. 10459–10468.

Yamamoto, M., Toya, Y., Schwencke, C., Lisanti, M. P. *et al.* (1998) 'Caveolin is an activator of insulin receptor signaling', *J. Biol. Chem.*, **273**(41), pp. 26962–26968.

Yamamoto, M., Toya, Y., Jensen, R. A. and Ishikawa, Y. (1999) 'Caveolin is an inhibitor of platelet-derived growth factor receptor signaling', *Exp. Cell Res.*, **247**(2), pp. 380–388.

Yancopoulos, G. D., Davis, S., Gale, N. W., Rudge, J. S. *et al.* (2000) 'Vascular-specific growth factors and blood vessel formation', *Nature*, **407**(6801), pp. 942–948.

Yang, G., Cai, K. Q., Thompson-Lanza, J. A., Bast, R. C. Jr. and Liu, J. (2004) 'Inhibition of breast and ovarian tumor growth through multiple signaling pathways by using retrovirus-mediated small interfering RNA against Her-2/neu gene expression', *J. Biol. Chem.*, **279**(6), pp. 4339–4345.

Yu, Y. and Sato, J. D. (1999) 'MAP kinases, phosphatidylinositol 3-kinase, and p70 S6 kinase mediate the mitogenic response of human endothelial cells to vascular endothelial growth factor', *J. Cellular Physiol.*, **178**(2), pp. 235–246.

Zamecnik, P. C. and Stephenson, M. L. (1978) 'Inhibition of Rous sarcoma virus replication and cell transformation by a specific oligodeoxynucleotide', *Proc. Natl. Acad. Sci. USA*, **75**(1), pp. 280–284.

Zambrowicz, B. P. and Sands, A. T. (2003) 'Knockouts model the 100 best-selling drugs–will they model the next 100?', *Nature Reviews. Drug Discovery*, **2**(1), pp. 38–51.

Zhang, L., Fogg, D. K. and Waisman, D. M. (2004) 'RNA interference-mediated silencing of the S100A10 gene attenuates plasmin generation and invasiveness of Colo 222 colorectal cancer cells' *J. Biol. Chem.*, **279**(3), pp. 2053–2062.

Zhou, X., Murphy, F. R., Gehdu, N., Zhang, J. *et al.* (2004) 'Engagement of $\alpha v \beta 3$ integrin regulates proliferation and apoptosis of hepatic stellate cells', *J. Biol. Chem.*, **279**(23), pp. 23996–24006.

9
Implantation of sponges and polymers

Silvia P. Andrade, Monica A. N. D. Ferreira and **Tai-Ping Fan**

Abstract

Cell recruitment, inflammation, angiogenesis and matrix deposition are key elements in chronic diseases (rheumatoid arthritis, psoriasis, cancer). These components can be induced using a variety of implantation techniques and biomaterials providing a microenvironment analogous to the process of repair that follows injury. The response to injury represented by the implant can be modulated by a range of potential therapeutic compounds and assessed by a number of variables including the formation of new blood vessels, infiltration of inflammatory cells, assay of relevant cytokines and determination of extracellular matrix components. In addition, implantation techniques have been used to host a variety of normal or diseased cells for the identification of early events associated with disease progression. As a direct consequence of recent advances in genetic manipulation, mouse models (i.e. knockouts, SCID, nude) have provided resources to delineate the mechanisms regulating the healing associated with implants. Here we outline methods of sponge and polymer implantation techniques and the usefulness of this methodology to study tumour or inflammatory angiogenesis in normal and genetically modified experimental animals.

Keywords
sponge implant; polymers; blood flow

9.1 Introduction

Research on angiogenesis was initiated with the development of several techniques permitting direct observations of the microvasculature in the living

Angiogenesis Assays Edited by Carolyn A. Staton, Claire Lewis and Roy Bicknell
© 2006 John Wiley & Sons, Ltd

animal. The response observed after the introduction of an appropriate stimulus, such as mechanical injury or injection of neoplastic or normal tissue implants, has allowed the cataloguing of a number of molecules and cells involved in the vascularization of normal repair or neoplastic tissue. A variety of implanted devices have been used to induce fibrovascular tissue. Such implants are particularly useful because they offer scope for modulating the environment within which angiogenesis occurs. The materials utilized have included stainless steel mesh chambers (Schilling *et al.*, 1959), hollow chambers with porous walls (Sprugel *et al.*, 1987; Dvorak *et al.*, 1987) and synthetic sponge matrix (polyvinyl alcohol, cellulose acetate, polyester and polyurethane) implants.

The realization that advances in angiogenesis research depended on making the assays more quantitative, and reproducible *in vitro* and *in vivo*, led to the development of new techniques and improvement of old or current models to comply with such requirements. For example, a laminin-rich reconstituted matrix extracted from Engelbreth-Holm-Swarm tumour composed of basement membrane components (Kleinman *et al.*, 1987) was developed as a Matrigel-plug assay to induce and/or to study angiogenesis (Passaniti *et al.*, 1992). Recent improvements to the assay involve encapsulation of the Matrigel in a plexiglass chamber or with flexible plastic tubing before subcutaneous implantation (Kragh *et al.*, 2003; Ley *et al.*, 2004; Baker *et al.*, 2006). This model was further improved by using alginate implants encapsulating tumour cells and injecting FITC-dextran into the animals. Quantification of the FITC-dextran found within the implant has been considered an improvement over the more traditional method, in which haemoglobin content was used to determine angiogenesis (Plunkett and Hailey, 1990; Hoffmann *et al.*, 1997).

Taking advantage of a wealthy body of information provided by implantation technique this experimental system model has further been optimized and adapted to characterize essential components and their roles in blood vessel formation in a variety of physiological and pathological conditions. A cannulated sponge model was described in 1987 (Andrade *et al.*, 1987) in which direct blood flow measurement was achieved by a ^{133}Xe clearance technique (Figure 9.1) This technique was later modified to assess inflammatory and tumour angiogenesis in wild and genetically modified mice by means of a fluorimetric diffusion method (Andrade *et al.*, 1997; Lage and Andrade, 2000; Ferreira *et al.*, 2004). In addition, at present, there is considerable interest in implantation techniques and scaffolds for tissue engineering due to their potential use in controlling tissue regeneration and repair. The National Institute of Health has estimated that 8–10 per cent of the American population has permanent medical implants (Kidd *et al.*, 2001). For this type of material, inflammation, angiogenesis and fibrosis must be determined for their biocompatibility. It is essential that they permit, and even stimulate host cells to rebuild functional tissue but, at the same time, do not modify the endogenous healing process.

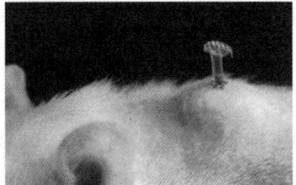

Implant in situ

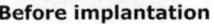

Before implantation   After implantation

Figure 9.1   Frontal view of a cannulated sponge disc *in situ* note that the cannula is exteriorized from, and immobilized perpendicularly to dorsal skin by sutures. Substances to be tested are injected directly into the sponge via the cannula. The cannulated sponge disc is shown before implantation (empty) and 14 days after implantation (fibrovascular tissue). (A colour reproduction of this figure can be viewed in the colour section towards the centre of the book).

9.2   Materials used in implantation techniques

A variety of implanted sponges and polymer matrices have been used to induce and study inflammatory angiogenesis . These include polyether polyurethane foam (Vitafoam Ltd., Manchester, UK), polyvinyl sponge (Rippey, Eldorado Hills, California, USA), polyvinyl acetyl (M-Pact Wordwide Management Corp, Eudora, KA, USA). As an experimental system, these materials provide an environment of defined dimensions allowing the invasion of various repair cells and the formation of fibrovascular tissue. However, the implants commonly become encapsulated and elicit the formation of fused macrophages known as foreign body giant cells. These cells, in turn, secrete a number of cytokines that markedly influence angiogenesis in the inflamed environment. Pre-treatment of the animals or implants may selectively inhibit this unwanted inflammatory response. The use of alginate polymers has also been extensively used to study inflammatory as well as tumour angiogenesis (Plunkett and Hailey, 1990; Hoffmann *et al.*, 1997). Matrigel, a widely used implantation matrix, is by its nature composed of key elements of the basement membrane matrix (collagen IV, laminin, entactin and heparin sulphate proteoglycan) providing a substrate on which endothelial cells can migrate, proliferate and form blood vessels. In addition, a range of growth factors have been identified in Matrigel (fibroblast growth factor, epidermal growth factor, insulin-like growth factor, platelet derived growth factors, and transforming

growth factor). Thus, extrapolation of results obtained from the analysis of a
Matrigel plug should be made with caution (Vukicevic *et al.*, 1992).

9.3 Implantation technique for assessment of inflammatory processes

Subcutaneous implantation of synthetic matrices to induce an inflammatory
response and to study repair processes is an old surgical procedure. Grindlay
and Waugh (1951) and later (Edwards *et al.*, 1960) were the first to use polyvinyl
alcohol sponge implants, in dogs and rabbits, respectively, as a framework for the
in-growth of vascularized connective tissue. These studies provided valuable
qualitative information on neovascularization and wound healing. However, this
technique was not limited to qualitative analysis and the sponge matrix model was
also used to determine biochemical variables of the fibrovascular tissue including
collagen metabolism (Paulini *et al.*, 1974), fibronectin deposition and proteoglycan
turnover (Bollet *et al.*, 1958). Additionally, the technique was employed to
characterize the sequence of histological changes in granulation tissue formation
(Holund *et al.*, 1979) and to monitor the kinetics of cellular proliferation
(Davidson *et al.*, 1985).

The sponge implant technique has also been modified to study more chronic
inflammatory responses and the evolution of the granulation tissue (Ford-
Hutchinson *et al.*, 1977). It is also possible to measure the extent of neutrophil
and macrophage accumulation in the sponge compartment by assaying the
inflammatory enzymes, myeloperoxidase (MPO) and N-acetylglucosaminidase
(NAG) (Bailey, 1988; Belo *et al.*, 2004a; Ferreira *et al.*, 2004). This model of
acute inflammation is particularly useful, allowing the collection and examination
of both cellular and fluid phases of the exudate formed within the sponge. The
effects of various anti-inflammatory agents on leukocyte migration and production
of inflammatory mediators, angiogenic factors, cytokines and chemokines have
been determined in fluids of the cannulated sponge implant (Table 9.1). In addition,
the potential inflammatory effect of a variety of agents has been determined by
soaking the implants prior or after host implantation (Iuvone *et al.*, 1994).

9.4 Implantation technique for assessment of systemic pathological conditions

The intensity and speed of the tissue response elicited by the sponge depends on
several factors including the size and type of sponge material (Salvatore *et al.*,
1961; Holund *et al.*, 1979). Factors such as species, strain, age, sex and nutritional
state of the animal have also been demonstrated to influence the tissue response by
the sponge implants (Noble and Boucek, 1955; Holund *et al.*, 1979; Reed *et al.*,
1998). More recently, sponge implants have been used to assess the influence of

Table 9.1 Inflammatory mediators, angiogenic factors, cytokines and chemokines detected in sponge fluids

Prostaglandins
Interleukin 1
Interleukin 6
Interleukin 8
Tumour necrosis factor-α
Vascular endothelial growth factor
Fibroblast growth factor
CXCL1-3/KC
CCL2/JE
Transforming growth factor- beta

diabetes and tumour growth on healing processes (Teixeira and Andrade, 1999; Belo *et al.*, 2004b). Using this approach, a relationship between the magnitude of hyperglycaemia and the severity of angiogenesis inhibition and granulation tissue formation was clearly demonstrated in subcutaneous implanted sponges of streptozotocin-induced diabetic rats (Teixeira and Andrade, 1999). Experiments examining the influence of a growing tumour over a proliferating fibrovascular tissue induced by a sponge have shown that this interaction resulted in suppression of inflammatory angiogenesis and altered production of chemokines in the implant microenvironment (Belo *et al.*, 2004b).

9.5 Implantation technique for assessment of tumour angiogenesis

Sponge implantation has also been used as a framework to host rodent and human cell lines (Mahadevan *et al.*, 1989, Andrade *et al.*, 1992; Lage and Andrade, 2000). The advantage of implantation techniques for the purpose of investigating tumour-induced angiogenesis is that the assessment of the relative contributions of the tumour cells to early changes in the implant blood-flow can be detected even before visible growth of the tumour mass is evident. Using the ^{133}Xe clearance technique the development of Colon 26 or melanoma B16 tumours has been determined. By using the fluorescein diffusion method the development of Ehrlich tumour has been evaluated regarding solid tumour haemodynamic features. The functional parameter expressed in terms of half-time ($t_{1/2}$; time taken for the fluorescence to reach 50 per cent of the peak in the systemic circulation), which is inversely proportional to blood flow, showed that in the tumour-free implants $t_{1/2}$ values decreased from 11. 55 ± 1 min at day 1 to 5. 7 ± 0. 44 min by day 14. In the tumour-bearing implants, this process was accelerated and maximum vascularization was achieved by day 7 (3 days after tumour cell inoculation). Increases in $t_{1/2}$ values were observed at days 10 and 14, which paralleled the tumour growth as indicated by wet weight. The results of these experiments indicate the suitability of the fluorescein diffusion technique for analysing angiogenesis induced by tumour cells and several haemodynamic features of Ehrlich tumour growth in conscious animals.

9.6 Implantation technique for assessment of inflammatory angiogenesis in genetically modified mice

The development of genetic animal models with selective expression or deletion of a range of essential proteins provides an experimental tool for assessing the contribution of individual molecules regulating the healing associated with implants. Taking advantage of the well established angiogenic and fibrogenic responses to a variety of implant material in wild-type mice, the sponge model has been adopted to investigate the role of the matricellular molecule, thrombospondin 2 (Kyriakides *et al.*, 2001), of platelet endothelial cell adhesion molecule-1 (PECAM-1, CD31) (Solowiej *et al.*, 2003), of the inflammatory mediator platelet activating factor (Ferreira *et al.*, 2004), and of the cytokine tumour necrosis factor alpha (Barcelos *et al.*, 2005). In addition, the sponge implant model has also been used to study gene transfer techniques to stimulate therapeutic angiogenesis (Wang *et al.*, 2000). The results of these experiments are sometimes puzzlingly or even contradictory with the well established role of the molecules indicating the need for more than one method of assessment.

9.7 Applications of cannulated sponge model for testing angiogenesis modulators

Over the past 20 years, we have used the cannulated sponge model to study the roles of a variety of inflammatory mediators, cytokines and growth factors in angiogenesis. In addition, the model offers a robust system for testing potential angiogenesis modulators (Table 9.2). It has become clear that the presence of the cannula is an important feature for successful neovascularization perhaps by providing a cut surface in the skin and thus a chronic wound. This is apart from its convenience to make localized injections accurately. Most of the compounds (cytokines) injected in the sponge compartment have been described to be involved in or to be of potential therapeutic value to either inflammatory or tumor angiogenesis.

9.8 Techniques for assessment of the vascularization in implants

Quantitative analysis of various components of the neovasculature in the implants relies on determining the number of endothelial cells and supporting vascular

Table 9.2 Substances shown to be angiogenic or anti-angiogenic in the cannulated sponge model

Test substance	References
ANGIOGENIC	
Peptides	
Substance P	Fan *et al.*, 1993
$[Sar^9, Met(O_2)^{11}]$substance P (NK_1 agonist)	
Bradykinin	Hu and Fan, (1993)
Angiotensin II	Andrade *et al.*, 1996; Walsh *et al.*, 1997
Vasoactive intestinal peptide	Hu *et al.*, 1996
Calcitonin gene related peptide	Hu *et al.*, 1996
Interleukin 1-α	Hu *et al.*, (1994)
Interleukin-8	Hu *et al.*, 1993
Tumour necrosis factor-α	Hu *et al.*, 1994
Vascular endothelial growth factor	Hu and Fan, 1995b
Basic fibroblast growth factor	Hu and Fan, 1995b; Andrade *et al.*, 1997
Hepatocyte growth factor	Sengupta *et al.*, 2003a
Laminin and fibrinogen	Mahadevan *et al.*, 1989
Fatty acids	
Platelet activating factor	Andrade *et al.*, 1992
Enzymes	
Thymidine phosphorylase	Moghaddam *et al.*, 1995
Polysaccharides	
Oligosaccharides of hyaluronan (1 to 4 kDa)	Lees *et al.*, 1995
Plant-derived materials	
Ginsenoside Rg_1	Sengupta *et al.*, (2004)
ANTI-ANGIOGENIC	
Neurokinin NK_1 receptor antagonists: selective peptide NK_1 receptor antagonist, L-668,169 selective non-peptide NK_1 receptor antagonist, RP 67580 and CP-96,345	Fan *et al.*, 1993
Angiotensin AT_1 receptor antagonist – losartan	Walsh *et al.*, 1997
Bradykinin B_1 receptor antagonist – $[Leu^8]des-Arg^9$-BK	Hu and Fan, 1993
Peptides	
IL-1 receptor antagonist	Hu *et al.*, 1994
VEGF-related molecules	
VEGF antibody	Hull *et al.*, 2003
Soluble truncated VEGF receptor (flt-1)	
Plant-derived substances	
Ginsenoside Rb_1	Sengupta *et al.*, 2004
Small molecular mass compounds	
Suramin and related polyanions	Braddock *et al.*, 1994
Protein kinase C inhibitor – calphostin C	Hu and Fan, 1995a
Tyrosine kinase inhibitor – lavendustin A	Hu and Fan, 1995b
Dexamethasone, hydrocortisone	Hori *et al.*, 1996; Lage and Andrade, 2000

(continued)

Table 9.2 *(Continued)*

Test substance	References
Angiostatic steroids: Tetrahydro-S, U-24067	
COX-2-selective inhibitors: Meloxicam, NS398	Sengupta *et al.*, 2003b
Clotrimazole	Belo *et al.*, 2004a
Thalidomide	
Rosiglitazone	
NON-ANGIOGENIC	
Histamine	Hu *et al.*, 1996
Endothelin 1	Hu *et al.*, (1996)
Endothelin 3	
Oligosaccharides of hyaluronan (33 kDa)	Lees VC *et al.*, 1995

elements, such as fibroblasts, pericytes, smooth muscle cells and the products of their activation, e.g. growth factors, all of which provide valuable information about the vascular component induced by the implant. However, information on the functional state of the neovasculature, the development of vasoactive regulatory systems and pharmacological reactivity of the newly formed blood vessels has been considered particularly relevant for therapeutic analysis.

Functional assessment of blood vessels

133Xe clearance

This technique is based on the principle that the amount of a locally deposited radioactive tracer decreases at a rate proportional to the blood flow at the site of the injection. The decrease in radioactivity is exponential and $t_{1/2}$ (time taken for the radioactivity to fall to 50 per cent of its original value) for the washout is inversely related to the local blood flow (Andrade *et al.*, 1987; Mahadevan *et al.*, 1990). Further studies showed a good correlation between ^{133}Xe clearance technique and the measurement of absolute blood flow in sponges using ^{113}Sn microspheres, with the amount of the neovasculature by the carmine red dye method, with the amount of haemoglobin in the implants and with the vascular density assessed by morphometric analysis (Hu *et al.*, 1995). A particular advantage of the ^{133}Xe method is that it allows non-destructive, and thus repeated, measurements of blood flow in the same animal over the period of neovascularization of the sponge. As a result it requires fewer animals and also allows an estimate of variability in individual animals. The radioactive nature of the tracer and cost are the main disadvantages of this technique.

Fluorescein diffusion method in conscious mice

The use of fluorescent dyes has long been employed in the clinic, especially in examination of ocular neovascularization based on the property of these vessels to be particularly leaky (Auerbach *et al.*, 1991). Using fundus fluorescein leakage angiography, ocular angiogenesis in a primate model has been investigated (Miller *et al.*, 1994). Furthermore, Lentner and Wienert (1996) have successfully replaced radioactive tracers to quantify veno-lymphatic drainage using the outflow rate of fluorescein dye injected intradermally.

The principle underlying the fluorimetric assay is that measurement of fluoro-chrome-generated emission in the bloodstream following the application of a fluorescent dye in a vascular compartment reflects the local blood flow (Andrade *et al.*, 1997). Using this technique, angiogenesis and tumour growth in conscious mice have been determined in subcutaneous implanted sponges bearing tumour cells (Lage and Andrade, 2000). It is relevant to point out that experiments for evaluation of blood flow performed in conscious animals is of particular advantage since anaesthesia has been considered a major problem in analysing tumour blood flow.

Histology, morphometric assessment, and quantitative *in vitro* receptor autoradiography of implant fibrovascular tissue

A range of histological techniques used in many different tissues have been adapted to study various events during the development of the fibrovascular tissue induced by the implants. For some types of implants (Matrigel, polyvinyl acetyl), pan-endothelial markers, such as CD31 and ICAM-2 have been used and successfully correlated with other parameters of vascularization (Solowiej *et al.*, 2003; Baker *et al.*, 2006). Other types of implants cross-react with the immuno-histochemistry procedure/reagents preventing the use of endothelial cell markers. Using AgNOR staining, a marker for proliferation and cellular activity, Picrosirius staining for collagen determination and TUNEL for apoptosis, we have character-ized the kinetics of the fibrovascular tissue infiltration in the process of implant host-interface (Figure 9.2).

Morphometric analysis of marked structures has allowed more objective quantification of several components (blood vessels, collagen, laminin, inflamma-tory cells) of the fibrovascular tissue induced by the implants. The sequential development of vasoactive regulatory systems (innervation, neurokinin and angio-tensin receptors) in sponge implants has also been determined using quantitative *in vitro* receptor autoradiography (Walsh *et al.*, 1996; Walsh *et al.*, 1997).

These analysis methods were used to follow the innervation and neurokinin receptors during sponge-induced angiogenesis in the rat (Walsh *et al.*, 1996). The fraction of endothelial cells immunoreactive for proliferating cell nuclear antigen, endothelial fractional area, and ^{133}Xe clearance were used as measures of endothelial proliferation, neovascularization and blood flow, respectively. Endothelial proliferation occurred predominantly in tissues surrounding the sponge, and peaked before neovascularization of sponge stroma and the establishment of sponge blood flow. Substance P-containing sensory nerves and specific, high affinity substance P binding sites with characteristics of neurokinin receptors of the NK$_1$ subclass, were localized to microvessels surrounding the sponge at all time points. Lower density substance P binding sites were localized to newly formed microvessels within the sponge stroma, progressively increasing in density from day 4 to day 14. Nerve fibres were observed in the stroma of only two of six sponges at day 14, and none at earlier time points. These data support the hypothesis that substance P-enhanced

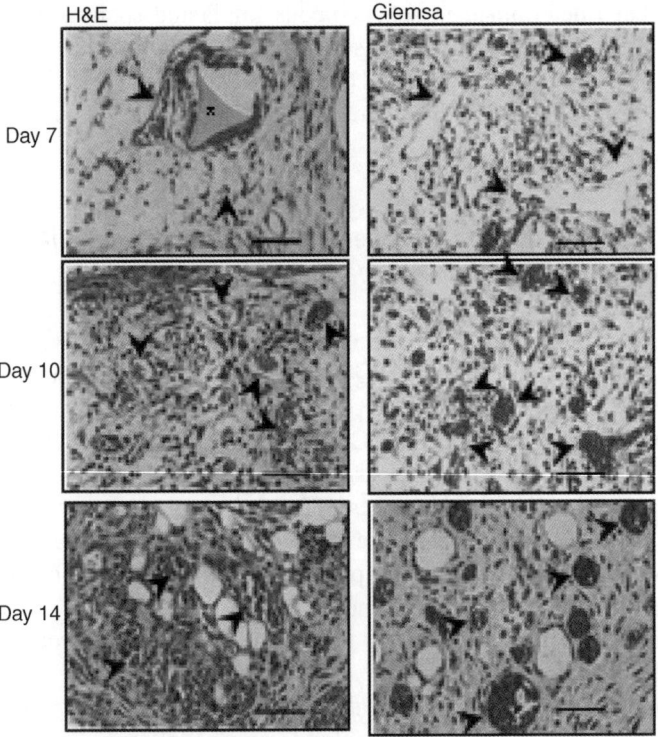

Figure 9.2 Granulation tissue induced by cannulated sponge implants representative histological sections (5 μm) of fibrovascular granulation tissue induced by cannulated sponge implants at days 7, 10 and 14 staining with H&E, Giemsa (collagen and fibrovascular infiltration), Dominici (mast cells), TUNEL (apoptosis), AgNOR (proliferation) Picrosirius (thin collagen –green, and thick collagen –red). The pores of the sponge matrix, seen as triangular shapes, are initially filled with a fibrinous network and numerous polymorphonuclear leukocytes. The matrix is progressively infiltrated by inflammatory cells, spindle-shaped like fibroblasts, endothelial cells and blood vessels embedded in a dense organized matrix. Bar = 45 μm; arrows, blood vessels; * matrix; triangles, mast cells. (A colour reproduction of this figure can be viewed in the colour section towards the centre of the book).

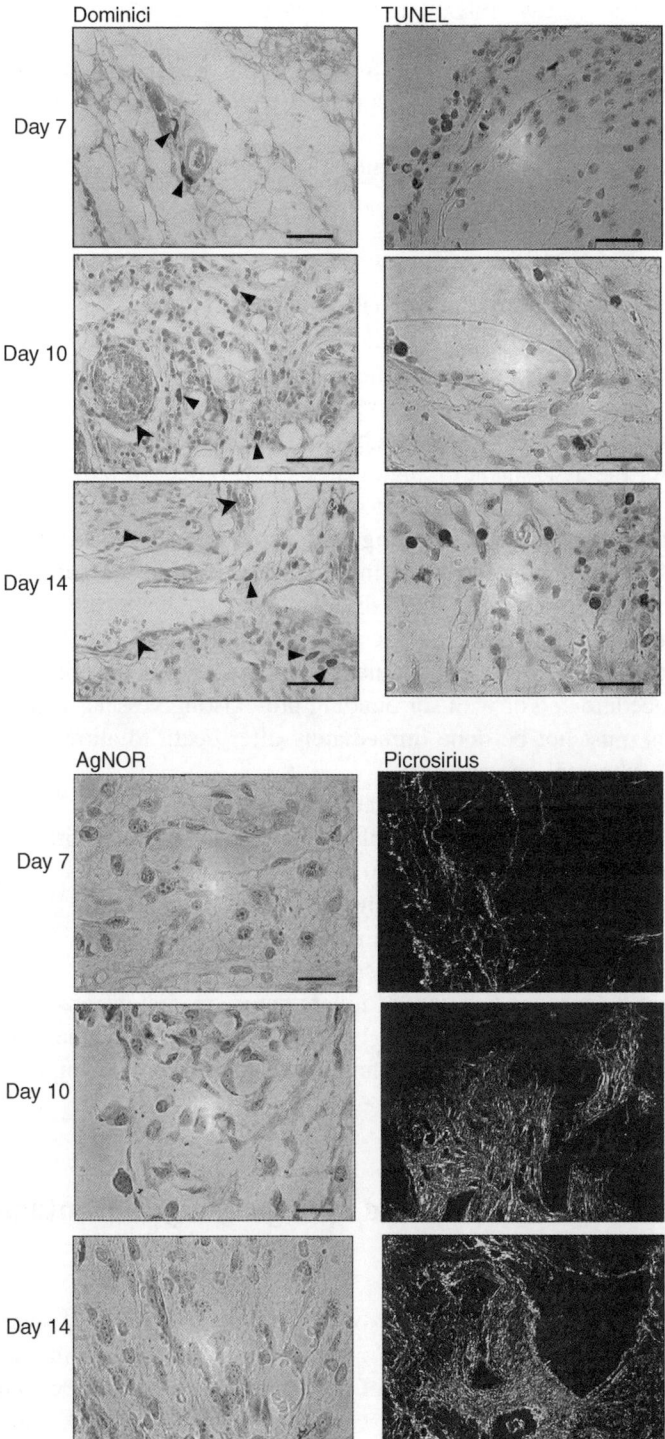

Figure 9.2 (Continued)

angiogenesis in this model results from a direct action on microvascular NK_1 receptors. Neovascularization is a sequential process, with early endothelial proliferation followed by new vessel formation and increased blood flow, with maturation of endogenous neurovascular regulatory systems occurring late in this process in inflamed tissues.

Biochemical analysis of implant fibrovascular tissue

One of the simplest and inexpensive measurements of neovascularization is analysis of the haemoglobin content of the implanted material. Since its development by Plunkett and Hailey (1990) this technique has been modified and extensively employed for the assessment of angiogenesis in a variety of biosynthetic matrices (Passaniti et al., 1992; Hu et al., 1995; Ferreira et al., 2004; Barcelos et al., 2005). The technique was validated following demonstration of good correlation between the haemoglobin content and other more specific angiogenic parameters (vessel number, vessel markers). The main disadvantages of this technique are possible overestimation of vascularization due to the vasodilatory nature of newly formed blood vessels and possible 'contamination' of blood spilled during and after the surgical procedure and/or with surrounding pre-existing vessels. Thus, removal of the implants must not be done immediately after death to allow time for blood coagulation (personal observation).

Other components of the fibrovascular tissue (collagen, glycosaminoglycans, fibronectin, laminin) induced by implants have been used to corroborate and/or indicate the ability of the matrices to induce relevant key stoma components in which blood vessels are laid after injury (Paulini et al., 1974; Davidson et al., 1985; Andrade et al., 1996; Solowiej et al., 2003).

In another series of experiments, using ELISA assays we have documented the levels and time-course of a variety of inflammatory mediators, angiogenic factors, cytokines and chemokines (Table 9.1). These data illustrate the complexity of chemical mediators in the process of angiogenesis, and provide important baseline data for future experiments.

9.9   Summary of cannulated sponge assay: advantages and disadvantages

There are limitations in the study of angiogenesis using the sponge model. Determination of blood flow *in vivo* is limited to fixed time points and damage of the neovasculature after injection of the tracers often occurs. Depending on the material, the inflammatory response can cause excessive matrix deposition and unwanted fibrosis. Because of the variety of the materials used (size, structure, composition, porosity) the pattern of the response varies widely. The maintenance of the animals bearing the implants can be a problem because they have to be kept individually. However, despite these limitations, since the pioneer description of the implantation technique, the experimental *in vivo* model has been modified and improved and its use extended to assess a variety of physiological and pathological conditions enabling detailed analysis of the angiogenic responses induced by a range of endogenous and exogenous molecules. The incorporation of quantitative *in vitro* receptor autoradiography, immunohistochemistry, morphometry and more recently, fluorimetric methods to assess blood flow in the fibrovascular tissue, have made the cannulated sponge model a simple, yet versatile and powerful tool to elucidate the cellular and molecular mechanisms of angiogenesis and to evaluate the efficacy of chemicals capable of modulating angiogenesis, leading to the discovery of novel therapeutic agents.

Acknowledgements

This work was supported by CNPq- Brazil, The Wellcome Trust, the British Heart Foundation and Hong Kong Research Grants Committee.

References

Andrade, S. P., Fan, T.-P. D. and Lewis, G. P. (1987) 'Quantitative *in vivo* studies on angiogenesis in a rat sponge model', *Br. J. Exp. Path*, **68**, pp. 755–766.

Andrade, S. P., Bakhle, Y. S., Hart, I. and Piper, P. J. (1992) 'Effects of tumour cells and vasoconstrictor responses in sponge implants in mice', *Br. J. Cancer*, **66**, pp. 821–826.

Andrade, S. P., Cardoso, C. C., Machado, R. D. and Beraldo, W. T. (1996) 'Angiotensin-II-induced angiogenesis in sponge implants in mice', *Int. J. Microcirc. Clin. Exp.*, **16**, pp. 302–307.

Andrade, S. P., Machado, R. D., Teixeira, A. S., Belo, A. V. *et al.* (1997) 'Sponge-induced angiogenesis in mice and the pharmacological reactivity of the neovasculature quantitated by a fluorimetric method', *Microvasc. Res.*, **54**, pp. 253–261.

Auerbach, R., Auerbach, W. and Polakowski, I. (1991) 'Assays for angiogenesis: A review', *Pharmacol. Ther.*, **51**, pp. 1–11.

Bailey, P. J. (1988) 'Sponge implants as models', *Meth. Enzym.*, **162**, pp. 327–334.

Baker, J. H. E., Huxham, L. A., Kyle, A. H., Lam, K. K. and Minchiton, A. I. (2006) 'Vascular-specific quantification in an *in vivo* Matrigel chamber angiogenesis assay', *Microvasc. Res.*,

Barcelos, L. S., Talvani, A., Teixeira, A. S., Vieira, L. Q. *et al.* (2005) 'Impaired inflammatory angiogenesis, but not leukocyte influx, in mice lacking TNFR1', *J. Leukoc. Biol.*, **78**, pp. 352–358.

Belo, A. V., Barcelos, L. S., Teixeira, M. M., Ferreira, M. A. and Andrade, S. P. (2004a) 'Differential effects of antiangiogenic compounds in neovascularization, leukocyte recruitment, VEGF production, and tumour growth in mice', *Cancer Invest.*, **22**, pp. 723–729.

Belo, A. V., Barcelos, L. S., Ferreira, M. A. N. D., Teixeira, M. M. and Andrade, S. P. (2004b) 'Inhibition of inflammatory angiogenesis by distant subcutaneous tumor in mice', *Life Sci.*, **74**, pp. 2827–2837.

Bollet, A. J., Goodwin, J. F., Simpson, W. F. and Anderson, D. V. (1958) 'Mucopolysaccharide, protein and DNA concentration of granulation tissue induced by polyvinyl sponges', *Proc. Soc. Exp. Biol. Med.*, **99**, pp. 418–421.

Braddock, P. S., Hu, D. E., Fan, T. P., Stratford, I. J. *et al.* (1994) 'A structure-activity analysis of antagonism of the growth factor and angiogenic activity of basic fibroblast growth factor by suramin and related polyanions', *Br. J. Cancer*, **69**, pp. 890–898.

Davidson, J. M., Klagsbrun, M., Hill, K. E., Buckley, A. *et al.* (1985) 'Accelerated wound repair, cell proliferation and collagen accumulation are produced by a cartilage-derived growth factor', *J. Cell Biol.*, **100**, pp. 1219–1227.

Dvorak, H. F., Harvey, V. S., Estrella, P., Brown, L. F. *et al.* (1987) 'Fibrin containing gels induce angiogenesis. Implications for tumor stroma generation and wound healing', *Lab. Invest.*, **57**, pp. 673–686.

Edwards R. H., Sarmenta, S. S. and Hass, G. M. (1960) 'Stimulation of granulation tissue growth by tissue extracts; study by intramuscular wounds in rabbits', *Arch. Path.*, **69**, pp. 286–302.

Fan, T. P., Hu, D. E., Guard, S., Gresham, G. A. and Watling, K. J. (1993) 'Stimulation of angiogenesis by substance P and interleukin-1 in the rat and its inhibition by NK1 or interleukin-1 receptor antagonists', *Br. J. Pharmacol.*, **110**, pp. 43–49.

Ferreira, M. A., Barcelos, L. A., Barcelos, L. A., Campos, P. P., Vasconcelos A. C., Taxeira M. M., Andrade, S. P. (2004) 'Sponge-induced angiogenesis and inflammation in PAF receptor-deficient mice (PAFR-KO)', *Br. J. Pharmacol.*, **141**(7), pp. 1185–1192.

Ford-Hutchinson, A. W., Walker, J. A. and Smith, J. A. (1977) 'Assessment of anti-inflammatory activity by sponge implantation techniques', *J. Pharmacol. Meth.*, **1**, pp. 3–7.

Grindlay, J. H. and Waugh, J. M. (1951) 'Plastic sponge which acts as a frame work for living tissue', *Arch. Surg.*, **63**, pp. 288–297.

Hoffmann, J., Schirner, M., Menrad, A. and Schneider, M. R. (1997) 'A highly sensitive model for quantification of *in vivo* tumor angiogenesis induced by alginate-encapsulated tumor cells', *Cancer Res.*, **57**, pp. 3847–3851.

Holund, B., Junker, P., Garbarsch, C., Christiffersen, P. and Lorenzen, I. (1979) 'Formation of granulation tissue in subcutaneously implanted sponges in rats', *Acta Path. Microbiol. Scand. Sect. A*, **87**, pp. 367–374.

Hori, Y., Hu, D. E., Yasui, K., Smither, R. L. *et al.* (1996) 'Differential effects of angiostatic steroids and dexamethasone on angiogenesis and cytokine levels in rat sponge implants', *Br. J. Pharmacol.*, **118**, 1584–1591.

Hu, D. E. and Fan, T. P. (1993) '[Leu8]des-Arg9-bradykinin inhibits the angiogenic effect of bradykinin and interleukin-1 in rats', *Br. J. Pharmacol.*, **109**, pp. 14–17.

Hu, D. E. and Fan, T. P. (1995a) 'Protein kinase C inhibitor calphostin C prevents cytokine-induced angiogenesis in the rat', *Inflammation*, **19**, pp. 39–54.

Hu, D. E. and Fan, T. P. (1995b) 'Suppression of VEGF-induced angiogenesis by the protein tyrosine kinase inhibitor, lavendustin A', *Br. J. Pharmacol.*, **114**, pp. 262–268.

Hu, D. E., Hori, Y. and Fan, T. P. (1993) 'Interleukin-8 stimulates angiogenesis in rats', *Inflammation*, **17**, pp. 135–143.

Hu, D. E., Hori, Y., Presta, M., Gresham, G. A. and Fan, T. P. (1994) 'Inhibition of angiogenesis in rats by IL-1 receptor antagonist and selected cytokine antibodies', *Inflammation*, **18**, pp. 45–58.

Hu, D. E., Hiley, C. R., Smither, R. L., Gresham, G. A. and Fan, T. P. (1995) 'Correlation of ^{133}Xe clearance, blood flow and histology in the rat sponge model for angiogenesis. Further studies with angiogenic modifiers', *Lab. Invest.*, **72**, pp. 601–610.

Hu, D. E., Hiley, C. R. and Fan, T. P. (1996) 'Comparative studies of the angiogenic activity of vasoactive intestinal peptide, endothelins-1 and -3 and angiotensin II in a rat sponge model', *Br. J. Pharmacol.*, **117**, pp. 545–551.

Hull, M. L., Charnock-Jones, D. S., Chan, C. L., Bruner-Tran, K. L. *et al.* (2003) 'Antiangiogenic agents are effective inhibitors of endometriosis', *J. Clin. Endocrinol. Metab.*, **88**, pp. 2889–2899.

Iuvone, T., Carnuccio, R. and Di Rosa, M. (1994) 'Modulation of granuloma formation by endogeous nitric oxide', *Eur. J. Pharmacol.*, **265**, pp. 89–92.

Kidd, K. R., Dal Ponte, D. B., Kellar, R. S. and Williams, S. K. (2001) 'A comparative evaluation of the tissue responses associated with polymeric implants in the rat and mouse', *J. Biomed. Mat. Res.*, **59**, pp. 682–689.

Kleinman, H. K., Kleinman, H. K., Graf, J., Iwamoto, Y. *et al.* (1987) 'Role of basement membranes in cell differentiation', *Ann. NY Acad. Sci.*, **513**, pp. 134–145.

Kragh, M., Hjarnaa, P. -J. V., Bramm, E., Kristjansen, P. E. G. *et al.* (2003) '*In vivo* chamber angiogenesis assay: an optimized Matrigel plug assay for fast assessment of anti-angiogenic activity', *Int. J. Oncol.*, **22**, pp. 305–311.

Kyriakides, T. R., Zhu, Y-H., Yang, Z., Huynh, G. and Bornstein, P. (2001) 'Altered extracellular matrix remodeling and angiogenesis in sponge granulomas of thrombos-pondin 2-null mice', *Am. J. Pathol.*, **159**, pp. 1255–1262.

Lage, A. P. and Andrade, S. P. (2000) 'Assessment of angiogenesis and tumor growth in conscious mice by a fluorimetric method', *Microvasc. Res.*, **59**, pp. 278–285.

Lees, V. C, Fan, T. P. and West, D. C. (1995) 'Angiogenesis in a delayed revascularization model is accelerated model is accelerated by angiogenic oligosaccharides of hyaluronan', *Lab. Invest.*, **73**, pp. 259–266.

Lentner, A. and Wienert, V. (1996) 'Influence of medical compression stockings drainage in phlebologically healthy test persons and patients with chronic venous insufficiency', *Int. J. Microcirc. Clin. Exp.*, **16**, pp. 320–324.

Ley, C. D., Olsen, M. W. B., Lund, E. L. and Kristjansen, P. E. G. (2004) 'Angiogenic synergy of bFGF and VEGF is antagonized by Angiopoietin-2 in a modified *in vivo* Matrigel assay', *Microvasc. Res.*, **68**, pp. 161–168.

Mahadevan, V., Hart, I. R. and Lewis, G. P. (1989) 'Factors influencing blood supply in wound granuloma quantitated by a new *in vivo* technique', *Cancer Res.*, **49**(2), pp. 415–419.

Mahadevan, V., Malik, S. T., Meager A., Fiers W., Lewis, G. P. and Hart, I. R. (1990) 'Role of tumour necrosis factor in flavone acetic acid-induced tumour vasculature shutdown', *Cancer Res.*, **50**(17), pp. 5537–5542.

Miller, J. C., Adamis, A. P., Shima, D. T., D'Amore, P. A. *et al.* (1994) 'Vascular endothelial growth factor/Vascular permeability factor is temporally and spacially correlated with ocular angiogenesis in a primate model', *Am. J. Pathol.*, **145**, pp. 574–584.

Moghaddam, A., Zhang, H. T., Fan, T. P., Hu, D. E. *et al.* (1995) 'Thymidine phosphorylase is angiogenic and promotes tumor growth', *Proc. Natl. Acad. Sci. USA*, **92**, pp. 998–1002.

Noble, N. L. and Boucek, R. J. (1955) 'Lipids of the serum and connective tissue of the rat and rabbit', *Circ. Res.*, **3**, pp. 344–350.

Passaniti, A., Taylor, A. M., Pili, R., Guo, Y. *et al.* (1992) 'A simple, quantitative method for assessing angiogenesis and antiangiogenic agents using reconstituted basement membrane, heparin, and fibroblast growth factor', *Lab. Invest.*, **67**, pp. 519–528.

Paulini, K., Korner, B., Beneke, G. and Endres, R. (1974) 'A quantitative study of the growth of connective tissue: Investigations on polyester-polyurethane sponges', *Connect. Tissue Res.*, **2**, pp. 257–264.

Plunkett, M. L. and Hailey, J. A. (1990) 'An *in vivo* quantitative angiogenesis model using tumor cells entrapped in alginate', *Lab. Invest.*, **62**(4): pp. 510–517.

Reed, M. J., Pendergrass, W., Penn P., Sage, E. H. and Abrass, I. B. (1998) 'Neovascularization in aged mice: delayed angiogenesis is coincident with decreased levels of transforming growth factors beta 1 and type 1 collagen', *Am. J. Pathol.*, **152**, pp. 113–123.

Salvatore, J. E., Gilmer, W. S. Jr, Kashgarian, M. and Barbee, W. R. (1961) 'An experimental study of the influence of pore size of implanted polyurethane sponges upon subsequent tissue formation', *Surg. Gynecol. Obstet.*, **112**, pp. 463–468.

Schilling, J. A., Joel, W. and Shurley, H. M. (1959) 'Wound healing: A comparative study of the histochemical changes in granulation tissue contained stainless steel wire mesh and polyvinyl cylinders', *Surgery*, **46**, pp. 702–710.

Sengupta, S, Sellers, L. A., Cindrova, T., Skepper, J. *et al.* (2003) 'Cyclooxygenase-2-selective nonsteroidal anti-inflammatory drugs inhibit hepatocyte growth factor/scatter factor-induced angiogenesis', *Cancer Res.*, **63**, pp. 8351–8359.

Sengupta, S., Gherardi, E., Sellers, L. A., Wood, J. M. *et al.* (2003) 'Hepatocyte growth factor/scatter factor can induce angiogenesis independently of vascular endothelial growth factor', *Arterioscler. Thromb. Vasc. Biol.*, **23**, pp. 69–75.

Sengupta, S., Toh, S. A., Sellers, L. A., Skepper, J. N. *et al.* (2004) 'Modulating angiogenesis: the yin and the yang in ginseng', *Circulation*, **110**, pp. 1219–1225.

Solowiej, A., Biswas, P., Graesser, D. and Madri, J. A. (2003) 'Lack of platelet endothelial cell adhesion molecule-1 attenuates foreign body inflammation because of decreased angiogenesis', *Am. J. Pathol.*, **162**, pp. 953–962.

Sprugel, K. M., McPherson, J. M., Clowes, A. W. and Ross, R. (1987) 'Effects of growth factors *in vivo*. I. Cell ingrowth into porus subcutaneous chambers', *Am. J. Path.*, **129**, pp. 601–613.

Teixeira, A. S. and Andrade, S. P. (1999) 'Aminoguanidine prevents impaired healing and deficient angiogenesis in diabetic rats', *Inflammation*, **23**, pp. 569–581.

Vukicevic, S., Kleinman H. K, Luyten, F. P., Roberts, A. B. *et al.* (1992) 'Identification of multiple active growth factors in basement membrane Matrigel suggests caution in interpretation of cellular activity related to extracellular matrix components', *Exp. Cell Res.*, **202**, pp. 1–8.

Walsh, D. A., Hu, D. E., Mapp, P. I., Polak, J. M. *et al.* (1996) 'Innervation and neurokinin receptors during angiogenesis in the rat sponge granuloma', *Histochem. J.*, **28**, pp. 759–769.

Walsh, D. A., Hu, D. E., Wharton, J., Catravas, J. D. *et al.* (1997) 'Sequential development of angiotensin receptors and angiotensin I converting enzyme during angiogenesis in the rat subcutaneous sponge granuloma', *Br. J. Pharmacol.*, **120**, pp. 1302–1311.

Wang, H., Gordon, D., Olszweski, B., Song, Y. L., Kovesdi I., Keiser, J. A. (2000) 'Rat sponge implant model: a new system for evaluating angiogenic gene transfer', *Int. J. Mol. Med.*, **6**(6), pp. 645–653.

10

Angiogenesis assays in the chick

Andries Zijlstra, David Mikolon and **Dwayne G. Stupack**

Abstract

The chorioallantoic membrane (CAM) of the chick presents an accessible, simple system in which to study angiogenesis *in vivo*. The use of the CAM requires relatively inexpensive tools, permitting the assay to be established in any laboratory. Moreover, the system is highly flexible, accommodating complex systems such as cells or tissues as well as purified pro- and anti-angiogenic compounds. The fact that CAM is amenable to genetic manipulation, serial imaging and systemic interventions which allow the controlled study of neovascularization in a complex physiological setting, combined with the relatively high throughput of the system, account for the increasing importance of this assay in the study of neovascularization.

Keywords

angiogenesis; assay; chick; chorioallantois; egg

10.1 Introduction

The chick embryo represents one of few current *in vivo* models developed prior to the 20th century, yet which promises to be increasingly useful even as we advance into the 21st. Since the chick is a living system, it provides powerful and relevant data that cannot be generated through *in vitro* assays. This is in part due to the complex nature of living systems, which produce a number of unrecognized, but critical factors, which are simply absent *in vitro* systems. Although 'simplistic' in design, the chick embryo model permits a more physiological system for *in vivo* analysis of cells, pathogens and pharmacological or biological reagents. For example, the chick embryo has been an instrumental host system for the growth and characterization of many viruses (Kotwal and Abrahams, 2004). The chick model remains in widespread use for the production of attenuated virus vaccines,

Angiogenesis Assays   Edited by Carolyn A. Staton, Claire Lewis and Roy Bicknell
© 2006 John Wiley & Sons, Ltd

most notably influenza, despite the advent of recombinant molecular techniques (Cox, 2005). During the past four decades the chick embryo has become a model of choice for the analysis of biological model systems including angiogenesis (Brooks *et al.*, 1999) and tumour cell metastasis (Kim *et al.*, 1998).

The chick embryo is an intricate model that properly represents the complexity of biological systems in the animal kingdom. Although the embryo is complex in nature, the extra-embryonic membrane known as the chorioallantoic membrane (CAM) is readily accessible outside the embryo, and provides a technically simple portal into complex biological systems. Although a number of investigations have been performed using the chorioallantoic membrane to characterize inflammatory responses (Leibovich *et al.*, 1987), the chick is nevertheless relatively immunoto-lerant. Therefore, unlike adult mice or chickens, the chick embryo is unable to reject transplanted tissue or cells. This property of the CAM has been exploited to sustain pathogens and cross-species xenografts for extended periods of study. Such 'xenografts' include mammalian tissue explants (Preminger *et al.*, 1981; Ausprunk and Folkman, 1976), tumours (Vogel and Berry, 1975) and even cultured human cancer cells (Auerbach *et al.*, 1975). The capacity of the CAM to maintain these diverse xenografted tissues is a result of the high vascular density of this respiratory organ, its ability to vascularize transplants, and its immunotolerance. The CAM initially forms when the mesodermal layer of the allantois fuses with the mesoder-mal layer of the chorion. Within the CAM, a dense vascular network of arterioles, venules and capillaries forms (Figure 10.1), with significant vascular expansion occurring by day 5–6 (Rizzo *et al.*, 1995) and maturation of the vessels is essentially complete by day 10–11 (Brooks *et al.*, 1999). The CAM encompasses a lumen that has direct contact with the abdominal cavity and also functions as a reservoir for liquid waste products. While the role of the CAM in gas exchange was defined in early studies of the chick embryo, the ability of the CAM to function as a model system for angiogenesis was not appreciated until the latter half of the 20th century.

The role of angiogenesis, or the growth of neovasculature from pre-existing vessels, was demonstrated in tumour xenograft growth on the CAM during the pioneering studies of Judah Folkman. Folkman demonstrated that addition of specific cytokines to the CAM could promote vessel growth beyond normal physiological parameters (Kessler *et al.*, 1976). Concordantly, he showed that either this (supra-physiological) neovascularization, or the normal developmental neovascularization of the chorioal-lantois could be blocked by selected chemical or biological agents (Brem and Folkman, 1975; Taylor and Folkman, 1982). These compounds were the first of what is now a broad class of agents termed anti-angiogenics.

Within the last quarter century or so, it has been recognized that tumours themselves secrete pro-angiogenic factors that act on local blood vessels, inducing endothelial cell de-differentiation, proliferation and invasion toward the angio-genic stimulus. Partial maturation and lumenization of these cells leads to neovessel formation, promoting neovascularization of the tumour. This neovascu-larization response is accomplished by usurping the normal physiological mechanisms that regulate blood vessel density in response to hypoxia/ischaemia

(a)   CAM Topical View

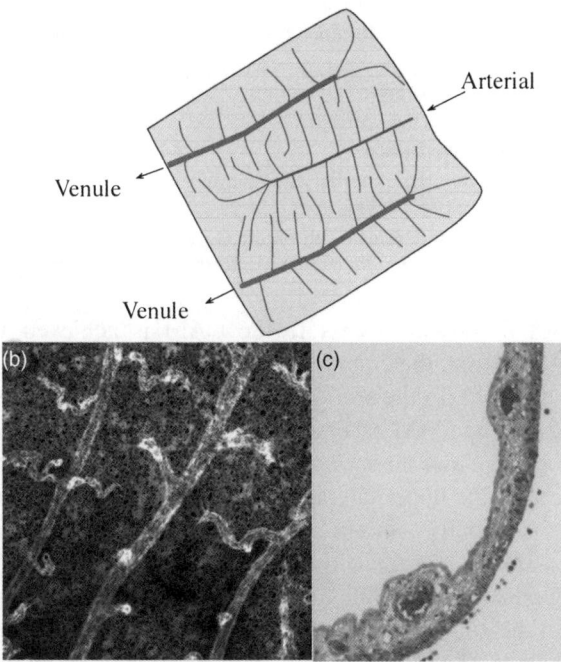

Figure 10.1 Vasculature within the chick CAM (a) A schematic view of the surface of the CAM, showing the interdigitating arterioles and venules present. (b) View of the vasculature, showing both arterioles and venules after injection of rhodamine-conjugated lectin (PHA). There is an extensive capillary network, indicative of the role of the CAM in mediating respiration. (c) An H&E stain of a transverse section through the CAM. In addition to the blood vessels (some of which contain nucleated avian red blood cells), the image clearly shows the two epithelial layers and the interstitial matrix.

and injury. Perhaps not surprisingly, it has been shown that non-tumour tissues that establish viable xenografts on the surface of the CAM produce pro-angiogenic factors. As might be expected, agents that block neovascularization also prevent engraftment (tumour or non-tumour) and growth on the CAM (Brooks *et al.*, 1994; Iruela-Arispe *et al.*, 1999).

Importantly, pro-angiogenic cytokines, such as basic fibroblast growth factor (bFGF) or vascular endothelial cell growth factor (VEGF), will induce angiogenesis when introduced to the surface of the mature CAM irrespective of whether there is an accompanying tissue or matrix that requires vascularization (Tsukamoto and Sugino, 1979; D'Amore *et al.*, 1981; Friedlander *et al.*, 1995). In this case, increased vascularization is observed within the CAM itself. This increased vascularization provides a convenient and reproducible response which can be used to assess the potency of agents that promote or inhibit angiogenesis.

Table 10.1   Minimal materials required for the basic CAM assay

---

1. Egg incubator
2. Light source for Egg candling
3. Rotary drill tool with drill/burr bit and cutting wheel (or manual egg pick)
4. Dissecting scissors
5. Curved dissection forceps
6. Stereo microscope (for scoring)
7. Safety glasses and Dust mask
8. Day 10, embryonated, specific pathogen-free eggs
9. Filter paper (similar to Whatman type I)
10. Pro-angiogenic agent (Basic Fibroblast Growth Factor)

For these types of assays, access to the CAM is achieved via one of two principal ways. In the first, the embryo is cultured in a sterile dish, sans shell, after the embryo is 3 days old. In the second method, the embryo is allowed to develop within the shell, and the CAM accessed via a small window cut in the shell. The window is cut above a 'false air sack' formed via transfer of the natural air sack at the base of the egg to the upper lateral surface of the egg (described below). Both techniques are minimally invasive, do not require significant equipment (Table 10.1) and do not significantly alter the local development of the CAM relative to untreated controls. While the 'in the shell' technique requires much less maintenance than the *ex ovo* ('Petri dish') method, and can be maintained to later stages of development, the out of shell technique provides better access to the test site improving the ability to repetitively treat, manipulate and image the angiogenic tissue.

10.2   Overview of the basic 'in shell' CAM assay

This overview describes the study of angiogenesis within mature CAM (day 11 and beyond) in which developmental angiogenesis has ceased. Prior to this time, active angiogenesis occurs within the CAM in response to endogenous angiogenic factors. It is possible to examine the effect of pro- or anti-angiogenic agents during this initial developmental period, however, the results obtained come with the caveat that the interaction with endogenous factors remains ambiguous and thus can complicate the interpretation. In most cases, it is therefore preferable to wait until day 11, maintaining the eggs in a humidified incubator, and then to examine angiogenesis in response to a defined, exogenously added factor. Recombinant cytokines or protein growth factors are typically added in a physical reservoir that permits them to be released slowly into the CAM itself, as well as to retain the reagent within a localized test area of the CAM. Typically, 6–7 mm discs of filter paper saturated with 15 μl of growth factor in a diluent (saline or phosphate buffered saline) are used. A standard hole punch can be used to generate these filter discs, which are then sterilized in ethanol and allowed to dry under UV light prior to application to the CAM.

Although the CAM will tolerate xenografts from a variety of different species, physical or chemical irritation of the CAM can trigger a strong inflammatory response. Since a number of inflammatory mediators also have pro-angiogenic activity, it is desirable to minimize irritation of the CAM. In fact, even the initial transfer of the air sac (described in more detail below) or the simple placement of the cytokine-saturated filter disc can be sufficient physical irritation to promote an inflammatory response that can result in 'non-specific' angiogenesis. For this reason, an anti-inflammatory agent is commonly added to the filter disks prior to their use. For example, hydroxycortisone acetate may be dissolved in 95 per cent ethanol which is diluted to 75 per cent and placed on the sterile discs. The solution is allowed to dry, thus seeding the disc with an anti-inflammatory compound that will slowly be released during the time period of the angiogenesis assay. While the addition of cortisone will not block all inflammation, it should be taken into consideration that some angiogenic stimuli work in concert with the inflammatory response (Leibovich *et al.*, 1987) and therefore overt suppression of inflammation may be undesirable.

Placement of the filter disc

Prior to placement of the disc, the egg must be prepared by transfer of the air sack, and the creation of a small access window (Figure 10.2). The transfer of the air sac is accomplished by placing the egg horizontally, and drilling two small holes, one on the 'upper' or 'dorsal' surface of the egg (as viewed horizontally) and one at the base of the egg, directly into the natural air sack cavity. These holes are generated with an egg-pick or, more conveniently, with a rotary tool and a 'burr' bit. After the second hole is drilled, the CAM may naturally 'drop,' as air exits the air sack and is drawn in via the hole on the 'dorsal' surface. However, the association between the CAM and the shell is often firm enough to prevent the transfer from occurring spontaneously, and a gentle vacuum may be applied to evacuate the natural air sack and initiate the transfer (or 'drop'). After transfer of the air sack, a small window is created in the upper surface of the egg, large enough to permit inspection and placement of the filter disc without damaging the underlying CAM.

The window can be cut using a rotary tool, or can be made 'piecemeal' by serially removing small sections of the shell with forceps; in fact, forceps are sufficient and preferred for this purpose. Although the rotary tool will make a 'neat' window, the forceps method eliminates the generation of eggshell dust that can promote irritation and inflammation of the underlying CAM. The entire process of dropping the CAM and creating a small window ($\sim$1 cm in diameter) takes a little more than a minute. Thus, several dozen eggs may be prepared in this manner each hour. After the CAMs are dropped and the windows made, the eggs are replaced in a humidified incubator for an hour to allow them to equilibrate. This period allows excess moisture on the CAM to evaporate, while the humidified environment ensures that the CAM does not dry out.

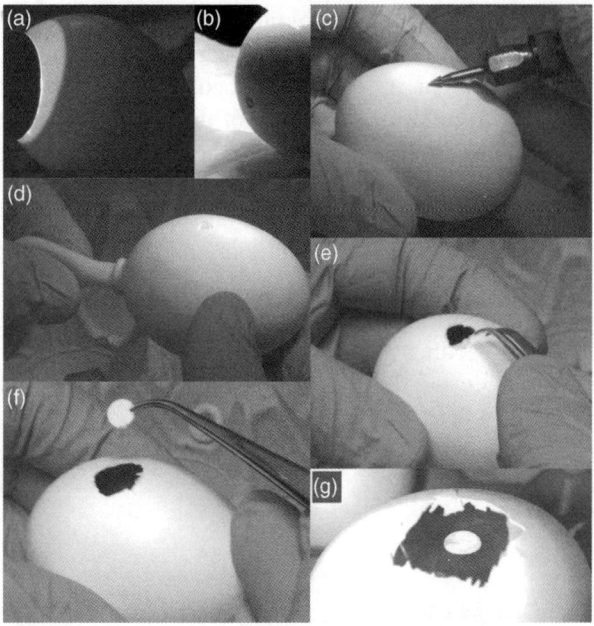

Figure 10.2 Preparation of the egg prior to disc placement (a) The egg is candled to confirm the position of the air sac and a hole is then made by drilling into the air sac (b) and on the dorsal surface of the egg (c). Candling also exposes the position of blood vessels, which is useful for intravenous injections (as depicted in Figure 10.4). Note that for the second hole, the side of the drill is used, and care is taken not to rupture the underlying CAM. Rounded, or 'burr' type bits also work very well for this purpose. (d) If the air sac does not spontaneously transfer from the end to just below the upper air hole, a pipette bulb may be used to evacuate the air sac. The upper hole will permit entry of air, functionally 'transferring' the location of the air sac. (e) Using forceps, the upper hole is increased in size to accommodate the disc or onplant. Shells removed are discarded and (f) the disc, saturated with diluent containing pro-angiogenic compounds (or diluent alone as a control) is then placed on the CAM, as shown in (g). The upper window can now be sealed with transparent tape.

The cytokine-soaked discs are then added to the CAM. Typically, the dry disks may be saturated just prior to placement on the CAM with 15 μl of pyrogen-free saline containing 200 ng of bFGF or VEGF (kept on ice). Once placed, the discs should not move. If the discs slide, the CAM is too moist, and resultant rapid diffusion of the cytokine away from the disc may compromise the assay. The disc should remain in place and slowly release the cytokine to promote neovascularization specifically in the CAM tissue beneath the disc. The window is subsequently closed with transparent tape, to seal the chamber. However, the window may be re-opened at any time to add additional reagents during the course of the assay. Anti-angiogenic compounds are usually not added until a day has passed (24 h), to permit the angiogenic response time to become established, although agents which take time to establish effects, such as a retroviral construct encoding an anti-angiogenic protein, might be added after only 6–12 h. These compounds are added in a total of about 5–10 µl directly to the upper surface of the filter disc.

Quantification of angiogenesis

Angiogenesis is typically allowed to proceed for 3 days, but a longer time-line can be applied. If the assay is allowed to progress longer than 3 days, a second addition of pro-angiogenic growth factors is typically required on the third day. Neovascularization is assessed by first euthanizing the embryos by transferring the eggs to ice for 30 min, and then subsequently harvesting the filter disc and the surrounding CAM with surgical scissors. The CAM is then rinsed briefly in PBS, inverted, and the visible blood vessels quantified by direct counting using a dissecting microscope. Vessels are quantified as individual segments, thus, each time a vessel branches, each new segment is counted regardless of whether it is an equal bifurcation of a major vessel, or simply an ancillary branch (Figure 10.3) (Brooks *et al.*, 1999; Storgard *et al.*, 2005). All filter discs should be scored by the same observer, although it is common practice to have each counted by more than one person. While individual counts may vary to some degree, similar trends are consistently observed. Thus, CAM blood vessels are quantitated in a blinded fashion, and the counts of two or more observers averaged.

This description highlights several advantages of the CAM assay. The technique is simple and rapid, thus allowing relatively high throughput. Moreover, the technological resources required are modest and relatively inexpensive (Table 10.1), requiring a pick or rotary tool, an egg incubator and a darkroom. A dissecting microscope is required only for the vessel counts on the CAMs at the termination of the assay. This should be optimized ergonomically as vessel counting can be somewhat tedious, particularly to a scientist new to the CAM system. However, vessel counting becomes easier as experience with the system is gained, permitting large number of compounds to be screened or allowing a wide range of concentrations of agents to be assessed within a single experiment. As an alternative, for very large screens it may be easier to simply score the vascularity of the CAM underlying the filter disc on a scale of 0–4. This method is also done in a blinded fashion by multiple observers, although it lacks the sensitivity of direct counts. Nevertheless, this semi-quantitative method does provides the capacity to score very large numbers of eggs on a weekly basis.

10.3   Variations on the CAM assay

The CAM assay allows for significant flexibility in approach, depending upon the nature of the scientific question posed. In addition to the general overview described above, we include here additional variations that permit one to address additional specific scientific questions.

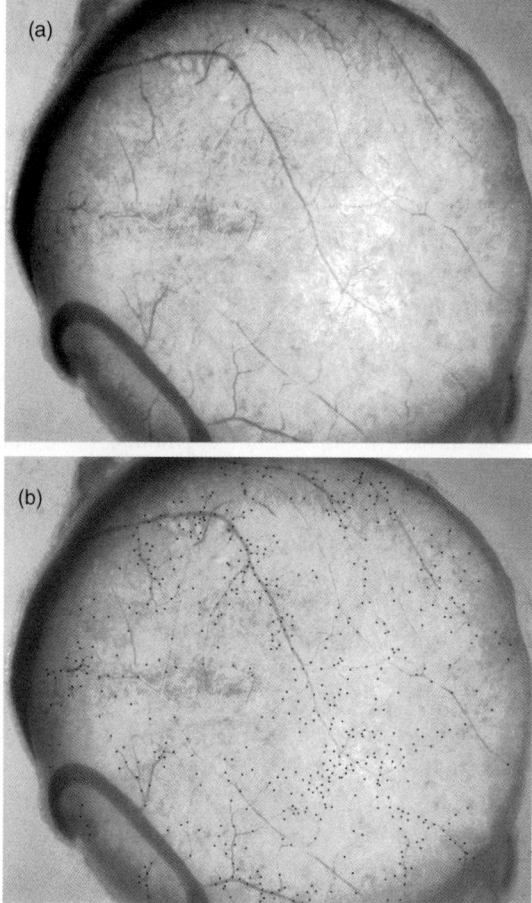

Figure 10.3   Example of branch point counting (a) The disc and the underlying CAM have been resected and inverted for examination under a dissecting microscope. (b) For clarity, each branch has been annotated with a black dot. While taking images of CAMs is recommended for archival purposes, the images captured rarely capture the same depth of field as a human observer, thus leading to a loss of sensitivity and accuracy. Therefore, CAMs should be quantitated immediately upon harvest, whenever possible.

Experimental treatment of chick embryos

The timing, concentration, location, and mode of treatment administration is critical in all animal experiments regardless of species. Therefore, the treatment strategy for any *in vivo* model must be able to accommodate limitations in experimental design. With respect to this, the chick embryo has proven to be extremely flexible in terms of treatment strategies. The test-site itself (local) is commonly treated with pro- or anti-angiogenic agents (as described above).

However, the animal as a whole (systemic) can also be subjected to treatment using various methods of administration. In many experimental designs, the local and systemic approaches are combined. For example, one might treat locally with a pro-angiogenic cytokine such as VEGF, and subsequently administer an anti-angiogenic compound systemically. The local and systemic approaches have specific concerns associated with them, briefly listed here.

Local treatment

Local treatment of the angiogenesis test site, as described in the basic overview above, is frequently desired because test compounds are in limited supply, are required at high concentrations or must be limited primarily to the test-site itself. This is readily achieved by incorporating the treatment (being pro- or anti-angiogenic) into the disc (or onplant, as described below). This ensures that the treatment is restricted primarily to the test site and is present from the start of the assay. Incorporation of the treatment also simplifies the experimental design and reduces the number of times an animal or a test site has to be manipulated. This approach is particularly useful for pharmacological compounds dissolved in organic solvents or small amounts of organic reagents such as growth factors and antibodies. When testing growth factors for their ability to induce angiogenesis, it is common practice to perform a titration. Due to variations in poultry stock, environmental conditions, growth factor batches and the release rate of the disc material, the optimal concentration for angiogenesis induction will often vary somewhat from laboratory to laboratory. Moreover, it should be kept in mind that the concentration of cytokines required to observe angiogenesis in any *in vivo* assay will vary from the concentrations of angiogenic growth factors used to stimulate endothelial cells in *in vitro* angiogenesis and cell signalling assays.

In some cases it may be preferential to initiate treatment some time after placement of the disc, or to treat the test-site repeatedly with the same (or sequentially with different) compounds. In such circumstances, treatment can be added to the test site hours and even days after the discs have been placed. In general, the test compounds are prepared in isotonic buffers such as PBS and 2.5–10 µl is added to each treatment site. However, multiple manipulations of each animal can increase the variability and decrease embryo viability. Consequently, the *ex vivo* assay, described below, may be more suitable for multiple manipulations.

Systemic treatment

In some cases local treatment is not desirable or feasible. Often, the test compounds cannot be suitably integrated with the disc or onplant material and must thus be introduced systemically. Moreover, in some cases the treatment may require delivery through the established blood flow in order to observe effect, or the treatment target is systemic or distant from the test site. In many cases both local and systemic treatments should be tested in order to optimize the experimental approach.

Intravenous injection of the CAM. The chick model does permit direct intravenous injection. Unlike mammalian systems in which the vasculature is essentially always easily accessible, the chick model requires special preparation at the onset of the experiment. The egg must be candled (Figure 10.2b) by transillumination to identify regions containing appropriately sized, CAM-anchored blood vessels, and the shell subsequently removed from these areas to permit access. However, the chick offers a significant advantage in return. Since clearance of test agents is not a significant factor, most agents need only a single dose, and will then remain in circulation for the duration of the assay.

Identification of blood vessels requires that the egg be candled in a darkened room. A bright light source is placed on the end of the egg, illuminating the egg and permitting easy identification of blood vessels of appropriate size for injection. A box is drawn surrounding the proposed injection site, typically on a straight and unbranched region of the blood vessel. The vessel should not move when the egg is rolled, since only those vessels anchored in the CAM are readily amenable to injection. A rotary tool is then used to score the shell along the lines of the box, penetrating just through the shell, but not breaking through the underlying membranes (Figure 10.4). The shell is left in place until the time of injection, typically on the second day of the assay. Small regions of the CAM exposed by scoring the shell may dry out somewhat, promoting a tight association with the shell. Therefore, a few drops of mineral oil are often applied to the top corners of the cuts, and allowed a few seconds to seep under the shell covering the injection window. This lubrication permits the shell to be gently removed using forceps, without damage to the underlying CAM. After the shell is removed, a further drop of mineral oil is added to the exposed surface of the CAM to render the membrane transparent, clearly exposing the underlying vessel.

The egg is top-lit through the upper window to back-light the blood vessel for injection. A 31-gauge needle (insulin syringe) is typically used on a syringe appropriate for 50 µl injections. The test compound, suspended in pyrogen-free saline, should be injected carefully, minimizing any movement to the egg by positioning the egg in a holder while freeing both hands to stabilize the syringe during injection. Introduction of the needle to the vessel is performed slowly at a shallow angle, in the direction of blood flow, and test compound administered slowly. Brief clearing of the vessel will obvious during the injection, which confirms good intravenous position. After removal of the needle from the vessel, a small piece of sterile gauze is used to staunch any bleeding which ensues. The window created for the injection is then sealed with adhesive tape, and the egg returned to the incubator.

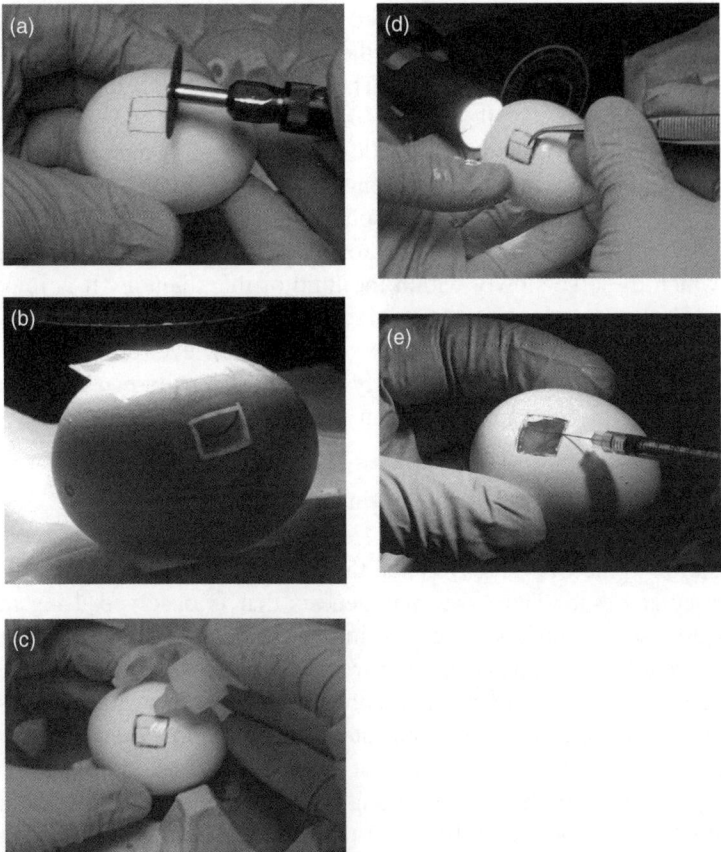

Figure 10.4 Preparing an egg for IV injection these events occur prior to the placement of a disc or onplant. During the candling process outlined in Figure 10.2, a vessel is selected for injection that is relatively straight and unbranched. (a) A box drawn around the vessel is used as a template. The egg is cut with a rotary tool, with care taken not to damage the underlying CAM. (b) The central portion of shell is left in place. This toplit image shows the exposed, undamaged CAM surrounding the central shell. Also visible in this image are the top window, covered with tape, and the initial hole drilled to transfer the air sack. (c) To remove the window, a few drops of mineral oil are allowed to wick under the shell. (d) Forceps are then used gently to remove the shell. An additional drop of mineral oil is added to clarify the underlying CAM for injection. (e) The egg is then injected carefully at a shallow angle. Visual cues are used for guidance, as there is no discernible difference in the tactile resistance of a blood vessel (relative to other regions of the CAM) as there might be when injecting a mouse intravenously, for example.

Intra-allantoic injection. Intravenous injection is an effective way to introduce material systemically. Unfortunately, I.V. injections are not suitable for all purposes. Intraperitoneal injections frequently supplant I.V. injections for the systemic administration of treatments. While the chick embryo can be injected directly, this frequently leads to reduced viability. In contrast the allantois is a suitable cavity in direct contact with the internal organs of the chick and the expansive vascular bed of the CAM. The allantois can be directly injected by stabbing a small gauge needle directly through the CAM and injecting into the fluid below. Larger volumes (100–500 µl) are readily injected making this extremely suitable for circumstances in which the soluent must be diluted or where the test compound is poorly soluble and must be introduced in large volumes. In some cases it is possible to inject insoluble compounds as homogenates which dissolve slowly within the fluid of the allantois. It is important to place the injection site in the upper surface of the CAM because leakage of fluid may occur if the injection site was below the ambient level of fluid in the CAM. This method of treatment is particularly useful for the *ex ovo* assay described below, since I.V. injection in that system is particularly challenging.

Use of onplants and tissues for angiogenesis assays

One limitation to the CAM assay is that the angiogenesis observed does not occur in an avascular tissue. Thus, the angiogenesis that is observed using the CAM assay, as assessed through a dissecting microscope, represents a combination of neovessel growth with expansion or growth of pre-existing vessels. This type of angiogenesis is sometimes contrasted with that seen in other angiogenesis assays, such as the Matrigel plug/sponge (Chapter 9) or the corneal angiogenesis assay (Chapter 11) where the inserted onplant serves as an avascular matrix for the invading vasculature. In these types of angiogenic models, all observed vessels are neovessels. However, it should be noted that similar studies can be performed in the chick CAM, using adapted methods.

Onplant assay

A variation of the CAM assay involves the placement of collagen gels supported by a nylon grid directly onto the CAM in place of the filter disc (Seandel *et al.*, 2001). The collagen is impregnated with angiogenic growth factors, which are subsequently released into the CAM. In response to these growth factors, neovessels grow up out of the CAM and into the collagen gel from beneath. This capacity for vessel outgrowth is precisely the reason that the CAM has been used in xenograft studies. In this case, one can exploit the avascular nature of the collagen gel, thus ensuring that all observed vessels invading the matrix are true neovessels (rather than pre-existing capillaries that have undergone growth or maturation), should this be important to the nature of the study. The approach is elegant, in that the supporting nylon mesh provides easy reference points that allow the user to score the area of grids for vascularization rather than each vessel individually. These studies may require somewhat higher magnification to permit the vessel counts.

Cell-based angiogenesis assays

The capacity of the CAM to support tumour cell growth and the formation of tumours *in situ* also permits one to use the CAM to examine the angiogenic potential of the tumour cells. For example, tumour cells can be seeded directly into the CAM and a tumour allowed to form for a week. The tumour can then be readily excised, weighed and the relative vascularity of the tumour determined.

To seed the tumours, the surface of the CAM is first abraded gently with a sterile swab. This is important, as it compromises the surface epithelium and allows direct access of the tumour cells into the CAM tissue. Next, 40 μl of cell suspension (12.5 $\times$ 10^8cells/ml, total of 5×10^6 cells) are introduced and should seed as a single bead on the abraded region of the CAM. The cells will develop into a tumour ranging from $\sim$0.1–1.0 g within a week. During this time, the developing tumour will elicit a vasculature through co-option of the existing CAM vasculature as well as through formation of neovessels. The tumour-associated vasculature can then be evaluated, permitting one to assess differences in the angiogenic potential of different cell lines, or differences within the same cell line after genetic manipulations (e.g. gene knockdown or expression of transgenes), or in response to topically applied or intravenously injected agents.

The extent of the vasculature can be measured via several different methods. The tumour can be resected and sectioned. Immunohistochemical or immuno-fluorescent staining of the sections can then be performed, and vessels quantitated. While this can be an involved process, if digital images can be captured, the availability of automated signal thresholding and pixel counting in most commercial image analysis software can greatly improve sensitivity and accuracy. Many images can be thus evaluated for the presence of vessels in a blinded manner. A major advantage of this approach is that it allows analysis of additional factors, such as the tumour histology and the expression of markers on the tumour or the associated angiogenic endothelium. However, this is a sample-based approach, and some investigators will prefer to assess neovascularization within the tumour as a whole using alternative approaches, discussed below.

Alternative quantitation methods

The collagen onplant method described above offers some benefits relative to the filter disc assay, yet both rely on direct vessel counting to quantify angiogenesis. However, rather than actually counting the individual vessels or branch points, it is also possible to automate quantification using proxy readouts such as haemoglobin or the presence of intravenously injected fluorophores to indicate the presence of blood vessels. In either case, a biopsy punch of the same or smaller diameter than the filter disc is used to obtain an equal area of the CAM from each sample. The samples are digested and the quantity of haemoglobin is assessed, typically with a commercial kit. Conversely, the presence of blood-borne fluorophores (e.g.

FITC-albumin) can be determined according to their respective excitation and emission spectra. This may seem a more objective manner to assess the presence of blood vessels, yet suffers from certain disadvantages.

The most obvious limitation to this approach is that the filter disc may sometimes be inadvertently placed over a pre-existing blood vessel of moderate size. Such a vessel might have few neovessels branching off it, yet would account for a significant blood volume in a given area of the CAM. Thus, such an area might be clearly non-angiogenic when quantitated using the dissecting microscope, but would appear as though it were angiogenic (a false-positive result) using the haemoglobin assay, or surrogate markers such as fluorescein-albumin or fluorescein-dextran. The use of endothelial specific, fluorophore-conjugated lectins can attenuate this problem somewhat, and is often used to assess angiogenesis. The use of these lectins permits specific binding to the surface of the endothelium, and thereby reduces the severity of 'false positives' since they measure vascular surface area rather than volume. However, fluorophores themselves suffer from a couple of limitations. The first is the presumption that they will bind relatively equally to blood vessels of different sizes and maturation status. The second disadvantage to a fluorophore approach is that they require intravenous injection. A significant amount of time is required for the I.V. injection, lectin circulation and subsequent tissue processing. While possible, I.V. injection of the chick embryo can be technically challenging, and may eliminate any potential time-saving associated with automated angiogenesis assessments relative to simple direct counting. However, in some cases direct counting is not a viable option. For example, while the angiogenesis assay described above uses defined growth factors to induce neovascularization, it will sometimes be desirable to assess angiogenesis in complex systems where an intact tissue or tumour cells are xenografted onto the CAM. In such cases, automated assessments may be the best option for quantification.

For these purposes, the chick is injected intravenously with fluorescent lectin (such as PHA), as described above, and the lectin is allowed to circulate and bind to the endothelium for 30 min prior to tumour harvest. Upon harvest, the tumour is trimmed and weighed, placed in a tube of PBS with 2 per cent SDS (2 µl/mg) and heated to 60°C to extract the endothelial cell-bound lectin. The fluorescence of the normalized tumours is then quantitated using a fluorometer. This approach permits an assessment of total tumour vascularization, thus avoiding pitfalls associated with regions of intense angiogenesis vs. regions that are avascular, and provides the additional advantage that results may be normalized from tumour to tumour, and against normal tissues. Nevertheless, this assay system lacks the flexibility of tissue sectioning, and possesses the normal limitations of assays which quantitate blood vessels via surrogate reporters, such as haemoglobin or lectins, as discussed above.

In ovo vs. *ex ovo*

While the majority of experimental application with the chick embryo is done *in ovo*, meaning within an intact egg, the eggshell itself can be a barrier that interferes with the experimental design. For this reason, removal of the egg shell (*ex ovo*) and *ex ovo* cultivation of the animal can be very beneficial. Specifically, the absence of the eggshell improves access to the CAM both for visual purposes including intravital imaging, as well as local treatment and tissue manipulations. To cultivate the chick embryo successfully *ex ovo*, the embryo must be decanted from the eggshell before the fifth day of development. Egg 'cracking' is preferably done on day 4 of development but the actual timeline may vary depending upon the temperature at which the animals have been maintained (i.e. temperature fluctuations during storage or shipment may alter embryo development). The embryo is decanted into a round bottom dish (such as sterilized weighboats, with square Petri dish lids) of approximately 100 ml with a lid and cultivated in a humidified 37°C incubator until the start of the experiment. Angiogenesis can be initiated when the CAM has matured, i.e. between developmental day 10 and 11 (Figure 10.5). At

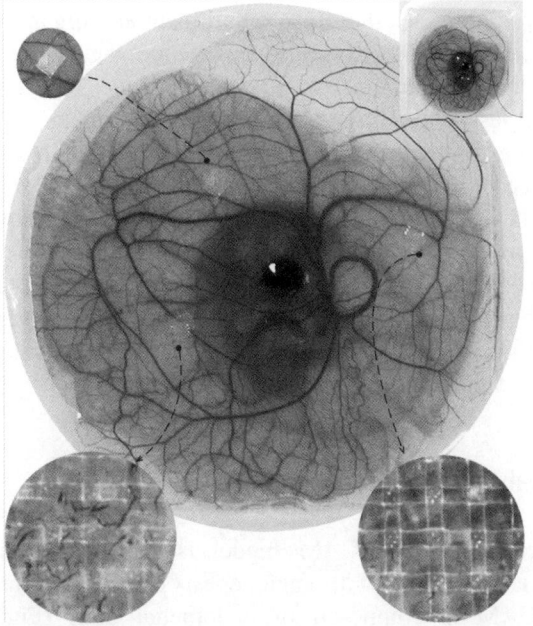

Figure 10.5 Use of the *ex ovo* model with onplants to assess angiogenesis in the *ex ovo*, or shell-less assay, 3-day-old eggs are 'cracked' and transferred to sterile dishes, then grown in a humidified incubator until they are 11 days old. In this case, sterile weighboats, as seen in the upper right inset, have been used, and are covered with square culture dish lids. An enhanced image of the upper left onplant (collagen/grid) is shown in the upper left inset, while three other onplants (nearly transparent) can be discerned in the main image. The use of four onplants here nicely illustrates the increased CAM surface available for angiogenesis studies in the *ex ovo* model. Quantification of vascularization following treatment with vascular endothelial growth factor (lower left inset) or diluent only (lower right inset) is shown, as scored via direct microscopic analysis. Typically, the number of grid squares supporting angiogenesis is scored. In this case, the grid on the left was scored with an angiogenic index of ~81 per cent, while that on the right was ~12.5 per cent.

this time the CAM will have spread over almost the entire surface of the culture dish ($\sim$100 cm^2). Both the disc method with branch-point scoring and the collagen-onplant assay can be applied to the *ex ovo* system, however, the system is most routinely used for the latter. The *ex ovo* embryo system forces the CAM to develop as a flat expansive membrane that is far larger and more accessible than the counterpart *in ovo*. This large, accessible surface is able to support multiple test-sites and generally 4–5 discs or grids are placed on each animal. Combined with the relative high number of animals that are routinely used in chick assays (6–10 animal/test), an *ex ovo* assay can provide a large number of data points for each assay (>20) making this assay a high-throughput system relative to the classic mouse assays. In addition to increasing the sample throughput, the large open CAM is readily accessible which greatly facilitates local treatment and manipulation of the test site as well as imaging of the angiogenic response.

There are a few issues to consider associated with the *ex ovo* assay. The need to decant the embryos increases the time commitment and organization required for the *ex ovo* assay. However, with a little experience, the 'egg cracking' becomes routine and requires little more than 30 s per animal. Additionally, intravenous injection of shell-less animals is complicated because the CAM is no longer associated with the rigid eggshell. As a consequence the vessels are difficult to penetrate and bleed more than their *in ovo* counterparts. The use of extremely small gauge needles or using intra-allantois injection can circumvent this complication. Furthermore, removal of the eggshell eliminates the calcium source required for proper bone development, thereby preventing full-term development of the animal. This proves problematic for late-stage animals (developmental day 19–21), although most assays are generally completed by developmental day 14–16, avoiding this complication.

Biochemical analysis of the CAM

A particular strength of the CAM model is the relative ease with which biochemical events associated with angiogenesis can be monitored. A significant component of CAM is composed of endothelial cells (Figure 10.1b), thus providing a readily accessible pool to perform such studies. Rapid resection and extraction of the CAM permits analysis of transcriptional regulation and/or protein production, as well as cell signalling events. The capacity to perform these studies *in vivo* is valuable, since these events often vary *in vivo* relative to results obtained *in vitro* using cultured endothelial cells.

For example, to study the kinetics of signalling events downstream of vascular endothelial growth factor (VEGF) *in vivo*, one might treat the CAM of 11-day-old chicks with VEGF-saturated filter discs for increasing periods of time. As controls, a different growth factor or diluent alone could be used (preferably both). At the appropriate times, the CAM underlying the filter disc is resected, and the tissue rinsed twice briefly in PBS, and then mechanically homogenized in RIPA buffer containing sodium vanadate (2 mM) and sodium fluoride (50 mM). The lysate is clarified by centrifugation (15 000 $g \times 5$ minutes). Proteins of interest (for example, Flk-1) can then be immunoprecipitated, typically from about 0.5 mg of protein. The immunoprecipitate is resolved by SDS-PAGE and the phosphorylation status of the precipitated proteins determined by immunoblotting using phospho-specific antibodies. This approach can be used to examine signalling events which occur very rapidly, as phosphorylation of proteins can be observed after only a few minutes (Alavi *et al.*, 2003).

Similarly, this approach could be adapted to answer other questions, such as the relative abundance of a particular metalloproteinase within angiogenic tissues. In this case, the kinetics of the study might follow an order of hours, rather than minutes. Depending upon the extraction used, RT-PCR analysis of mRNA transcripts or even zymography could be performed. Overall, this capacity to easily employ biochemical methods in analysis of angiogenesis in the CAM is a considerable strength of the assay. Aside from providing a means to investigate novel molecular events *in vivo*, it affords a means to evaluate and/or confirm whether test or known therapeutic anti-angiogenic agents (such as kinase or protease inhibitors) are actually influencing their putative 'targets' *in vivo*.

10.4 Summary of the CAM assay: disadvantages and advantages

There are limitations to the study of angiogenesis in the chick. As an embryonic tissue, the CAM is only amenable to studies for a limited period and is typically utilized for no more than a week or so. While this is better than or on a par with most *in vitro* angiogenesis assays, it is significantly shorter than mouse models, which can be extended for several weeks, if necessary. Moreover, murine models offer the advantage of a number of established different genetic backgrounds. Thus, the roles of specific genes in angiogenesis may be assessed directly through the use of knockout animals. In contrast, in the avian system one is dependent upon the use of avian retroviruses (RCAS) to transduce exogenous genes into the chick CAM (Bell and Brickell, 1997). The murine system is also a mammalian system, and as such may be considered a somewhat closer mimic of human physiology. Nevertheless, the chick embryo properly represents the complexity of biological systems involved in angiogenesis and can be effectively applied for the study of neovascularization.

As discussed, the chick CAM offers several advantages to studying angiogenesis. One important factor for many laboratories is that the assay is relatively inexpensive, and requires only basic laboratory tools to execute. Large numbers of eggs can be rapidly prepared for the CAM assay, and the short length of the assay permits a relatively high throughput. The assay is simple to set up and maintain in a laboratory, yet represents a significant step above *in vitro* assays for several reasons. It is a live assay system, thus the forming neovessels are, by definition, functional, and the ability of these neovessels to support a biological process, such as tumour growth can be readily assessed. Furthermore, like many *in vitro* assays the CAM is amenable to biochemical analysis, and extracts of CAMs have been used to demonstrate signalling events *in vivo*. This is, in part, due to the large proportion of endothelial cells which comprise the CAM tissue, and represents a significant advantage of the CAM, relative to other *in vivo* assays, such as the corneal pocket assay or the Matrigel assay. Like the corneal assay, the CAM assay can be serially assessed, or even imaged intravitally as angiogenesis occurs when the shell-less system is used. The simplicity, flexibility and short duration of the assay in fact provides the chick assay as the only semi high-throughput angiogenesis model currently available.

The chick system also offers some subtle advantages relative to other *in vivo* systems that may not be immediately obvious. For example, while murine anti-human antibodies typically do not recognize murine antigens, many recognize chick antigens, permitting greater experimental and analytical options than in the murine system. The blood volume of a chick embryo is about the same as a mouse, but the lack of excretion permits test reagents to be maintained in circulation for extended periods, in addition, the accessibility of the CAM permits serial application of topical agents to be easily performed. The physiology of the chick permits one to use the system to study additional angiogenesis dependent events, such as tumour metastasis. The chick CAM thus has the advantage of presenting a single model system which can support angiogenesis studies via a number of different mechanisms. As such, it represents a powerful, yet highly flexible tool for laboratories that study neovascularization.

References

Alavi, A., Hood, J. D., Frausto, R., Stupack, D. G. and Cheresh, D. A. (2003) 'Role of Raf in vascular protection from distinct apoptotic stimuli', *Science*, **301**(5629), pp. 94–96.

Auerbach, R., Arensman, R., Kubai, L. and Folkman, J. (1975) 'Tumor-induced angiogenesis: lack of inhibition by irradiation', *Int. J. Cancer*, **15**(2), pp. 241–245.

Ausprunk, D. H. and Folkman, J. (1976) 'Vascular injury in transplanted tissues. Fine structural changes in tumor, adult, and embryonic blood vessels', *Virchows Arch. B Cell Pathol.*, **18**(1), pp. 31–44.

Bell, E. J. and Brickell, P. M. (1997) 'Replication-competent retroviral vectors for expressing genes in avian cells in vitro and in vivo', *Mol. Biotechnol.*, **7**(3), pp. 289–298.

Brem, H. and Folkman, J. (1975) 'Inhibition of tumor angiogenesis mediated by cartilage', *J. Exp. Med.*, **141**(2), pp. 427–439.

Brooks, P. C., Montgomery, A. M. Rosenfeld, M., Reisfeld, R. A. *et al.* (1994) 'Integrin alpha v beta 3 antagonists promote tumor regression by inducing apoptosis of angiogenic blood vessels', *Cell*, **79**(7), pp. 1157–1164.

Brooks, P. C., Montgomery, A. M. and Cheresh, D. A. (1999) 'Use of the 10-day-old chick embryo model for studying angiogenesis', *Methods Mol. Biol.*, **129**, pp. 257–269.

Cox, M. M. (2005) 'Cell-based protein vaccines for influenza', *Curr. Opin. Mol. Ther.*, **7**(1), pp. 24–9.

D'Amore, P. A., Glaser, B. M., Brunson, S. K. and Fenselau, A. H. (1981) 'Angiogenic activity from bovine retina: partial purification and characterization.' *Proc. Natl. Acad. Sci. USA*, **78**(5), pp. 3068–3072.

Friedlander, M. Brooks, P. C., Shaffer, R. W., Kincaid, C. M. *et al.* (1995) 'Definition of two angiogenic pathways by distinct alpha v integrins', *Science*, **270**(5241), pp. 1500–1502.

Iruela-Arispe, M. L. Lombardo, M., Krutzsch, H. C., Lawler, J. and Roberts, D. D. (1999) 'Inhibition of angiogenesis by thrombospondin-1 is mediated by 2 independent regions within the type 1 repeats.' *Circulation*, **100**(13), pp. 1423–1431.

Kessler, D. A., Langer, R. S., Pless, N. A. and Folkman, J. (1976) 'Mast cells and tumor angiogenesis', *Int. J. Cancer*, **18**(5), pp. 703–709.

Kim, J., Yu, W., Kovalski, K. and Ossowski, L. (1998) 'Requirement for specific proteases in cancer cell intravasation as revealed by a novel semiquantitative PCR-based assay', *Cell*, **94**(3), pp. 353–362.

Kotwal, G. J. and Abrahams, M. R. (2004) 'Growing poxviruses and determining virus titer', *Methods Mol. Biol.*, **269**, pp. 101–112.

Leibovich, S. J. Polverini, P. J., Shepard, H. M., Wiseman, D. M., Shively, V. and Nuseir, N. (1987) 'Macrophage-induced angiogenesis is mediated by tumour necrosis factor-alpha', *Nature*, **329**(6140), pp. 630–632.

Preminger, G. M., Koch, W.F., Fried, F.A. and Mandell, J. (1981) 'Chorioallantoic membrane grafting of the embryonic murine kidney. An improved in vitro technique for studying kidney morphogenesis', *Invest. Urol.*, **18**(5), pp. 377–381.

Rizzo, V., Kim, D., Duran, W. N. and DeFouw, D. O. (1995) 'Ontogeny of microvascular permeability to macromolecules in the chick chorioallantoic membrane during normal angiogenesis', *Microvasc. Res.*, **49**(1), pp. 49–63.

Seandel, M., Noack-Kunnmann, K., Zhu, D., Aimes, R. T. and Quigley, J. P. (2001) 'Growth factor-induced angiogenesis in vivo requires specific cleavage of fibrillar type I collagen', *Blood*, **97**(8), pp. 2323–2332.

Storgard, C., Mikolon, D. and Stupack, D. G. (2005) 'Angiogenesis assays in the chick CAM', *Methods Mol. Biol.*, **294**, pp. 123–136.

Taylor, S. and Folkman, J. (1982) 'Protamine is an inhibitor of angiogenesis', *Nature*, **297**(5864), pp. 307–312.

Tsukamoto, K. and Sugino, Y. (1979) 'Tumor angiogenesis activity in clonal cells transformed by bovine adenovirus type 3', *Cancer Res.*, **39**(4), pp. 1305–1309.

Vogel, H. B. and Berry, R. G. (1975) 'Chorioallantoic membrane heterotransplantation of human brain tumors', *Int. J. Cancer*, **15**(3), pp. 401–408.

11

Corneal angiogenesis assay

Siqing Shan and Mark W. Dewhirst

Abstract

New vessel growth in the avascular and transparent cornea occurs under a variety of pathological conditions and is readily distinguishable. Therefore, the corneal neovascularization (CNV) assay has become a widely used *in vivo* model for angiogenesis research. Many techniques, including chemical cauterization and mechanical manipulations have been developed. This chapter describes the background, physiology, induction techniques and image analysis of CNV with emphasis on the most frequently applied micropocket assay. In this assay, angiogenesis is induced by placing a polymer pellet containing an angiogenic factor, such as bFGF, into a surgically created stromal pocket and allowing for its sustained release. A putative anti- or pro-angiogenic substance is incorporated into the pellet, or given locally or systemically. The degree of suppression or enhancement of angiogenesis is quantified by image analysis. CNV assay offers a quantitable and precise means of screening for angiogenesis inhibitors, but also has limitations. It is desirable to combine CNV with other *in vivo* assays to investigate angiogenesis.

Keywords

cornea; neovascularization; rodents; micropocket assay

11.1 Introduction

Corneal neovascularization (CNV) has been an important research subject for two major reasons. First, CNV is a common histopathological feature of corneal diseases. A study with 1278 human corneal bottoms obtained by keratoplasty showed 19.9 per cent with angiogenesis (Cursiefen *et al.*, 1998). CNV remains a severely disabling ophthalmologic condition that causes loss of the corneal transparency and visual impairment. Secondly, the cornea is the only avascular

Angiogenesis Assays   Edited by Carolyn A. Staton, Claire Lewis and Roy Bicknell
© 2006 John Wiley & Sons, Ltd

transparent tissue in the body, so any vessels growing in the cornea, induced by different techniques, are newly formed, readily visible and can be quantified. Therefore, CNV is an ideal model for quantitative investigation of angiogenesis and the evaluation of substances with potential pro- or anti-angiogenic properties.

11.2   Anatomy and histology

The cornea is the anterior outer fibrous tunic of the eye. It consists of an external layer of non-keratinized, stratified squamous epithelium overlying a prominent basement membrane; a stroma (substantia propria) composed of collagen fibres, sparse elastic fibres, and scattered fibroblasts; Descemet's membrane; and an internal layer of endothelium (Figure 11.1). The corneal epithelial stem cells, located at the limbal basal layer, are the ultimate source for corneal epithelial cell proliferation and differentiation (Chen and Tseng, 1990). The cornea does not contain blood vessels.

The maintenance of corneal avascularity and transparency depends on many cellular and molecular mechanisms. An early study showed that the corneal cells isolated from adult rabbits inhibited lymphocyte-induced angiogenesis in intradermal injection sites (Kaminski and Kaminska, 1978). HCO_3^--dependent fluid secretion by the corneal endothelium controls corneal hydration and maintains corneal transparency (Sun and Bonanno, 2002). Several cornea-specific gene products have been reported including cornea-derived transcript 6 (CDT6), which encodes for a protein homologous to the angiopoietins, and was detected only in the corneal

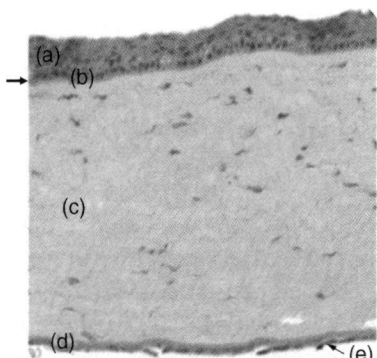

Figure 11.1 Histology of the rat cornea (a) the non-keratinized, stratified squamous epithelium; (b) basement membrane; (c) substantia propria (stroma) composed of collagen fibres, sparse elastic fibres, and scattered fibroblasts; (d) Descemet's membrane; (e) endothelium.

stroma but not other adult human tissues examined (Peek *et al.*, 1998). Interferon-alpha (IFN-α) appears to have an important role in regulating transcription of the human CDT6 promoter (Liu and Wilson, 2001). Other molecules such as proletin (PRL)-like molecules (Duenas *et al.*, 1999), thrombospondin (TSP)-1 and -2 (Cursiefen *et al.*, 2004), and matrilysin, matrix metalloproteinase (MMP)-7 (Kure *et al.*, 2003), probably also play important roles in maintaining corneal avascularity (angiogenic privilege).

11.3   Brief history of CNV assays

It has been observed that many pathological conditions cause vessel growth in the normally avascular cornea. Experimental manipulation with different methods in rabbit cornea was used to investigate the CNV process in the early 1970s (Gimbrone *et al.*, 1973; Gimbrone *et al.*, 1974; Auerbach *et al.*, 1975). Keratectomy-induced capillary endothelial cell proliferation at the limbus of the eye was inhibited by irradiation in canines, and radiation dose responses reported in the 1970s represented the radiosensitivity of neovasculature (Gillette *et al.*, 1975; Fike and Gellette, 1978).

Different tissues were implanted into surgically created corneal stromal micro-pockets in rabbits to study their effects on CNV. It was found that ovarian corpus luteum induced CNV, but follicles and cartilage did not (Brem and Folkman, 1975; Gospodarowicz and Thakral, 1978). Because of the pivotal role of angiogenesis in malignancy development and metastasis, tumour tissue or tumour cells were implanted into the corneal micropocket to study tumour-induced angiogenesis (Gimbrone *et al.*, 1974). It was found that the new vessel induction was related to the aggressiveness of tumour type (Chaudhury *et al.*, 1980) and less related to tumour cell viability (Auerbach *et al.*, 1975; Ryu and Albert, 1979). Tumour extracts or tumour cell conditioned-medium also induced similar neovasculariza-tion (Fournier *et al.*, 1981; Polverini and Leibovich, 1984; Galardy *et al.*, 1994). These results were among the first to demonstrate that tumour induced angiogen-esis is dependent on pro-angiogenic factors secreted by tumour cells.

Many techniques have been developed to induce CNV experimentally in the past three decades, and a variety of pro-angiogenic and anti-angiogenic reagents have been tested in CNV models. Among these, the corneal micropocket assay, with the implantation of a sustained-release polymer pellet containing purified known growth factors, is the most commonly used at present. Rabbits were the first animal model used for the micropocket assays and rats and mice have been the most frequently employed since the late 1970s (Muthukkaruppan and Auerbach, 1979; Fournier *et al.*, 1981).

Recent advances in CNV using transgenic or knockout mouse models allow investigators to evaluate the roles of particular genes and their expressions in corneal angiogenesis. For example, CD18 and intracellular adhesion molecule-1 (ICAM-1)-deficient mice developed significantly lower corneal neutrophil counts, vascular endothelial growth factor (VEGF) mRNA levels and corneal neovascularization than strain-specific normal controls, suggesting that CD18 and ICAM-1 are mediators of the inflammatory and VEGF-dependent CNV (Moromizato *et al.*, 2000). Small leucine-rich proteoglycans play important roles in the organization of the extracellular matrix as well as the regulation of cell behaviour, two biological processes that are essential for angiogenesis. Compared with wild-type mice, in decorin (DCN)-deficient mice, the growth of vessels was significantly diminished, suggesting that DCN is an angiogenesis regulator (Schonherr *et al.*, 2004). In lethally irradiated FVB mice that had been transplanted with bone marrow mononuclear cells from transgenic mice constitutively expressing beta-galactosidase (β-gal), bone marrow-derived endothelial progenitor cells (BM-EPCs) showing β-gal staining comprised a significant portion of the endothelium in VEGF-induced CNV (Murayama *et al.*, 2002). Genetic ablation of chemokine receptors CCR2 or CCR5, which are involved in leukocyte and endothelial chemotaxis, inhibits the development of CNV (Ambati *et al.*, 2003). Matrilysin (matrix metalloproteinase (MMP)-7) deficient mice develop significantly higher CNV compared with wild-type littermates and the differences seem unrelated to the bFGF and VEGF levels in the corneal epithelium, indicating matrilysin's important role in maintaining corneal avascularity during wound healing (Kure *et al.*, 2003).

11.4   The process of corneal neovascularization

A wide variety of pathological conditions, including chemical or mechanical injuries, viral or bacterial infections, parasite infestation and immune-mediated keratitis, can cause CNV, the process of ingrowth of new vessels originated from sclero-corneal limbus vessels. Although different causes or induction techniques may induce corneal angiogenesis in different ways, the principal processes of vessel growth involve the upregulation of pro-angiogenic factors and/or down-regulation of anti-angiogenic regulators. There are complex interactions of cytokines, growth factors and changes in the corneal cellular elements and stroma that interrupt the balance between mechanisms maintaining corneal avascularity and pro-angiogenic factors, thus causing the eventual proliferation and migration of vessel endothelial cells of limbal venules and sprouting in corneal stroma toward the injury site or implanted stimulus, forming a complex new blood vessel network. The front tips of this network are the most angiogenic, from where new sprouts form and the tips migrate further inward (Asahara *et al.*, 1998; Tong and Yuan, 2001). Figure 11.2 summarizes the major cellular and humoral elements involved in corneal angiogenesis.

Chemical cauterization induces cellular responses in epithelium, endothelium and fibroblasts of the cornea, as well as in the vascular cells in the sclero-limbal vessels. The first mitoses in vascular endothelial cells and pericytes were observed 36 h after corneal injury, and the initial capillary sprouts appeared at 39 h (Burger and Klintworth, 1981).

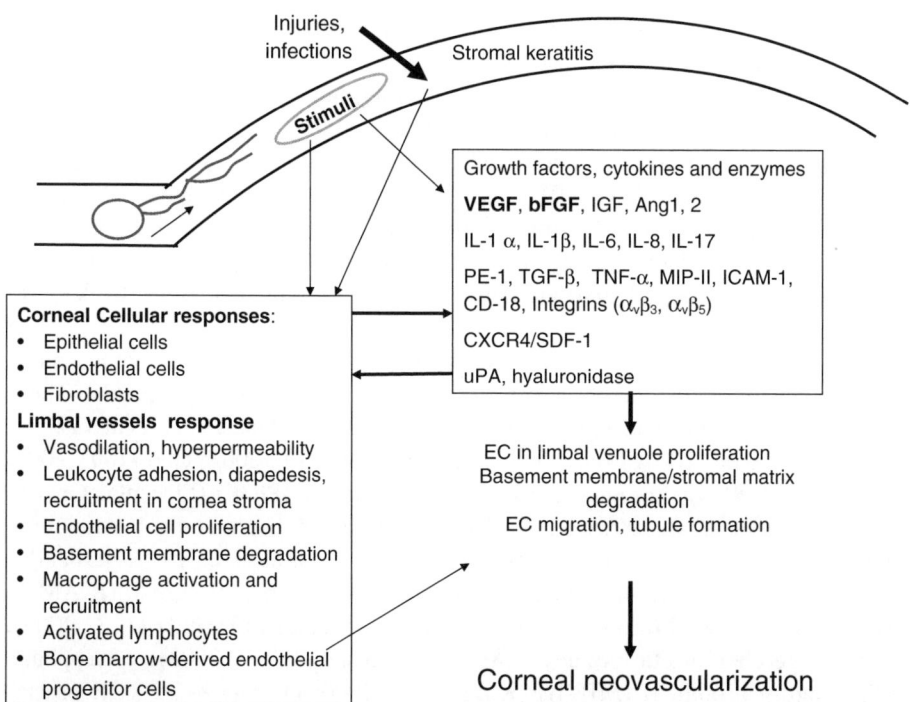

Figure 11.2 A schematic diagram of the corneal neovascularization ang, angiopoietin; bFGF, basic fibroblast growth factor; COX-2, cyclooxygenase-2; CXCR4, chemokine receptors CXCR4 and its ligand SDF; EC, endothelial cells; ICAM-1, intracellular adhesion molecule-1; IGF, insulin-like growth factor; IL, interleukins; MIP-II, macrophage inflammatory protein II; PGE, prostaglandin E; TGF-β, transforming growth factor β; TNF-α, tumour necrosis factor-α; TXB-2, thromboxane B2; uPA-urokinase plasminogen activator; VEGF, vascular endothelial growth factor.

Leukocytes play an important role in CNV. Fluorescent angiography shows that increased leukocyte adhesion in the limbal vessels occurs 24 h after pellet implantation. Margination/diapedesis of leukocytes from the limbal venules was shown 2 days post-implantation. Treatment with anti-ICAM-1 antibody resulted in reduced leukocyte sticking and rolling, and significantly diminished the area covered by new blood vessels (Becker *et al.*, 1999). Polymorphonuclear neutrophils (PMNs) produce pro-angiogenic cytokines, such as VEGF, tumour necrosis factor-alpha (TNF-α), and interleukins (IL-1, IL-6, and IL-8). In addition, PMN-derived proteinases promote endothelial cell migration. CNV in response to fibroblast growth factor-2 (FGF-2; a.k.a. basic fibroblast growth factor, bFGF) is diminished by PMN depletion using serial intraperitoneal injections of monoclonal antibody to Gr-1 (Shaw *et al.*, 2003).

VEGF produced by leukocytes and upregulated by hypoxia after injury/infection is the major cytokine regulating corneal angiogenesis. VEGF elevation correlates temporally and spatially with cautery-induced angiogenesis in the rat cornea. Both inflammatory products and hypoxia appear sufficiently to increase VEGF expression near the cautery lesion to increase vascular permeability of limbal vessels and induce endothelial cell migration and proliferation (Edelman *et al.*, 1999). Increased VEGF, bFGF, and other growth factors by neutrophils promote proliferation and migration of endothelial cells of the limbal vessels. Upregulated matrix metaloproteinases (MMPs) and plasminogen activators induce basement membrane breakdown and facilitate endothelial migration and sprout formation. VEGF also mobilizes bone marrow-derived endothelial progenitor cells (BM-EPCs) into the circulation thereby further augmenting CNV (Takahashi *et al.*, 1999).

In addition to VEGF and bFGF, the two most potent angiogenic factors, many other cytokines, growth factors and enzymes are also involved in angiogenesis. These include angiopoietin 1 and 2 and their receptor Tie-2 (Lin *et al.*, 1997; Asahara *et al.*, 1998; Otani *et al.*, 2001; Lobov *et al.*, 2002; White *et al.*, 2003; Cho *et al.*, 2004; Oliner *et al.*, 2004); Placental-derived growth factor-1 (PlGF) (Placental Growth Factor-1) (Ziche *et al.*, 1997); insulin-like growth factor (Grant *et al.*, 1993a; Grant *et al.*, 1993b); cyclooxygenase-2 (Daniel *et al.*, 1999; Pourtau *et al.*, 1999; Hernandez *et al.*, 2001; Leahy *et al.*, 2002); intracellular adhesion molecule-1 (ICAM-1) (Becker *et al.*, 1999; Moromizato *et al.*, 2000), interleukins (Yoshida *et al.*, 2003); macrophage chemotactic protein-1 (MCP-1) (Yoshida *et al.*, 2003); macrophage inflammatory protein II (MIP-II) (King *et al.*, 2000; Qian *et al.*, 2002); prostaglandin E (PGE) (Frucht and Zauberman, 1984; Ziche *et al.*, 1989; Leahy *et al.*, 2002; Kuwano *et al.*, 2004), all of which have been shown to modify CNV.

11.5   Experimental induction of corneal neovascularization

Almost all types of corneal injury induce neovascularization due, to some extent, to inflammation and therefore many experimental techniques have been employed in the development of corneal angiogenesis assays.

Chemical cauterization

Chemical cautery with silver/potassium nitrate or sodium hydroxide on rabbit or rat cornea elicits neovascularization as capillary sprouts extend centripetally from the corneo-scleral limbus to the cautery site. Typically, the eyes of deep-anaesthetized rat or rabbit are cauterized by applying the tip of silver nitrate applicators (75 per cent silver nitrate 25 per cent potassium nitrate) to the surface of the cornea for 5 s. The detailed techniques can be found in many references (Burger and Klintworth, 1981; Proia *et al.*, 1988; Lai *et al.*, 2002; Kon *et al.*, 2003). Alternatively, thermal cautery can be used (Robin *et al.*, 1985). The chemical cautery technique is easy to perform and useful for the evaluation of compounds with potential inhibitory effects on angiogenesis. Therefore, it is still a preferred model used by some investigators. Its shortcoming is that it is not an ideal model to test locally implanted pro-angiogenic biological substances.

Intrastromal injection

Intrastromal injection of biomaterials, such as human serum albumin (Verbey *et al.*, 1988), bovine albumin (Damms *et al.*, 1997), rabbit serum (Slegers *et al.*, 2000), specific parasitic antigens (Gallin *et al.*, 1988; Verbey and van Haeringen, 1990; Pearlman *et al.*, 1998), lipid hydroperoxide (Conrad *et al.*, 1994) and interleukin-2 (Epstein *et al.*, 1991) induce immuno-inflammatory responses and CNV. The model is mostly used for immune-mediated stromal keratitis and consequent corneal neovascularization.

Mechanical debridement or denudation

Scraping of the corneal and limbal epithelium (Fike and Gellette, 1978; Joussen *et al.*, 2001; Moore *et al.*, 2002; Poulaki *et al.*, 2004) can induce CNV in canine, rabbit and rat eyes. It is simple to perform, and therefore can be used as a screening method for evaluation of potential inhibitors of angiogenesis. The disadvantage may be its inability to generate quantitative stimulation for CNV as well as variations caused in size and depth of scraping between test animals.

Intracorneal suture

Several techniques using single or multiple sutures with 11-0 to 7-0 silk or nylon in rabbit, rat or mouse corneas have been developed to induce CNV (Harvey and Cherry, 1983; Corrent *et al.*, 1989; Friling *et al.*, 1996). Such mechanical manipulation and foreign material existence in the cornea causes inflammatory responses and new blood vessel in-growth from sclero-corneal limbus toward suture site. These techniques are simple and easy to perform, and are good screening methods to investigate the efficacy of photothrombosis therapy for corneal pathological angiogenesis and to evaluate compounds with potential anti-angiogenic effects.

Surgical grafts

Allograft rejection is one of the leading causes of corneal graft failure and one of the most serious complications of perforating keratoplasty in man when the cornea is heavily vascularized. Therefore, seeking effective measures, systemic or topical, to inhibit neovascularization after corneal transplantation is an important research area in vision science (Stark, 1981; Slegers *et al.*, 2000; Qian *et al.*, 2002; Murthy *et al.*, 2003). Allografts and xenografts in rabbits, rats, guinea pigs and mice have also been used as models in studies of angiogenesis in general (Hill and Maske, 1988; Young *et al.*, 1989; Benelli *et al.*, 1997). However, the demanding technical skill for this procedure prevents its wide application as an angiogenesis assay for screening purposes.

Micropocket with implantation of slow-release polymer pellets

Implantation of tumour fragments or tumour cell pellets into a surgically created rabbit corneal pocket to investigate tumour induced neovascularization was first described in 1974 (Gimbrone *et al.*, 1974). This technique has also been applied to rats (Fournier *et al.*, 1981), guinea pigs (Hori, 1990) and mice (Muthukkaruppan and Auerbach, 1979; Chen *et al.*, 1995). A wide variety of materials have been examined for angiogenic stimulatory or inhibitory activities in the cornea. These include tumour tissue (Gross *et al.*, 1981), tumour cells, as well as tumour extracts, concentrated tumour conditioned medium (Conn *et al.*, 1980; Fournier *et al.*, 1981) and other tissues and cells.

As more angiogenic growth factors and cytokines have been discovered and become commercially available, purified growth factors/cytokines, alone or in combination, incorporated into controlled slow release polymers as stimuli, have been widely used (Table 11.1). VEGF and bFGF are the most commonly used angiogenic inducers, both in rats and in mice, due to their potent angiogenic effects and commercial availability. In this chapter, we discuss the micropocket assay in detail, including the preparation of controlled release polymer containing bFGF or VEGF, the surgical procedure and methods of quantification of CNV.

Table 11.1 Angiogenesis stimulators tested in corneal micropocket assays

---

Tissue implants:
  Tumour tissues/cells
    Rat Walker carcino-sarcoma 256, rabbit V2 carcinoma, mouse B16 melanoma, mouse teratoma, retinoblastoma, mammary carcinoma, Lewis lung carcinoma and fibrosarcoma M4, VX2 carcinoma, human endometrial adenocarcinoma, glioma, mouse sarcoma180
  Other tissues and cells
    Ovarian corpus luteum, stimulated T-cells, tumour associated macrophages, macrophages, skeleton muscle extract, human coronary atherectomy specimen, rat self-bone marrow cells
Tumour extract/conditioned medium in polymers
Cytokines in polymers (single or in combination)
    Prostaglandin E1, basic fibroblast growth factor (FGF-2), insulin-like growth factor (IGF), transforming growth factor-beta (TGF-β), VEGF

Preparation of controlled slow-release polymer pellet

Polymer solution.   There are several non-inflammatory polymers that can be used as a sustained release pellet for the CNV micropocket assay. The two most commonly used are poly-2-hydroxylethylmethacrylate (Hydron®, Sigma) and ethylene-vinyl acetate copolymer (Evax®), and both polymers have been tested for their biocompatibility. 'Empty' pellets with a control solution such as PBS did not induce inflammation and CNV for an observation period of up to 3 weeks (Klotz *et al.*, 2000). A recent study has reported an alternative biodegradable vehicle for bFGF, namely an acidic gelatin hydrogel, which yields consistent and reproducible results in induction of CNV (Yang *et al.*, 2000).

We routinely use Hydron® for CNV assays by dissolving Hydron polymer in absolute ethanol (12 per cent W/V) in a sterile tube that is placed on a rotator overnight at 37°C or 3 days at room temperature. The resulting solution can be stored at room temperature before pellet making.

Preparation of testing reagents.   Media conditioned by cultured cells of particular interest are the preferred form of angiogenic biological material to be assayed in many laboratories, because such medium may contain a naturally occurring cohort of pro- and anti-angiogenic molecules. With tumour cells, the balance of these factors is toward promotion of tumour angiogenesis. Details of preparation of conditioned media are available in papers by Polverini, Lingen and colleagues (Polverini and Leibovich, 1984; Lingen *et al.*, 1996).

To simplify the procedure and control the amount of angiogenic stimulus more accurately, we prefer to use commercially available growth factors, commonly bFGF and VEGF, to test putative antiangiogenic compounds. The doses of bFGF and VEGF vary in the literature. We prefer to use bFGF at 50 ng per pellet for rats, and 25 ng per pellet for mice. It has been shown from previous reports (D'Amato *et al.*, 1994) and from our studies that sucrose aluminum sulfate (Sucralfate) protects bFGF from degradation, slows its release and produces consistent aggressive angiogenesis. Therefore, we routinely incorporate Sucralfate into the pellets.

A stock solution of bFGF (recombinant human bFGF, R&D Systems, Inc., Minneapolis, MN, USA) at a concentration of 200 ng/µl is made by dissolving the lyophilized product with sterile PBS containing 0.1 per cent bovine serum albumin. Then 5 µl or 10 µl aliquots of the stock bFGF solution are made and stored at −80°C. Sucralfate (Sigma, St Louis, MO, USA) stock solution is made by suspending it in sterile PBS at 100 µg/µl and storing at 4°C until use.

Constitution of casting gel and pellet making.   The pellets are made the day before the corneal surgery in a laminar flow hood under sterile conditions. Depending on the total number of pellets needed and pellet size (usually 2–5 µl) the total amount of casing gel (Hydron® solution plus mixture of testing agents) is calculated, half consisting of 12 per cent Hydron solution and the other half a mixture of bFGF/Sucralfate and a putative testing bio-substance (if desired) in PBS. This casting gel is promptly pipetted on a sterile PTFE (polytetrafluoroethylene, Teflon®) sheet in a Petri dish. Sucralfate tends to precipitate quickly and Hydron is viscous, so caution should be taken to vortex the stock suspension and the mixture containing Sucralfate during process. The Petri dish containing the pellets is placed in a 4°C refrigerator overnight to allow for polymerization. Dried discs of uniform size (about 2 mm in diameter) are used.

Surgical procedure

We routinely use Fischer 344 rats for the corneal assay. The surgical procedure is modified from that described by Polverini and colleagues (Koch *et al.*, 1991). Following anaesthesia with sodium pentobarbital (Abbott Laboratories, North Chicago, IL, USA) given i.p. at 50 mg/kg body weight, a rat is placed under a dissecting microscope and a drop of Ophthaine is applied on the cornea of one eye. The eye is proptosed by stretching sutures on the upper and lower eyelids. A 3 mm incision is made at the cornea centre and through the outer layer of cornea stroma. A micropocket is created by separating the lamella of the stroma toward the limbus of the outer canthus with a modified iris spatula (No. 10092-12, Fine Science Tools, Inc., Belmont, CA, USA). The distance between the bottom of the micropocket and the limbus is 1 to 1.2 mm. Too deep an incision (near the inter layer of stroma) or the deeper pocket (the bottom closer than 1 mm from the limbus) will cause non-specific inflammation and vessel overgrowth. A prepared sterile pellet is placed near the incision and is rehydrated, so that the pellet can be folded and inserted into the micropocket and positioned at the bottom. The area occupied by the pellet should be no more than half of the pocket; otherwise, the spontaneous complete or partial expulsion of the implant will occur due to transient corneal oedema after surgery. Since the partial expulsion of the implant would interfere with the sealing of the pocket and incision healing, resulting in non-specific inflammation and vigorous neovascularization regardless of the implant property, we prefer to use a small pellet with 2 μl of final casting gel, if possible. The cornea is covered with gentamycin ophthalmic ointment (Alleran, Inc., Irvine, CA, USA) after surgery to maintain the corneal moisture and to provide antibacterial action.

   The polymer pellet and surgical procedure for the mouse micropocket assay has been reported by Chen *et al.* (1995). In this laboratory, we use a modified smaller spatula with similar techniques as for rats to create a corneal pocket. For a more accurate amount of testing protein or stimuli in the implants, we prefer to prepare the pellet as described above for rats, most commonly with 100 ng bFGF and 40 μg Sucralfate per pellet. During surgery, the pellet disc will be cut into four quarters and each quarter will therefore contain about 25 ng bFGF and 10 μg Sucralfate.

Anti-angiogenic treatment for CNV assay

There are several ways, alone or in combination, to administer test compounds for evaluation of their effects on angiogenesis. The selection of administration routes depends on the biological characteristics, bioavailability, pharmacodynamics and pharmacokinetics of test compounds.

Systemic administration

This route is used when a testing material can reach the cornea in adequate amounts via the circulation systemic administration (Kwak *et al.*, 2002; Yoon *et al.*, 2004). Chronic osmotic minipumps implanted in appropriate sites can be used to deliver small amounts of testing materials at a constant rate for up to 2 weeks (Ambati *et al.*, 2002). Systemic administration is more clinically relevant and is ideal to test drugs in preclinical investigations (Danesi *et al.*, 1997; Volpert *et al.*, 1998; Teicher *et al.*, 2002; Shaw *et al.*, 2003).

Topical application

Some compounds have been shown inhibition on CNV induced by different methods with topical application in eye drops (Verbey *et al.*, 1988; Benelli *et al.*, 1997; Siefert *et al.*, 1999), or by direct subconjunctival injection of (Phillips *et al.*, 1992; Lee *et al.*, 2002; Yoshida *et al.*, 2003; Pan *et al.*, 2004). These methods are especially useful for evaluation of anti-angiogenic and anti-inflammatory drugs for ophthalmologic diseases.

Combined or parallel pellet

To test pro- or anti-angiogenic effects of expensive biological materials with limited availability and quantity, such as antibodies, cytokines, recombinant proteins and RNA aptamers, a combined polymer pellet can be used, in which a known angiogenic stimulus plus a test putative biological inhibitor are mixed and incorporated into the casting gel. A known biomaterial with a similar structure to the test substance but without pro- or anti-angiogenic properties plus the same stimulus are incorporated into another pellet and implanted into the opposite eye as a control (Lin *et al.*, 1997; Klotz *et al.*, 2000; White *et al.*, 2003). Some authors use separate pellets for inducer and putative inhibitor, because of the concern that direct interaction between the test bio-substance and the angiogenic stimulator may bias the results. They can be placed into the same pocket (Kim *et al.*, 1999; Simonini *et al.*, 2000) or into two pockets parallel to each other (Shin *et al.*, 2000).

Monitoring and scoring of corneal neovascularization

A variety of methods are used to monitor and score CNV as detailed below.

Slit lamp microscopy

This method allows monitoring and scoring of corneal neovascularization in living animals at different time points. Measurements of the vascularized area and vessel length can be carried out from photographic images. One report describes a technique coupling video data acquisition with computerized analysis of the video images, recording the entire pattern of CNV over time in individual animals. A stereotactic holding and positioning device allows alignment of the cornea such that it is viewed in a known and repeatable way. Contrast between blood vessels and corneal stroma in the images is enhanced by illuminating the cornea with monochromatic light centred on the peak absorption of haemoglobin (Conrad *et al.*, 1994). These CNV scoring techniques require costly facilities, and the new vessels in eyes may not be as easily quantified in comparison with a flattened whole mount (see below) in a two-dimensional image.

Whole mounts of flat cornea

Five to 7 days after pellet implantation, animals receive either fluorescent dye for corneal vessel endothelium staining, or water resistant colloidal carbon suspension to visualize the corneal blood vessels. The corneas are removed and flattened onto slides, which are then imaged at low power on a microscope. Digitized images of the slide mount corneas are further analysed for quantitative or qualitative endpoints.

Fluorescein angiography.   Fluorescein angiography clearly permits visualization of corneal vessels (Asahara *et al.*, 1998; Yang *et al.*, 2000; Sawa *et al.*, 2004) and allows the study of sclero-limbal vascular responses to angiogenic stimuli such as leukocyte adhesion (Becker *et al.*, 1999). VEGF mediated increases in vascular permeability, such as fluorescein leakage can be scored from limbal vessels. For example, the anti-angiogenic VEGF receptor (Flk-1/KDR) inhibitor, SU5416, showed significantly lower leakage scores compared with controls (Takeda *et al.*, 2003).

Endothelial-specific fluorescein-conjugated lectin has been used for staining corneal vessels (Joussen *et al.*, 2001) (Figure 11.3). Anaesthetized animals are given lectin i.v, and 30 min later, the eyes are harvested and fixed with 10 per cent neutral buffered formalin for 24 h. The corneas are subsequently dissected, flat-mounted on glass slides and observed with a fluorescence microscope. Neovascularization is quantified by setting a threshold level of fluorescence,

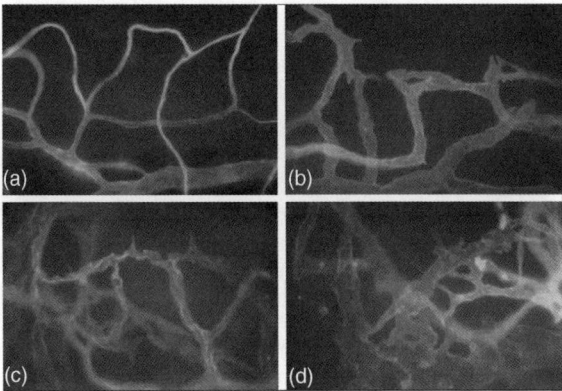

Figure 11.3 Lectin staining of corneal vessels on day 7 after corneal scraping in C57BL6 mice (a) Untreated eyes with short vascular loops at the limbus. (b) Eyes of the TNP-470-injected group in contrast had a shorter vessel length as well as reduced vascular density. (c) Topical treatment resulted in less inhibition. (d) Eyes of the control group showed a dense capillary network. (Joussen, A. M. *et al.*, Investigative Ophthalmology & Visual Science, 42:2510–2516, 2001. Reproduced by permission of IOVS).

above which only vessels are captured. The tissue itself is autofluorescent, which necessitates the shareholding procedure. The total neovascularization area is divided by the total corneal area to yield the percentage of visualization.

Colloidal carbon- perfusion. The most frequent method used to quantitate CNV is image analysis from whole mounts of flat cornea following perfusion with India ink (Proia *et al.*, 1988; Polverini *et al.*, 1989; Shan *et al.*, 2001; Sarayba *et al.*, 2005). CNV scoring is usually done 5–7 days postoperatively. Following anaesthesia with sodium pentobarbital (50 mg/kg for rats), rats are administered heparin 300 IU in 0.3 ml via the tail vein. Five minutes later, the thoracic cavity is opened and the descending aorta is cannulated retrogradely. An incision is made in the right auricle for drainage of blood and saline during flushing. Between 40–60 ml of warm (35°C) saline is perfused at a pressure of 100 mmHg to wash out the blood from the vessels in the upper part of the body. The pressure is maintained by elevating the solution container to a set distance above the animal's body and allowing gravity to provide the flushing pressure. Perfusion is continued until the eyes, ears and nose become pale. Then the major vessels just above the heart are blocked with a haemostat, and 10 ml of Higgins waterproof India ink (Sanford, Bellwood, IL, USA) is injected via the tubing for one minute at a constant rate until the eyes, ears and front paws become black. Some authors suggest adding gelatin to India ink and to freeze the eyes using compressed dichlorodifluoromethane immediately after ink perfusion to solidify the ink in vessels (Proia *et al.*, 1988).

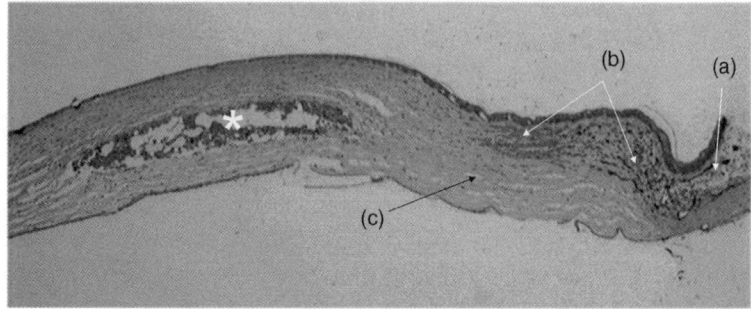

Figure 11.4 bFGF-induced neovascularization in the rat cornea a hydron pellet containing 50 ng bFGF and 20 µg Sucralfate was placed into the surgically created stromal micropocket (*). On postoperative day 5, India ink was perfused through the rat thoracic aorta and the eye was enucleated and fixed in formalin. Histology shows that numerous carbon particles filled capillaries growing centrally from the sclero-corneal limbal venules (a) into the external half of corneal stroma (b). Separation of collagen fibres in the stroma indicates oedema (c).

The eyes are enucleated and fixed in 10 per cent neutralized buffered formalde-hyde overnight. The next day the cornea with a 1 mm rim of adjacent scleral tissue is dissected from the rest of the globe. Three peripheral (about one-third of radium) radial cuts (120° apart) are made and the cornea is gently flattened between two slides. Some authors prefer to dehydrate the flat-mounted cornea before imaging for better visualization of corneal vessels (Proia *et al.*, 1988). Images of flat-mounted corneas are taken using a microscope with a CCD camera connected to a PC equipped with an image acquisition software, such as Scion Image and a frame grabber (Scion Corporation, Frederick, MD). Figure 11.4 shows a corneal histological section of a rat perfused with India ink 5 days after the implantation of a bFGF-containing pellet into the stromal pocket. Example images of inhibited CNV after treatment and control are shown in Figure 11.5.

Image analysis and quantitation of CNV

Some authors regard the corneal bioassay as a qualitative assay and score the neovascularization as positive or negative based on certain criteria (Polverini *et al.*, 1991). However, inhibitory effects and decreased CNV often does not show as absence of vessel growth. Some suggest semi-quantitative scoring on a scale from 0 to 6+ according to the distances from the limbus to the end points of new vessels (Conrad *et al.*, 1994). Alternatively, many other investigators prefer to quantitate different parameters and analyse patterns of vessel growth using more sophisti-cated image analysis.

To quantitate the CNV, several parameters have been determined from digital images of flat-mounted corneas, such as the vascularized area (expressed either as square mm or as the percentage of the whole cornea), the circumference (clock

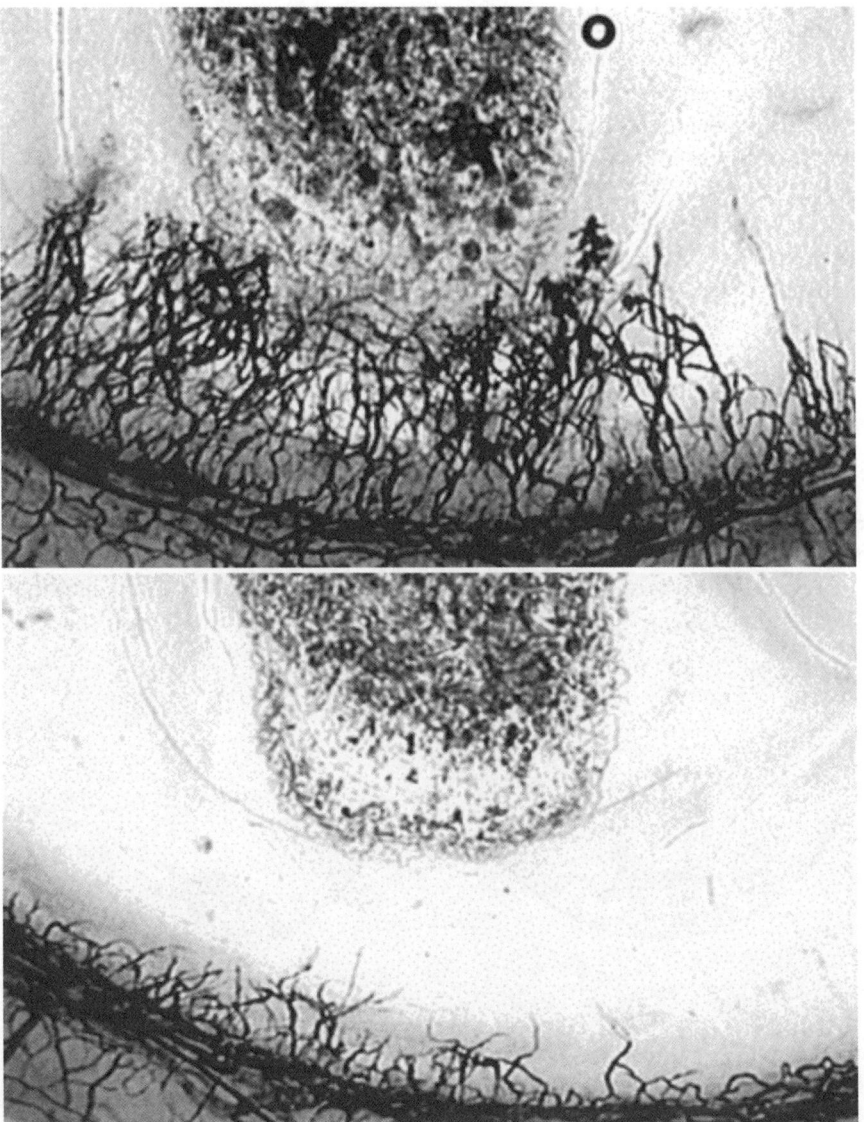

Figure 11.5 The flat-mount corneal images of neovascularization induced by bFGF a polymer pellet was implanted into the corneal micropocket (the *oval dark area*) in one eye of a Fischer 344 rat. On day 7, the corneal neovasculature was visualized by ink angiography. Abundant new vessels grew from the limbus toward the pellet in a control animal (a); whereas vessels in the cornea of a rat treated with SU5416 were scarce (b).

hour), vessel length and vascular density. Detailed vascular patterns such as capillary diameter and vessel tip sprouts can be also determined, and a variety of image analysis software is available (Proia *et al.*, 1988; Asahara *et al.*, 1998; Bocci *et al.*, 1999; Shan *et al.*, 2001; Sarayba *et al.*, 2005). Sometimes, changes in these parameters are not proportional. For example, one might observe significant reduction in circumference of angiogenesis involvement, while observing no difference in vascular density. This may be because the determinants for each response parameter are multifactoral. Some studies show that different angiogenic stimuli induce different patterns of the CNV (Asahara *et al.*, 1998; Ekstrand *et al.*, 2003; Kirchmair *et al.*, 2004). Indeed, a study with mathematic modelling of CNV in rat cornea for quantitative analysis of various biological events involved in angiogenesis reported that redistribution and uptake of angiogenic factors during angiogenesis can have significant effects on the structure of vascular networks (Tong and Yuan, 2001). A decrease in the uptake rate of bFGF results in increases in vessel density, self-loop formation and rapid front migration speed. The randomness in the direction of sprout formation determined the curvature of vessels, whereas the probability of sprout formation from a vessel segment had a significant effect on the total number of vessels in vascular networks (Figure 11.6).

Because of the unpredictability of any one endpoint, we analyse images in several ways. First, we examine India-ink perfused flat-slide mounted cornea images at low magnification to determine the overall quality of the preparation and to measure whole corneal area and diameter. CNV induced by implanted pellet containing angiogenic growth factor usually appears in a reversed fan shape with

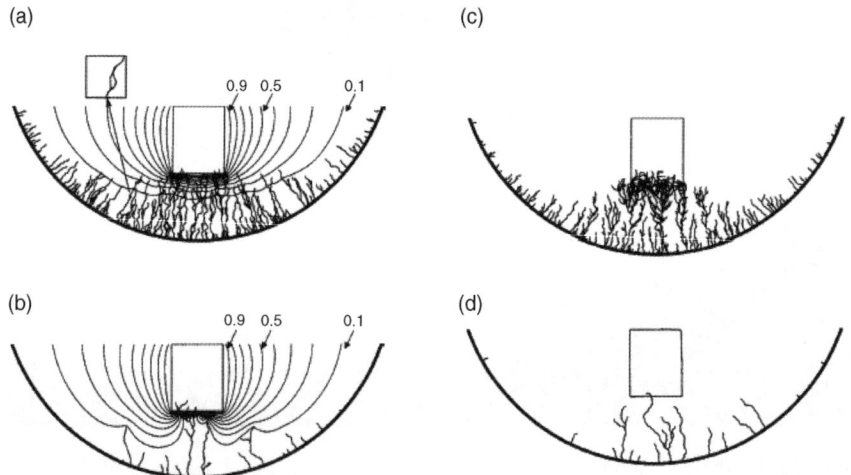

Figure 11.6 A mathematical model of CNV in a mathematical model, diffusion of angiogenic factors, uptake of these factors by endothelial cells, and randomness in the rate of sprout formation and the direction of sprout growth were analysed. Examples show typical vascular networks with different rates of bFGF uptake and different rates of sprout formation, (a) at low uptake rate and (b) at high uptake rate. The concentration distribution in the cornea is shown as isoconcentration lines, (c) at high rate of sprout formation, and (d) at low rate of sprout formation. A brush-border structure formed near the front edge of vascular network at high rate of sprout formation (Tong and Yuan, Microvascular Research **61**, 14–27, 2001. Reproduced by permission of Elsevier).

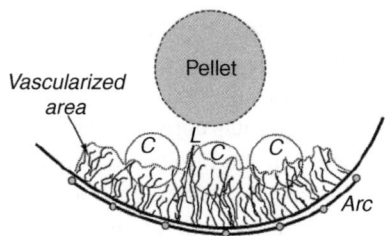

Figure 11.7   Method for quantitation of corneal vascularization the arc (*Arc*) along the limbus for the vascularized area is drawn, and its length was measured for calculation of the circumference of the neovascularization area (i.e. the arc angle). The marked arc was evenly divided into six sections, distances between vessel tips and the limbus at the five intersection points of the arc were measured, and the average vessel length was calculated. The perpendicular distance between the tip of the longest vessel and the limbus (longest vessel length, *L*) was also measured. Three circles (*C*) indicate fields for measuring the vascular area density. The area enclosed by the dotted line and the arc was measured and presented as total vascularized area (Shan *et al.*, Clin. Cancer Res. 7, 2590-6, 2001. Reproduced by permission of the American Association for Cancer Research).

the bottom at the sclero-corneal limbus. However, when a pellet is partially expulsed and dislocated within the corneal incision wound, in addition to vigorous CNV on the pellet side, there is obvious CNV on the other side of the cornea due to non-specific inflammation around the centrally located wound. We exclude such preparations from final analyses. This artifact rarely occurs as long as adequate technical skills are used and the appropriate pellet size is applied.

Images at higher magnification only include the vascularized area for detailed measurement. Figure 11.7 shows our method for digital image analysis using Scion Image or Image J. Both can be downloaded from websites (http://www.scioncorp.com, and http://rsb.info.nih.gov/ij/index.html). By default, all images are 640×480 pixels in size. The shortest distance between the pellet and the limbus is marked as the 'mid-line', and the length is measured. The arc along the limbus for the entire vascularized area is drawn and the length of the arc is measured. The circumference of this area (i.e. the arc angle) is calculated by using the following equation:

$$\text{Circumference (in clock hours)} = \text{Arc length} \times 360/\pi D/30,$$

where D is the corneal diameter.

The marked arc is evenly divided into a few sections. The perpendicular distances between vessel tips and the limbus at the five intersection points of the arc are measured, and the *average vessel length* is calculated. The perpendicular distance between the tip of the longest vessel and the limbus (*longest vessel length*) is also measured. To measure the *vascular area density*, three circles are located between the pellet and the limbus. The total area (pixel numbers) of each circle (Area-T) is determined. With a threshold setting at which level the vascular images are closest to those without threshold, the total vessel area (area-V) represented by total black pixel numbers in each circle is determined. The vascular area density is defined as the ratio of Area-V to Area-T. An average value from these three circles is obtained for each cornea. The reason we use three sampled circles located in the area between the implanted pellet and the limbus, instead of total vascularized area, is because the zone near the limbus even in non-vascular spots often turns black above the threshold, thus the vascular area density is overestimated. Total *vascularized corneal area* is also measured by marking the perimeter of the vascularized area. We also measure the individual implanted pellet area and the distance between pellet and the limbus for quality control. All parameters in pixels are converted to mm or mm^2 by calibration against a stage micrometer image taken at the same magnification and zoom parameters. The total vascularized area can also be expressed as a percentage of the whole cornea.

Histology and biochemistry assay of CNV

The CNV assay allows visual monitoring of the physical changes associated with new vessel growth, and also provides tissue for histology, immunohistochemistry and biochemical studies to elucidate mechanisms that regulate the growth and remodelling of the new vessels at tissue, cellular and molecular levels (Addison *et al.*, 2000; Hamano *et al.*, 2000; Thompson *et al.*, 2003; Gabison *et al.*, 2004).

11.6   Discussion and conclusions

The corneal neovascularization model is probably the most widely used *in vivo* assay for angiogenesis studies. The normally avascular and transparent cornea provides the cleanest tissue in which to observe newly formed blood vessels under a variety conditions in a living animal. There are many options for evaluating the effects of materials of interest with different properties using different administration approaches, either locally, or systemically. It presents an ideal model for investigation of angiogenic mechanisms at morphological and molecular levels, and provides a system for evaluation of anti-angiogenic compounds in pre-clinical trials.

Many methods regarding the induction and scoring of CNV have been developed during extensive research in past decades. It is impossible and unnecessary to standardize this assay, because the purposes of studies, investigator's experiences, preference, technical ability and facilities available are different. Obviously, chemical cauterization, mechanical debridement or intracorneal suture are faster to perform, and are therefore suitable for screening potential angiogenic inhibitors. The micropocket assay, using implantation of slow release polymer pellet(s) containing a known angiogenic factor or testing reagent, offers more flexibility for controlling pro- and anti-angiogenic stimuli and because of this is more easily quantitated.

In our experience, the micropocket assay in rats is easier than mice surgically, and the amounts of reagent in each pellet are more accurate. However, mouse models are more economical and more biological reagents are commercially available, as well as a greater number of tumour lines that grow in mice rather than rats. In addition, transgenic or knockout mouse models represent unique environments and opportunities that are currently not available in rats.

Successful and reproducible micropocket assay implementation requires meticulous surgical skills and practice to minimize corneal trauma. Obtaining a precise incision at the proper depth, and the distance of the micropocket to the sclerocorneal limbus is important. Pre-surgical inspection of the eyes to select animals with healthy corneas is required. We found that ovariectomized rats tend to have corneal scars (unpublished data), probably due to lacrimal gland regression and decreased tear secretion as a consequence of oestrogen/androgen deficiency (Azzarolo *et al.*, 2003). Most image analyses are based on ink or lectin perfused flat mount corneas, so the consistency of perfusion technique, specimen preparation and image acquisition are critically important.

The complex processes of CNV involve numerous molecular and cellular interactions within the cornea. It is possible that there are different pathways of CNV after different injuries or experimental inductions. A study showed that in the rat cornea, alpha (v) beta (3) ligation (alpha (v)-integrin antagonists LM609 and cRGDfV) inhibited bFGF-induced neovascularization, but did not block CNV induced by chemical burn. This suggests an angiogenic pathway independent of alpha(v)-integrins (Klotz *et al.*, 2000). Therefore, a combination with other *in vivo* assays or using two different CNV inducers for a given testing compound may strengthen the evidence of its effects on angiogenesis.

Heterogeneity of the extent of angiogenesis induced by exogenous growth factors may be determined by genetic influences. One study showed the strain-related influences on naive resting limbal vessel phenotype and gene expression whereby bFGF induced CNV was significantly altered in different mouse species (Chan *et al.*, 2004). Therefore, genetic differences in angiogenic responses among different strains must be considered.

Angiogenesis in different tissues is not the same, and CNV is not a universal angiogenesis assay. For example, tumour angiogenesis has its own characteristics that are different from CNV. Evaluation of anti-angiogenic pro-drugs for cancer therapy should therefore not rely on CNV alone. It is desirable that CNV should be combined with other *in vivo* assays to investigate tumour angiogenesis, such as skin/tissue window models (Lin *et al.*, 1997; Blackwell *et al.*, 2000; Shan *et al.*, 2001).

Although the corneal micropocket assay is an excellent *in vivo* angiogenesis model, there are many technical factors that affect the angiogenesis quantitation. It is also relatively time consuming, especially the image analysis of the ink-perfused flat mount corneas, and thus limits its application in large-scale screening. It is necessary to develop an accurate and more efficient automated image analysis computation to facilitate the evaluation of rapidly increasing numbers of potential pro-drugs for treatment of angiogenesis dependent diseases.

Acknowledgements

The authors apologize for not being able to cite many excellent original papers on this subject. The field has been in existence for more than three decades, so the depth and breadth of material is large. We have done our best to provide a broad overview of the subject and have cited example publications that illustrate the various points that we wished to make.

This work was supported in part by the Duke SPORE for breast cancer. The authors appreciate the assistance of Dr Peter Polverini who originally taught us this method and Drs Kathleen L. Johnson and Rachel Richardson who assisted in the editorial preparation of the manuscript.

References

Addison, C. L., Daniel, T. O., Burdick, M. D., Liu, H. *et al.* (2000) 'The CXC chemokine receptor 2, CXCR2, is the putative receptor for ELR+ CXC chemokine-induced angiogenic activity', *J. Immunol.*, **165**(9), pp. 5269–5277.

Ambati, B. K., Joussen, A. M., Ambati, J., Moromizato, Y. *et al.* (2002) 'Angiostatin inhibits and regresses corneal neovascularization', *Arch. Ophthalmol.*, **120**(8), pp. 1063–1068.

Ambati, B. K., Anand, A., Joussen, A. M., Kuziel, W. A. *et al.* (2003) 'Sustained inhibition of corneal neovascularization by genetic ablation of CCR5', *Invest. Ophthalmol. Visual Sci.*, **44**(2), pp. 590–593.

Asahara, T., Chen, D., Takahashi, T., Fujikawa, K. *et al.* (1998) 'Tie2 receptor ligands, angiopoietin-1 and angiopoietin-2, modulate VEGF-induced postnatal neovascularization.[see comment]', *Circ. Res.*, **83**(3), pp. 233–240.

Auerbach, R., Arensman, R., Kubai, L. and Folkman, J. (1975) 'Tumour-induced angiogenesis: lack of inhibition by irradiation', *Int. J. Cancer*, **15**(2), pp. 241–245.

Azzarolo, A. M., Eihausen, H. and Schechter, J. (2003) 'Estrogen prevention of lacrimal gland cell death and lymphocytic infiltration', *Exp. Eye Res.*, **77**(3), pp. 347–354.

Becker, M. D., Kruse, F. E., Azzam, L., Nobiling, R. *et al.* (1999) 'In vivo significance of ICAM-1-dependent leukocyte adhesion in early corneal angiogenesis', *Invest. Ophthalmol. Visual Sci.*, **40**(3), pp. 612–618.

Benelli, U., Ross, J. R., Nardi, M. and Klintworth, G. K. (1997) 'Corneal neovascularization induced by xenografts or chemical cautery. Inhibition by cyclosporin A', *Invest. Ophthalmol. Visual Sci*, **38**(2), pp. 274–282.

Blackwell, K. L., Haroon, Z. A., Shan, S., Saito, W. *et al.* (2000) 'Tamoxifen inhibits angiogenesis in estrogen receptor-negative animal models', *Clin. Cancer Res.*, **6**(11), pp. 4359–4364.

Bocci, G., Danesi, R., Benelli, U., Innocenti, F. *et al.* (1999) 'Inhibitory effect of suramin in rat models of angiogenesis in vitro and in vivo', *Cancer Chemotherapy Pharmacol.*, **43**(3), pp. 205–212.

Brem, H. and Folkman, J. (1975) 'Inhibition of tumour angiogenesis mediated by cartilage', *J. Exp. Med.*, **141**(2), pp. 427–439.

Burger, P. C. and Klintworth, G. K. (1981) 'Autoradiographic study of corneal neovascularization induced by chemical cautery', *Lab. Invest.*, **45**(4), pp. 328–335.

Chan, C. K., Pham, L. N., Chinn, C., Spee, C. *et al.* (2004) 'Mouse strain-dependent heterogeneity of resting limbal vasculature', *Invest. Ophthalmol. Visual Sci.*, **45**(2), pp. 441–447.

Chaudhury, T. K., Lerner, M. P. and Nordquist, R. E. (1980) 'Angiogenesis by human melanoma and breast cancer cells', *Cancer Letters*, **11**(1), pp. 43–49.

Chen, C., Parangi, S., Tolentino, M. J. and Folkman, J. (1995) 'A strategy to discover circulating angiogenesis inhibitors generated by human tumours', *Cancer Res.*, **55**(19), pp. 4230–4233.

Chen, J. J. and Tseng, S. C. (1990) 'Corneal epithelial wound healing in partial limbal deficiency', *Invest. Ophthalmol. Visual Sci.*, **31**(7), pp. 1301–1314.

Cho, C. H., Kammerer, R. A., Lee, H. J., Steinmetz, M. O. *et al.* (2004) 'COMP-Ang1: a designed angiopoietin-1 variant with nonleaky angiogenic activity', *Proc. Natl. Acad. Sci. USA*, **101**(15), pp. 5547–5552.

Conn, H., Berman, M., Kenyon, K., Langer, R. and Gage, J. (1980) 'Stromal vascularization prevents corneal ulceration', *Invest. Ophthalmol. Visual Sci*, **19**(4), pp. 362–370.

Conrad, T. J., Chandler, D. B., Corless, J. M. and Klintworth, G. K. (1994) 'In vivo measurement of corneal angiogenesis with video data acquisition and computerized image analysis', *Lab. Invest.*, **70**(3), pp. 426–434.

Corrent, G., Roussel, T. J., Tseng, S. C. and Watson, B. D. (1989) 'Promotion of graft survival by photothrombotic occlusion of corneal neovascularization', *Arch. Ophthalmol.*, **107**(10), pp. 1501–1506.

Cursiefen, C., Kuchle, M. and Naumann, G. O. (1998) 'Angiogenesis in corneal diseases: histopathologic evaluation of 254 human corneal buttons with neovascularization', *Cornea*, **17**(6), pp. 611–613.

Cursiefen, C., Masli, S., Ng, T. F., Dana, M. R. *et al.* (2004) 'Roles of thrombospondin-1 and -2 in regulating corneal and iris angiogenesis', *Invest. Ophthalmol. Visual Sci.*, **45**(4), pp. 1117–1124.

D'Amato, R. J., Loughnan, M. S., Flynn, E. and Folkman, J. (1994) 'Thalidomide is an inhibitor of angiogenesis', *Proc. Natl. Acad. Sci. USA*, **91**(9), pp. 4082–4085.

Damms, T., Ross, J. R., Duplessie, M. D. and Klintworth, G. K. (1997) 'Intracorneal bovine albumin: an immunologic model of corneal angiogenesis', *Graefes Arch. Clin. Exp. Ophthalmol.*, **235**(10), pp. 662–666.

Danesi, R., Agen, C., Benelli, U., Paolo, A. D. *et al.* (1997) 'Inhibition of experimental angiogenesis by the somatostatin analogue octreotide acetate (SMS 201-995)', *Clin. Cancer Res.*, **3**(2), pp. 265–272.

Daniel, T. O., Liu, H., Morrow, J. D., Crews, B. C. and Marnett, L. J. (1999) 'Thromboxane A2 is a mediator of cyclooxygenase-2-dependent endothelial migration and angiogenesis', *Cancer Res.*, **59**(18), pp. 4574–4577.

Duenas, Z., Torner, L., Corbacho, A. M., Ochoa, A. *et al.* (1999) 'Inhibition of rat corneal angiogenesis by 16-kDa prolactin and by endogenous prolactin-like molecules', *Invest. Ophthalmol. Visual Sci.*, **40**(11), pp. 2498–2505.

Edelman, J. L., Castro, M. R. and Wen, Y. (1999) 'Correlation of VEGF expression by leukocytes with the growth and regression of blood vessels in the rat cornea', *Invest. Ophthalmol. Visual Sci.*, **40**(6), pp. 1112–1123.

Ekstrand, A. J., Cao, R., Bjorndahl, M., Nystrom, S. *et al.* (2003) 'Deletion of neuropeptide Y (NPY) 2 receptor in mice results in blockage of NPY-induced angiogenesis and delayed wound healing', *Proc. Natl. Acad. Sci. USA*, **100**(10), pp. 6033–6038.

Epstein, R. J., Hendricks, R. L. and Harris, D. M. (1991) 'Photodynamic therapy for corneal neovascularization', *Cornea*, **10**(5), pp. 424–432.

Fike, J. R. and Gillette, E. L. (1978) '60Co gamma and negative pi meson irradiation of microvasculature', *Int. J. Radiat. Oncol. Bio. Phys.*, **4**(9–10), pp. 825–828.

Fournier, G. A., Lutty, G. A., Watt, S., Fenselau, A. and Patz, A. (1981) 'A corneal micropocket assay for angiogenesis in the rat eye', *Invest. Ophthalmol. Visual Sci.*, **21**(2), pp. 351–354.

Friling, R., Yassur, Y., Levy, R., Kost, J. *et al.* (1996) 'A role of transforming growth factor-beta 1 in the control of corneal neovascularization', *In Vivo*, **10**(1), pp. 59–64.

Frucht, J. and Zauberman, H. (1984) 'Topical indomethacin effect on neovascularisation of the cornea and on prostaglandin E2 levels', *Br. J. Ophthalmol.*, **68**(9), pp. 656–659.

Gabison, E., Chang, J. H., Hernandez-Quintela, E., Javier, J. *et al.* (2004) 'Anti-angiogenic role of angiostatin during corneal wound healing', *Exp. Eye Res.*, **78**(3), pp. 579–589.

Galardy, R. E., Grobelny, D., Foellmer, H. G. and Fernandez, L. A. (1994) 'Inhibition of angiogenesis by the matrix metalloprotease inhibitor N-[2R-2-(hydroxamidocarbony-methyl)-4-methylpentanoyl)]-L-tryptophan methylamide', *Cancer Res.*, **54**(17), pp. 4715–4718.

Gallin, M. Y., Murray, D., Lass, J. H., Grossniklaus, H. E. and Greene, B. M. (1988) 'Experimental interstitial keratitis induced by Onchocerca volvulus antigens', *Arch. Ophthalmol.*, **106**(10), pp. 1447–1452.

Gillette, E. L., Mauer, G. D. and Severin, G. A. (1975) 'Endothelial repair of radiation damge following beta irradioation', *Radiology*, **116**(1), pp. 175–177.

Gimbrone, M. A., Jr, Leapman, S. B., Cotran, R. S. and Folkman, J. (1973) 'Tumour angiogenesis: iris neovascularization at a distance from experimental intraocular tumours', *J. Natl. Cancer Inst.*, **50**(1), pp. 219–228.

Gimbrone, M. A., Jr, Cotran, R. S., Leapman, S. B. and Folkman, J. (1974) 'Tumour growth and neovascularization: an experimental model using the rabbit cornea', *J. Natl. Cancer Inst.*, **52**(2), pp. 413–427.

Gospodarowicz, D. and Thakral, K. K. (1978) 'Production a corpus luteum angiogenic factor responsible for proliferation of capillaries and neovascularization of the corpus luteum', *Proc. Natl. Acad. Sci. USA*, **75**(2), pp. 847–851.

Grant, M. B., Mames, R. N., Fitzgerald, C., Ellis, E. A. *et al.* (1993a) 'Insulin-like growth factor I acts as an angiogenic agent in rabbit cornea and retina: comparative studies with basic fibroblast growth factor', *Diabetologia*, **36**(4), pp. 282–291.

Grant, M. B., Mames, R. N., Fitzgerald, C., Ellis, E. A. *et al.* (1993b) 'Insulin-like growth factor I as an angiogenic agent. In vivo and in vitro studies', *Ann. NY Acad. Sci.*, **692** (Aug. 27), pp. 230–242.

Gross, J., Azizkhan, R. G., Biswas, C., Bruns, R. R. *et al.* (1981) 'Inhibition of tumour growth, vascularization, and collagenolysis in the rabbit cornea by medroxyprogesterone', *Proc. Natl. Acad. Sci. USA*, **78**(2), pp. 1176–1180.

Hamano, K., Li, T. S., Kobayashi, T., Kobayashi, S. *et al.* (2000) 'Angiogenesis induced by the implantation of self-bone marrow cells: a new material for therapeutic angiogenesis', *Cell Transplant.*, **9**(3), pp. 439–443.

Harvey, P. T. and Cherry, P. M. (1983) 'Indomethacin v. dexamethasone in the suppression of corneal neovascularization', *Can. J. Ophthalmol.*, **18**(6), pp. 293–295.

Hernandez, G. L., Volpert, O. V., Iniguez, M. A., Lorenzo, E. *et al.* (2001) 'Selective inhibition of vascular endothelial growth factor-mediated angiogenesis by cyclosporin A:

roles of the nuclear factor of activated T cells and cyclooxygenase 2', *J. Exp. Med.*, **193**(5), pp. 607–620.

Hill, J. C. and Maske, R. (1988) 'An animal model for corneal graft rejection in high-risk keratoplasty', *Transplantation*, **46**(1), pp. 26–30.

Hori, S. (1990) '[Pathophysiology of intraocular neovascularization]', *Nippon Ganka Gakkai Zasshi Acta Societatis Ophthalmologicae Japonicae*, **94**(12), pp. 1103–1121.

Joussen, A. M., Beecken, W. D., Moromizato, Y., Schwartz, A. *et al.* (2001) 'Inhibition of inflammatory corneal angiogenesis by TNP-470', *Invest. Ophthalmol. Visual Sci*, **42**(11), pp. 2510–2516.

Kaminski, M. and Kaminska, G. (1978) 'Inhibition of lymphocyte-induced angiogenesis by enzymatically isolated rabbit cornea cells', *Archivum Immunologiae et Therapiae Experimentalis*, **26**(1–6), pp. 1079–1082.

Kim, J. H., Kim, J. C., Shin, S. H., Chang, S. I. *et al.* (1999) 'The inhibitory effects of recombinant plasminogen kringle 1–3 on the neovascularization of rabbit cornea induced by angiogenin, bFGF, and VEGF', *Exp. Molec. Med.*, **31**(4), pp. 203–209.

King, W. J., Comer, R. M., Hudde, T., Larkin, D. F. and George, A. J. (2000) 'Cytokine and chemokine expression kinetics after corneal transplantation', *Transplantation*, **70**(8), pp. 1225–1233.

Kirchmair, R., Gander, R., Egger, M., Hanley, A. *et al.* (2004) 'The neuropeptide secretoneurin acts as a direct angiogenic cytokine in vitro and in vivo', *Circulation*, **109**(6), pp. 777–783.

Klotz, O., Park, J. K., Pleyer, U., Hartmann, C. and Baatz, H. (2000) 'Inhibition of corneal neovascularization by alpha(v)-integrin antagonists in the rat', *Graefes Arch. Clin. Exp. Ophthalmol.*, **238**(1), pp. 88–93.

Koch, A. E., Burrows, J. C., Polverini, P. J., Cho, M. and Leibovich, S. J. (1991) 'Thiol-containing compounds inhibit the production of monocyte/macrophage-derived angiogenic activity', *Agents and Actions*, **34**(3–4), pp. 350–357.

Kon, K., Fujii, S., Kosaka, H. and Fujiwara, T. (2003) 'Nitric oxide synthase inhibition by N(G)-nitro-L-arginine methyl ester retards vascular sprouting in angiogenesis', *Microvascular Res.*, **65**(1), pp. 2–8.

Kure, T., Chang, J. H., Kato, T., Hernandez-Quintela, E. *et al.* (2003) 'Corneal neovascularization after excimer keratectomy wounds in matrilysin-deficient mice.[erratum appears in Invest Ophthalmol Vis Sci. 2003 Feb;44(3):960. Note: Gosheh Faris [corrected to Ghosheh Faris]]'. *Invest. Ophthalmol. Visual Sci.*, **44**(1), pp. 137–144.

Kuwano, T., Nakao, S., Yamamoto, H., Tsuneyoshi, M. *et al.* (2004) 'Cyclooxygenase 2 is a key enzyme for inflammatory cytokine-induced angiogenesis', *FASEB J.*, **18**(2), pp. 300–310.

Kwak, D. H., Kim, J. K., Kim, J. Y., Jeong, H. Y. *et al.* (2002) 'Anti-angiogenic activities of Cnidium officinale Makino and Tabanus bovinus', *J. Ethnopharmacol.*, **81**(3), pp. 373–379.

Lai, C. M., Spilsbury, K., Brankov, M., Zaknich, T. and Rakoczy, P. E. (2002) 'Inhibition of corneal neovascularization by recombinant adenovirus mediated antisense VEGF RNA', *Exp. Eye Res.*, **75**(6), pp. 625–634.

Leahy, K. M., Ornberg, R. L., Wang, Y., Zweifel, B. S. *et al.* (2002) 'Cyclooxygenase-2 inhibition by celecoxib reduces proliferation and induces apoptosis in angiogenic endothelial cells in vivo', *Cancer Res.*, **62**(3), pp. 625–631.

Lee, S., Zheng, M., Deshpande, S., Eo, S. K. *et al.* (2002) 'IL-12 suppresses the expression of ocular immunoinflammatory lesions by effects on angiogenesis', *J. Leukocyte Biol.*, **71**(3), pp. 469–476.

Lin, P., Polverini, P., Dewhirst, M., Shan, S. *et al.* (1997) 'Inhibition of tumour angiogenesis using a soluble receptor establishes a role for Tie2 in pathologic vascular growth', *J. Clin. Invest.*, **100**(8), pp. 2072–2078.

Lingen, M. W., Polverini, P. J. and Bouck, N. P. (1996) 'Inhibition of squamous cell carcinoma angiogenesis by direct interaction of retinoic acid with endothelial cells', *Lab. Invest.*, **74**(2), pp. 476–483.

Liu, J. J. and Wilson, S. E. (2001) 'Characterization of human and mouse angiopoietin-like factor CDT6 promoters', *Invest. Ophthalmol. Visual Sci.*, **42**(12), pp. 2776–2783.

Lobov, I. B., Brooks, P. C. and Lang, R. A. (2002) 'Angiopoietin-2 displays VEGF-dependent modulation of capillary structure and endothelial cell survival in vivo', *Proc. Natl. Acad. Sci. USA*, **99**(17), pp. 11205–11210.

Moore, J. E., McMullen, T. C., Campbell, I. L., Rohan, R. *et al.* (2002) 'The inflammatory milieu associated with conjunctivalized cornea and its alteration with IL-1 RA gene therapy', *Invest. Ophthalmol. Visual Sci.*, **43**(9), pp. 2905–2915.

Moromizato, Y., Stechschulte, S., Miyamoto, K., Murata, T. *et al.* (2000) 'CD18 and ICAM-1-dependent corneal neovascularization and inflammation after limbal injury', *Am. J. Pathol.*, **157**(4), pp. 1277–1281.

Murayama, T., Tepper, O. M., Silver, M., Ma, H. *et al.* (2002) 'Determination of bone marrow-derived endothelial progenitor cell significance in angiogenic growth factor-induced neovascularization in vivo', *Exp. Hematol.*, **30**(8), pp. 967–972.

Murthy, R. C., McFarland, T. J., Yoken, J., Chen, S. *et al.* (2003) 'Corneal transduction to inhibit angiogenesis and graft failure', *Invest. Ophthalmol. Visual Sci.*, **44**(5), pp. 1837–1842.

Muthukkaruppan, V. and Auerbach, R. (1979) 'Angiogenesis in the mouse cornea', *Science*, **205**(4413), pp. 1416–1418.

Oliner, J., Min, H., Leal, J., Yu, D. *et al.* (2004) 'Suppression of angiogenesis and tumour growth by selective inhibition of angiopoietin-2', *Cancer Cell*, **6**(5), pp. 507–516.

Otani, A., Takagi, H., Oh, H., Koyama, S. and Honda, Y. (2001) 'Angiotensin II induces expression of the Tie2 receptor ligand, angiopoietin-2, in bovine retinal endothelial cells', *Diabetes*, **50**(4), pp. 867–875.

Pan, X., Wang, Y., Zhang, M., Pan, W. *et al.* (2004) 'Effects of endostatin-vascular endothelial growth inhibitor chimeric recombinant adenoviruses on antiangiogenesis', *World J. Gastroenterol.*, **10**(10), pp. 1409–1414.

Pearlman, E., Hall, L. R., Higgins, A. W., Bardenstein, D. S. *et al.* (1998) 'The role of eosinophils and neutrophils in helminth-induced keratitis.[erratum appears in Invest Ophthalmol Vis Sci 1998 Aug; 39(9):1641]'. *Invest. Ophthalmol. Visual Sci.*, **39**(7), pp. 1176–1182.

Peek, R., van Gelderen, B. E., Bruinenberg, M. and Kijlstra, A. (1998) 'Molecular cloning of a new angiopoietinlike factor from the human cornea', *Invest. Ophthalmol. Visual Sci.*, **39**(10), pp. 1782–1788.

Phillips, G. D., Whitehead, R. A. and Knighton, D. R. (1992) 'Inhibition by methylprednisolone acetate suggests an indirect mechanism for TGF-B induced angiogenesis', *Growth Factors*, **6**(1), pp. 77–84.

Polverini, P. J. and Leibovich, S. J. (1984) 'Induction of neovascularization in vivo and endothelial proliferation in vitro by tumour-associated macrophages', *Lab. Invest.*, **51**(6), pp. 635–642.

Polverini, P. H., Schmizu, K. and Solt, D. B. (1989) 'Genetic control of expression of the angiogenic phenotype', *Northwestern Dental Res.*, **1**(1), pp. 4–7.

Polverini, P. J., Bouck, N. P. and Rastinejad, F. (1991) 'Assay and purification of naturally occurring inhibitor of angiogenesis', *Methods Enzymol.*, **198**, pp. 440–450.

Poulaki, V., Mitsiades, N., Kruse, F. E., Radetzky, S. *et al.* (2004) 'Activin A in the regulation of corneal neovascularization and vascular endothelial growth factor expression', *Am. J. Pathol.*, **164**(4), pp. 1293–1302.

Pourtau, J., Mirshahi, F., Li, H., Muraine, M. *et al.* (1999) 'Cyclooxygenase-2 activity is necessary for the angiogenic properties of oncostatin M', *FEBS Lett.*, **459**(3), pp. 453–457.

Proia, A. D., Chandler, D. B., Haynes, W. L., Smith, C. F. *et al.*(1988) 'Quantitation of corneal neovascularization using computerized image analysis', *Lab. Invest.*, **58**(4), pp. 473–479.

Qian, Y., Hamrah, P., Boisgerault, F., Yamagami, S. *et al.* (2002) 'Mechanisms of immunotherapeutic intervention by anti-CD154 (CD40L) antibody in high-risk corneal transplantation', *J. Interferon Cytokine Res.*, **22**(12), pp. 1217–1225.

Robin, J. B., Regis-Pacheco, L. F., Kash, R. L. and Schanzlin, D. J. (1985) 'The histopathology of corneal neovascularization. Inhibitor effects', *Arch. Ophthalmol.*, **103**(2), pp. 284–287.

Ryu, S. and Albert, D. M. (1979) 'Evaluation of tumour angiogenesis factor with the rabbit cornea model', *Invest. Ophthalmol. Visual Sci.*, **18**(8), pp. 831–841.

Sarayba, M. A., Li, L., Tungsiripat, T., Liu, N. H. *et al.* (2005) 'Inhibition of corneal neovascularization by a peroxisome proliferator-activated receptor-gamma ligand', *Exp. Eye Res.*, **80**(3), pp. 435–442.

Sawa, M., Awazu, K., Takahashi, T., Sakaguchi, H. *et al.* (2004) 'Application of femtosecond ultrashort pulse laser to photodynamic therapy mediated by indocyanine green', *Br. J. Ophthalmol.*, **88**(6), pp. 826–831.

Schonherr, E., Sunderkotter, C., Schaefer, L., Thanos, S. *et al.* (2004) 'Decorin deficiency leads to impaired angiogenesis in injured mouse cornea', *J. Vasc. Res.*, **41**(6), pp. 499–508.

Shan, S., Lockhart, A. C., Saito, W. Y., Knapp, A. M. *et al.* (2001) 'The novel tubulin-binding drug BTO-956 inhibits R3230AC mammary carcinoma growth and angiogenesis in Fischer 344 rats', *Clin. Cancer Res.*, **7**(8), pp. 2590–2596.

Shaw, J. P., Chuang, N., Yee, H. and Shamamian, P. (2003) 'Polymorphonuclear neutrophils promote rFGF-2-induced angiogenesis in vivo', *J. Surg. Res.*, **109**(1), pp. 37–42.

Shin, S. H., Kim, J. C., Chang, S. I., Lee, H. and Chung, S. I. (2000) 'Recombinant kringle 1-3 of plasminogen inhibits rabbit corneal angiogenesis induced by angiogenin', *Cornea*, **19**(2), pp. 212–217.

Siefert, B., Pleyer, U., Muller, M., Hartmann, C. and Keipert, S. (1999) 'Influence of cyclodextrins on the in vitro corneal permeability and in vivo ocular distribution of thalidomide', *J. Ocular Pharmacol. Ther.*, **15**(5), pp. 429–438.

Simonini, A., Moscucci, M., Muller, D. W., Bates, E. R. *et al.* (2000) 'IL-8 is an angiogenic factor in human coronary atherectomy tissue', *Circulation*, **101**(13), pp. 1519–1526.

Slegers, T. P., van Rooijen, N., van Rij, G. and van der Gaag, R. (2000) 'Delayed graft rejection in pre-vascularised corneas after subconjunctival injection of clodronate liposomes', *Curr. Eye Res.*, **20**(4), pp. 322–324.

Stark, W. J. (1981) 'Transplantation antigens and keratoplasty', *Dev. Ophthalmol.*, **5**, pp. 33–40.

Sun, X. C. and Bonanno, J. A. (2002) 'Expression, localization, and functional evaluation of CFTR in bovine corneal endothelial cells', *Am. J. Physiol. Cell Physiol.*, **282**(4), pp. C673–683.

Takahashi, T., Kalka, C., Masuda, H., Chen, D. *et al.* (1999) 'Ischemia- and cytokine-induced mobilization of bone marrow-derived endothelial progenitor cells for neovascularization', *Nature Med.*, **5**(4), pp. 434–438.

Takeda, A., Hata, Y., Shiose, S., Sassa, Y. *et al.* (2003) 'Suppression of experimental choroidal neovascularization utilizing KDR selective receptor tyrosine kinase inhibitor', *Graefes Arch. Clin. Exp. Ophthalmol.*, **241**(9), pp. 765–772.

Teicher, B. A., Alvarez, E., Menon, K., Esterman, M. A. *et al.* (2002) 'Antiangiogenic effects of a protein kinase Cbeta-selective small molecule', *Cancer Chemotherapy Pharmacol.*, **49**(1), pp. 69–77.

Thompson, L. J., Wang, F., Proia, A. D., Peters, K. G. *et al.* (2003) 'Proteome analysis of the rat cornea during angiogenesis', *Proteomics*, **3**(11), pp. 2258–2266.

Tong, S. and Yuan, F. (2001) 'Numerical simulations of angiogenesis in the cornea', *Microvascular Res.*, **61**(1), pp. 14–27.

Verbey, N. L., van Haeringen, N. J. and de Jong, P. T. (1988) 'Modulation of immunogenic keratitis in rabbits by topical administration of inhibitors of lipoxygenase and cyclooxygenase', *Curr. Eye Res.*, **7**(4), pp. 361–368.

Verbey, N. L. and van Haeringen, N. J. (1990) 'The influence of a fish oil dietary supplement on immunogenic keratitis', *Invest. Ophthalmol. Visual Sci.*, **31**(8), pp. 1526–1532.

Volpert, O. V., Fong, T., Koch, A. E., Peterson, J. D. *et al.* (1998) 'Inhibition of angiogenesis by interleukin 4', *J. Exp. Med.*, **188**(6), pp. 1039–1046.

White, R. R., Shan, S., Rusconi, C. P., Shetty, G. *et al.* (2003) 'Inhibition of rat corneal angiogenesis by a nuclease-resistant RNA aptamer specific for angiopoietin-2', *Proc. Natl. Acad. Sci. USA*, **100**(9), pp. 5028–5033.

Yang, C. F., Yasukawa, T., Kimura, H., Miyamoto, H. *et al.* (2000) 'Experimental corneal neovascularization by basic fibroblast growth factor incorporated into gelatin hydrogel', *Ophthal. Res.*, **32**(1), pp. 19–24.

Yoon, S. C., Kim, J. K., Kwak, D. H., Ko, J. J. *et al.* (2004) 'Antitumour activity of Soamsan, a traditional Korean medicine, via suppressing angiogenesis and growth factor transcription', *J. Ethnopharmacol.*, **93**(2–3), pp. 403–408.

Yoshida, S., Yoshida, A., Matsui, H., Takada, Y. and Ishibashi, T. (2003) 'Involvement of macrophage chemotactic protein-1 and interleukin-1beta during inflammatory but not basic fibroblast growth factor-dependent neovascularization in the mouse cornea', *Lab. Invest.*, **83**(7), pp. 927–938.

Young, E., Olkowski, S. T., Dana, M., Mallette, R. A. and Stark, W. J. (1989) 'Pretreatment of donor corneal endothelium with ultraviolet-B irradiation', *Transplant. Proc.*, **21**(1) Pt 3, pp. 3145–3146.

Ziche, M., Maglione, D., Ribatti, D., Morbidelli, L. *et al.* (1997) 'Placenta growth factor-1 is chemotactic, mitogenic, and angiogenic', *Lab. Invest.*, **76**(4) pp. 517–531.

Ziche, M., Alessandri, G. and Gullino, P. M. (1989) 'Gangliosides promote the angiogenic response', *Lab. Invest.*, **61**(6), pp. 629–634.

12

Dorsal air sac model

Sei Yonezawa, Tomohiro Asai and Naoto Oku

Abstract

The dorsal air sac (DAS) model is regarded as an *in vivo* model of angiogenesis, and is especially useful to for studying early phases of tumour-induced angiogenesis. This method characteristically uses tumour cell-secreted angiogenic factors for the induction of angiogenesis. For making a DAS, tumour cells are loaded into a chamber ring covered with 0.45 μm pore filters, and the ring is implanted into the dorsal interspace of mouse, which is formed by an air injection under the hind skin of the animal. In this model, angiogenic factors released from tumour cells pass though the filters, and induce angiogenesis on the skin attached to the ring. In this chapter, we present several examples for evaluating anti-angiogenic therapy by using the DAS model. Finally, we describe a unique usage of the DAS model for obtaining an angiogenic specific peptide which was used for neovessel-targeting probe in a drug delivery system.

Keywords

dorsal air sac; DAS; photodynamic therapy; PDT; anti-angiogenic therapy; anti-neovascular therapy; phage displayed random peptide library; drug delivery system; DDS

12.1 Introduction

Angiogenesis is an indispensable event in tumour progression. Moreover, the newly developed vessels play an important role in blood-borne tumour metastasis. Therefore, Shimizu and Oku (2004) introduced the notion that a number of anti-angiogenic therapies have been developed for the treatment of cancer. *In vivo* assay systems for angiogenesis have also been developed including chick chorioallantoic membrane (CAM) assay, corneal angiogenesis assay and chamber assays as

Angiogenesis Assays   Edited by Carolyn A. Staton, Claire Lewis and Roy Bicknell
© 2006 John Wiley & Sons, Ltd

detailed in other chapters of this book. Oikawa *et al.* (1997) first reported that the DAS model is one of the most easily prepared without requiring adept techniques as the simple implantation of a chamber ring loaded with tumour cells causes angiogenic vessel formation on the murine skin attached to the ring. Therefore, this method has been widely used for studying the mechanisms of tumour-inducing angiogenesis and for evaluating the characteristics of various anti-angiogenic agents. In this chapter we detail how the DAS model is prepared and used by showing several experimental examples including anti-angiogenic photodynamic therapy (PDT) and angiogenesis suppression by tea polyphenol.

12.2   Preparation of the DAS model

The scheme of preparing DAS model is shown in Figure 12.1. The chamber ring is composed of a plastic 1 cm diameter ring (Millipore) and nitrocellulose membrane filters (0.45 μm pore size, Millipore), which are secured on both sides of the ring by using CEMENT (Millipore), and the chamber is then sterilized by gamma irradiation or ethylene oxide. Then, cultured tumour cells are diluted with saline or cell culture medium (1×10^7 cells/0.15 ml), and loaded into the chamber ring through a hole in the side (see Figure 12.1), which is then sealed using the supplied plastic bar (Millipore). For the negative control, a chamber ring loaded with saline or medium is also prepared, by soaking an empty sealed ring in saline before implantation.

Mice are first anaesthetized by intraperitoneal injection of nembutal (Dainippon Pharmaceutical Co., Ltd. Japan), for example. Then the dorsal skin is lifted by the tips of thumb and fingers, pricked with a needle subcutaneously, and air (about 8 ml) is pumped into the back. After cutting the skin of the back enough to insert the chamber ring, it is implanted into the interspace. The cut area is then sutured.

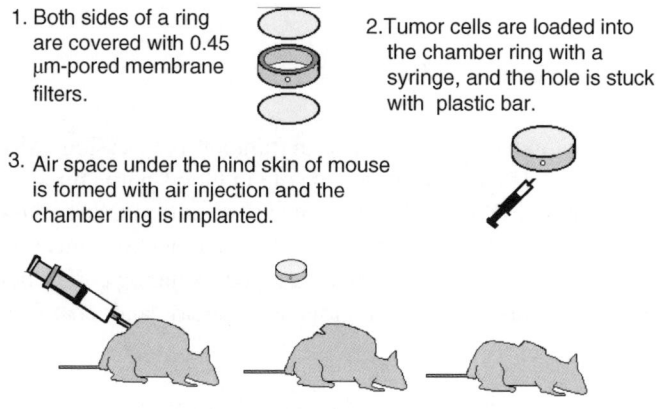

1. Both sides of a ring are covered with 0.45 μm-pored membrane filters.

2. Tumor cells are loaded into the chamber ring with a syringe, and the hole is stuck with plastic bar.

3. Air space under the hind skin of mouse is formed with air injection and the chamber ring is implanted.

Figure 12.1   Preparation of DAS model.

Because of the transplant surgery under general anaesthesia, heat retention is recommended until mice recover from anaesthesia.

Tumour-induced angiogenesis on the skin attached to the chamber ring is obvious by day 4 or 5 after implantation. To observe the neovessels, the mouse is sacrificed under anaesthesia, and a portion of the skin including the chamber ring is cut off and the chamber ring carefully detached from the skin. Since the chamber ring is attached firmly to the back skin, and the newly produced blood vessels are fragile, partial bleeding is often observed on the skin.

12.3   Evaluation of anti-angiogenic photodynamic therapy using a DAS model

This section details an example where the DAS model has been used. PDT is a potential cancer treatment method that uses a combination of photosensitizer and tissue-penetrating laser light. Dolmans *et al.* (2003), Peng *et al.* (2001) and Reuther *et al.* (2001) conclude that when cancer patients, or experimental animals, are treated with photosensitizers and laser light with an appropriate wavelength at the tumour site, it results in the topical production of activated oxygen which effects tumour destruction. Kurohane *et al.* (2001) previously observed that laser irradiation at 15 min post-injection of photosensitizer (15-min PDT), strongly suppressed tumour growth due to haemostasis, compared with conventional scheduling of PDT (laser irradiation at 3 h post-injection, 3-h PDT). Ichikawa *et al.* (2004) examined the effect of 15-min PDT on angiogenic vessels in comparison with that of 3-h PDT in mice using the DAS model. In this experiment angiogenic vessels formed in the DAS model were irradiated with laser at 15 min or 3 h post-injection of Visudyne®, a photosensitizing drug that is used clinically for the treatment of age-related macular degeneration (liposomal drug of benzo-porphyrin derivative monoacid ring A, BPD-MA).

In this study, a chamber ring loaded with Meth A sarcoma cells was implanted dorsally into 5-week-old BALB/c male mice (Japan SLC Inc., Shizuoka, Japan). PDT treatment was preformed on day 4 after ring implantation. Fifteen min or 3 h after intravenous administration of Visudyne®, skins on the chamber rings were irradiated with a laser light (wave length; 689 nm). At 24 h after PDT treatment, the mice were sacrificed and the skin attached to the chamber ring was examined in the same manner as Takeuchi *et al.* (2002), Asai *et al.* (2002) and Oku *et al.* (2002).

Figures 12.2a and 12.2b show the typical images of the skin attached to the chamber ring preloaded with saline or tumour cells, respectively. In saline rings only pre-existing blood vessels are seen (Figure 12.2a), while many angiogenic vessels, as well as pre-existing blood vessels, are observed in rings pre-loaded with tumour cells (Figure 12.2b). Some of these angiogenic vessels were haemolysed since they were fragile. After 15-min PDT treatment, the haemorrhagic

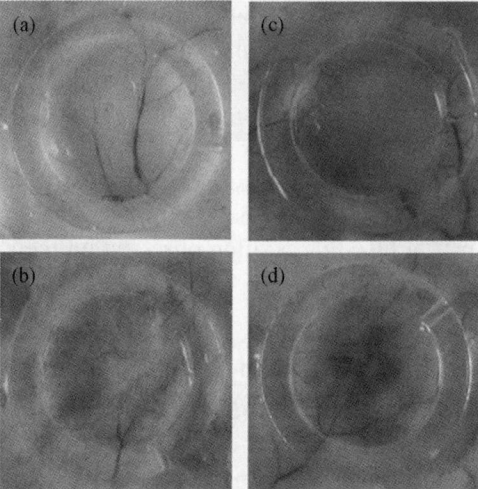

Figure 12.2 Skin angiogenesis in the DAS model usage of a DAS model in PDT is shown. Saline-loaded (a) or Meth-A sarcoma (1 × 10^7 cells/0.15 ml)-loaded (b–d) chamber rings were dorsally implanted into BALB/c mice. At 4 days after implantation of the chamber ring, PDT treatment was performed by an i.v. injection of saline (a, b) or Visudyne®, 0.25 mg/kg in terms of BPD-MA (c, d). The animals were exposed to a laser light of 689 nm with 150 J/cm^2 of fluence at 15 min (c) or 3 h (d) post-injection of Visudyne®. At 24 h after PDT treatment, the mice were sacrificed; and the neovascularized dorsal skin was resected for observation. Data reproduced with kind permission from *Cancer Letters*, Ichikawa *et al.*, 2004.

photodamage of the angiogenic site was dramatically increased (Figure 12.2c). In contrast, no significant change was observed after 3-h PDT (Figure 12.2d). These results suggest that the enhanced anti-tumour activity after 15-min PDT is mainly introduced by endothelial cell degeneration, leading to vascular haemorrhage rather than by tumour cell damage, although Visudyne® taken up by tumour cells as early as 15 min post-injection simultaneously may cause the cell death after laser irradiation.

12.4 Determination of blood volume or blood flow in the angiogenic site using the DAS model

The DAS method has recently been modified to measure blood volume in the angiogenesis-induced skin as determined by the amount of ^{51}Cr-labelled erythrocytes circulating in the skin (Funahashi *et al.*, 1999). Radiolabelling of erythrocytes is carried out by a simple incubation of erythrocytes with [^{51}Cr] sodium chromate. Kurohane *et al.* (2001) utilized this method to observe the effects of 15-min PDT on blood volume. In brief, erythrocytes freshly prepared from BALB/c mice were incubated for 30 min at 37°C with 9.25 MBq of [^{51}Cr] sodium chromate (Daiichi Pure Chemical Co., Ltd., Tokyo, Japan). After washing, the radiolabelled erythrocytes were suspended in saline. The chamber-ring-bearing mice were prepared as described previously and allowed to develop for 4 days prior to PDT treatment. Fifteen minutes or 3 h after intravenous injection of liposomal BPD-MA (a sensitizer similar to Visudyne®, with a different liposomal composition) into the DAS model mice, skins on the chamber rings were irradiated by a laser light. At 1, 6 and 24 h after PDT treatment, the mice were injected via the tail vein with ^{51}Cr-labelled erythrocytes (1×10^8 cells/0.2 ml). Control mice, namely those without treatment, those injected with saline and then irradiated, and those injected with liposomal BPD-MA without laser irradiation were also injected with ^{51}Cr-labelled erythrocytes. At 30 min after erythrocyte injection, the mice were sacrificed with diethylether anaesthesia, and the skin attached the chamber ring was cut away. The radioactivity of ^{51}Cr in the skin was measured using a gamma counter. ^{51}Cr accumulated in angiogenic skins 1 h after PDT (see Figure 12.3), indicating the erythrocytes were unable to flow away through the vessels, suggesting that there was endothelial cell damage. The flow of ^{51}Cr-labelled erythrocytes into the skin was recovered to the control level 24 h after 3-h PDT. On the other hand, complete suppression of blood flow was observed 24 h after 15-min PDT, suggesting that anti-angiogenic PDT causes haemostasis.

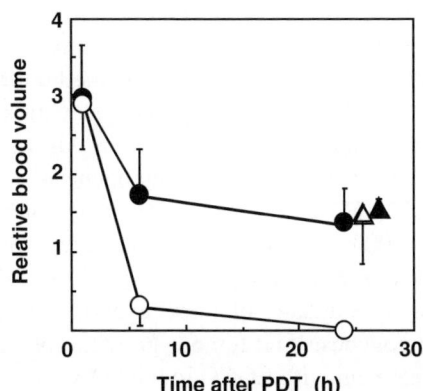

Figure 12.3 Blood volume change at the PDT-treated neovascular site meth a sarcoma cell-loaded chamber rings (1×10^7 cells/0.15 ml) were implanted into the dorsal air sac of mice. Day 4 after the implantation, the mice were injected via a tail vein with 2 mg/kg of liposomal BPD-MA. Then the mice were kept in the dark for 15 min (open circles) or 3 h (closed circles) before irradiation with 689 nm laser light (150 J/cm^2). At 1, 6, or 24 h after the irradiation, the mice were injected via a tail vein with ^{51}Cr-labelled erythrocyte (1×10^8 cells/0.2 ml). Thirty minutes post-injection, the PDT-treated neovascularized skin was removed and the radioactivity was then counted with a gamma counter. Blood volume is presented as ratio of treated to control values. Each point represents the mean ± S.D. (n = 4). Open triangle, laser irradiation alone; closed triangle, liposomal BPD-MA injection without laser irradiation. Data reproduced with kind permission from *Cancer Letters*, Kurohane *et al.*, 2001.

12.5   Quantitative or semi-quantitative analysis of angiogenesis in the DAS model

For qualitative analysis of angiogenesis, the skin attached to the chamber ring of a DAS model is examined with naked eyes or under a microscope. Imaging software enables quantification of vessel density using photographs of angiogenic skins, however, it is sometimes difficult to subtract pre-existing vessels from this analysis. In contrast, the Evans blue method is quite easy and useful for semi-quantitative evaluation of angiogenesis. This method utilizes the fact that the endothelial cell layer of angiogenic neovessels is characteristically weak and leaky. Since angiogenic vessels are leaky enough to allow outflow of the dye, Evans blue accumulates in the interstitial spaces around the neovessels, but does not leak out from pre-existing vessels. Therefore, the amount of accumulated dye between angiogenic skin and control skin is quite different.

The Evans blue is used by intravenous injection of 0.2 ml of 1 per cent (w/v) Evans Blue (Wako Pure Chemical, Osaka, Japan) in phosphate-buffered saline (PBS), pH 7.4, into the DAS bearing mice via the tail vein, 4–5 days after ring implantation. One minute after injection, the mice are sacrificed under anaesthesia and the angiogenic skin is detached. The skin is then frozen on dry ice and cut along the chamber ring. The pigment in the skin attached to the chamber ring is extracted using 0.15 per cent Na_2SO_4-70 per cent acetone for 1 h at room temperature. The extracted solution is then centrifuged at 12 000 x g for 1 min, and the absorbance of the supernatant is determined at 620 nm.

Yamakawa *et al.* (2004) applied this Evans blue method for evaluating the anti-angiogenic effect of a tea polyphenol, (–)-epigallocatechin gallate (EGCG), in the DAS model. Matrix metalloproteinases (MMPs) are known to play an important role in endothelial cell invasion during angiogenesis and we reported that membrane type-1 MMP (MT1-MMP) activity was potentially suppressed by EGCG in Oku *et al.* (2003). Therefore, we examined the effect of EGCG on invasion and tube formation of endothelial cells *in vitro*, and on angiogenesis and tumour growth *in vivo*. The results of these experiments indicated that EGCG actually suppressed angiogenesis and tumour growth, potentially via the suppression of MT1-MMP. Treatment with EGCG in the DAS model showed a significant suppression of angiogenesis (Figure 12.4). In this experiment a chamber ring loaded with HT1080 human fibrosarcoma cells was prepared and implanted dorsally into 5-week-old BALB/c male mice. EGCG (40 mg/kg, 0.2 ml/mouse/day) or PBS was injected intravenously into a tail vein for 4 days from the day of implantation. At day 4 after ring implantation, the skin attached to the chamber ring was surgically removed, and the degree of neovascularization was semi-quantified by Evans blue method. Negative controls were obtained utilizing mice implanted with medium-loaded chamber rings. A strong induction of angiogenesis was observed when HT1080 fibrosarcoma cells were loaded into the chamber rings (Figure 12.4a), in contrast to the controls where implantation of

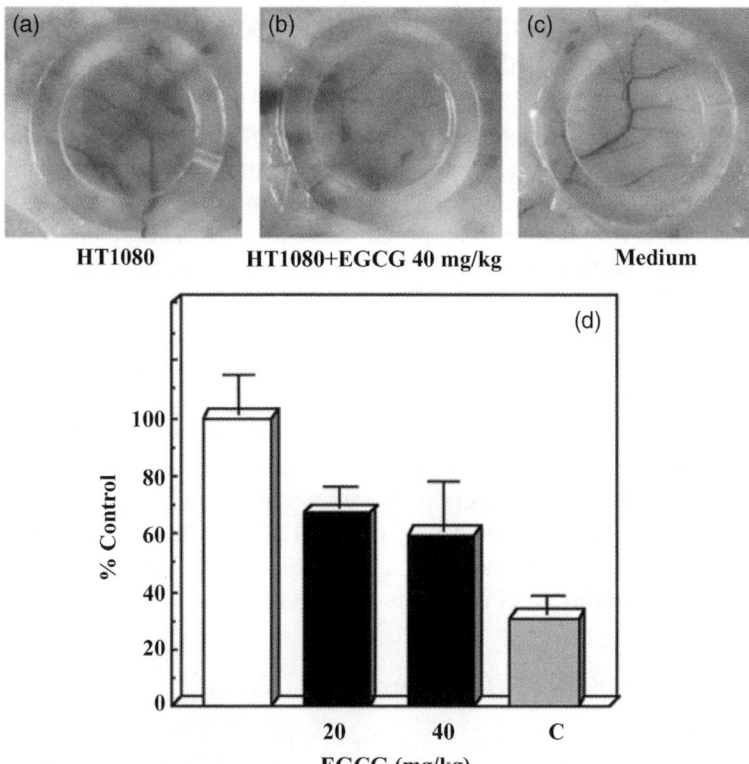

Figure 12.4   Suppression of angiogenesis with EGCG treatment a chamber ring loaded with HT1080 fibrosarcoma cells (a, b) or with medium alone (c) was dorsally implanted into mice. The mice were injected with 0.2 ml of EGCG (40 mg/kg, b) or PBS (a, c) everyday from day 0 to day 3 after implantation of the chamber ring. At day 4, mice were sacrificed and the attached skin on the chamber ring was observed. d, A chamber ring loaded with HT1080 fibrosarcoma or with medium alone was similarly implanted as negative control (shadow bars). The mice were injected with 0.2 ml of EGCG (closed bars) or PBS (open bars) everyday from day 0 to day 3 after implantation of the chamber ring. At day 4, Evans blue was injected into the mice and at 1 min after the injection, they were sacrificed and the attached skin on the chamber ring was cut out. The pigment was extracted with 0.15 per cent Na_2SO_4-70 per cent acetone and measured absorbance at 620 nm. Data are presented as mean $+/-$ SEM. Data reproduced with kind permission from *Cancer Letters*, Yamakawa *et al.*, 2004.

a medium-loaded chamber ring did not induce angiogenesis (Figure 12.4c). EGCG treatment (40 mg/kg/day) resulted in suppression of angiogenesis (Figure 12.4b). Corresponding to these observations, accumulation of Evans blue in the angiogenic skin was reduced in EGCG-treated group (40.6 per cent reduction in this experiment) compared with PBS controls (Figure 12.4d).

12.6   Isolation and application of neovessel-targeting probe using DAS model

As tumours are dependent upon angiogenesis, disruption of this process proves an attractive target for the development of potential therapies. Current anti-cancer drugs cause damage to growing cells thereby affecting the neovessels as well as the tumour cells. We previously proposed that anti-neovascular therapy, whereby neovessel-targeted drug carriers are used to deliver anti-cancer drugs to angiogenic endothelial cells, may prove effective in cancer treatment via indirect lethal damage to tumour cells through damage to neovessels, which would block the supply of oxygen and nutrients to the tumour cells. For this purpose, we designed liposomes modified with a neovessel specific probe as drug carriers, and we iso-lated peptides specific for tumour angiogenic vasculature as the probe molecules by using a phage-displayed peptide library, described in Oku *et al.* (2002). *In vivo* biopanning of phage-displayed peptide library was performed in angiogenic model mice prepared by a DAS method (see Figure 12.5). In this experiment, a DAS model was quite advantageous because it enabled us to isolate specific phage clones having the ability to bind only to angiogenic vessels, not to tumour cells.

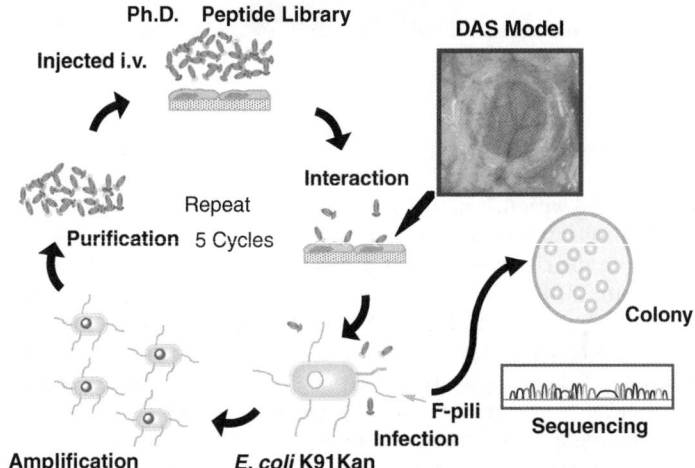

Figure 12.5 *In vivo* biopanning by using a DAS model the DAS model mice were prepared by using highly metastatic murine B16BL6 melanoma cells (1×10^7 cells/ring). Five days after implantation, these mice were used for *in vivo* biopanning. A phage-displayed random peptide library expressing pentadecapeptides at the N terminus of pIII coat protein of M13 phage was injected into the DAS model mice via a tail vein. Four minutes after injection, the phages that had accumulated in angiogenic vessels were recovered and titrated. Biopanning steps were repeated for five cycles. After selected phages were cloned, sequences of the presented peptides were determined. The purified peptides obtained by this method potentially associate with neovessels. (A colour reproduction of this figure can be viewed in the colour section towards the centre of the book).

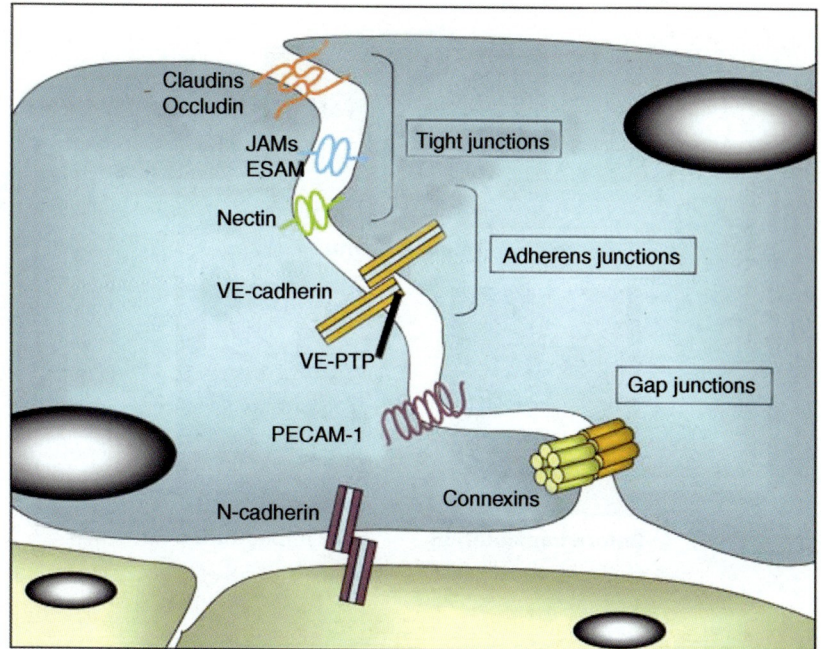

Figure 1.1

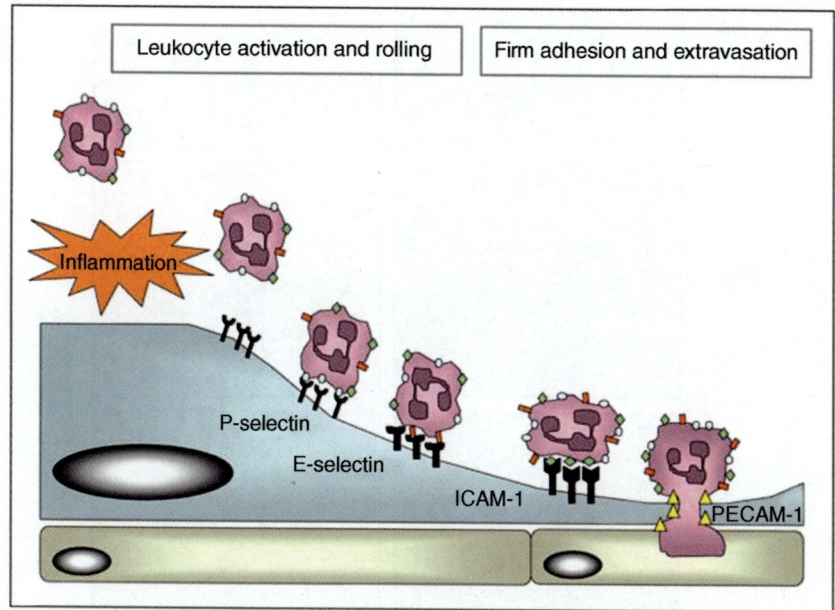

Figure 1.2

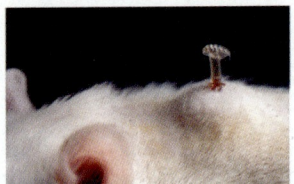

Implant in situ

Before implantation

After implantation

Figure 9.1

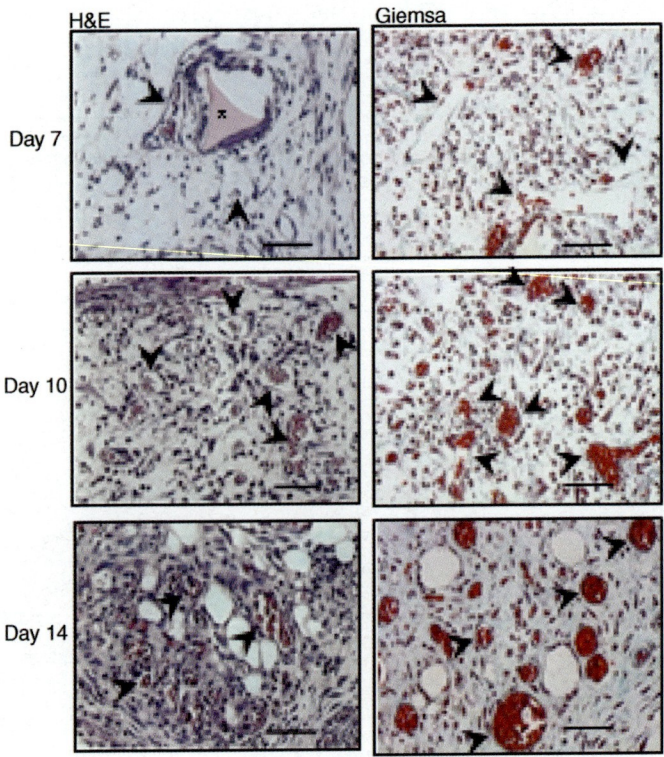

Figure 9.2

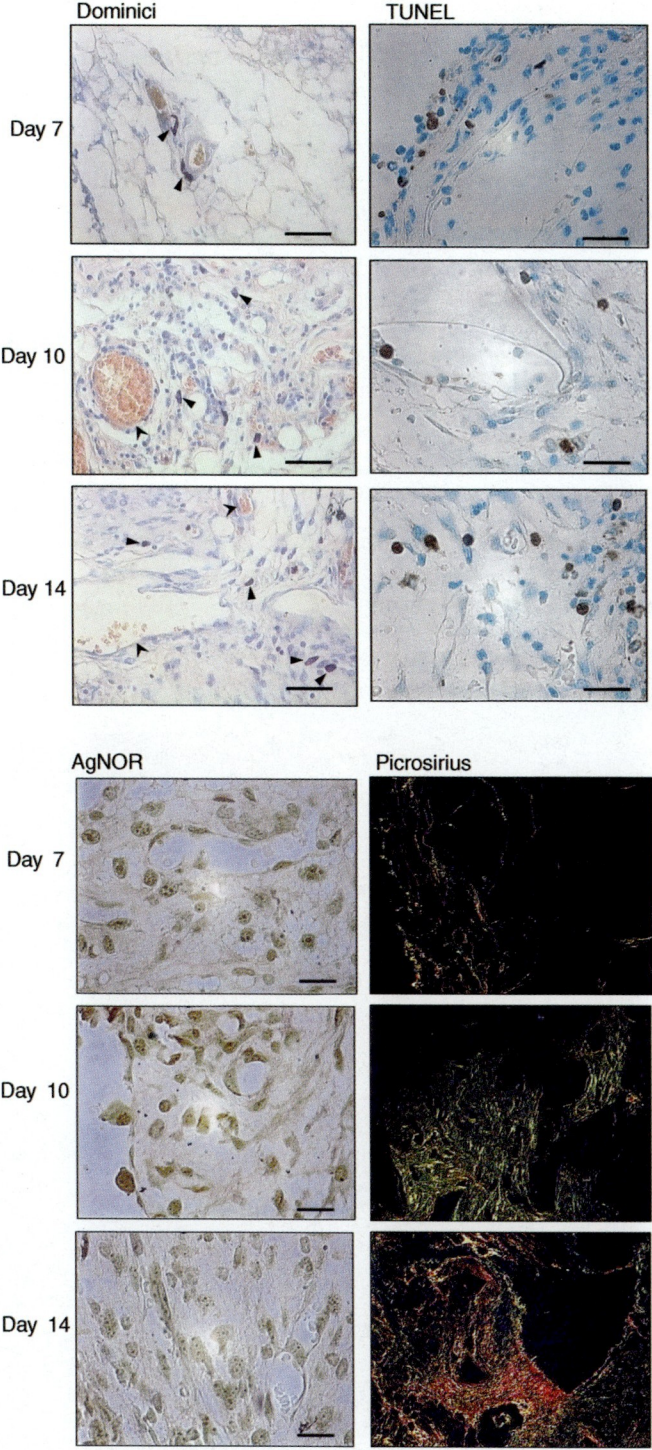

Figure 9.2 (*Continued*)

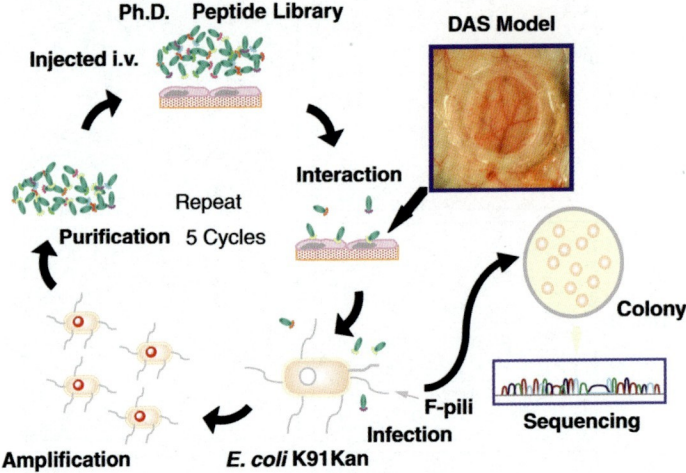

Figure 12.5

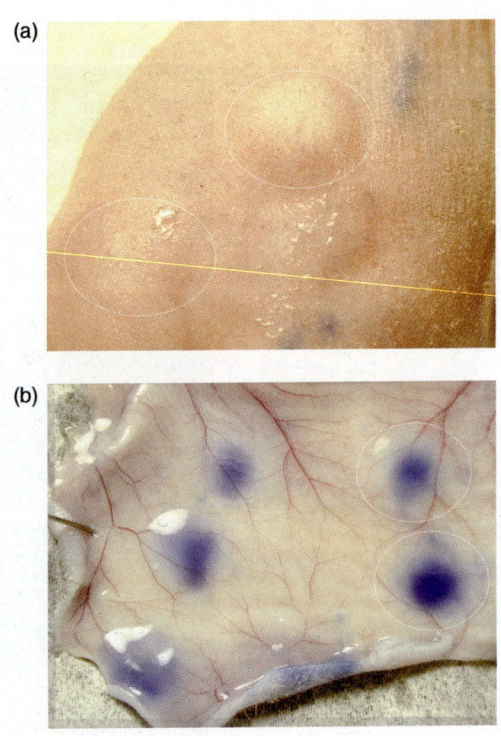

Figure 14.1

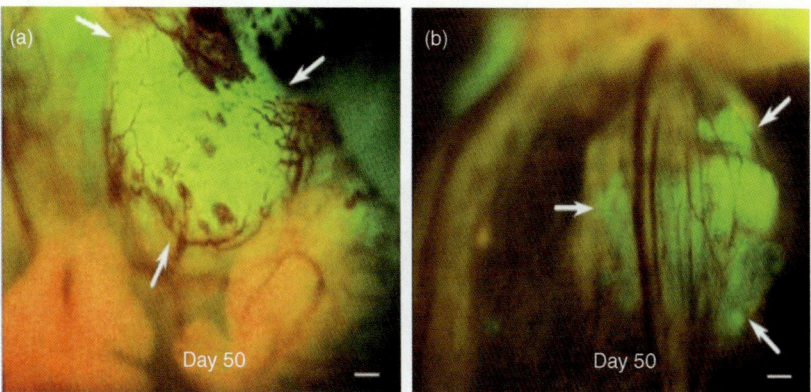

Figure 15.1

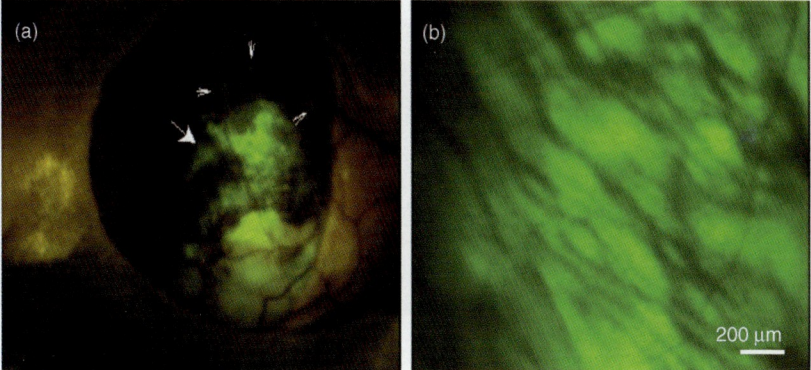

Figure 15.2

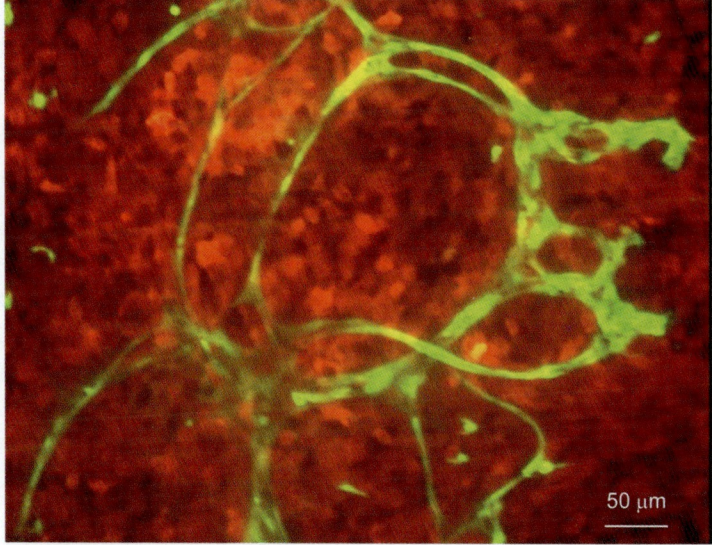

Figure 15.3

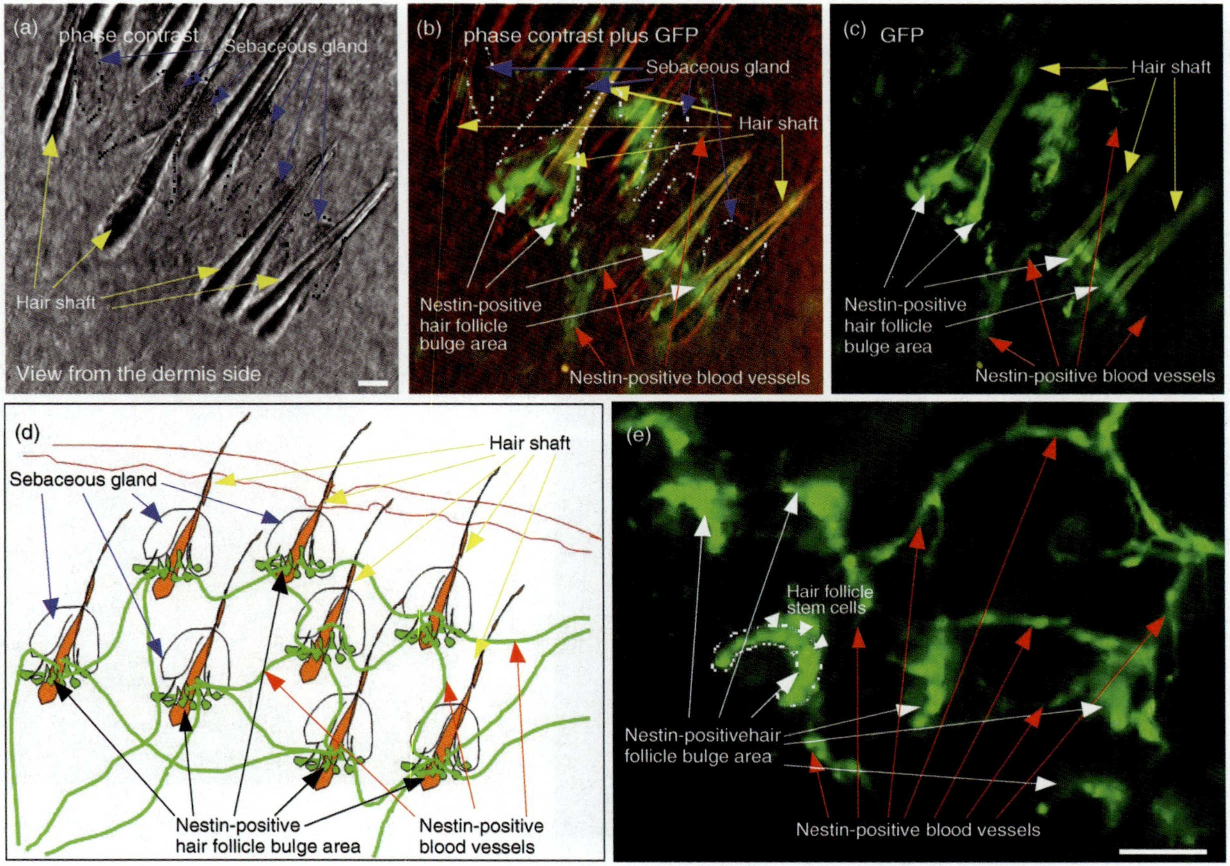

(a) phase contrast
Sebaceous gland
Hair shaft
View from the dermis side

(b) phase contrast plus GFP
Sebaceous gland
Hair shaft
Nestin-positive hair follicle bulge area
Nestin-positive blood vessels

(c) GFP
Hair shaft
Nestin-positive hair follicle bulge area
Nestin-positive blood vessels

(d)
Hair shaft
Nestin-positive blood vessels
Nestin-positive hair follicle bulge area
Sebaceous gland

(e)
Hair follicle stem cells
Nestin-positive hair follicle bulge area
Nestin-positive blood vessels

Figure 15.4

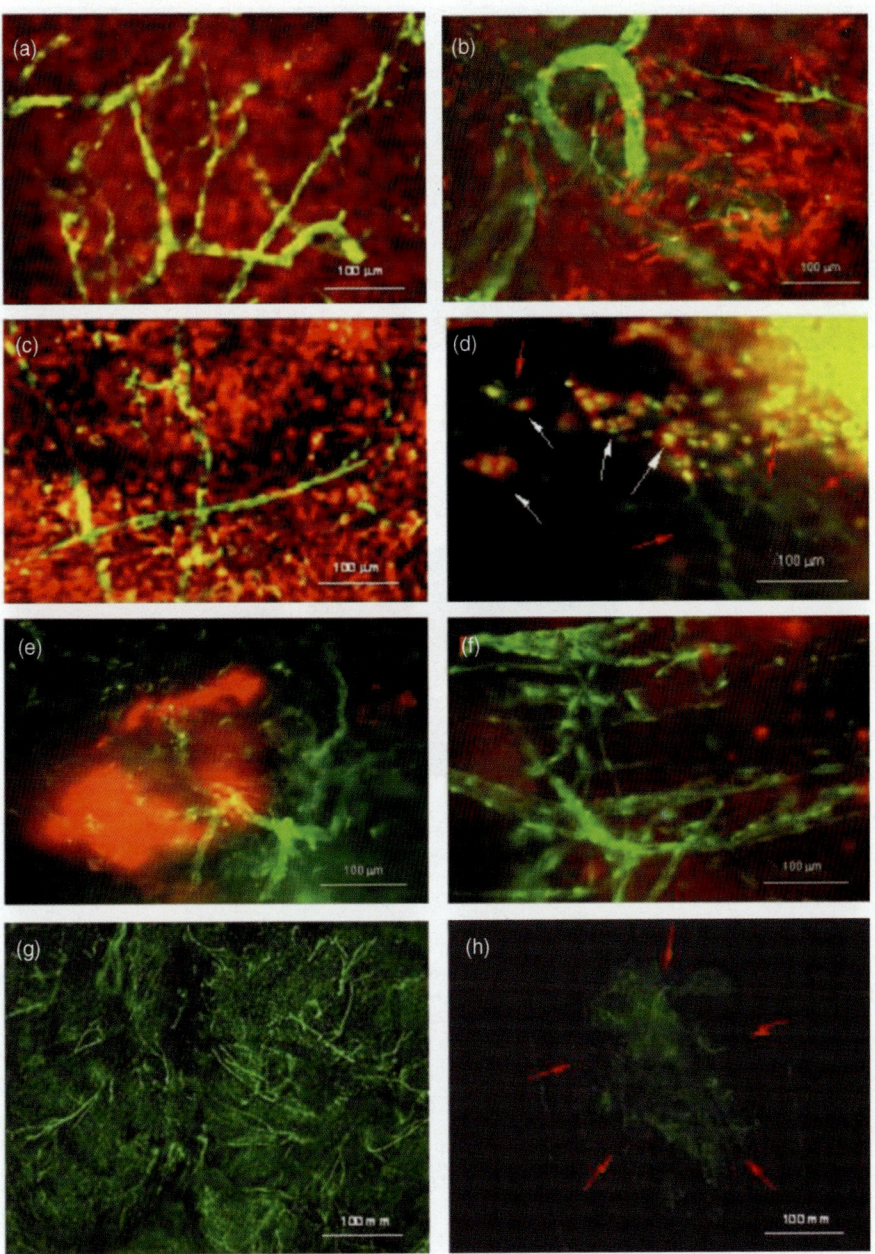

Figure 15.5

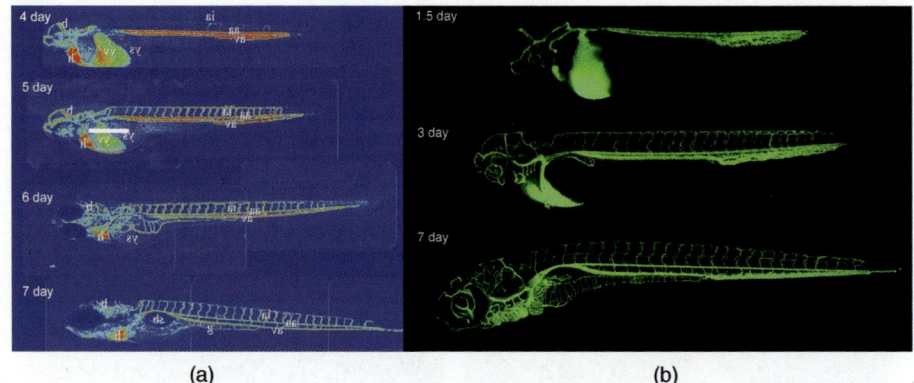

(a)   (b)

Figure 16.2

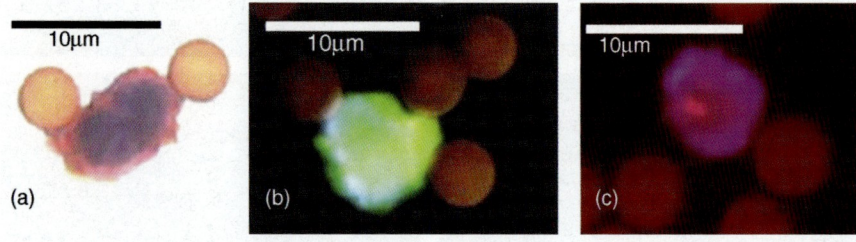

(a)   (b)   (c)

Figure 17.1

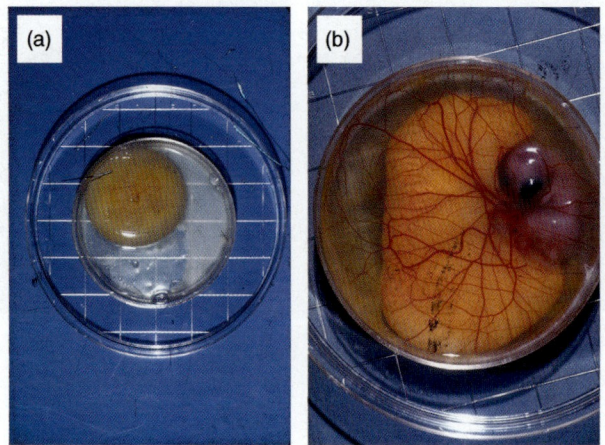

Figure 19.3

After determination of the epitope sequences of the obtained phage clones, APRPG-modified liposome was revealed to be favourable for the active targeting of the angiogenic endothelium. Furthermore, Maeda *et al.* (2004) reported that anti-neovascular therapy using this type of liposome as a carrier of adriamycin markedly suppressed tumour growth possibly through a cytotoxic effect of the drug against angiogenic endothelial cells.

12.7 Conclusions

The advantages of a DAS model are that this approach is simply carried out without special skills, tumour-induced angiogenesis is reproduced in natural environment, and tumour cells derived from different species, including human, can be used in conventional mice since the cells loaded in a chamber ring would not be recognized by the host immune system. In contrast, variability in the amount of pre-existing vessels where the chamber ring is implanted affects the evaluation of angiogenesis such as blood flow determination. In some cases, it is difficult to evaluate the degree of angiogenesis by qualitative observations, although advances in the use of image analysis software to view vessels are making interpretation of results easier. It is possible actually to observe or count vessels by eye, but angiogenic vessels vary in diameter and length as well as number so quantification by eye minimizes the effects seen with individual treatments. Although the methods described above are indirect, they allow for greater interpretation of results. The activity of anti-angiogenic substances is easily examined by using DAS models and this methodology is useful for development of anti-angiogenic drugs or anti-angiogenic therapies. Moreover, DAS model can be modified for analysis of the mechanisms of angiogenesis. For example, implantation of a chamber ring loaded with tumour cells manipulated to over- or under-express certain factors will help reveal the role of the relevant test factor in angiogenesis.

A recent development in the DAS method has taken this one step further by pre-loading the ring with a single test peptide rather than tumour cells, which will secrete a number of different factors. Hamada *et al.* (2004) used this method to compare the effects of a test peptide with VEGF, a known pro-angiogenic protein. This novel use of the DAS assay might give new insight for understanding the mechanisms of angiogenesis stimulated by single test factors.

References

Asai, T., Shimizu, K., Kondo, M., Kuromi, K. *et al.* (2002) 'Anti-neovascular therapy by liposomal DPP-CNDAC targeted to angiogenic vessels', *FEBS Letters*, **250**(1–3), pp. 167–170.

Dolmans, D. E., Fukumura, D. and Jain, R. K. (2003) 'Photodynamic therapy for cancer', *Nature Rev. Cancer*, **3**(5), pp. 380–387.

Funahashi, Y., Wakabayashi, T., Semba, T., Sonoda, J. *et al.* (1999) 'Establishment of a quantitative mouse dorsal air sac model and its application to evaluate a new angiogenesis inhibitor', *Oncol. Res.*, **11**(7), pp. 319–329.

Hamada, Y., Yuki, K., Okazaki, M., Fujitani, W. *et al.* (2004) 'Osteopontin-derived peptide SVVYGLR induces angiogenesis in vivo', *Dental Materials J.*, **23**(4), pp. 650–655.

Ichikawa, K., Takeuchi, Y., Yonezawa, S., Hikita, T. *et al.* (2004) 'Antiangiogenic photo-dynamic therapy (PDT) using Visudyne causes effective suppression of tumor growth', *Cancer Lett.*, **205**(1), pp. 39–48.

Kurohane, K., Tominaga, A., Sato, K., North, J. R. *et al.* (2001) 'Photodynamic therapy targeted to tumor-induced angiogenic vessels', *Cancer Lett.*, **167**(1), pp. 49–56.

Maeda, N., Takeuchi, Y., Takada, M., Sadzuka, Y. *et al.* (2004) 'Anti-neovascular therapy by use of tumor neovasculature-targeted long-circulating liposome', *J. Controlled Release*, **100**(1), pp. 41–52.

Oikawa, T., Sasaki, M., Inose, M., Shimamura, M. *et al.* (1997) 'Effects of cytogenin, a novel microbial product, on embryonic and tumor cell-induced angiogenic responses in vivo', *Anticancer Res.*, **17**(3C), pp. 1881–1886.

Oku, N., Asai, T., Watanabe, K., Kuromi, K. *et al.* (2002) 'Anti-neovascular therapy using novel peptides homing to angiogenic vessels', *Oncogene*, **21**(17), pp. 2662–2669.

Oku, N., Matsukawa, M., Yamakawa, S., Asai, T. *et al.* (2003) 'Inhibitory effect of green tea polyphenols on membrane-type 1 matrix metalloproteinase, MT1-MMP', *Biol. Pharm. Bull.*, **26**(9), 1235–1238.

Peng, Q., Warloe, T., Moan, J., Godal, A. *et al.* (2001) 'Antitumor effect of 5-aminolevulinic acid-mediated photodynamic therapy can be enhanced by the use of a low dose of photofrin in human tumor xenografts', *Cancer Res.*, **61**(15), pp. 5824–5832.

Reuther, T., Kubler, A. C., Zillmann, U., Flechtenmacher, C. and Sinn, H. (2001) 'Compar-ison of the in vivo efficiency of photofrin II-, mTHPC-, mTHPC-PEG- and mTHPCnPEG-mediated PDT in a human xenografted head and neck carcinoma', *Lasers Surg. Med.*, **29**(4), pp. 314–322.

Shimizu, K. and Oku, N. (2004) 'Cancer anti-angiogenic therapy', *Biol. Pharm. Bull.*, **27**(5), pp. 599–605.

Takeuchi, Y., Kurohane, K., Ichikawa, K., Yonezawa, S. *et al.* (2002) 'Induction of intensive tumor suppression by antiangiogenic photodynamic therapy using polycation-modified liposomal photosensitizer', *Cancer*, **97**(8), pp. 2027–2034.

Yamakawa, S., Asai, T., Uchida, T., Matsukawa, M. *et al.* (2004) '(–)-Epigallocatechin gallate inhibits membrane-type 1 matrix metalloproteinase, MT1-MMP, and tumor angiogenesis', *Cancer Lett.*, **210**(1), pp. 47–55.

13

Chamber assays

Michael D. Menger, Matthias W. Laschke and Brigitte Vollmar

Abstract

Angiogenesis plays a pivotal role in tumour growth and metastasis, and is a hallmark of both inflammatory and ischaemic diseases. Accordingly, major focus has been given to angiogenesis research during the last two decades. The most frequently used tool to investigate blood vessel formation is the histological cross-sectional preparation, however, this technique lacks information on the dynamic process of both blood vessel development and microhaemodynamic perfusion. Because it is essential to know whether a newly formed blood vessel is perfused, thereby contributing to tissue oxygenation, *in vivo* imaging of the development of new blood vessels has been introduced for experimental studies in angiogenesis research. While acute preparations do not allow us to follow the prolonged process of angiogenesis, chamber assays are capable of investigating blood vessel formation and microvascular networking over time periods of two to three weeks. With the dorsal skinfold chamber, vascularization during tumour growth, wound healing, tissue ischaemia, transplant engraftment, biomaterial incorporation and tissue engineering can be studied quantitatively, including the elucidation of microhaemodynamic, cellular and molecular mechanisms of the vascularization process. Thus, chamber assays may have the potential substantially to contribute to the development of novel therapeutic strategies in counteracting neovascularization in tumour growth and in supporting rapid blood vessel development during healing of normoxic and hypoxic tissues.

Keywords

dorsal skinfold chamber; intravital microscopy; *in vivo* angiogenesis assays; neovascularization; tissue healing

13.1 Introduction

Angiogenesis plays a pivotal role in tumour growth and metastasis, and is a hallmark of both inflammatory and ischaemic diseases (Carmeliet and Jain, 2000;

Angiogenesis Assays   Edited by Carolyn A. Staton, Claire Lewis and Roy Bicknell

Jain, 2001). In malignant and inflammatory diseases, angiogenesis is considered pathological in nature, propagating the progression of the disease. In wound healing, angiogenesis contributes to the process of regeneration, while in ischaemic diseases it counteracts tissue hypoxia by development of collateral circulation. Accordingly, the clinical interest is diverse, searching for both sophisticated anti-angiogenic strategies to inhibit tumour development, growth and spreading, and effective angiogenic stimulation therapies to improve tissue oxygenation in ischaemic myocardial and peripheral occlusive arterial diseases.

Physiological angiogenesis occurs primarily during embryonal development, and is rarely observed in adults. Experimental studies analysing the microvasculature of intact skin during a whole life span of mice convincingly demonstrated changes in haemodynamics and vascular reactivity during maturation and ageing, however, complete lack of angiogenesis and new vessel formation (Vollmar et al., 2000). The only organ which presents with angiogenesis in adults is the female reproductive system (Fraser and Lunn, 2000). Accordingly, clinical interest focuses on the mechanisms of new blood vessel formation in folliculogenesis and therapeutic approaches in follicle maturation failure.

With the view that the formation of new blood vessels represents an interesting target for the treatment of disease states, major focus has been given to angiogenesis research during the last two decades. The most frequently used tool to investigate blood vessel formation is the histological cross-sectional preparation, which is stained with antibodies specifically directed against epitopes of endothelial cells such as CD31 (platelet-endothelial cell adhesion molecule-1; PECAM-1) and von Willebrand factor. This technique allows quantitation of the number of newly formed blood vessels per area of tissue, however, it lacks information on the dynamic process of both blood vessel development, i.e. budding and sprouting, and microhaemodynamic perfusion, i.e. onset of blood flow within the newly formed microvascular network. Because it is essential to know whether a newly formed blood vessel is perfused, thereby contributing to the oxygenation of the tissue (Vajkoczy et al., 1999a), in vivo imaging of the development of new blood vessels is preferred for experimental studies in angiogenesis research.

13.2   *In vivo* imaging of angiogenesis and microcirculation

In vivo imaging of the microcirculation is performed by light-transillumination or fluorescent epi-illumination microscopy. If the thickness of the tissue preparation does not exceed 300 µm, as known for the mesentery (Zweifach, 1954), the omentum (Intaglietta et al., 1970) the cremaster muscle (Prewitt and Johnson, 1976) and the cheek pouch preparation (Duling, 1973), transillumination techniques can be used. In contrast, solid organs, such as the heart (Huwer et al., 1998) and the liver (Menger et al., 1991a), may be studied by epi-illumination using fluorescent techniques and specific dyes. Meanwhile, the microvasculature and microhaemodynamics of almost all organs can be visualized and investigated by intravital microscopy, including brain (Kawamura *et al.*, 1990), heart (Huwer *et al.*, 2001), lung (Schneider *et al.*, 2001), liver (Vollmar *et al.*, 1998), spleen (McCuskey and McCuskey, 1977), pancreas (Menger *et al.*, 1996), pancreatic islets (Vollmar *et al.*, 1999), stomach (Menger, 1994), small and large intestine (Gonzalez *et al.*, 1994; Maßberg *et al.*, 1998; Riaz *et al.*, 2002), urinary bladder (Bajory *et al.*, 2002), bone (Albrektsson and Albrektsson, 1978), periosteum (Rücker *et al.*, 1998), muscle (Tyml and Budreau, 1991), subcutis (Rücker *et al.*, 2000) and skin (Barker *et al.*, 1989). The technique allows investigation of the individual geometry of the microvascular networks, including feeding arterioles, nutritive capillaries and post-capillary and collecting venules, as well as microcirculatory haemodynamics, i.e. red blood cell velocity, blood vessel diameters, microvascular blood flow and wall shear rates.

The use of these preparations for angiogenesis studies, however, is limited due to the fact that most of these preparations are acute in nature. Because angiogenesis is a process which lasts over days or weeks, repetitive surgical interventions would be necessary to document the individual steps of new vessel formation *in vivo*. Surgical interventions, however, and the repeated preparation of the tissue in particular, represent major stimuli to induce inflammation and angiogenesis *per se* (Fiebig *et al.*, 1991; Thaler *et al.*, 2005; Greene *et al.*, 2005), and will thus influence the analysis of specific angiogenic stimuli associated with distinct disease states. Only the skin, which is exposed without surgical preparation, may be studied for angiogenesis repeatedly over time without being affected by surgical preparations (Vollmar *et al.*, 2000; Barker *et al.*, 1989). This advantage is counteracted by the fact that skin does not provide ideal transparency for intravital microscopic studies and varies tremendously depending on pigmentation and age of the animal.

To overcome these limitations, chamber assays may be used. They also require a surgical preparation to expose subcutis, muscle, brain or bone within the chamber, however, a recovery period of 48 h to 72 h before start of the experiment allows normalization of the microcirculation. The chamber tissue preparations may then provide ideal translucency for intravital microscopy, and perfectly serve as host tissue for implantation experiments.

13.3 Chamber assays

Rabbit ear chamber

In 1924, Sandison (1928) was the first to implant a transparent observation chamber into the rabbit ear. This enabled him to study the growth and behaviour of living cells by intravital microscopy. Later on, this rabbit ear chamber technique underwent modifications, aimed at facilitating implantation and improving image quality (Algire, 1943; Arfors *et al.*, 1970). In 1943, Algire adapted the transparent chamber technique to the mouse, which allowed, in particular, the study of growth and vascularization of tumour implants. A major drawback of these early used chamber models, however, was that the tissue under investigation consisted primarily of newly formed granulation tissue. Accordingly, studies on angiogenesis of tumour implants were influenced by additional processes of neovascularization associated with granulation tissue formation during wound healing.

Dorsal skinfold chamber

Later developments modified the chamber technique so as to provide a preparation of a preformed tissue which is not affected by the implantation procedure. Using the dorsal skinfold as chamber implantation site, those techniques were introduced during the last three decades in rats (Papenfuss *et al.*, 1979), hamsters (Endrich *et al.*, 1980) and mice (Lehr *et al.*, 1993). In these models, the host tissue for microscopic observation within the skinfold preparation includes striated muscle, i.e. the panniculus carnosus, subcutis and skin. The chambers can be used for intravital microscopic studies over a time period of 3 to 4 weeks (Menger and Lehr, 1993), and do not provoke angiogenesis and granulation tissue formation. Thus, angiogenesis of cell implants within these chambers can be studied without being affected by neovascularization events of the host tissue itself.

Today, the most common chronic model to study the dynamics of angiogenesis and microcirculation by repetitive intravital fluorescence microscopy is the dorsal skinfold chamber. The chamber consists of two symmetrical titanium frames that are implanted into the dorsal skinfold of the animals so as to sandwich the extended double layer of skin (Figure 13.1). One layer is completely removed in a

Figure 13.1 Syrian golden hamster skinfold chamber titanium chamber mounted in the skinfold of the dorsum of a Syrian golden hamster.

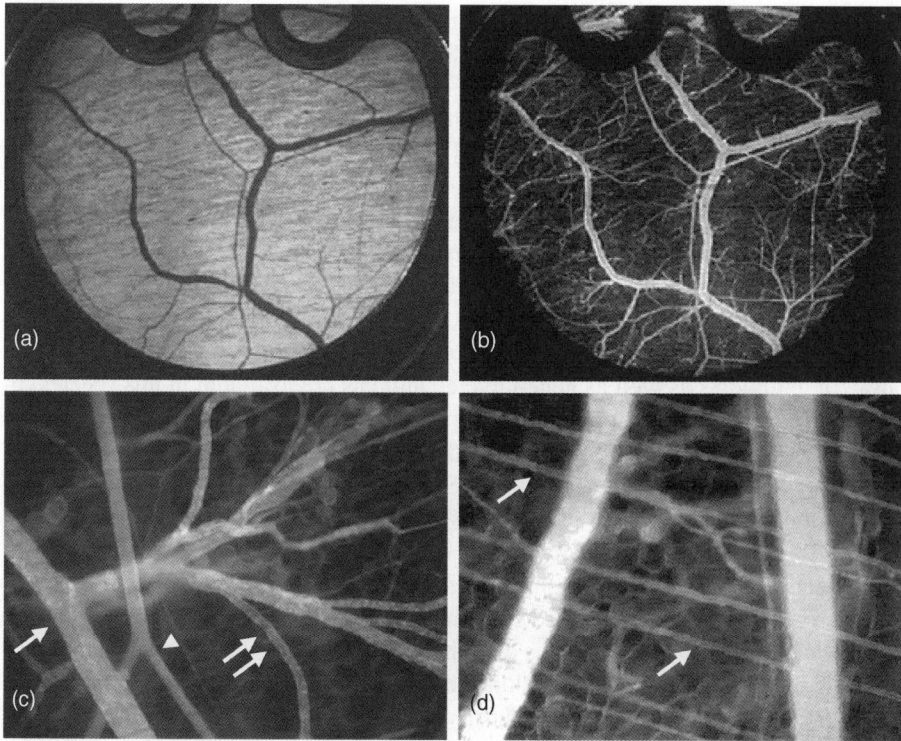

Figure 13.2 Microscopic imaging of the microcirculation in the dorsal skinfold chamber dorsal skinfold chamber in hamsters as documented by intravital transillumination (a) or fluorescence epi-illumination microscopy (b). Higher magnification allows for quantitative analysis of the microcirculation, including terminal arterioles (c, arrow head), nutritive capillaries (d, arrows), and post-capillary (c, double arrow) and collecting venules (c, arrow). Contrast enhancement is achieved by i.v. injection of fluorescein-isothiocyanate (FITC)-dextran 150 000. Magnifications ×6 (a,b), ×60 (c), and ×120 (d).

circular area of 15 mm in diameter and the remaining layer, consisting of striated muscle, subcutaneous tissue and epidermis, is covered with a glass cover slip incorporated into one of the frames (Figure 13.2) (Menger and Lehr, 1993). A recovery period of 48 h is ideally allowed to overcome microcirculatory changes due to the surgical trauma of the preparation procedure. The cover slip can temporarily be removed, so that tumour cells, transplants or tissue engineering constructs can be implanted into the tissue, which are then exposed for the study of neovascularization (Laschke *et al.*, 2006a; Menger *et al.*, 1994; Leunig *et al.*, 1992). In contrast to previously used chamber models, in which the tissue is sandwiched by two cover slips, which restrict growth of tissue to two dimensions, the back site of the presently used preparation is not covered by a glass slide, facilitating three-dimensional growth of the implants.

The major advantage of the dorsal skinfold preparation is that microcirculatory studies can be performed repeatedly in unanaesthetized animals over a prolonged period of time, including analysis of identical microvessels of arteriolar, capillary and venular nature (Menger *et al.*, 2002) (Figure 13.2). The chamber can be mounted in hamsters, rats and mice. The use of hamsters may be preferred, because in their dorsal skinfold the retractor muscle is only loosely attached to the underlying panniculus carnosus of the subcutis. This guarantees ideal removal of the retractor muscle without surgical alterations to the underlying tissue, which serves for later microscopic analysis. In mice, but particularly in rats, the retractor muscle is more firmly attached to the panniculus carnosus of the subcutis, including multiple vascular connections between the two tissue layers, which have to be dissected when removing the retractor muscle. This increases the surgical trauma to the tissue preparation. On the other hand, the use of mice may be preferred, because a large array of antibodies as well as gene-targeted animals are available in this species. In addition, the use of immunocompromised animals, such as nude or SCID mice (Lehr *et al.*, 1993; Leunig *et al.*, 1992), gives the advantage of studying angiogenesis in xenografts, particularly human tumour cells.

A further advantage of the use of hamsters is that the panniculus carnosus of the hamster is considerably thinner than that of the mouse and the rat, providing better translucency and thus microscopic image quality (Figure 13.2). The thin panniculus carnosus in the hamster also guarantees that there is only one layer of striated muscle capillaries. This facilitates quantitative analysis of the functional capillary density in the hamster dorsal skinfold preparation, while in mice and rats quantitative analysis of capillary density requires focusing through the different layers of capillaries in order to include all of the microvessels within the panniculus carnosus (Menger *et al.*, 2002).

The use of transillumination techniques restricts the analysis to the study of microvascular and microcirculatory parameters, such as microvessel diameter, density and blood flow. The implementation of epi-illumination fluorescence techniques enables additional simultaneous *in vivo* analysis of distinct cellular and molecular aspects, such as leukocyte and platelet adhesion (Menger and Lehr, 1993; Vollmar *et al.*, 2003), vascular permeability (Menger *et al.*, 1992a), microlymphatics (Menger *et al.*, 2003), gene expression (Fukumura *et al.*, 2001; Brown *et al.*, 2001), cell migration (Vajkoczy *et al.*, 1999b), tissue oxygenation (Rücker *et al.*, 2000; Torres Filho and Intaglietta, 1993), and cellular injury, including both apoptosis (Westermann *et al.*, 1999) and necrosis (Harris *et al.*, 1997). Leukocyte–endothelial cell interactions are studied by staining white blood cells *in vivo* with acridine orange or rhodamine 6G (Menger *et al.*, 1992a). In contrast, the analysis of platelet–endothelial cell interaction requires *ex situ* isolation and staining of the platelets with rhodamine or 2'-7'-bis-(2-carboxyethyl)-5-(and-6)-carboxyfluorescein (BCECF) (Vollmar *et al.*, 2003). Vascular permeability is analysed by measuring the extravasation of a high molecular weight fluorescence marker such as fluorescein-isothiocyanate (FITC)-dextran 150 000 or FITC-dextran labelled bovine serum albumin after intravenous application (Menger *et al.*, 1992a). Accordingly, microlymphatics can be studied when injecting those high molecular weight fluorescence markers not intravenously but locally into the interstitial space (Schacht *et al.*, 2004).

To monitor *in vivo* gene expression of distinct growth factors which may be involved in angiogenesis, for example vascular endothelial growth factor (VEGF), cells may be engineered that express green fluorescent protein (GFP) under the control of the VEGF promoter (Fukumura *et al.*, 2001). These cells can be implanted into the chamber, and VEGF promoter activity is then assessed by GFP imaging (Brown *et al.*, 2001). Further, the use of GFP-transfected cells or cells stained *ex vivo* with other fluorescent dyes allows quantitative analysis of migration activity (Vajkoczy *et al.*, 1999b). Oxygen transport within the microvasculature, and to tissues, is studied by phosphorescence quenching (Kerger *et al.*, 1996) or NADH imaging techniques (Rücker *et al.*, 2000). Finally, individual necrotic cell death can be analysed by the *in vivo* use of propidium iodide (Harris *et al.*, 1997), while apoptotic cell death may be detected by intravital microscopic imaging of nuclear condensation, margination and fragmentation using bis-benzimide staining (Westermann *et al.*, 1999).

Due to this wide spectrum of parameters which can be investigated, the skinfold chamber is an ideal tool to study mechanisms of angiogenesis under physiological conditions (Vollmar *et al.*, 2001), as well as in tumour growth (Endrich *et al.*, 1982), endometrial proliferation (Laschke *et al.*, 2005a), wound healing, ischaemia and hypoxia (Harder *et al.*, 2004), organ transplantation (Menger *et al.*, 1989), synthetic material and biomaterial incorporation (Menger *et al.*, 1990a; Menger *et al.*, 1992b), and tissue engineering (Laschke *et al.*, 2006a).

Cranial window preparation

In order to provide an orthotopic brain tumour model, the cranial window preparation was used for tumour implantation (Yuan *et al.*, 1994; Foltz *et al.*, 1995). To create the cranial window, a 6 × 7 mm bone flap is elevated over the frontal and parietal regions of the skull. Care has to be taken to free the flap from the underlying dura and sagittal sinus. The dura overlying the hemispheres is then removed, avoiding any damage to the sinus and bridging veins. The tumour specimens are directly placed onto the pial surface of either hemisphere and the window is sealed with a glass cover slip using a histocompatible glue. The model was initially created by transplanting large tumour specimens, which, however, only allowed study of the fully developed tumour microcirculation, and not the individual steps of angiogenesis. This drawback was overcome by orthotopically implanting glioma cell suspensions or spheroids (Vajkoczy *et al.*, 2002).

A variety of experimental studies have demonstrated that C6 gliomas vascularize quite similarly compared with those implanted heterotopically into the skinfold chamber. Detailed analysis of the function of the newly developed microvasculature, however, revealed that gliomas within the cranial window, but not within dorsal skinfold chambers, develop blood–brain barrier characteristics (Yuan *et al.*, 1994). It is still a matter of controversy whether the cranial window is more appropriate for the study of glioma vascularization than the dorsal skinfold chamber, because the cranial window implants are also not placed intraparenchy-mally. In fact, ultrastructural analyses of morphology and permeability have demonstrated that the newly developed microvasculature differs significantly from that of host vessels but not from each other (Coomber *et al.*, 1988). This view is also supported by vascular corrosion cast studies of subcutaneously implanted tumour cells, demonstrating that the individual tumour cell and not the host tissue is the driving force in determining the characteristics of individual types of tumour microvasculature (Konerding *et al.*, 1999).

Of interest, the cranial window preparation may not only be used for the study of neovascularization of orthotopically implanted brain tumours, but may also serve to study angiogenesis of orthotopically implanted primary and secondary bone tumours. Recently, Klenke and co-workers (2005) demonstrated that in a cranial window preparation of SCID mice growth and neovascularization of solid human osteosarcoma as well as A459 lung carcinoma and PC-3 prostate carcinoma can be analysed by intravital microscopy when the calvaria serves as the site of orthotopic engraftment.

Bone chamber preparations

Other chamber techniques have been introduced to study the process of vasculariza-tion during bone healing. A bone chamber implant which allows intravital microscopic observations of the vasculature of bone and bone marrow was first introduced by Branemark in 1959 using a rabbit model. Later modifications to this chronic approach used the hollow titanium screw to access the bone tissue (McCuskey *et al.*, 1971; Albrektsson and Albrektsson, 1978; Winet, 1996). A different modification was recently reported by Hansen-Algenstaedt and co-workers (Hansen-Algenstaedt *et al.*, 2005), using a sandwich principle for chamber creation. This technique allows the use of the method in mice, in which the hollow screen technique did not provide an adequate observation area for representative studies. This recent model bears some additional advantages, including that 1) both pre-existing and newly growing blood vessels can be observed, 2) the bone within the observation area remains biomecha-nically active, and 3) tumour and metastasis studies can be performed.

13.4 Physiological angiogenesis

In adult tissue, angiogenesis is a characteristic of pathological conditions, such as tumour growth, wound healing and inflammation. Non-pathological angiogenesis is rare and restricted to the female reproductive tissues (Reynolds *et al.*, 1992). During the cycle of follicle development and corpus luteum formation, vascular changes are tightly regulated in that angiogenesis is turned on for brief periods and then completely inhibited. Thus, folliculogenesis offers a unique system to study not only the induction of angiogenesis, but also the maturation and regression of blood vessels (Goede *et al.*, 1998).

After isolation from donor ovaries, follicles with a size from 250 to 1000 μm are implanted into the dorsal skinfold chamber (Vollmar *et al.*, 2001). Before transplantation, the follicles may be stained by fluorescent cytoplasmatic or nuclear dyes. The latter have the advantage that nuclear changes, such as apoptosis-associated condensation, margination and fragmentation can also be observed *in vivo* during the process of follicle neovascularization. Of interest, small pre-antral follicles lacking a theca interna fail to vascularize, and should therefore not be used (Vollmar *et al.*, 2001).

Within the first three days after implantation, follicles provoke an angiogenic response within the host tissue. This is primarily observed in capillaries and post-capillary venules, and is characterized by protrusion of buds and sprouts (Figure 13.3). After 6 days, sprouting capillaries connect with each other, and after

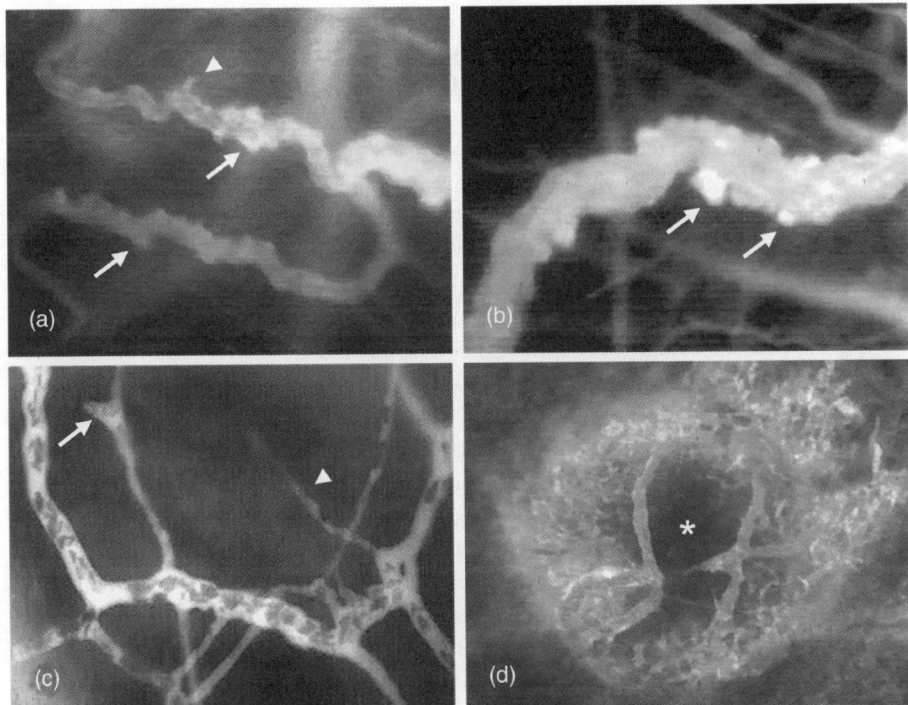

Figure 13.3 Neovascularization of ovarian follicle post-capillary venules of host tissue in the hamster dorsal skinfold chamber at day 3 after transplantation of a syngeneic ovarian follicle (a,b). Note the capillary buds (arrows) and sprouts (arrow head), originating from these venules, demonstrating the first signs of angiogenesis. At day 6 after transplantation (c) a network of newly formed microvessels, which vary tremendously in diameter, has developed. The fluorescent dye FITC-dextran 150 000 within the microvessels indicates onset of perfusion. The newly built network further grows, as can be recognized by additional capillary buds (arrow) and sprouts (arrow head). At day 14, a densely meshed network of newly formed capillaries supplies the transplanted follicle, which is drained by two central venular systems (d). Magnifications x40 (d), and x120 (a,b,c).

10 to 14 days a network of newly formed microvessels is developed, which completely vascularizes the implanted follicle (Figure 13.3) (Vollmar *et al.*, 2001). Of interest, ovariectomy of the recipient animal was shown to accelerate angiogenesis and to improve the microcirculation of the transplanted follicles (Laschke *et al.*, 2002). In addition, cryopreservation of the isolated follicles before implantation into the skinfold chamber does not affect the process of vascularization (Laschke *et al.*, 2003).

13.5 Tumour angiogenesis

Using the modified dorsal skinfold chambers, angiogenesis and vascularization of a variety of different tumours have been analysed, including melanoma (Endrich *et al.*, 1980), colon adenocarcinoma (Leunig *et al.*, 1992), rhabdomysarcoma (Borgstrom *et al.*, 1996), prostate cancer (Borgstrom *et al.*, 1998), renal cell carcinoma (Heuser *et al.*, 2003), pancreatic cancer (Guba *et al.*, 2005) and high-grade glioma (Vajkoczy *et al.*, 1998).

In highly vascularized tumours, first signs of angiogenesis, i.e. dilatation and tortuous elongation of host vessels with microvascular budding and sprouting, can be observed at day 2 to 4 after cell implantation (Vajkoczy *et al.*, 1999a; Vajkoczy *et al.*, 1998). The newly formed microvessels subsequently branch and interconnect with each other, forming complex microvascular networks. The microvessels are extremely heterogeneous in diameter and chaotic in network structure (Leunig *et al.*, 1992), but present with initial red blood cell perfusion. By day 10, the tumour mass is completely vascularized and starts to grow steadily (Vajkoczy *et al.*, 2000). With the increase in size during the second to the third week, the microvasculature remodels along a gradient, with high angiogenic activity in the peritumoural and low activity in the central area of the tumour mass (Vajkoczy *et al.*, 1999a; Vajkoczy *et al.*, 1998). Thus, the multi-step process of tumour angiogenesis involves three phases: first, an avascular phase (day 0 to day 6) characterized by initial tumour angiogenesis originating from host vessels, but lag of tumour growth; second, an early phase of tumour vascularization (day 6 to day 14) with development of tumour microcirculation, and, consequently, initiation of tumour growth; and third, an advanced phase of tumour vascularization (day 14 to day 22) with maintained high peritumoural angiogenic activity, but microvascular remodelling within the mature tumour (Vajkoczy *et al.*, 2000). This time course is the one of glioma growth, and may vary to a considerable extent in tumour cell lines of different origin.

Tumour neovascularization may be quantified by determining the microvessel density, representing the length of all newly formed microvessels per tumour area (Leunig *et al.*, 1992). However, because not all of the newly formed microvessels are perfused with red cells, and because this information is essential to learn about nutritional blood flow and tumour metabolism, it seems mandatory to determine also the *functional* vascular density, which includes only the length of all red cell perfused tumour microvessels per tumour area (Vajkoczy *et al.*, 2000). The advantage of the chamber technique in combination with intravital microscopic analysis is that both angiogenesis and microcirculation can be investigated over time. This is of particular interest, because the microcirculation may strongly influence tumour oxygenation and metabolism (Helmlinger *et al.*, 1997; Alexandrakis *et al.*, 2004), drug delivery (Jain, 1998), cell–cell and cell–matrix interaction (Vajkoczy *et al.*, 2001) as well as immune cell invasion (Melder *et al.*, 1996). In addition, besides angiogenesis (Jain, 2001), the microcirculation of the established neovasculature may represent an additional target for therapeutic tumour destruction, as demonstrated by microvascular thrombus formation in pancreatic tumours after mTOR-inhibitor treatment (Guba *et al.*, 2005).

13.6 Angiogenesis in endometriosis

Angiogenesis is thought to be a determinant for the development and progression of endometriosis. However, little is known on its mechanisms in this disease, which is probably due to the lack of adequate experimental models. Recently, it was shown that endometriosis can be mimicked by transplanting endometrial tissue into dorsal skinfold chambers, allowing quantitative analysis of angiogenesis and neovascularization by intravital fluorescence microscopy (Figure 13.4) (Laschke *et al.*, 2005a). These studies revealed that combined inhibition of vascular endothelial growth factor (VEGF), fibroblast growth factor and platelet-derived growth factor, but not inhibition of VEGF alone, effectively suppresses angiogenesis and vessel maturation in endometriotic lesions (Laschke *et al.*, 2006b). With the use of this model, further studies may be capable of elucidating the effects of other anti-angiogenic drugs on new vessel formation in endometriosis-like lesions.

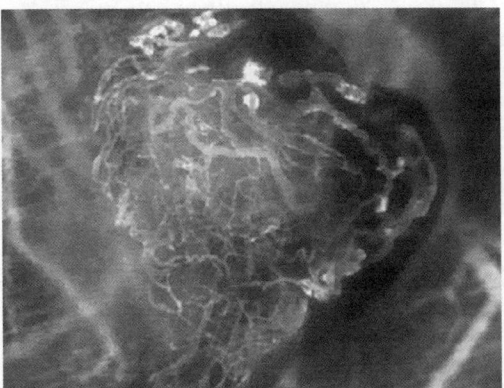

Figure 13.4 Vascularization of endometriosis-like lesion vascularization of an endometriosis-like lesion at day 10 after transplantation into a dorsal skinfold chamber. Note the chaotically arranged dense network of newly formed microvessels. Contrast enhancement is achieved by i.v. injection of fluorescein-isothiocyanate (FITC)-dextran 150 000. Magnification x40.

13.7 Angiogenesis in wound and bone healing

The study of wound healing in rodents is limited in their comparability to that of humans due to the fact that healing of wound defects in rodents are primarily driven by wound contraction but not granulation tissue formation. Thus, in rodents only models which naturally limit wound contraction, such as the ear of the hairless mouse, may be used for analysis of angiogenesis and vascularization in wound healing (Bondar *et al.*, 1991; Vollmar *et al.*, 2002). For primary wound healing, however, chamber models may be suitable. Tissue can be dissected and sutured for studying the healing process over time. Of interest, such studies may clarify whether primary wound healing is mainly dependent on angiogenesis, i.e. new vessel formation, or on inosculation, i.e. re-anastomosis of preformed blood vessels.

There are some early studies analysing the process of angiogenesis and neovascularization of both cortical and spongy bone using skinfold chambers (Funk *et al.*, 1986; Leunig *et al.*, 1994). In contrast to cortical bone, spongy bone shows early induction of angiogenesis at day 3 after implantation (Figure 13.5). This is associated with petechial bleeding within the angiogenic front, probably due to massive permeabilization of the microvasculature by high levels of VEGF (vascular permeability factor). Within a 14-day period spongy bone is completely vascularized with a densely meshed network of newly formed capillaries (Figure 13.5).

Bone healing *in situ*, however, may be preferably studied using hollow screw chambers in long bones in rabbits (Branemark, 1959) or sandwich chambers in long bones of mice and rats (Hansen-Algenstaedt *et al.*, 2005). While the latter technique was only recently introduced, and, thus, little experience with this model is reported, the hollow screw titanium chamber is known to enable vital microscopic studies of bone tissue *in situ*. This model was used to show that fractured and orthotopically grafted bone regains its circulation mainly via ingrowth of newly developed vessels of host origin, starting at 5–8 days after grafting (Albrektsson, 1980a). Bone remodelling then occurs after 3 weeks. Of interest, however, there is also an additional mode of blood flow recovery, involving blood vessels of more than 30 μm in diameter, which interconnect by end-to-end anastomoses between host and pre-existing graft vessels. In this case bone remodelling starts as early as 1 week after fracture and transplantation (Albrektsson, 1980a).

A major advantage of this model is that it allows study of the bone microvasculature for prolonged periods of time of up to one year. If bone healing reveals early vascularization (5–9 days after grafting), a vital bone and a vivid blood circulation is achieved after one year. In contrast, if bone healing is associated with late onset of angiogenesis and vascularization (17–29 days after grafting), only parts of the bone are revascularized after one year, containing islands of dead tissue (Albrektsson and Linder, 1981). Major surgical trauma to both the graft and

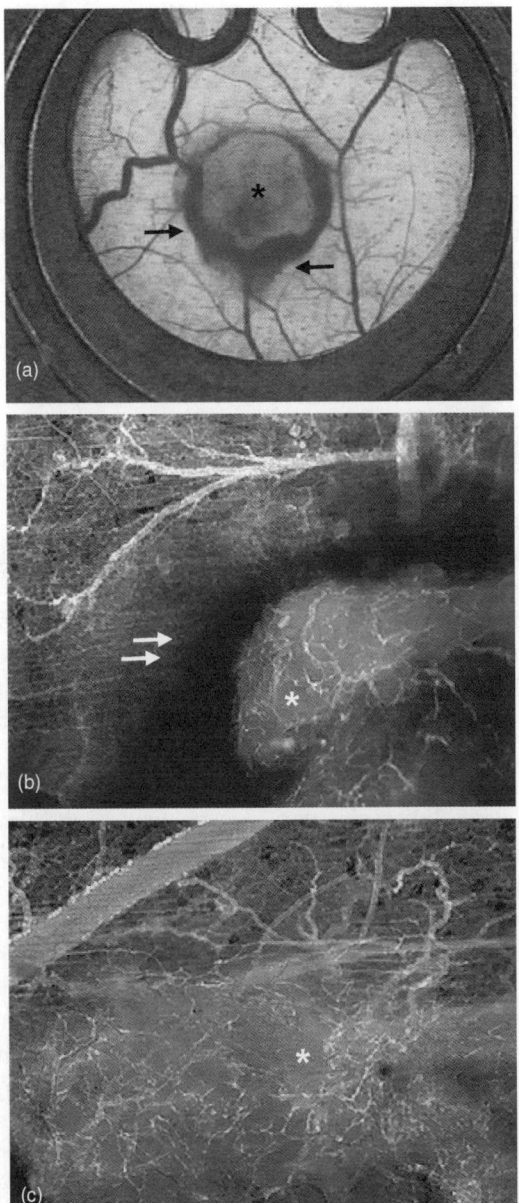

Figure 13.5 Vascularization of spongy bone graft dorsal skinfold chamber at day 3 after transplantation of spongy bone tissue (a, asterisk). Note the petechial bleeding within the surrounding margin (arrows), indicating the angiogenic front. Panel b displays higher magnification of a spongy bone transplant at day 6, indicating neovascularization (asterisk). Of interest, petechial bleedings can still be found within the margin of the bone transplant (arrows). Panel c shows a fully revascularized spongy bone graft (asterisk) at day 14 after transplantation. The petechial bleedings at the margin of the transplant are completely resolved. Contrast enhancement is achieved by i.v. injection of fluorescein-isothiocyanate (FITC)-dextran 150 000. Magnifications x6 (a), and x40 (b and c).

the host bed alters early revascularization and healing, and therefore should meticulously be avoided (Albrektsson, 1980b).

With recent developments of bone substitute materials for treating large bone defects after trauma or tumour surgery, analysis of their capability of revascularization becomes of major interest. Using the hamster dorsal skinfold chamber model, those substitutional materials, such as nano-scaled hydroxyapatite, implanted into muscle and subcutaneous tissue, provoke rapid angiogenesis and complete neovascularization within a time period of 10 to 14 days (Figure 13.6).

13.8   Angiogenesis in ischaemia and hypoxia

In chronic ischaemic diseases, angiogenesis is thought to be the only way to overcome the hypoxic conditions of the compromised tissue. To mimic chronic ischaemic conditions within the dorsal skinfold chamber tissue, feeding arterioles may be ligated (Menger *et al.*, 1988). This procedure, however, is complicated due to multiple preformed collateral connections within the subcutaneous tissue. In the dorsal skinfold chamber model in mice, Harder and co-workers (Harder *et al.*, 2004) have recently introduced a modification of this technique by creating a flap, which becomes necrotic in its distal part due to deprivation of nutritional blood flow. The elevation of the flap includes the transsection of both the deep circumflex iliac artery and the lateral thoracic artery. After extending the elevated tissue to the back side of the dorsal part of the titanium frame, the flap's lateral sides are sutured back to the surrounding tissue within the chamber. Of interest, however, chronic ischaemia does not primarily provoke an angiogenic response to prevent manifestation of necrosis. Instead, a non-vascularized zone, which is potentially salvageable, develops distally to a hyperaemic zone and demarcates the necrotic tissue (Harder *et al.*, 2005a). Further conditioning of the ischaemic tissue prevents the development of necrosis (Harder *et al.*, 2005b). This, however, is neither achieved by angiogenesis and new vessel formation nor by increasing the ischaemic tolerance of the tissue, but simply by improving the nutritive microcirculation through the pre-existing microvasculature (Harder *et al.*, 2005b). Ongoing experiments have to clarify whether the lack of angiogenesis is due to failure of growth factor production, and whether this can be overcome by transplanting genetically modified cells producing VEGF, fibroblast growth factor and platelet-derived growth factor.

13.9   Angiogenesis in transplantation

While most of the organs for transplantation are revascularized immediately after grafting by anastomotic suturing of the main feeding and draining blood vessels,

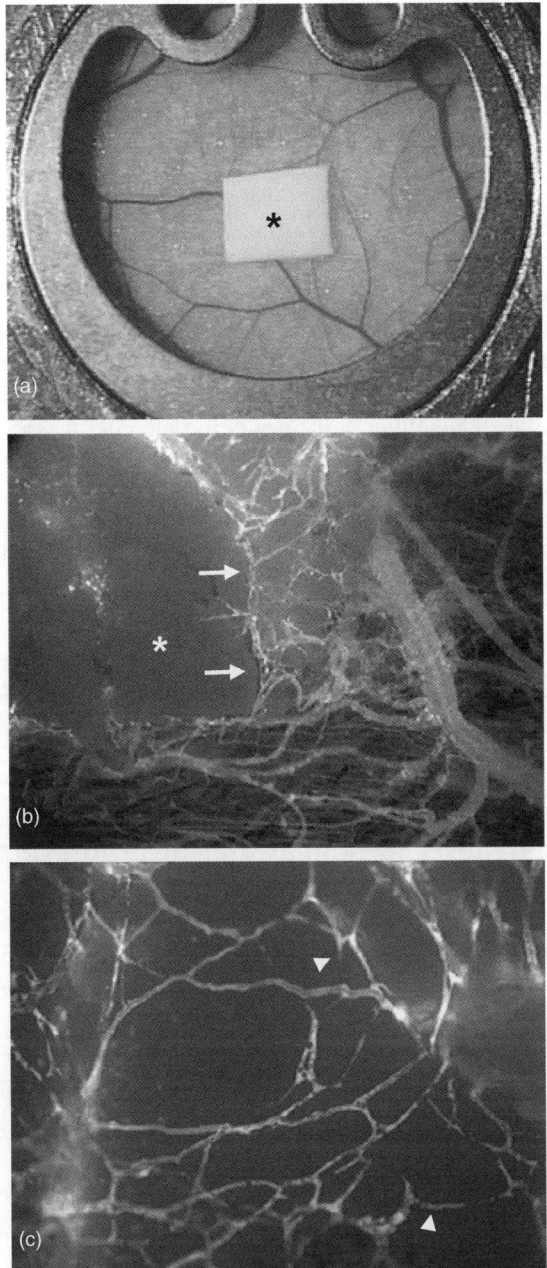

Figure 13.6 Vascularization of bone substitute material dorsal skinfold chamber at day 0 after implantation of a bone substitute consisting of nano-scaled hydroxyapatite (a, asterisk). At day 6 after implantation, note the angiogenic response at the margin of the bone substitute, characterized by the formation of new capillary-like blood vessels (arrows). Panel c displays a complete network of newly formed microvessels, overgrowing the bone substitute. Nonetheless, additional capillary sprouts can be seen (arrow heads). Contrast enhancement is achieved by i.v. injection of fluorescein-isothiocyanate (FITC)-dextran 150 000. Magnifications x6 (a), and x60 (b and c).

including heart, lung, liver, pancreas and kidney grafts, some organs, such as parathyroids and isolated pancreatic islets are avascular at the time of transplantation, and are dependent on the process of angiogenesis to rebuild a microvascular network sufficient for an adequate blood supply (Menger *et al.*, 2001). This process of angiogenesis can be ideally analysed using the dorsal skinfold chamber model. Accordingly, the technique of islet grafting (Menger *et al.*, 1989; Menger *et al.*, 1990b) and, more recently, the methods for heterotopic parathyroid transplantation (Strieth *et al.*, 2005) have been introduced.

Intravital microscopy showed that the avascular, freely transplanted islets, and even pseudo-islets, develop a new microvasculature, similar to that observed in pancreatic islets *in situ* (Menger *et al.*, 1994; Beger *et al.*, 1998). This is of particular interest, because the endothelium of the newly formed microvessels originates from the vasculature of the host tissue (Vajkoczy *et al.*, 1995). The purity of the isolated islets is a determinant for successful vascularization (Heuser *et al.*, 2000). Cryopreservation of the isolated islets before transplantation and hyperglycaemia of the recipient do not alter the angiogenic process (Menger *et al.*, 2001; Menger *et al.*, 1992c; Menger *et al.*, 1992d). In contrast, immunosuppressive agents, such as cyclosporine, deoxyspergualin or RS-61443 inhibit adequate vascularization (Menger *et al.*, 1991b; Beger and Menger, 1997; Vajkoczy *et al.*, 1999c), while additional anti-oxidative treatment with vitamin E effectively reduces graft rejection without affecting the process of vascularization (Vajkoczy *et al.*, 1997).

13.10   Angiogenesis in biomaterial incorporation

Due to the interest in using foreign materials in surgery, it is necessary to study the inflammatory and angiogenic response after foreign material implantation. For this purpose, the use of the chamber model in the hamster is ideal, because the model not only allows for repeated microscopic analysis over a time period of weeks, but also provides perfect visualization due to its high translucency. Experiments testing synthetic, biosynthetic and biological prosthesis materials have demonstrated that engraftment and neoangiogenesis are mainly dependent on material porosity and material composition (Menger *et al.*, 1990a; Menger *et al.*, 1990c; Menger *et al.*, 1992b). Other studies, analysing metal implants used in trauma and orthopaedic surgery indicated a distinct leukocytic response by silver implants, and a reduction of microvascular inflammation by the use of stainless steel, nickel-reduced alloys, and, in particular, titanium (Kraft *et al.*, 2000; Kraft *et al.*, 2001). Recent studies, investigating synthetic mesh materials used in hernia surgery showed that these meshes are rapidly vascularized by newly formed microvessels (Laschke *et al.*, 2005b). This vascularization is associated with a leukocytic inflammatory response, which, however, disappears within a 14-day period after material implantation.

13.11   Angiogenesis in tissue engineering

Long-term function of three-dimensional tissue constructs depends on adequate vascularization after implantation. During the last few years, research in tissue engineering has therefore focused on the analysis of angiogenesis. For this purpose, the dorsal skinfold chamber technique has been introduced, allowing a more detailed analysis of angiogenic dysfunction and engraftment failure of three-dimensional tissue constructs (Figure 13.7) (Laschke *et al.*, 2006a). Druecke and colleagues (2004) used the murine dorsal skinfold chamber in order to study neovascularization and biocompatibility of acellular poly(ether ester) block-co-polymer scaffolds of different pore sizes. This type of biomaterial can be easily tailored, is biodegradable and has been proven to support in-growth of different

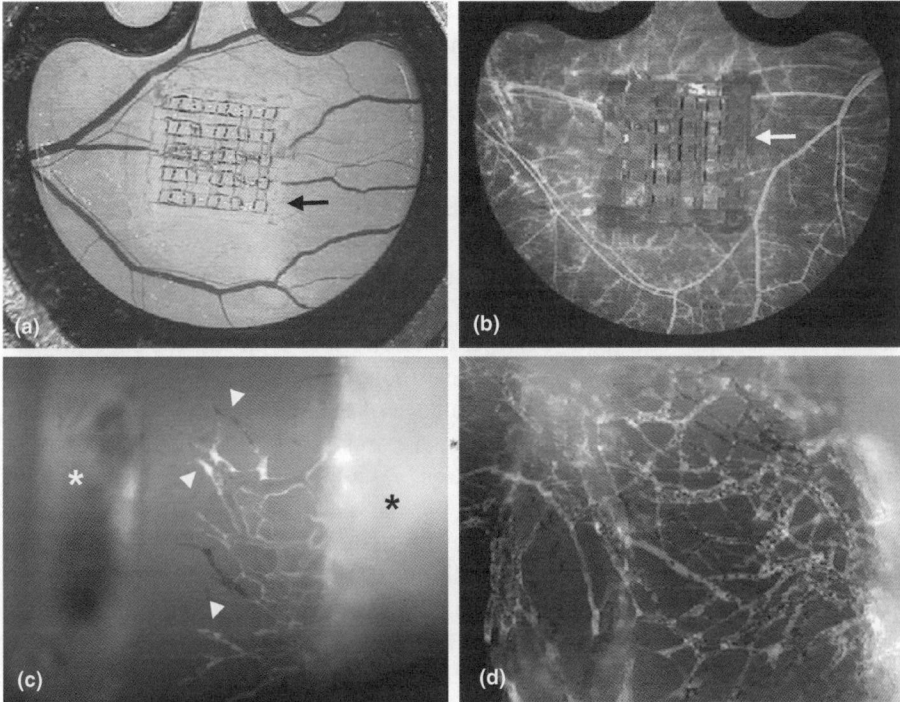

Figure 13.7 Vascularization of 3D scaffold for tissue engineering transillumination microscopy of a dorsal skinfold chamber at day 0 after implantation of a three-dimensional scaffold with a grid-like structure used for tissue engineering constructs (a, arrow). Panel b displays a similar scaffold, visualized by epi-illumination fluorescence microscopy. At day 10 after implantation, note the angiogenic response with capillary sprouts (c, arrow heads) in between the grid structures (asterisks), originating from the host tissue. Of interest, at day 14, a densely meshed network of capillaries is formed, completely overgrowing the scaffold material (d). Contrast enhancement is achieved by i.v. injection of fluorescein-isothiocyanate (FITC)-dextran 150 000. Magnifications x6 (a and b), and x60 (c and d).

tissues. The results of the study showed that new blood vessel in-growth into the scaffolds is influenced by pore size and is most pronounced in the scaffolds with the largest pores. This is in line with reports of new vessel formation in synthetic vascular graft materials (Menger *et al.*, 1990c), and underlines the importance of the biomaterial properties for the vascularization process of implanted tissue constructs.

To achieve appropriate vascularization, several approaches are currently under investigation. These include the modification of biomaterial properties of scaffolds as well as the stimulation of blood vessel development and maturation by different growth factors using slow release devices through pre-encapsulated microspheres (Laschke *et al.*, 2006a). Moreover, new microvascular networks in tissue substitutes can be engineered by the use of endothelial cells and stem cells or the creation of arteriovenous shunt loops. However, it is still a matter of debate, whether these techniques can accelerate the process of angiogenesis in such a way that it is sufficient for early vascularization of the constructs. Rather, it may be necessary to create the microvascular networks within three-dimensional constructs *in vitro* before implantation, or to co-stimulate angiogenesis and parenchymal cell proliferation in order to engineer the vascularized tissue substitute *in situ*.

13.12   Conclusions and perspectives

Angiogenesis is a complex multi-step process involved in the pathogenesis of a variety of diseases. The mechanisms of new vessel formation are partly understood, however, they still require further research using sophisticated experimental models. Chamber models are indeed ideal tools to study the individual steps of angiogenesis over time *in vivo*. The advantage of the chamber models is that they provide detailed information on functional parameters, such as vessel growth, microvascular perfusion, tissue oxygenation, vascular permeability and vasomotor response. This advantage is opposed by the need of sophisticated surgical preparation and a 48 h period to allow the tissue to recover from the surgical stress to achieve adequate baseline conditions. Further, quantitation of functional parameters requires time consuming analyses with computer-assisted dynamic image analysis systems. However, the fact that the combined use of intravital microscopy simultaneously allows for the study of cellular, molecular and functional parameters provides an approach which can increase our understanding of the links between angiogenesis and gene expression, cell–cell interaction, cell migration, cell proliferation and cell death. Thus, the use of chamber models in biomedical research will further contribute to the development of novel therapeutic strategies, counteracting tumour growth, endometriosis and chronic ischaemic diseases, as well as wound and bone healing failure. Of utmost interest for future developments is the question as to whether research with these chamber models can also contribute to an improvement of vascularization of foreign materials and three-dimensional scaffolds in tissue engineering and organ replacement.

References

Albrektsson, T. (1980a) '*In vivo* studies of bone grafts. The possibility of vascular anastomoses in healing bone', *Acta Orthopaedica Scand.*, **51**, pp. 9–17.

Albrektsson, T. (1980b) 'The healing of autologous bone grafts after varying degrees of surgical trauma. A microscopic and histochemical study in the rabbit', *J. Bone Joint Surg. (British)*, **62**, pp. 403–410.

Albrektsson, T. and Albrektsson, B. (1978) 'Microcirculation in grafted bone. A chamber technique for vital microscopy of rabbit bone transplants', *Acta Orthopaedica Scand.*, **49**, pp. 1–7.

Albrektsson, T. and Linder, L. (1981) 'Intravital, long-term follow-up of autologous experimental bone grafts', *Arch Orthopaedic Trauma Surg.*, **98**, pp. 189–193.

Alexandrakis, G., Brown, E. B., Tong, R. T., McKee, T. D. *et al.* (2004) 'Two-photon fluorescence correlation microscopy reveals the two-phase nature of transport in tumours', *Nat. Med.*, **10**, pp. 203–207.

Algire, G. H. (1943) 'An adaptation of the transparent chamber technique to the mouse', *J. Natl. Cancer Inst.*, **4**, pp. 1–11.

Arfors, K. E., Jonsson, J. A. and McKenzie, F. N. (1970) 'A titanium rabbit ear chamber: assembly, insertion, and results', *Microvasc. Res.*, **2**, pp. 516–519.

Bajory, Z., Hutter, J., Krombach, F. and Messmer, K. (2002) 'New method: the intravital videomicroscopic characteristics of the microcirculation of the urinary bladder in rats', *Urol. Res.*, **30**, pp. 148–152.

Barker, J. H., Hammersen, F., Bondar, I., Uhl, E. *et al.* (1989) 'The hairless mouse ear for *in vivo* studies of skin microcirculation', *Plastic Reconstructive Surg.*, **83**, pp. 948–959.

Beger, C. and Menger, M. D. (1997) 'RS-61443 prevents microvascular rejection of pancreatic islet xenografts', *Transplantation*, **63**, pp. 577–582.

Beger, C., Cirulli, V., Vajkoczy, P., Halban, P. A. and Menger, M. D. (1998) 'Vascularization of purified pancreatic islet-like cell aggregates (pseudoislets) after syngeneic transplantation', *Diabetes*, **47**, pp. 559–565.

Bondar, I., Uhl, E., Barker, J. H., Galla, T. J. *et al.* (1991) 'A new model for studying microcirculatory changes during dermal wound healing', *Res. Exp. Med.*, **191**, pp. 379–388.

Borgstrom, P., Bourdon, M. A., Hillan, K. J., Sriramarao, P. and Ferrara, N. (1998) 'Neutralizing anti-vascular endothelial growth factor antibody completely inhibits angiogenesis and growth of human prostate carcinoma micro tumours *in vivo*', *Prostate*, **35**, pp. 1–10.

Borgstrom, P., Hillan, K. J., Sriramarao, P. and Ferrara, N. (1996) 'Complete inhibition of angiogenesis and growth of microtumours by anti-vascular endothelial growth factor neutralizing antibody: novel concepts of angiostatic therapy from intravital videomicroscopy', *Cancer Res.*, **56**, pp. 4032–4039.

Branemark, P. I. (1959) 'Vital microscopy of bone marrow in rabbit', *Scand. J. Clin. Lab. Investigation*, **11**, pp. 1–82.

Brown, E. B., Campbell, R. B., Tsuzuki, Y., Xu, L. *et al.* (2001) '*In vivo* measurement of gene expression, angiogenesis and physiological function in tumours using multiphoton laser scanning microscopy', *Nature Med.*, **7**, pp. 864–868.

Carmeliet, P. and Jain, R. K. (2000) 'Angiogenesis in cancer and other diseases', *Nature*, **407**, pp. 249–257.

Coomber, B. L., Stewart, P. A., Hayakawa, E. M., Farrell, C. L. and Del Maestro, R. F. (1988) 'A quantitative assessment of microvessel ultrastructure in C6 astrocytoma spheroids transplanted to brain and to muscle', *J. Neuropathol. Exp. Neurol.*, **47**, pp. 29–40.

Druecke, D., Langer, S., Lamme, E., Pieper, J. *et al.* (2004) 'Neovascularization of poly(ether ester) block-copolymer scaffolds *in vivo*: long-term investigations using intravital fluorescent microscopy', *J. Biomed. Mat. Res. A*, **68**, pp. 10–18.

Duling, B. R. (1973) 'The preparation and use of the hamster cheek pouch for studies of the microcirculation', *Microvasc. Res.*, **5**, pp. 423–429.

Endrich, B., Asaishi, K., Goetz, A. and Messmer, K. (1980) 'Technical report – a new chamber technique for microvascular studies in unanesthetized hamsters', *Res. Exp. Med.*, **177**, pp. 125–134.

Endrich, B., Hammersen, F., Götz, A. and Messmer, K. (1982) 'Microcirculatory blood flow, capillary morphology and local oxygen pressure of the hamster amelanotic melanoma A-Mel-3', *J. Natl. Cancer Inst.*, **68**, pp. 475–485.

Fiebig, E., Ley, K. and Arfors, K. E. (1991) 'Rapid leukocyte accumulation by "spontaneous" rolling and adhesion in the exteriorized rabbit mesentery', *Int. J. Microcirculation: Clin. Exp.*, **10**, pp. 127–144.

Foltz, R. M., McLendon, R. E., Friedman, H. S., Dodge, R. K. *et al.* (1995) 'A pial window model for the intracranial study of human glioma microvascular function', *Neurosurgery*, **36**, pp. 976–984.

Fraser, H. M. and Lunn, S. F. (2000) 'Angiogenesis and its control in the female reproductive system', *Br. Med. Bull.*, **56**, pp. 787–797.

Fukumura, D., Xu, L., Chen, Y., Gohongi, T. *et al.* (2001) 'Hypoxia and acidosis independently up-regulate vascular endothelial growth factor transcription in brain tumours *in vivo*', *Cancer Res.*, **61**, pp. 6020–6024.

Funk, W., Endrich, B. and Messmer, K. (1986) 'A novel method for follow-up studies of the microcirculation in non-malignant tissue implants', *Res. Exp. Med.*, **186**, pp. 259–270.

Goede, V., Schmidt, T., Kimmina, S., Kozian, D. and Augustin, H. G. (1998) 'Analysis of blood vessel maturation processes during cyclic ovarian angiogenesis', *Lab. Invest.*, **78**, pp. 1385–1394.

Gonzalez, A. P., Sepulveda, S., Massberg, S., Baumeister, R. and Menger, M. D. (1994) '*In vivo* fluorescence microscopy for the assessment of microvascular reperfusion injury in small bowel transplants in rats', *Transplantation*, **58**, pp. 403–408.

Greene, A. K., Alwayn, I. P., Nose, V., Flynn, E. *et al.* (2005) 'Prevention of intra-abdominal adhesions using the antiangiogenic COX-2 inhibitor celecoxib', *Ann. Surg.*, **242**, pp. 140–146.

Guba, M., Yezhelyev, M., Eichhorn, M. E., Schmid, G. *et al.* (2005) 'Rapamycin induces tumour-specific thrombosis via tissue factor in the presence of VEGF', *Blood*, **105**, pp. 4463–4469.

Hansen-Algenstaedt, N., Schaefer, C., Wolfram, L., Joscheck, C. *et al.* (2005) 'Femur window – a new approach to microcirculation of living bone *in situ*', *J. Orthopaedic Res.*, **23**, pp. 1073–1082.

Harder, Y., Amon, M., Erni, D. and Menger, M. D. (2004) 'Evolution of ischaemic tissue injury in a random pattern flap: a new mouse model using intravital microscopy', *J. Surg. Res.*, **121**, pp. 197–205.

Harder, Y., Amon, M., Georgi, M., Banic, A. *et al.* (2005a) 'Evolution of a "falx lunatica" in demarcation of critically ischaemic myocutaneous tissue', *Am. J. Physiol.: Heart Circ. Physiol.*, **288**, pp. H1224–H1232.

Harder, Y., Amon, M., Schramm, R., Georgi, M. *et al.* (2005b) 'Heat shock preconditioning reduces ischaemic tissue necrosis by heat shock protein (HSP)-32-mediated improvement of the microcirculation rather than induction of ischaemic tolerance', *Ann. Surg.*, **242**, pp. 869–878.

Harris, A. G., Steinbauer, M., Leiderer, R. and Messmer, K. (1997) 'Role of leukocyte plugging and edema in skeletal muscle ischemia-reperfusion injury', *Am. J. Physiol.*, **273**, pp. H989–H996.

Helmlinger, G., Yuan, F., Dellian, M. and Jain, R. K. (1997) 'Interstitial pH and pO2 gradients in solid tumours *in vivo*: high-resolution measurements reveal a lack of correlation', *Nature Med.*, **3**, pp. 177–182.

Heuser, M., Wolf, B., Vollmar, B. and Menger, M. D. (2000) 'Exocrine contamination of isolated islets of Langerhans deteriorates the process of revascularization after free transplantation', *Transplantation*, **69**, pp. 756–761.

Heuser, M., Ringert, R. H., Zoeller, G. and Hemmerlein, B. (2003) 'Dynamic assessment of angiogenesis in renal cell carcinoma spheroids by intravital microscopy', *J. Urol.*, **169**, pp. 1267–1270.

Huwer, H., Nikoloudakis, N., Rissland, J., Vollmar, B. *et al.* (1998) '*In vivo* analysis of microvascular injury after myocardial cryothermia', *J. Surgical Res.*, **79**, pp. 1–7.

Huwer, H., Winning, J., Vollmar, B., Rissland, J. *et al.* (2001) 'Microvascularization and ventricular function after local alginate-encapsulated angiogenic growth factor treatment in a rat cryothermia-induced myocardial infarction model', *Microvasc. Res.* **62**, pp. 211–214.

Intaglietta, M., Tompkins, W. R. and Richardson, D. R. (1970) 'Velocity measurements in the microvasculature of the cat omentum by on-line method', *Microvasc. Res.*, **2**, pp. 462–473.

Jain, R. K. (1998) 'The next frontier of molecular medicine: delivery of therapeutics', *Nature Med.*, **4**, pp. 655–657.

Jain, R. K. (2001) 'Normalizing tumour vasculature with anti-angiogenic therapy: a new paradigm for combination therapy', *Nature Med.*, **7**, pp. 987–989.

Kawamura, S., Schürer, L., Goetz, A., Kempski, O. *et al.* (1990) 'An improved closed cranial window technique for investigation of blood–brain barrier function and cerebral vasomotor control in the rat', *Int. J. Microcirc.: Clin. Exp.*, **9**, pp. 369–383.

Kerger, H., Saltzman, D. J., Menger, M. D., Messmer, K. and Intaglietta, M. (1996) 'Systemic and subcutaneous microvascular PO2 dissociation during 4-h hemorrhagic shock in conscious hamsters', *Am. J. Physiol.*, **270**, pp. H827–H836.

Klenke, F. M., Merkle, T., Fellenberg, J., Abdollahi, A. *et al.* (2005) 'A novel model for the investigation of orthotopically growing primary and secondary bone tumours using intravital microscopy', *Lab. Animals*, **39**, pp. 377–383.

Konerding, M. A., Malkusch, W., Klapthor, B., van Ackern, C. *et al.* (1999) 'Evidence for characteristic vascular patterns in solid tumours: quantitative studies using corrosion casts', *Br. J. Cancer*, **80**, pp. 724–732.

Kraft, C., Hansis, M., Arens, S., Menger, M. D. and Vollmar, B. (2000) 'Striated muscle microvascular response to silver implants: A comparative *in vivo* study with titanium and stainless steel', *J. Biomed. Mat. Res.*, **49**, pp. 192–199.

Kraft, C. N., Burian, B., Perlick, L., Wimmer, M. A. *et al.* (2001) 'Impact of a nickel-reduced stainless steel implant on striated muscle microcirculation: A comparative *in vivo* study', *J. Biomed. Mat. Res.*, **57**, pp. 404–412.

Laschke, M. W., Menger, M. D. and Vollmar, B. (2002) 'Ovariectomy improves neovascularization and microcirculation of freely transplanted ovarian follicles', *J. Endocrinol.*, **172**, pp. 535–544.

Laschke, M. W., Menger, M. D. and Vollmar, B. (2003) 'Cryopreservation does not affect neovascularization of freely transplanted ovarian follicles'. *Fertility Sterility*, **79**, pp. 1458–1460.

Laschke, M. W., Elitzsch, A., Vollmar, B. and Menger, M. D. (2005a) '*In vivo* analysis of angiogenesis in endometriosis-like lesions by intravital fluorescence microscopy', *Fertility and Sterility*, **84**(Suppl. 2), pp. 1199–1209.

Laschke, M. W., Häufel, J. M., Thorlacius, H. and Menger, M. D. (2005b) 'New experimental approach to study host tissue response to surgical mesh materials *in vivo*', *J. Biomed. Mat. Res. A*, **74**, pp. 696–704.

Laschke, M. W., Elitzsch, A., Vollmar, B., Vajkoczy, P. and Menger, M.D. (2006b) 'Combined inhibition of vascular endothelial growth factor (VEGF), fibroblast growth factor and platelet-derived growth factor, but not inhibition of VEGF alone, effectively suppresses angiogenesis and vessel maturation in endometriotic lesions', *Human Reprod.*, **21**, pp. 262–268.

Laschke, M. W., Harder, Y., Amon, M., Martin, I. *et al.* (2006a) 'Angiogenesis in tissue engineering: breathing life into constructed tissue substitutes', *Tissue Eng.*, 12:2093–2104.

Lehr, H. A., Leunig, M., Menger, M. D., Nolte, D. and Messmer, K. (1993) 'Dorsal skinfold chamber technique for intravital microscopy in nude mice', *Am. J. Pathol.*, **143**, pp. 1055–1062.

Leunig, M., Yuan, F., Menger, M. D., Boucher, Y. *et al.* (1992) 'Angiogenesis, microvascular architecture, microhemodynamics, and interstitial fluid pressure during early growth of human adenocarcinoma LS174T in SCID mice', *Cancer Res.*, **52**, pp. 6553–6560.

Leunig, M., Yuan, F., Berk, D. A., Gerweck, L. E. and Jain, R. K. (1994) 'Angiogenesis and growth of isografted bone: quantitative *in vivo* assay in nude mice', *Lab. Invest.*, **71**, pp. 300–307.

Massberg, S., Eisenmenger, S., Enders, G., Krombach, F. and Messmer, K. (1998) 'Quantitative analysis of small intestinal microcirculation in the mouse', *Res. Exp. Med.*, **198**, pp. 23–35.

McCuskey, R. S. and McCuskey, P. A. (1977) '*In vivo* microscopy of the spleen', *Biblioteca Anatomica*, **16**, pp. 121–125.

McCuskey, R. S., McClugage, S. G. and Younker, W. J. (1971) 'Microscopy of living bone marrow *in situ*', *Blood*, **38**, pp. 87–95.

Melder, R. J., Koenig, G. C., Witwer, B. P., Safabakhsh, N. *et al.* (1996) 'During angiogenesis, vascular endothelial growth factor and basic fibroblast growth factor regulate natural killer cell adhesion to tumour endothelium', *Nature Med.*, **2**, pp. 992–997.

Menger, M. D. (1994) 'Microcirculation of gastric mucosa in pathogenesis of stomach ulcer', *Zentralblatt für Chirurgie*, **119**, pp. 1–10.

Menger, M. D. and Lehr, H. A. (1993) 'Scope and perspectives of intravital microscopy–bridge over from *in vitro* to *in vivo*', *Immunol. Today*, **14**, pp. 519–522.

Menger, M. D., Bonkhoff, H. and Vollmar, B. (1996) 'Ischemia-reperfusion-induced pancreatic microvascular injury. An intravital fluorescence microscopic study in rats', *Digestive Dis. Sci.*, **41**, pp. 823–830.

Menger, M. D., Hammersen, F., Barker, J., Feifel, G. and Messmer, K. (1988) 'Tissue PO2 and functional capillary density in chronically ischaemic skeletal muscle', *Adv. Exp. Med. Biol.*, **222**, pp. 631–636.

Menger, M. D., Jaeger, S., Walter, P., Feifel, G. *et al.* (1989) 'Angiogenesis and hemodynamics of microvasculature of transplanted islets of Langerhans', *Diabetes*, **38**(Suppl. 1), pp. 199–201.

Menger, M. D., Walter, P., Hammersen, F. and Messmer, K. (1990a) 'Quantitative analysis of neovascularization of different PTFE-implants', *Eur. J. Cardiothoracic Surg.*, **4**, pp. 191–196.

Menger, M. D., Jäger, S., Walter, P., Hammersen, F. and Messmer, K. (1990b) 'A novel technique for studies on the microvasculature of transplanted islets of Langerhans *in vivo*', *Int. J. Microcirculation: Clin. Exp.*, **9**, pp. 103–117.

Menger, M. D., Hammersen, F., Walter, P. and Messmer, K. (1990c) 'Neovascularization of prosthetic vascular grafts. Quantitative analysis of angiogenesis and microhemodynamics by means of intravital microscopy', *Thoracic Cardiovasc. Surgeon*, **38**, pp. 139–145.

Menger, M. D., Marzi, I. and Messmer, K. (1991a) '*In vivo* fluorescence microscopy for quantitative analysis of the hepatic microcirculation in hamsters and rats', *Eur. Surg. Res.*, **23**, pp. 158–169.

Menger, M. D., Wolf, B., Höbel, R., Schorlemmer, H. U. and Messmer, K. (1991b) 'Microvascular phenomena during pancreatic islet graft rejection', *Langenbecks Archiv für Chirurgie*, **376**, pp. 214–221.

Menger, M. D., Pelikan, S., Steiner, D. and Messmer, K. (1992a) 'Microvascular ischemia-reperfusion injury in striated muscle: significance of "reflow-paradox" ', *Am. J. Physiol.*, **263**, pp. H1901–H1906.

Menger, M. D., Hammersen, F. and Messmer K. (1992b) '*In vivo* assessment of neovascularization and incorporation of prosthetic vascular biografts', *Thoracic Cardiovasc. Surgeon*, **40**, pp. 19–25.

Menger, M. D., Pattenier, J., Wolf, B., Jäger, S. *et al.* (1992c) 'Cryopreservation of islets of Langerhans does not affect angiogenesis and revascularization after free transplantation', *Eur. Surg. Res.*, **24**, pp. 89–96.

Menger, M. D., Vajkoczy, P., Leiderer, R., Jäger, S. and Messmer K. (1992d) 'Influence of experimental hyperglycemia on microvascular blood perfusion of pancreatic islet isografts', *J. Clin. Invest.*, **90**, pp. 1361–1369.

Menger, M. D., Vajkoczy, P., Beger, C. and Messmer K. (1994) 'Orientation of microvascular blood flow in pancreatic islet isografts', *J. Clin. Invest.*, **93**, pp. 2280–2285.

Menger, M. D., Yamauchi, J. and Vollmar, B. (2001) 'Revascularization and microcirculation of freely grafted islets of Langerhans', *World J. Surg.*, **25**, pp. 509–515.

Menger, M. D., Laschke, M. W. and Vollmar, B. (2002) 'Viewing the microcirculation through the window: some twenty years experience with the hamster dorsal skinfold chamber', *Eur. Surg. Res.*, **34**, pp. 83–91.

Menger, M. D., Laschke, M. W., Amon, M., Schramm, R. *et al.* (2003) 'Experimental models to study microcirculatory dysfunction in muscle ischemia-reperfusion and osteomyocutaneous flap transfer', *Langenbecks Arch. Surg.*, **388**, pp. 281–290.

Papenfuss, H. D., Gross, J. F., Intaglietta, M. and Treese, F. A. (1979) 'A transparent access chamber for the rat dorsal skin fold', *Microvasc. Res.*, **18**, pp. 311–318.

Prewitt, R. L. and Johnson, P.C. (1976) 'The effect of oxygen on arteriolar red cell velocity and capillary density in the rat cremaster muscle', *Microvasc. Res.*, **12**, pp. 59–70.

Reynolds, L. P., Killilea, S. D. and Redmer, D. A. (1992) 'Angiogenesis in the female reproductive system', *FASEB J.*, **6**, pp. 886–892.

Riaz, A. A., Wan, M. X., Schaefer, T., Schramm, R. *et al.* (2002) 'Fundamental and distinct roles of P-selectin and LFA-1 in ischemia/reperfusion-induced leukocyte–endothelium interactions in the mouse colon', *Annals Surg.*, **236**, pp. 777–784.

Rücker, M., Roesken, F., Vollmar, B. and Menger, M. D. (1998) 'A novel approach for comparative study of periosteum, muscle, subcutis, and skin microcirculation by intravital fluorescence microscopy', *Microvasc. Res.*, **56**, pp. 30–42.

Rücker, M., Strobel, O., Vollmar, B., Roesken, F. and Menger, M. D. (2000) 'Vasomotion in critically perfused muscle protects adjacent tissues from capillary perfusion failure', *Am. J. Phys.: Heart Circ. Phys.*, **279**, pp. H550–H558.

Sandison, J. C. (1928) 'The transparent chamber of the rabbit's ear giving a complete description of improved techniques of construction and introduction and general account of growth and behavior of living cells and tissues seen with the microscope', *Am. J. Anat.*, **41**, pp. 447–472.

Schacht, V., Berens von Rautenfeld, D. and Abels, C. (2004) 'The lymphatic system in the dorsal skinfold chamber of the Syrian golden hamster *in vivo*', *Arch. Dermatol. Res.*, **295**, pp. 542–548.

Schneider,. P., Foitzik, T., Kahrau, S., Podufal, A. and Buhr, H. J. (2001) 'An experimental rat model for studying pulmonary microcirculation by *in vivo* videomicroscopy', *Microvasc. Res.*, **62**, pp. 421–434.

Strieth, S., von Johnston, V., Eichhorn, M. E., Enders, G. *et al.* (2005) 'A new animal model to assess angiogenesis and endocrine function of parathyroid heterografts *in vivo*', *Transplantation*, **79**, pp. 392–400.

Thaler, K., Mack, J. A., Berho, M., Grotendorst, G. *et al.* (2005) 'Coincidence of connective tissue growth factor expression with fibrosis and angiogenesis in postoperative peritoneal adhesion formation', *Eur. Surg. Res.*, **37**, pp. 235–241.

Torres Filho, I. P. and Intaglietta, M. (1993) 'Microvessel PO2 measurements by phosphorescence decay method', *Am. J. Physiol.*, **265**, pp. H1434–H1438.

Tyml, K. and Budreau, C. H. (1991) 'A new preparation of rat extensor digitorum longus muscle for intravital investigation of the microcirculation', *Int. J. Microcirc.: Clin. Exp.*, **10**, pp. 335–343.

Vajkoczy, P., Olofsson, A. M., Lehr, H. A., Leiderer, R. *et al.* (1995) 'Histogenesis and ultrastructure of pancreatic islet graft microvasculature: Evidence for graft vascularization by endothelial cells of host origin', *Am. J. Pathol.*, **146**, pp. 1397–1405.

Vajkoczy, P., Lehr, H. A., Hübner, C., Arfors, K. E. and Menger, M. D. (1997) 'Prevention of pancreatic islet xenograft rejection by dietary vitamin E', *Am. J. Pathol.*, **150**, pp. 1487–1495.

Vajkoczy, P., Schilling, L., Ullrich, A., Schmiedek, P. and Menger, M. D. (1998) 'Characterization of angiogenesis and microcirculation of high-grade glioma: an intravital multifluorescence microscopic approach in the athymic nude mouse', *J. Cerebral Blood Flow Metab.*, **18**, pp. 510–520.

Vajkoczy, P., Menger, M. D., Vollmar, B., Schilling, L. *et al.* (1999a) 'Inhibition of tumour growth, angiogenesis, and microcirculation by the novel Flk-1 inhibitor SU5416 as assessed by intravital multi-fluorescence videomicroscopy', *Neoplasia*, **1**, pp. 31–41.

Vajkoczy, P., Goldbrunner, R., Farhadi, M., Vince, G. *et al.* (1999b) 'Glioma cell migration is associated with glioma-induced angiogenesis *in vivo*', *Int. J. Dev. Neurosci.*, **17**, pp. 557–563.

Vajkoczy, P., Vollmar, B., Wolf, B. and Menger, M. D. (1999c) 'Effects of cyclosporine A on the process of vascularization of freely transplanted islets of Langerhans', *J. Molec. Med.*, **77**, pp. 111–114.

Vajkoczy, P., Ullrich, A. and Menger, M. D. (2000) 'Intravital fluorescence videomicroscopy to study tumour angiogenesis and microcirculation', *Neoplasia*, **2**, pp. 53–61.

Vajkoczy, P., Farhadi, M., Gaumann, A., Heidenreich, R. *et al.* (2002) 'Microtumour growth initiates angiogenic sprouting with simultaneous expression of VEGF, VEGF receptor-2, and angiopoietin-2', *J. Clin. Invest.*, **109**, pp. 777–785.

Vollmar, B., Siegmund, S. and Menger,. M. D. (1998) 'An intravital fluorescence microscopic study of hepatic microvascular and cellular derangements in developing cirrhosis in rats', *Hepatology*, **27**, pp. 1544–1153.

Vollmar, B., Janata, J., Yamauchi, J., Wolf, B. *et al.* (1999) 'Exocrine, but not endocrine, tissue is susceptible to microvascular ischemia/reperfusion injury following pancreas transplantation in the rat', *Transplant. Int.*, **12**, pp. 50–55.

Vollmar, B., Morgenthaler, M., Amon, M. and Menger, M. D. (2000) 'Skin microvascular adaptations during maturation and aging of hairless mice', *Am. J. Physiol.: Heart Circ. Physiol.*, **279**, pp. H1591–H1599.

Vollmar, B., Laschke, M. W., Rohan, R., Koenig, J. and Menger, M. D. (2001) '*In vivo* imaging of physiological angiogenesis from immature to preovulatory ovarian follicles', *Am. J. Pathol.*, **159**, pp. 1661–1670.

Vollmar, B., El-Gibaly, A. M., Scheuer, C., Strik, M. W. *et al.* (2002) 'Acceleration of cutaneous wound healing by transient p53 inhibition', *Lab. Invest.*, **82**, pp. 1063–1071.

Vollmar, B., Slotta, J. E., Nickels, R. M., Wenzel, E. and Menger, M. D. (2003) 'Comparative analysis of platelet isolation techniques for the *in vivo* study of the microcirculation', *Microcirculation*, **10**, pp. 143–152.

Westermann, S., Vollmar, B., Thorlacius, H. and Menger, M. D. (1999) 'Surface cooling inhibits tumour necrosis factor-alpha-induced microvascular perfusion failure, leukocyte adhesion and apoptosis in the striated muscle', *Surgery*, **126**, pp. 881–889.

Winet, H. (1996) 'The role of microvasculature in normal and perturbed bone healing as revealed by intravital microscopy', *Bone*, **19**(1 Suppl.), pp. 39S–57S.

Yuan, F., Salehi, H. A., Boucher, Y., Vasthare, U. S., *et al.* (1994) 'Vascular permeability and microcirculation of gliomas and mammary carcinomas transplanted in rat and mouse cranial windows', *Cancer Res.*, **54**, pp. 4564–4568.

Zweifach, B.W. (1954) 'Direct observation of the mesenteric microcirculation in experimental animals', *Anatomical Record*, **120**, pp. 277–291.

14
Tumour models: analysis of angiogenesis *in vivo*

Sven A. Lang and **Oliver Stoeltzing**

Abstract

Abundant studies have been conducted for elucidating the complex process of angiogenesis *in vivo*. In cancer research, tumour models of angiogenesis have a substantial role in the assessment of the efficacy of novel anti-angiogenic and anti-neoplastic therapy regimens in a preclinical setting. However, it has to be realized that these models also harbour some difficulties that eventually may lead to contradictory results. For example, tumour neovascularization and drug efficacy can be significantly affected by the tumour microenvironment. This in turn might result in a great heterogeneity of 'anti-angiogenic-activity' and chemosensitivity observed in models. Hence, for the development of promising anti-angiogenic therapy concepts, translational tumour models have to be designed appropriately in order best to reflect the complex tumour biology encountered in humans. This chapter will provide an overview on common tumour models with a focus on orthotopic models of tumour angiogenesis in mice and discuss their advantages and weaknesses, respectively.

Keywords

angiogenesis; tumour models; mice; tumour microenvironment

14.1 Introduction

Over the past decade, abundant experimental studies have been conducted for investigating the complex process of tumour angiogenesis *in vivo*. A great variety of tumour models have been developed and applied either to elucidate the exact role of certain angiogenic regulators *in vivo*, such as growth factor receptors and

Angiogenesis Assays   Edited by Carolyn A. Staton, Claire Lewis and Roy Bicknell
© 2006 John Wiley & Sons, Ltd

transcription factors, or to assess drug efficacy of novel anti-angiogenic agents. For the latter, the use of experimental tumour models has become a substantial component, as they allow sufficient monitoring of drug distribution, plasma levels and drug related toxicity (Fidler, 1986). Furthermore, anti-angiogenic effects can be studied in a relevant microenvironmental setting, as tumour angiogenesis involves the interaction of tumour endothelial cells (EC) with other tumour compartments such as the tumour cells themselves, peri-endothelial cells (pericytes), immune cells and stromal cells (e.g. fibroblasts). However, the initial 'failure' of many anti-angiogenic therapy concepts which have been transposed into clinical studies was, at least in part, created by an inappropriate use and interpretation of experimental tumour models (McCarty *et al.*, 2003). One of the key issues seems to be the phenomenon of organ-site specific angiogenesis which also determines efficacy of anti-angiogenic and vascular targeting agents (Keyes *et al.*, 2003; Cheung *et al.*, 1998; Jung *et al.*, 2000). In our experience, most anti-angiogenic agents elicit the strongest growth-inhibitory effects in subcutaneously grown (xenografted or syngenic) tumours, whereas a similar efficacy is much less frequently achieved when cells are implanted at the corresponding orthotopic site, such as the stomach or pancreas. Therefore, orthotopic tumour models should be used whenever possible in order to investigate the therapeutic effects of novel anti-angiogenic agents (in combination with standard chemotherapy) before subsequent initiation of clinical studies, as already proposed some time ago by Isaiah J. Fidler Ph.D., D.V.M. (Department of Cancer Biology; The University of Texas, MD Anderson Cancer Center, Houston, TX, USA) (Fidler, 2001b; Fidler, 2001a; Fidler, 1999).

In general, for studying neoplastic angiogenesis, tumour cells can either be implanted subcutaneously (syngenic or xenografted) or grown at an orthotopic site. In addition, tumour cells can also be directly implanted at a 'relevant orthotopic site of tumour metastasis', such as the liver, bone or lung. In order to help the researcher in choosing the 'right' tumour model, this chapter will provide an overview on experimental tumour models for analysing angiogenesis *in vivo* and discuss their advantages and disadvantages, respectively. In addition, various methods of assessing angiogenesis (and vascular permeability) in tumours and tissue sections are briefly described.

14.2   Tumour microenvironment and angiogenesis

As mentioned above, the process of angiogenesis is highly dependent on the organ microenvironment, resulting in organ-site specific tumour angiogenesis (Singh and Fidler, 1996; Fukumura *et al.*, 1997; Fidler, 2001b). This has great implication for researchers investigating the *in vivo* effects of 'vascular targeting agents' or certain anti-angiogenic substances, as observed results may differ dramatically among applied experimental tumour models, despite the same tumour cells/tumour entity being examined (Keyes *et al.*, 2003). Recent reports suggest that the tumour microenvironment may also determine efficacy of anti-angiogenic and anti-neoplastic agents (Jung *et al.*, 2000; McCarty *et al.*, 2003). In addition, for future vascular targeting strategies, it is important to realize that the tumour vasculature in primary tumour lesions may be quite distinct from that developed in their corresponding metastases.

The effect of the organ microenvironment on tumour biology has been recognized for more than 100 years, since Paget's 'seed and soil' hypothesis (Paget, 1989; Fidler, 2001b). It has been generally accepted that the organ/tumour microenvironment may significantly influence tumour biology *in vivo*. Great discrepancies in tumour vascularization, growth rate, metastatic potential, and the efficacy of systemic treatment between ectopic and orthotopic tumours have been reported in several experimental studies (Kitadai *et al.*, 1995; Paget, 1989; Fidler, 1995). The specific tumour microenvironment is composed of a variety of heterogeneous cell types as well as components of the extracellular matrix (ECM). Each component of this microenvironment interacts with one another, thereby leading to complex cellular responses. The microenvironment not only consists of organ specific cells such as hepatocytes or epithelial cells, but also endothelial cells, pericytes, inflammatory cells, fibroblasts and the ECM. The observation that blood vessel growth is controlled by the microenvironment rather than by some intrinsic genetic endothelial programme is also supported by transplantation studies in the quail chick system (Pardanaud *et al.*, 1987). Endothelial cells lose organ-specific characteristics when they are isolated from the organ environment and cultured *in vitro* (Jung *et al.*, 2000; Takahashi *et al.*, 1995). These studies highlight the importance of understanding the biological heterogeneity of tumour angiogenesis and the redundancy of angiogenic factor expression in specific organ microenvironments. Since neovascularization is essential for metastasis formation and their subsequent outgrowth (Folkman, 1990; Fidler and Ellis, 1994), organ site-specific microenvironmental regulation of angiogenesis is one of the most important determinants for the organ preference of metastasis formation. Only metastatic cells that express the appropriate and corresponding cell surface receptors will be able to implant and grow at the respective organ site.

The design of promising anti-angiogenic therapy concepts in the future also demands an appropriate design of translational–preclinical–tumour models that best reflect the complex tumour biology in humans. However, for reasons of feasibility and adequate timely fashion, certain compromises have to be considered and accepted in the design and use of experimental tumour models, which will be discussed in the following section.

14.3 Tumour models

Since the original proposal that targeting a tumour vasculature could be an efficient form of cancer therapy (Folkman, 1971), much data have accumulated to support the critical role of tumour-associated vessels in tumour growth and progression. In general, it has to be realized that in experimental tumour models for investigating angiogenesis *in vivo*, endothelial cells (and pericytes) in the neovasculature are derived from the host (e.g. mouse) and are not of human origin. Therefore we are investigating the effects of angiogenic growth factors or anti-angiogenic substances on a neoplastic neovasculature that may be quite distinct from that observed in humans (Goldbrunner *et al.*, 2004). Secondly, tumours – for example in murine models – are established within a few weeks after cell implantation, whereas human cancer develops over a period of several months or years, respectively. This fact is also reflected by histopathological morphology of human tumours. For example, in pancreatic cancer, in addition to a neoplastic microvasculature, tumour sections predominantly show a stromal tissue reaction compared with a relatively small extent of epithelial cancer cells. In contrast, in our experimental tumour models we basically observe tumour cells in tissue sections, as tumour cells have been implanted in a cell suspension bolus in order eventually to form subcutaneous or orthotopic tumours. Thus, the aspect of an influence of stromal cells on neoplastic angiogenesis is only insufficiently represented in most tumour models. Similarly, intra-tumoural lymphangiogenesis seems to be absent in subcutaneously grown and most orthotopically implanted tumours. However, tumour models are an essential part of current preclinical translational investigations, despite the above mentioned limitations. Thoroughly planned experimental studies with subsequent critical data interpretation will substantially contribute to elucidating the complex process of tumour angiogenesis and to evaluating the efficacy of novel anti-angiogenic therapy concepts. This section will provide an overview on common tumour models with a focus on models of tumour angiogenesis in mice, as these rodents are most frequently used in research laboratories.

Measuring tumour burden

The use of experimental animal models not only requires an extensive preparation effort, such as getting a detailed animal protocol approved, it also demands that certain guidelines for animal welfare are implemented throughout the entire study (UKCCCR, 1998). Thus, it is absolutely critical to gather as much information as possible whenever animal experiments are being conducted. Furthermore, for planning the group sizes (animal numbers) of investigational arms, it is important to realize that a sample size of 10 animals per group is required in order to detect a 30 per cent change in tumour growth parameters with a power of 80 per cent and 5 per cent two-sided significance.

In experimental tumour models, angiogenesis is mostly analysed by visualizing tumour vessels via immunohistochemical staining after the study has finally been terminated. In addition, various objective parameters of tumour growth (a consequence of angiogenesis) should be obtained before tumours are being sectioned and processed. *Tumour diameters* may be measured in order to calculate *tumour volumes* if grown tumours can not be excised easily or are being monitored in a subcutaneous tumour model. These volumes can easily be estimated by using the equation: *Length* (longest diameter) x *Width^2* (shortest diameter) x *0.5*. Another convincing parameter for assessing tumour burden is the measurement of final *tumour weight*. However, this method is only applicable for tumour models that lead to the development of single excisable tumours (e.g. subcutaneously grown tumours) (Stoeltzing *et al.*, 2004). Before excising tumours, photographic imaging of the tumour situation *in situ* should always be considered, as this may provide some additional valuable information. Furthermore, the development of distant organ metastases or loco-regional lymph node metastases needs to be examined in every animal. *Extent of metastasis* can be rated by either counting nodules (lymph nodes, peritoneal nodes, liver metastases, etc.) or by excising all tumour nodules and determining the overall weight of metastatic tumour masses in each animal (Stoeltzing *et al.*, 2002). Extent of organ metastases may also be estimated by measuring organ weight (liver, lung) (Stoeltzing *et al.*, 2003a). If abundant metastases are evident, as for example in peritoneal carcinomatosis, tumour nodules only exceeding a certain threshold diameter may be counted or excised. *Animal survival* is another important aspect, especially when new anti-angiogenic and anti-neoplastic substances are being evaluated (Reinmuth *et al.*, 2003; Stoeltzing *et al.*, 2003c). With respect to analysis of tumour angiogenesis, these models are difficult to interpret as mice are being individually sacrificed based on their health condition and not simultaneously in all groups.

As treatment with anti-angiogenic agents or chemotherapeutic substances may also exhibit drug-related organ toxicity, asservation of organ tissues (liver, kidney) is to be considered for detection of cellular alterations in selected studies. Blood samples can be obtained for measuring drug levels or circulating angiogenic growth factors. In addition, animal weights need to be measured intermittently and at the time of termination. Only an extensive acquisition of experimental *in vivo* data will ultimately allow an adequate interpretation of preclinical studies that potentially translate into promising clinical therapy regimens.

Cancer cells

Over the past decade, a large number of tumorigenic cancer cell lines of common human cancer entities have been investigated in experimental tumour models. In general, tumour cells to be used in animal models may either be syngenic [e.g. murine colon cancer cells CT26 (Reinmuth *et al.*, 2003)] (Samant *et al.*, 2005; Dell'Eva *et al.*, 2004; Peron *et al.*, 2004; Miller *et al.*, 2001) or of human origin, the latter requiring a model with immunodeficient animals, such as nude mice (Balb/c$^{nu/nu}$) (Stoeltzing *et al.*, 2004; Kern *et al.*, 1994; Presta *et al.*, 2000; Fidler, 2001a; Tang *et al.*, 2001). The advantage of using human cancer cells is that cells may display a more relevant molecular pathology in terms of angiogenic factor expression or chemosensitivity, which mirrors the tumour biology in patients more accurately. On the other hand, the use of human cancer cell lines also harbours some difficulties. For pancreatic cancer, approximately 40 cell lines derived from pancreatic ductal adenocarcinomas have been reported and widely used (Sipos *et al.*, 2003). However, the diversity of culture methods employed, together with long-term culture, have led to a considerable level of heterogeneity with undesired subcloning or partial loss of the initial molecular characteristics. This phenomenon, which is also valid for many other cancer cell lines, may in part lead to significant differences in growth patterns and degrees of tumour vascularization observed *in vivo*, despite the same cancer entity being studied. Fast growing tumour cells may elicit a higher degree of neovascularization (immature vessels) that respond better to anti-angiogenic therapy, than slower growing tumours with an 'average' degree of angiogenesis (more mature vessels with higher degree of pericyte coverage). Hence, selection of a cancer cell line for conducting an experimental tumour model remains a critical determinant.

With regard to syngenic cancer cells, their use can (in our opinion) only be recommended for certain studies that require an immunocompetent host system or for selected metastasis models. Otherwise these tumour cells tend to proliferate (and metastasize) at such high rates *in vivo* that experiments need to be terminated within a few weeks, leaving only a small window for therapy experiments. In addition, it is uncertain whether these cell lines express an angiogenic factor profile that is comparable to corresponding human tumour cells, not to mention alterations in the genetic profile.

Animals

For investigating tumour angiogenesis, a great variety of tumour models have been described that involve rodents. This chapter will mainly focus on mouse models of angiogenesis. However, certain assays might require the use of hamsters or rats, respectively, as some interesting syngenic cell lines are available for *in vivo* experiments in these animals or larger animals are needed for radiological (angiogenesis/perfusion) imaging studies (Heckl *et al.*, 2004). In our laboratory, we mainly investigate tumour growth and angiogenesis in mice. Interestingly, evidence is supported by several studies that angiogenesis in mice (or rats) may be distinct from that occurring in human tumours (Goldbrunner *et al.*, 2004; Sikder *et al.*, 2003). For example, over the past several years, data have been provided to support the hypothesis of a central role for the Id family of helix-loop-helix factors in the process of tumour angiogenesis (Benezra, 2001, Benezra *et al.*, 2001). Remarkably, mice lacking 1–3 *Id*1 and/or *Id*3 alleles have been demonstrated to resist growth of tumour implants due to defects in tumour angiogenesis (Lyden *et al.*, 1999; Lyden *et al.*, 2001). The tumour model system evaluated in those studies involved either an intradermal injection of xenografted tumour cells into host animals or the use of angiogenic stimuli in Matrigel plugs. In a recent study, Sikder and colleagues showed that skin tumours initiated by classical two-step chemical carcinogenesis do not demonstrate altered angiogenesis in Id1 null animals and the authors proposed a model for tumour angiogenesis that may occur via Id-dependent or Id-independent mechanisms (Sikder *et al.*, 2003).

In addition, we observed mouse strain-specific xenografted tumour growth rates and drug absorption. This suggests that mouse strains should be investigated in preliminary experiments for determining optimal drug dosing (toxicity) and tumour growth parameters. Subsequent experimental studies should be performed using one strain consistently throughout the experiments.

14.4   Subcutaneous tumour models

The subcutaneous (s.c.) implantation of tumour cells is one of the easiest approaches for investigating tumour growth and angiogenesis *in vivo* which has been used widely (Saadoun *et al.*, 2005; Schmid *et al.*, 2005; Graepler *et al.*, 2005; Qian *et al.*, 2004). Cells can be implanted into the subcutis (right or left flank) of mice by means of a 27- or 30-gauge needle. Usually, preliminary experiments are necessary for determining the optimal amount of cells to be implanted (approx. 1–2×10^6) and if cell suspensions need to be supplemented with matrices (e.g. Matrigel, collagen) in order to support tumour cell implantation and initial angiogenesis (Yamamura *et al.*, 1993; Bonfil *et al.*, 1994). For most cell lines, the tumour take is about 90–100 per cent in a s.c. tumour model, which makes this system quite reliable, so that overall animal numbers can be minimized. Critical for this s.c. model is to avoid intradermal tumour cell inoculation (unless skin cancer is being investigated), as this will rapidly lead to necrotic skin regions with subsequent tumour perforation. In addition, injection into the fat pad, which is typically located ventrally of the hind limb, should be avoided as this promotes angiogenesis and tumour growth *in vivo* which might result in tremendous growth heterogeneity in the s.c. model.

Despite the very well known differences in microenvironment of subcutaneously grown tumours versus growth at an orthotopic site (Keyes *et al.*, 2003) and the consequences for angiogenesis and tumour growth *in vivo*, this is a very valuable and suitable experimental animal model. Tumour growth can be easily assessed by measuring tumour diameters every other day and weighing excised tumours at the end of the experiment. The process of angiogenesis cannot be visualized or monitored on a daily basis with this model, but if animal numbers are high enough, subgroups of randomized animals may be sacrificed sequentially (i.e. day 0, day 2, day 10; including animals from the control groups) for imaging effects of anti-angiogenic therapy on neovasculature (e.g. EC apoptosis, which occurs in a time dependent fashion) (Bruns *et al.*, 2000b).

In our laboratory, the effects of anti-angiogenic and/or anti-neoplastic substances are first investigated in s.c. tumour models. As already mentioned earlier, sub-cutaneously grown tumours tend to respond more effectively to (anti-angiogenic) therapy or molecular studies (for example mediated by siRNA) and results will be obtained in a timely fashion. Hence, studies involving an expensive and time consuming orthotopic tumour model would only be conducted if preliminary data support the stated hypothesis. However, if s.c. tumour experiments show promising results, results should be confirmed in an orthotopic tumour model whenever feasible (if cells can be grown at an orthotopic site). In contrast, experimental studies for vessel morphology and analysis of molecular parameters of neoplastic vasculature (EC, pericytes) should only be examined in an appropriate tumour/organ microenvironment.

14.5 Orthotopic tumour models

As new anti-tumour and anti-angiogenic compounds reach clinical trials, one challenge that arises is whether tumours at several anatomical sites will respond similarly to the same agent (Jung *et al.*, 2000). As mentioned above, preclinical models have shown that the same tumour implanted into different sites elicit differences in response to chemotherapy (Wilmanns *et al.*, 1992; Staroselsky *et al.*, 1990; Donelli *et al.*, 1967). Thus organ microenvironment plays a critical role for models investigating tumour angiogenesis/vessel morphology and experimental animal models should reflect a typical clinical situation, e.g. colon cancer cells metastasize to the liver and not into the subcutis (Fidler, 1986). A great variety of orthotopic tumour models have been developed for most common human cancer entities and some selected models will be described briefly in this section (Manzotti *et al.*, 1993).

Orthotopic models harbour some specific difficulties such as 1) measuring or monitoring tumour growth *in vivo* (this might require the use of GFP transfected cells); 2) visualizing angiogenesis *in vivo*; 3) cancer cell lines may not be tumorigenic at an orthotopic site or at an organ-site which is typically used for 'artificial' implantation of metastases (e.g. liver, bone, lung); 4) final tumour take (= development of tumour mass) may be quite low (50–70 per cent), demanding higher numbers of animals per experimental group; and 5) tumour cells might not lead to the expected formation of typical metastases *in vivo* (lymph nodes, liver, lung, etc.) (Dore *et al.*, 1987; Bruns *et al.*, 1999). The latter is a relevant problem for most human cancer cell lines that are being investigated in immunodeficient mice (Dore *et al.*, 1987). To overcome this problem, tumour cells can be transformed into a more metastatic phenotype by repetitive injection and re-culturing of some metastasized cells, as previously described for human pancreatic cancer cells (Bruns *et al.*, 1999).

Similar to subcutaneous tumour cell implantation, inoculation of cancer cells at an orthotopic location might also require the growth support of co-injected matrices, such as collagen or Matrigel (see above) in order to facilitate implantation and growth of (human) cancer cells. Some orthotopic models use excised tumour tissues (from subcutaneously grown tumours or human tissues) for 'cell implantation', as tumour angiogenesis may develop slower and more realistically to a human situation, which involves a compartment with stromal cells (Schmidt *et al.*, 2000; Hotz *et al.*, 2001; Furukawa *et al.*, 1993a).

In contrast to s.c. tumour models, injection techniques for orthotopic models are much more difficult, as the respective organ might have to be exposed for cell implantation. This procedure will require appropriate anaesthesia of animals and an adequate environment for operation (light, microscope, sterile instruments, etc.). This means that in addition to the problem of a lower 'tumour take' rate, operative deaths have to be taken into account for planning overall animal numbers. Another very interesting aspect is that anaesthesia may also affect the efficiency of a tumour model, as certain metabolites of pentobarbital (nembutal) or analgesics might act as anti-metastatics *in vivo*, resulting in a significant reduction in liver metastases formation in metastasis models (Thaker *et al.*, 2005; Gaspani *et al.*, 2002). Thaker *et al.* demonstrated that GABA receptors are present on human colon cancer cell lines (KM12SM, HT29, RKO). In their study, continuous exposure to Nembutal (0.1–500 μg/ml) resulted in an IC50 level of 58 μg/ml for the KM12SM cells and 120 μg/ml for the HT29 cells. In addition, Nembutal reduced cellular cAMP concentration in colon cancer cells and resulted in a dose and time dependent decrease in MMP-2 and MMP-9 levels. In *in vivo* experiments, the incidence of liver metastases in the nembutal group was zero compared with eight out of nine in the metaphane group (Thaker *et al.*, 2005).

An alternative to investigating tumour growth and angiogenesis of an orthotopically implanted 'primary' tumour, is the examination of tumour growth at an organ site which represents a typical site for metastases formation, such as the liver, lung or peritoneum (Stoeltzing *et al.*, 2002; Goto *et al.*, 2002; Reinmuth *et al.*, 2003; Herynk *et al.*, 2003). As mentioned earlier, orthotopically implanted cancer cells might not lead to formation of metastases in the manner in which they would be expected to occur in humans. Therefore we have used models of direct cancer cell implantation into the liver or into the spleen of mice for development of colorectal cancer hepatic tumours (Stoeltzing *et al.*, 2003a). Similarly, cancer cells can be injected via the tail vein in order to produce lung 'metastases' (Goto *et al.*, 2002). Most importantly, wherever cancer cells are being injected (either into primary tumour or metastasis model), the developing tumours must not affect the health condition of the animals, by inducing bowel obstruction or jaundice, for example. Considering the rapid growth rates of most cancer models, these complications might occur in an early phase of the experimental study, requiring premature termination of the entire study.

In the following section, various orthotopic tumour models and implantation techniques are briefly described and discussed.

Gastrointestinal cancers

Oesophageal cancer

No models have been reported for oesophageal cancer. Obstruction with consecutive dysphagia is the major problem for this model. In contrast, orthotopic models for head and neck cancer (squamous cell carcinoma) have been developed

(Myers *et al.*, 2002; Holsinger *et al.*, 2003). To produce tumours, authors used human JMAR cells (in 50 µl of HBSS) and injected them into the tongue of athymic nude mice, using a 30-gauge hypodermic needle and tuberculin syringe. Based on their experience with this model, a tumour take rate of at least 67 per cent was estimated.

Gastric cancer

Several human cancer cell lines (adenocarcinomas) have been studied in ortho-topic gastric cancer models (Yashiro *et al.*, 1996; Fujihara *et al.*, 1998). However, most tested cell lines were not tumorigenic in our hands. Cell implantation into the gastric wall can be achieved by two methods. First, cells can be directly injected (40–50 µl) into the gastric wall (greater curvature), remote of the pylorus in order to avoid obstruction (Stoeltzing *et al.*, 2004; McCarty *et al.*, 2004a; McCarty *et al.*, 2004b), and tumours will develop over 4–6 weeks. With TMK-1 cells we observed a tumour take rate of 60–70 per cent with this model. Secondly, tumour tissues can be directly implanted into the stomach of animals. Authors reported on using tissues derived from subcutaneously grown human gastric cancer specimen for this model (Furukawa *et al.*, 1993a; Furukawa *et al.*, 1993b; Lu *et al.*, 2004). Lu *et al.* grew BGC-823 cell subcutaneously and removed the tumour at a diameter of 2–2.5 cm. Pieces of fresh tumour tissue (1 mm^3) near the margin were subsequently used for orthotopic transplantation.

Colon cancer

An abundant number of human or syngenic colon cancer cell lines have been reported to be used in experimental models. For orthotopic tumour formation (caecal tumours) cells (for example HT29 cells) can be injected as single cell suspensions (in 50 µl HBSS) into the caecal wall of nude mice (Morikawa *et al.*, 1988, Yokoi *et al.*, 2005). After a period of 5–7 weeks, mice will have developed tumours and lymph node metastases to such an extent that the experiment needs to be terminated (Yokoi *et al.*, 2005). From a clinical perspective, this model appears not to be that relevant, as primary colon cancer (not rectum cancer) can be resected in most patients. In contrast, treatment of distant metastases (liver metastases) is the great challenge. Therefore we prefer investigating growth of human colon cancer cells in our experimental models only at organ sites that are relevant for this malignancy (i.e. liver) (Stoeltzing *et al.*, 2003a).

Pancreatic cancer

Similar to gastric cancer, a variety of cell lines have been used for orthotopic pancreatic cancer (Sipos *et al.*, 2003), that can be injected directly into the tail of the pancreas (50 μl volumes). The numbers of cells to be injected vary from 250 000 to one million, depending on the cell line. Usually tumours form quite rapidly and the experiments need to be terminated within 35 days (Stoeltzing et al., 2003b; Bruns et al., 2000b; Schniewind et al., 2004). Injection into the pancreatic head should be avoided as this may cause bowel obstruction and jaundice leading to severe morbidity and mortality. Other groups have also implanted tissue fragments of syngenic ductal adenocarcinoma into the pancreas of rats in order to establish an authentic model of pancreatic tumour growth and angiogenesis (Hotz et al., 2000; Hotz et al., 2001; Hotz et al., 2003). Hotz and colleagues first injected cancer cells into the subcutis of rats to allow tumour formation. Tumours were then harvested under strict aseptic conditions and minced by a scalpel into small fragments (1.0 mm^3). To avoid necrotic tissue from central tumour areas, only macroscopically viable tumour tissue from the outer part of the donor tumours was used for subsequent orthotopic implantation. For operative procedures, rats were anaesthetized and animals' abdomens opened through a midline incision. The spleen with the tail of the pancreas was exteriorized and 3–5 tissue pockets prepared in the pancreatic parenchyma, serving as an implantation bed. One donor tumour fragment was then placed into each pancreatic tissue pocket in a way that the tumour tissue was completely surrounded by pancreatic parenchyma. In their experiments, no sutures or fibrin glue were used to fix the tumour fragments to the recipient pancreas. With this model, a tumour take of 75 per cent was achieved and a duration of experiments of 16 weeks (Hotz et al., 2001). In a similar model, this author group used human pancreatic cancer tissue implants (MiaPaCa-2) in an experimental mouse model (Hotz et al., 2003).

Breast cancer

Although most human breast cancer cells injected into the mammary fat pad of athymic (female) nude mice elicit metastatic properties (lung, lymph nodes) in vivo, some cell lines have been developed (selected) that also metastasize to the skeletal systems or brain which resembles the situation in humans (Yoneda et al., 2001; Hoffman, 1999). Bone metastasis thus far has only been achieved by injecting breast cancer cells into the heart (left ventricle) (Sasaki et al., 1995) or directly into the bone of nude mice (Wang and Chang, 1997a), respectively. In addition, the use of a syngenic breast cancer line (4T1) has been described (Samant et al., 2005; Hiraga et al., 2005).

A number of highly metastatic breast cancer cell lines have been used for orthotopic tumour models in the past that can be injected as cell suspension into mammary fat pad of animals, preferably into female animals in (Kim and Price, 2005).

For orthotopic growth and metastasis formation, human breast cancer cells (MDA-MB-435) (ER-) have been inoculated into mammary fat pads of female athymic nude mice (ages 4 to 5 weeks) via 5 mm incision (in the skin over the lateral thorax) after animals had been anaesthetized. In a recently described experiment, a number of 1×10^6 cells/0.2 ml have been used for implantation into the tissue by means of a 27-gauge needle (Lang *et al.*, 2005). In this experiment the animals survived for 10 weeks. As these cells do metastasize to bone, lung and lymph nodes, mice have to be examined thoroughly at the end of the experiment. Some other models of tumour growth and angiogenesis involve the use of transgenic mice in order to set up a realistic tumour system with sequential carcinogenesis, which are described later in this section.

Prostate cancer

Understanding the mechanism of prostate cancer metastasis is essential to the design of a more effective therapy which in part depends on the use of a clinically relevant *in vivo* model. Similarly to breast cancer, prostate cancer growth, angiogenesis and metastasis can be investigated in orthotopic models by injecting cells directly into the prostate of male nude mice, as described by research groups in the I. J. Fidler laboratory (Stephenson *et al.*, 1992; Pettaway *et al.*, 1996; Huang *et al.*, 2002).

In these experimental studies, investigators used LNCaP prostate cancer cells to produce tumours in the prostate (Stephenson *et al.*, 1992). Interestingly, this human cancer cell line grew only in the prostate. In contrast, enhanced tumorigenicity at the orthotopic site was found for PC-3M cells, where lymph node metastases were observed in all mice given an injection of PC-3M cells in the prostate. Bilateral orchiectomy did not alter the tumorigenicity of either PC-3M or LNCaP cells or the incidence of lymph node metastasis by PC-3M cells. In subsequent studies, Pettaway *et al.* determined whether the implantation of human prostate cancer cells into the prostates of nude mice and their subsequent growth could be used to select variants with increasing metastatic potential (Pettaway *et al.*, 1996). For this purpose, PC-3M and LNCaP cells were injected into the prostates of athymic mice and tumours from the prostate or lymph nodes were harvested, and cells were consecutively reinjected into the prostate. This cycle was repeated three to five times. Eventually, PC-3M-LN4 cells produced enhanced regional lymph node and distant organ metastasis as compared with PC-3M-Pro4 or PC-3M cells. This model is now being used to study preclinical efficacy of anti-angiogenic therapy regimens (Huang *et al.*, 2002).

Non-small-cell-lung cancer (NSCLC)

For investigating tumour growth and angiogenesis of human lung cancer (NSCLC), a variety of approaches have been used to accomplish tumour implantation at an orthotopic site, however, development of better models has been an important research focus over the last few years. So far, reported models for NSCLC include either implantation of human cancerous tissues which were obtained by surgical resection (Hoffman, 1999), or injection of tumour cell suspensions into the rodent airways (McLemore *et al.*, 1987; Howard *et al.*, 1991; Hastings *et al.*, 2001), into pleural cavity (McLemore *et al.*, 1988; Kraus-Berthier *et al.*, 2000), or into the lung parenchyma after skin incision (Doki *et al.*, 1999), or eventually by a transthoracic approach via thoracotomy (Wang *et al.*, 1997b; Miyoshi *et al.*, 2000), respectively. In contrast, only two studies describe the use of orthotopic models to investigate *in vivo* growth of small-cell lung cancer (SCLC), which comprises about 20 per cent of all lung cancer cases (McLemore *et al.*, 1988; Kuo *et al.*, 1993).

Despite their availability, orthotopic models of human lung cancer are not widely used, and most of the research and development of novel therapeutics for lung cancer still relies upon s.c. tumour models, which are potentially less relevant for translation into clinical situations. Therefore, Onn and colleagues optimized the model for orthotopic implantation of NSCLC tumour cells (NCI-H358: bronchioloalveolar carcinoma, NCI-H1299 and A549: poorly differentiated NSCLC, PC14PE6: selected from human adenocarcinoma cell line PC14 to produce pleural effusion when injected into mice) for evaluating anti-angiogenesis treatment *in vivo* (Onn *et al.*, 2003; Onn *et al.*, 2004). In their initial description of this model, authors prepared equal volumes of cell suspensions for thoracic injection in a PBS and Matrigel stock, giving final dilution factor of approximately x60. Subsequently, mice were anaesthetized and placed in the right lateral decubitus position, and cell stock suspensions were injected (by means of 1 ml tuberculin syringes with 30-gauge needles) percutaneously into the left lateral thorax, at the lateral dorsal axillary line, approximately 1.5 cm above the lower rib line just below the inferior border of the scapula. After tumour injection, the mouse was turned to the left lateral decubitus position to allow cell distribution (Onn *et al.*, 2003). For establishing an aggressive phenotype of NCI-H226 cells, the authors re-cultured cells and re-injected them into the lungs of nude mice, as described for other models (Myers *et al.*, 2002; Bruns *et al.*, 1999). For a chemotherapy study, the experimental duration was 7–9 weeks for this model (Onn *et al.*, 2003). In addition, GFP transfection of cells can be performed to monitor tumour growth *in vivo*, however, monitoring of angiogenesis remains difficult.

14.6 Transgenic mouse models of tumour angiogenesis

Within the last two decades an increasing number of genetically engineered mouse models have been developed for studying carcinogenesis and ultimatively angiogenesis. In general, production of transgenic mice is done by targeted disruption of specific genes resulting in gene 'knock-in' (transgenic) or 'knock-out' animals (Lewandoski, 2001; Jonkers and Berns, 2002). These animals either overexpress the targeted oncogenes (knockin mice) or lack tumour suppressor genes (knockout mice), some of them in a tissue specific manner. The most important advantage of these models is that tumour growth takes place at a predictable age in their natural tissue environment and thereby negative impacts of ectopic tumour models like microenvironment and long-term cell culture are avoided. Moreover, tumour growth and angiogenesis can be examined through a series of pre-malignant lesions followed by carcinoma *in situ* until the emergence of invasive cancer, for example revealing in detail when during the process of multistep carcinogenesis angiogenesis becomes apparent. Furthermore, in 'knock-in' or 'knock-out' models the efficiency of anti-angiogenic therapy strategies can be tested on tumours with specific genetic background which commonly occurs in tumour development. While most of the models have a permanent knock-out of the targeted genes, new technology based on bacteria-derived tetracycline-inducible system (Gossen and Bujard, 1992) allows induction or suppression of expression of certain genes under control of a tetracycline-dependent transcriptional activator protein (Tet-Off and Tet-On systems). Activation and deactivation of specific genes can thereby be initiated at different time points and impact on tumour growth and angiogenesis can be studied. Therefore, genetically engineered mice are helpful for both evaluating the angiogenic process in general and assessing therapeutic effects of angiogenesis inhibitors alone or in combination with other chemotherapeutic agents. To date, more than 50 transgenic mouse models of cancer are in use with even more been under evaluation and some of them will be described briefly (Crnic and Christofori, 2004).

RIP-Tag transgenic mice

The prototype model of pancreatic islet β-cell tumour carcinogenesis is the RIP-Tag mouse model. Tumours within this model are induced by targeted over-expression of *SV40 Tag* oncogene under control of the insulin gene regulatory region (Hanahan, 1985). Through different stages of tumour formation, ranging from oncogene overexpression without consequences through hyperplastic islets, characterized as carcinoma *in situ*, these mice end up with formation of solid tumours. Analyses of the subset of hyperplastic islets that progress to solid tumours revealed that the onset of angiogenesis is required for tumour progression providing evidence for an angiogenic switch before the appearance of invasive malignancy (Hanahan *et al.*, 1996; Folkman *et al.*, 1989). The changes in the microvascular system during this particular tumorigenesis have been described more in detail by Ryschich and co-workers (2002). In addition, the functional involvement of FGFs has been shown in RIP-Tag model by inhibiting the angiogenic switch with soluble FGF receptors (Compagni *et al.*, 2000). Recently, the efficiency of therapies targeting tumour vasculature by kinase inhibitors for platelet-derived growth factor receptor (PDGFR), vascular endothelial growth factor receptor (VEGFR) and their combination with metronomic chemotherapy have also been demonstrated in RIP-Tag mouse models (Bergers *et al.*, 2003; Pietras and Hanahan, 2005).

BPV1.69 transgenic mice

Another murine model, with spontaneously developing tumours of the skin, involves mice which are transgenic for bovine papillomavirus (BPV1.69) (Adams and Cory, 1991). This model is characterized by multistage tumorigenesis of dermal fibrosarcomas occurring in distinct stages: normal skin, mild and aggressive fibromatosis and fibrosarcomas (Sippola-Thiele et al., 1989). The different stages of tumour growth are not only determined by histology but also by an associated expression of certain oncogenes (e.g. Jun-B) and transcription factors (e.g. AP-1) (Bossy-Wetzel et al., 1992). Again within tumour progression, there is a dramatic increase in vascular density during premalignancy, providing evidence for an angiogenic switch before invasive tumour growth onset (Hanahan et al., 1996).

HER2/neu transgenic mice

HER2/neu is an oncogene overexpressed in 25 to 30 per cent of human breast cancers, influencing prognosis and tumour biology (Choudhury and Kiessling, 2004). Moreover, HER2/neu signalling and overexpression stimulate the transcription of VEGF and thereby promoting the angiogenic progression (Kumar and Yarmand-Bagheri, 2001). Therefore, transgenic mice expressing the HER2/neu oncogene under tissue-specific transcriptional control of mouse mammary tumour virus promoter (MMTV-LTR) are a suitable model of mammary carcinogenesis (Guy et al., 1992). Moreover, this mouse model shows vessel recruitment, tissue invasion and gives rise to distant metastases (Sacco et al., 2000; Sacco et al., 1998). Tumour development again reveals a multistage process from atypical hyperplasia to carcinoma in situ and invasive carcinoma (Di Carlo et al., 1999). In addition, tumours arising in HER2/neu mice bear oestrogen receptors (ER) which are also present in a subset of human tumours. Several reports evaluated the effect of different anti-angiogenic treatment regimes alone or in combination with other chemotherapeutic agents or gene therapies, using this tumour model (Lee et al., 2002; Sacco et al., 2000; Sacco et al., 2003).

14.7 Monitoring vascular permeability

In addition to tumour angiogenesis, vascular permeability represents an important parameter of vessel function/dysfunction in malignant tissues. Most tumours express high amounts of vascular endothelial growth factor (VEGF), which was initially identified as vascular permeability factor (VPF) by Harold Dvorak and colleagues (Senger et al., 1983; Nagy et al., 1995; Feng et al., 2000). Importantly, increased vascular permeability may lead to high interstitial pressure levels in solid malignancies, a situation which consecutively leads to loco-regional perfusion deficiencies and tissue hypoxia. Thus, chemotherapy may not reach tumour cells effectively and chemoresistance may additionally develop. New anti-angiogenic therapy regimen have been shown to eradicate immature (hyper-permeable) tumour vessels which subsequently leads to tumour vessel normalization and lower interstitial pressures. Anti-angiogenic therapy may therefore render tumour cells susceptible to chemotherapy again. However, the parameter 'vascular permeability' is often not addressed in tumour models, which in part might be due to difficulties in monitoring and rating tumour permeability *in vivo*. Dvorak *et al.* described a method using FITC-labelled dextran (MW 250 000–1 000 000 Da) for perfusion of tumours and subsequent analysis of vessel morphology and distribution into surrounding tissue (Senger *et al.*, 1983).

Another valid assay for investigating vascular permeability is the Miles *in vivo* permeability assay, which can be performed on tumour bearing animals by injecting Evans' blue dye into the tail vein, as described below, and excising tumours after 30 min. The dye can subsequently be extracted and measured by spectophotometry. An alternative is the use of conditioned media derived from tumour cells in non-tumour bearing animals (Figure 14.1).

For this assay, conditioned media is collected after 48 h incubation in 1 per cent FBS-modified Eagle's medium at 80 per cent cell density, centrifuged for 5 min at $350 \times g$, and filtered through a 0.22 μm filter. Animals (e.g. nude mice) then receive an i.v. injection with sterile 0.5 per cent Evans blue dye (200 μl) via the tail vein. After 10 min, animals are given intradermal injections into the dorsal skin at different sites and killed 20 min later. The dorsal skin of each animal is then harvested to permit visualization of intradermal dye leakage. In order determine the relative degree of vascular permeability, two dimensions (*a* and *b*) of the elliptically appearing area of dye leakage can be measured at each injection site and the area can be calculated with the formula $a \times b \times \pi$. In addition, optical densitometry can be performed on digitally acquired images (Stoeltzing *et al.*, 2002; Stoeltzing *et al.*, 2003a).

14.8   Analysis of angiogenesis in tumours

In order to investigate neoplastic vascularization and vessel function in primary and metastatic tumour tissues, various approaches and methods have been applied to visualize and quantitate tumour angiogenesis *in vivo*, including: immunohistochemical analyses, ultrastructural analyses (electron microscopy), fluorescent *in vivo* imaging of tumour vessels, and many other experimental models of angiogenesis (Staton *et al.*, 2004).

Immunohistochemical staining of tumour vessels is one of the most commonly used procedures for evaluating effects of anti-angiogenic agents on angiogenesis in experimental models. Using current technology, it is necessary for this approach to obtain tissue for histological evaluation, which needs to be embedded accordingly. The standard for evaluation of vascularization in tumour specimen is to highlight

(a)

(b)

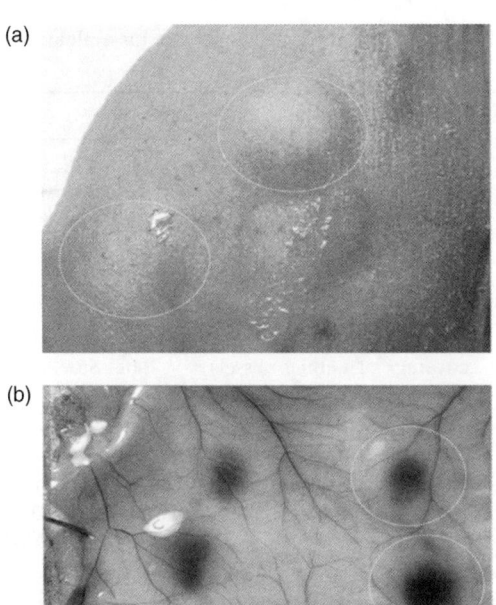

Figure 14.1 Miles *in vivo* permeability assay using conditioned media conditioned media (CM) was collected from tumour cells and processed, as described. Animals (nude mice) received an i.v. injection of sterile 0.5 per cent Evans blue dye (200 µl) via tail vein. (a) Ten minutes later, animals were given intradermal injections of CM into the dorsal skin at different sites (circled area) and killed 20 min later. (b) The dorsal skin of each animal was harvested to permit visualization of intradermal dye leakage. In order determine the relative degree of vascular permeability, the area can be calculated with the formula $a \times b \times \pi$. In addition, optical densitometry can be performed on digitally acquired images, as described (Stoeltzing *et al.*, 2002; Stoeltzing *et al.*, 2003a). (A colour reproduction of this figure can be viewed in the colour section towards the centre of the book).

the tumour endothelium with antibodies that differentiate ECs from other cells within the tumour. For this purpose, endothelial-specific antibodies such as FVIII-RA, CD31/PECAM-1, VE-cadherin and CD34 have been used in immuno-histochemical staining procedures (Table 14.1). However, the second and much more difficult step is the rating and appropriate interpretation of stained tumour sections. One of the most common parameters for rating tumour angiogenesis is calculation of microvessel density (MVD), which was considered to be the 'gold standard'. Various methods and guidelines for determining MVD in tumours have been described, based on the initial method reported by Weidner *et al.* (Takahashi *et al.*, 1996; Weidner *et al.*, 1991). For example, vessel numbers may be determined by choosing the five most vascularized areas within a tumour and counting the

Table 14.1 Overview on immunohistochemical analyses for evaluating angiogenesis in experimental models *in vivo*

Staining	Parameter	References
H&E	structural analysis, necrosis	(Huang *et al.*, 2000, Solorzano *et al.*, 2001, Bruns *et al.*, 2000b, Kedar *et al.*, 2002, McCarty *et al.*, 2004a)
CD31 (CD34)	vessel density, vessel area	(Bruns *et al.*, 2000a, Solorzano *et al.*, 2001, Goto *et al.*, 2002)
VE-cadherin	vessel density, vessel area	(Martin *et al.*, 2005)
CD31 + αSMA	vessel morphology, pericyte coverage of tumour vessels	(Terayama *et al.*, 1996, Reinmuth *et al.*, 2002, Stoeltzing *et al.*, 2004, Winkler *et al.*, 2004)
CD31 + desmin	vessel morphology, pericyte coverage of tumour vessels	(Stoeltzing *et al.*, 2004)
vWF	vessel density	(Terayama *et al.*, 1996)
TUNEL (+CD31)	apoptosis (EC apoptosis)	(Bruns *et al.*, 2000a, Bruns *et al.*, 2000b, Kim *et al.*, 2004b, Stoeltzing *et al.*, 2003c)
BrdU (+CD31)	cell proliferation (EC proliferation)	(Stoeltzing *et al.*, 2004)
PCNA	cell proliferation	(McCarty *et al.*, 2004a, Kim *et al.*, 2004a)

vessels/branches in these areas under higher magnification. Vessels can also be quantified using image analysis (densitometry: vessel area as parameter of functional vascular volume) or vessel density can be graded on a scale 0+ to 3+, with being 3+ the most vascular (mostly applied for human tissue specimens). In addition, the degree of angiogenesis can be assessed by counting numbers of branch-points or measuring inter-capillary distance. Other important aspects of neoplastic vascularization are 'vessel maturation' or 'vessel function'. Most tumour vessels appear to be immature in their structure, as they are lacking support by peri-endothelial cells (pericytes), consequently leading to increased vessel permeability and blood flow. Hence, analysis of vessel maturation has become an interesting aspect of angiogenesis research which either can be achieved by staining for peri-endothelial cell markers, such as α-smooth muscle actin (αSMA) (Baluk *et al.*, 2003; Stoeltzing *et al.*, 2003a) or by ultrastructural analysis (Caruso *et al.*, 2004; Wesseling *et al.*, 1995).

In the era of anti-angiogenic therapy, it has become even more difficult objectively to assess tumour vascularization. Due to effects like tumour shrinkage, reduction of interstitial pressure and 'vessel normalization' by anti-angiogenic and anti-neoplastic therapy regimen, the simple determination of MVD may not reflect these profound changes (Jain, 2001; Winkler *et al.*, 2004). Further studies are required to identify optimal biomarkers for monitoring tumour angiogenesis in patients receiving anti-angiogenic therapy.

14.9 Conclusions

Experimental models of tumour angiogenesis have become a substantial component for assessing drug efficacy of novel anti-angiogenic and anti-neoplastic agents in preclinical studies. Abundant models have been described for assaying neoplastic angiogenesis *in vivo*. However, it has to be realized that these models also harbour some difficulties that may lead to contradictive results in experimental studies. Tumour angiogenesis (and drug efficacy) may not only be influenced by the organ site or microenvironment (subcutaneous versus orthotopic), also mouse strain, or tumour cell line to be investigated may display a great heterogeneity in terms of angiogenesis and chemosensitivity. The most important aspect appears to be the organ microenvironment, suggesting that orthotopic models should be used whenever feasible. In preliminary studies, the use of subcutaneous tumour models may have the advantage of an easy practicability and efficient growth monitoring, so that it seems to be absolutely reasonable to start with this model.

In contrast, monitoring effects of anti-angiogenic or anti-neoplastic agents on angiogenesis *in vivo* is much more difficult and might require sequential termination of experiments, which in turn will require higher numbers of animals in experimental groups. Hence, in most tumour models, neoplastic vascularization is being investigated as an end-point analysis. Currently, new *in vivo* imaging techniques are being developed that potentially allow effective monitoring of angiogenesis or the impact of novel vascular targeting agents that may contribute to successful analysis of tumour models. However, the appropriate design of an experimental model will have the most impact on the success of any study.

References

United Kingdom Co-ordinating Committee on Cancer Research (UKCCCR) (1998) *Guidelines for the Welfare of Animals in Experimental Neoplasia* (2nd edn)', *Br. J. Cancer*, **77**(1), pp. 1–10.
Adams, J. M. and Cory, S. (1991) 'Transgenic models of tumor development', *Science*, **254**(5035), pp. 1161–1167.
Baluk, P., Morikawa, S., Haskell, A., Mancuso, M. and McDonald, D. M. (2003) 'Abnormalities of basement membrane on blood vessels and endothelial sprouts in tumors', *Am. J. Pathol.*, **163**(5), pp. 1801–1815.
Benezra, R. (2001) 'Role of Id proteins in embryonic and tumor angiogenesis', *Trends Cardiovasc. Med.*, **11**(6), pp. 237–241.
Benezra, R., Rafii, S. and Lyden, D. (2001) 'The Id proteins and angiogenesis', *Oncogene*, **20**(58), pp. 8334–8341.
Bergers, G., Song, S., Meyer-Morse, N., Bergsland, E. and Hanahan, D. (2003) 'Benefits of targeting both pericytes and endothelial cells in the tumor vasculature with kinase inhibitors', *J. Clin. Invest.*, **111**(9), pp. 1287–1295.

Bonfil, R. D., Vinyals, A., Bustuoabad, O. D., Llorens, A. *et al.* (1994) 'Stimulation of angiogenesis as an explanation of Matrigel-enhanced tumorigenicity', *Int. J. Cancer*, **58**(2), pp. 233–239.

Bossy-Wetzel, E., Bravo, R. and Hanahan, D. (1992) 'Transcription factors junB and c-jun are selectively up-regulated and functionally implicated in fibrosarcoma development', *Genes Dev.*, **6**(12A), pp. 2340–2351.

Bruns, C. J., Harbison, M. T., Kuniyasu, H., Eue, I. and Fidler, I. J. (1999) 'In vivo selection and characterization of metastatic variants from human pancreatic adenocarcinoma by using orthotopic implantation in nude mice', *Neoplasia*, **1**(1), pp. 50–62.

Bruns, C. J., Harbison, M. T., Davis, D. W., Portera, C. A. *et al.* (2000a) 'Epidermal growth factor receptor blockade with C225 plus gemcitabine results in regression of human pancreatic carcinoma growing orthotopically in nude mice by antiangiogenic mechanisms', *Clin. Cancer Res.*, **6**(5), pp. 1936–1948.

Bruns, C. J., Solorzano, C. C., Harbison, M. T., Ozawa, S. *et al.* (2000b) 'Blockade of the epidermal growth factor receptor signaling by a novel tyrosine kinase inhibitor leads to apoptosis of endothelial cells and therapy of human pancreatic carcinoma', *Cancer Res.*, **60**(11), pp. 2926–2935.

Caruso, R. A., Ieni, A., Fabiano, V., Basile, G. and Inferrera, C. (2004) 'Perivascular mast cells in advanced gastric adenocarcinomas: an electron microscopic study', *Anticancer Res.*, **24**(4), pp. 2257–2263.

Cheung, N., Wong, M. P., Yuen, S. T., Leung, S. Y. and Chung, L. P. (1998) 'Tissue-specific expression pattern of vascular endothelial growth factor isoforms in the malignant transformation of lung and colon', *Hum. Pathol.*, **29**(9), pp. 910–914.

Choudhury, A. and Kiessling, R. (2004) 'Her-2/neu as a paradigm of a tumor-specific target for therapy', *Breast Dis.*, **20**(25–31).

Compagni, A., Wilgenbus, P., Impagnatiello, M. A., Cotten, M. and Christofori, G. (2000) 'Fibroblast growth factors are required for efficient tumor angiogenesis', *Cancer Res.*, **60**(24), pp. 7163–7169.

Crnic, I. and Christofori, G. (2004) 'Novel technologies and recent advances in metastasis research', *Int. J. Dev. Biol.*, **48**(5–6), pp. 573–581.

Dell'Eva, R., Pfeffer, U., Vene, R., Anfosso, L. *et al.* (2004) 'Inhibition of angiogenesis in vivo and growth of Kaposi's sarcoma xenograft tumors by the anti-malarial artesunate', *Biochem. Pharmacol.*, **68**(12), pp. 2359–2366.

Di Carlo, E., Diodoro, M. G., Boggio, K., Modesti, A. *et al.* (1999) 'Analysis of mammary carcinoma onset and progression in HER-2/neu oncogene transgenic mice reveals a lobular origin', *Lab. Invest.*, **79**(10), pp. 1261–1269.

Doki, Y., Murakami, K., Yamaura, T., Sugiyama, S. *et al.* (1999) 'Mediastinal lymph node metastasis model by orthotopic intrapulmonary implantation of Lewis lung carcinoma cells in mice', *Br. J. Cancer*, **79**(7–8), pp. 1121–1126.

Donelli, M. G., Rosso, R. and Garattini, S. (1967) 'Selective chemotherapy in relation to the site of tumor transplantation', *Int. J. Cancer*, **2**(5), pp. 421–424.

Dore, J. F., Bailly, M. and Bertrand, S. (1987) 'Metastases of human tumors in experimental animals', *Anticancer Res.*, **7**(5B), pp. 997–1003.

Feng, D., Nagy, J. A., Dvorak, A. M. and Dvorak, H. F. (2000) 'Different pathways of macromolecule extravasation from hyperpermeable tumor vessels', *Microvasc. Res.*, **59**(1), pp. 24–37.

Fidler, I. J. (1986) 'Rationale and methods for the use of nude mice to study the biology and therapy of human cancer metastasis', *Cancer Metastasis Rev.*, **5**(1), pp. 29–49.

Fidler, I. J. (1995) 'Critical factors in the biology of human cancer metastasis', *Am. Surg.*, **61**(12), pp. 1065–1066.

Fidler, I. J. (1999) 'Critical determinants of cancer metastasis: rationale for therapy', *Cancer Chemother. Pharmacol.*, **43**(Suppl.), pp. S3–10.

Fidler, I. J. (2001a) 'Angiogenic heterogeneity: regulation of neoplastic angiogenesis by the organ microenvironment', *J. Natl. Cancer Inst.*, **93**(14), pp. 1040–1041.

Fidler, I. J. (2001b) 'Seed and soil revisited: contribution of the organ microenvironment to cancer metastasis', *Surg. Oncol. Clin. N. Am.*, **10**(2), pp. 257–269, vii–viiii.

Fidler, I. J. and Ellis, L. M. (1994) 'The implications of angiogenesis for the biology and therapy of cancer metastasis', *Cell*, **79**(2), pp. 185–188.

Folkman, J. (1971) 'Tumor angiogenesis: therapeutic implications', *N. Engl. J. Med.*, **285**(21), pp. 1182–1186.

Folkman, J. (1990) 'What is the evidence that tumors are angiogenesis dependent?' *J. Natl Cancer Inst.*, **82**(1), pp. 4–6.

Folkman, J., Watson, K., Ingber, D. and Hanahan, D. (1989) 'Induction of angiogenesis during the transition from hyperplasia to neoplasia', *Nature*, **339**(6219), pp. 58–61.

Fujihara, T., Sawada, T., Hirakawa, K., Chung, Y. S. *et al.* (1998) 'Establishment of lymph node metastatic model for human gastric cancer in nude mice and analysis of factors associated with metastasis', *Clin. Exp. Metastasis*, **16**(4), pp. 389–398.

Fukumura, D., Yuan, F., Monsky, W. L., Chen, Y. and Jain, R. K. (1997) 'Effect of host microenvironment on the microcirculation of human colon adenocarcinoma', *Am. J. Pathol.*, **151**(3), pp. 679–688.

Furukawa, T., Fu, X., Kubota, T., Watanabe, M. *et al.* (1993a) 'Nude mouse metastatic models of human stomach cancer constructed using orthotopic implantation of histologically intact tissue', *Cancer Res.*, **53**(5), pp. 1204–1208.

Furukawa, T., Kubota, T., Watanabe, M., Kitajima, M. and Hoffman, R. M. (1993b) 'A novel 'patient-like' treatment model of human pancreatic cancer constructed using orthotopic transplantation of histologically intact human tumor tissue in nude mice', *Cancer Res.*, **53**(13), pp. 3070–3072.

Gaspani, L., Bianchi, M., Limiroli, E., Panerai, A. E. and Sacerdote, P. (2002) 'The analgesic drug tramadol prevents the effect of surgery on natural killer cell activity and metastatic colonization in rats', *J. Neuroimmunol.*, **129**(1–2), pp. 18–24.

Goldbrunner, R. H., Bendszus, M. and Tonn, J. C. (2004) 'Models for angiogenesis in gliomas', *Cancer Treat. Res.*, **117**, pp. 115–135.

Gossen, M. and Bujard, H. (1992) 'Tight control of gene expression in mammalian cells by tetracycline-responsive promoters', *Proc. Natl. Acad. Sci. USA*, **89**(12), pp. 5547–5551.

Goto, H., Yano, S., Zhang, H., Matsumori, Y. *et al.* (2002) 'Activity of a new vascular targeting agent, ZD6126, in pulmonary metastases by human lung adenocarcinoma in nude mice', *Cancer Res.*, **62**(13), pp. 3711–3715.

Graepler, F., Verbeek, B., Graeter, T., Smirnow, I. *et al.* (2005) 'Combined endostatin/sFlt-1 antiangiogenic gene therapy is highly effective in a rat model of HCC', *Hepatology*, **41**(4), pp. 879–886.

Guy, C. T., Webster, M. A., Schaller, M., Parsons, T. J. *et al.* (1992) 'Expression of the neu protooncogene in the mammary epithelium of transgenic mice induces metastatic disease', *Proc. Natl. Acad. Sci. USA*, **89**(22), pp. 10578–10582.

Hanahan, D. (1985) 'Heritable formation of pancreatic beta-cell tumours in transgenic mice expressing recombinant insulin/simian virus 40 oncogenes', *Nature*, **315**(6015), pp. 115–122.

Hanahan, D., Christofori, G., Naik, P. and Arbeit, J. (1996) 'Transgenic mouse models of tumour angiogenesis: the angiogenic switch, its molecular controls, and prospects for preclinical therapeutic models', *Eur. J. Cancer,*. **32A**(14), pp. 2386–2393.

Hastings, R. H., Burton, D. W., Quintana, R. A., Biederman, E. *et al.* (2001) 'Parathyroid hormone-related protein regulates the growth of orthotopic human lung tumors in athymic mice', *Cancer*, **92**(6), pp. 1402–1410.

Heckl, S., Pipkorn, R., Nagele, T., Vogel, U. *et al.* (2004) 'Molecular imaging: Bridging the gap between neuroradiology and neurohistology', *Histol. Histopathol.*, **19**(2), pp. 651–668.

Herynk, M. H., Stoeltzing, O., Reinmuth, N., Parikh, N. U. *et al.* (2003) 'Down-regulation of c-Met inhibits growth in the liver of human colorectal carcinoma cells', *Cancer Res.*, **63**(11), pp. 2990–2996.

Hiraga, T., Hata, K., Ikeda, F., Kitagaki, J. *et al.* (2005) 'Preferential inhibition of bone metastases by 5'-deoxy-5-fluorouridine and capecitabine in the 4T1/luc mouse breast cancer model', *Oncol. Rep.*, **14**(3), pp. 695–699.

Hoffman, R. M. (1999) 'Orthotopic metastatic mouse models for anticancer drug discovery and evaluation: a bridge to the clinic', *Invest, New Drugs*, **17**(4), pp. 343–359.

Holsinger, F. C., Doan, D. D., Jasser, S. A., Swan, E. A. *et al.* (2003) 'Epidermal growth factor receptor blockade potentiates apoptosis mediated by Paclitaxel and leads to prolonged survival in a murine model of oral cancer', *Clin. Cancer Res.*, **9**(8), pp. 3183–3189.

Hotz, H. G., Hines, O. J., Foitzik, T. and Reber, H. A. (2000) 'Animal models of exocrine pancreatic cancer', *Int. J. Colorectal Dis.*, **15**(3), pp. 136–143.

Hotz, H. G., Reber, H. A., Hotz, B., Sanghavi, P. C. *et al.* (2001) 'Angiogenesis inhibitor TNP-470 reduces human pancreatic cancer growth', *J. Gastrointest. Surg.*, **5**(2), pp. 131–138.

Hotz, H. G., Reber, H. A., Hotz, B., Yu, T. *et al.* (2003) 'An orthotopic nude mouse model for evaluating pathophysiology and therapy of pancreatic cancer', *Pancreas*, **26**(4), pp. e89–98.

Howard, R. B., Chu, H., Zeligman, B. E., Marcell, T. *et al.* (1991) 'Irradiated nude rat model for orthotopic human lung cancers', *Cancer Res.*, **51**(12), pp. 3274–3280.

Huang, S., Robinson, J. B., Deguzman, A., Bucana, C. D. and Fidler, I. J. (2000) 'Blockade of nuclear factor-kappaB signaling inhibits angiogenesis and tumorigenicity of human ovarian cancer cells by suppressing expression of vascular endothelial growth factor and interleukin 8', *Cancer Res.*, **60**(19), pp. 5334–5339.

Huang, S. F., Kim, S. J., Lee, A. T., Karashima, T. *et al.* (2002) 'Inhibition of growth and metastasis of orthotopic human prostate cancer in athymic mice by combination therapy with pegylated interferon-alpha-2b and docetaxel', *Cancer Res.*, **62**(20), pp. 5720–5726.

Jain, R. K. (2001) 'Normalizing tumor vasculature with anti-angiogenic therapy: a new paradigm for combination therapy', *Nat. Med.*, **7**(9), pp. 987–989.

Jonkers, J. and Berns, A. (2002) 'Conditional mouse models of sporadic cancer', *Nat. Rev. Cancer*, **2**(4), pp. 251–265.

Jung, Y. D., Ahmad, S. A., Akagi, Y., Takahashi, Y. *et al.* (2000) 'Role of the tumor microenvironment in mediating response to anti-angiogenic therapy', *Cancer Metastasis Rev.*, **19**(1–2), pp. 147–157.

Kedar, D., Baker, C. H., Killion, J. J., Dinney, C. P. and Fidler, I. J. (2002) 'Blockade of the epidermal growth factor receptor signaling inhibits angiogenesis leading to regression of human renal cell carcinoma growing orthotopically in nude mice', *Clin. Cancer Res.*, **8**(11), pp. 3592–3600.

Kern, F. G., McLeskey, S. W., Zhang, L., Kurebayashi, J. *et al.* (1994) 'Transfected MCF-7 cells as a model for breast-cancer progression', *Breast Cancer Res. Treat.*, **31**(2–3), pp. 153–165.

Keyes, K. A., Mann, L., Teicher, B. and Alvarez, E. (2003) 'Site-dependent angiogenic cytokine production in human tumor xenografts', *Cytokine*, **21**(2), pp. 98–104.

Kim, L. S. and Price, J. E. (2005) 'Clinically relevant metastatic breast cancer models to study chemosensitivity', *Methods Mol. Med.*, **111**, pp. 285–295.

Kim, L. S., Huang, S., Lu, W., Lev, D. C. and Price, J. E. (2004a) 'Vascular endothelial growth factor expression promotes the growth of breast cancer brain metastases in nude mice', *Clin. Exp. Metastasis*, **21**(2), pp. 107–118.

Kim, S. J., Uehara, H., Yazici, S., Langley, R. R. *et al.* (2004b) 'Simultaneous blockade of platelet-derived growth factor-receptor and epidermal growth factor-receptor signaling and systemic administration of paclitaxel as therapy for human prostate cancer metastasis in bone of nude mice', *Cancer Res.*, **64**(12), pp. 4201–4208.

Kitadai, Y., Ellis, L. M., Takahashi, Y., Bucana, C. D. *et al.* (1995) 'Multiparametric in situ messenger RNA hybridization analysis to detect metastasis-related genes in surgical specimens of human colon carcinomas', *Clin. Cancer Res.*, **1**(10), pp. 1095–1102.

Kraus-Berthier, L., Jan, M., Guilbaud, N., Naze, M. *et al.* (2000) 'Histology and sensitivity to anticancer drugs of two human non-small cell lung carcinomas implanted in the pleural cavity of nude mice', *Clin. Cancer Res.*, **6**(1), pp. 297–304.

Kumar, R. and Yarmand-Bagheri, R. (2001) 'The role of HER2 in angiogenesis', *Semin. Oncol.*, **28**(5) Suppl 16, pp. 27–32.

Kuo, T. H., Kubota, T., Watanabe, M., Furukawa, T. *et al.* (1993) 'Site-specific chemo-sensitivity of human small-cell lung carcinoma growing orthotopically compared to subcutaneously in SCID mice: the importance of orthotopic models to obtain relevant drug evaluation data', *Anticancer Res.*, **13**(3), pp. 627–630.

Lang, J. Y., Chen, H., Zhou, J., Zhang, Y. X. *et al.* (2005) 'Antimetastatic effect of salvicine on human breast cancer MDA-MB-435 orthotopic xenograft is closely related to Rho-dependent pathway', *Clin. Cancer Res.*, **11**(9), pp. 3455–3464.

Lee, J. C., Kim, D. C., Gee, M. S., Saunders, H. M. *et al.* (2002) 'Interleukin-12 inhibits angiogenesis and growth of transplanted but not in situ mouse mammary tumor virus-induced mammary carcinomas', *Cancer Res.*, **62**(3), pp. 747–755.

Lewandoski, M. (2001) 'Conditional control of gene expression in the mouse', *Nat. Rev. Genet.*, **2**(10), pp. 743–755.

Lu, X. G., Zhan, L. B., Feng, B. A., Qu, M. Y. *et al.* (2004) 'Inhibition of growth and metastasis of human gastric cancer implanted in nude mice by d-limonene', *World J. Gastroenterol.*, **10**(14), pp. 2140–2144.

Lyden, D., Hattori, K., Dias, S., Costa, C. *et al.* (2001) 'Impaired recruitment of bone-marrow-derived endothelial and hematopoietic precursor cells blocks tumor angiogenesis and growth', *Nat. Med.*, **7**(11), pp. 1194–1201.

Lyden, D., Young, A. Z., Zagzag, D., Yan, W. *et al.* (1999) 'Id1 and Id3 are required for neurogenesis, angiogenesis and vascularization of tumour xenografts', *Nature*, **401**(6754), pp. 670–677.

Manzotti, C., Audisio, R. A. and Pratesi, G. (1993) 'Importance of orthotopic implantation for human tumors as model systems: relevance to metastasis and invasion', *Clin. Exp. Metastasis*, **11**(1), pp. 5–14.

Martin, T. A., Watkins, G., Lane, J. and Jiang, W. G. (2005) 'Assessing microvessels and angiogenesis in human breast cancer, using VE-cadherin', *Histopathology*, **46**(4), pp. 422–430.

McCarty, M. F., Liu, W., Fan, F., Parikh, A. *et al.* (2003) 'Promises and pitfalls of anti-angiogenic therapy in clinical trials', *Trends Mol. Med.*, **9**(2), pp. 53–58.

McCarty, M. F., Takeda, A., Stoeltzing, O., Liu, W. *et al.* (2004a) 'ZD6126 inhibits orthotopic growth and peritoneal carcinomatosis in a mouse model of human gastric cancer', *Br. J. Cancer*, **90**(3), pp. 705–711.

McCarty, M. F., Wey, J., Stoeltzing, O., Liu, W. *et al.* (2004b) 'ZD6474, a vascular endothelial growth factor receptor tyrosine kinase inhibitor with additional activity against epidermal growth factor receptor tyrosine kinase, inhibits orthotopic growth and angiogenesis of gastric cancer', *Mol. Cancer Ther.*, **3**(9), pp. 1041–1048.

McLemore, T. L., Liu, M. C., Blacker, P. C., Gregg, M. *et al.* (1987) 'Novel intrapulmonary model for orthotopic propagation of human lung cancers in athymic nude mice', *Cancer Res.*, **47**(19), pp. 5132–5140.

McLemore, T. L., Eggleston, J. C., Shoemaker, R. H., Abbott, B. J. *et al.* (1988) 'Comparison of intrapulmonary, percutaneous intrathoracic, and subcutaneous models for the propagation of human pulmonary and nonpulmonary cancer cell lines in athymic nude mice', *Cancer Res.*, **48**(10), pp. 2880–2886.

Miller, C. G., Krummenacher, C., Eisenberg, R. J., Cohen, G. H. and Fraser, N. W. (2001) 'Development of a syngenic murine B16 cell line-derived melanoma susceptible to destruction by neuroattenuated HSV-1', *Mol. Ther.*, **3**(2), pp. 160–168.

Miyoshi, T., Kondo, K., Ishikura, H., Kinoshita, H. *et al.* (2000) 'SCID mouse lymphogenous metastatic model of human lung cancer constructed using orthotopic inoculation of cancer cells', *Anticancer Res.*, **20**(1A), pp. 161–163.

Morikawa, K., Walker, S. M., Nakajima, M., Pathak, S. *et al.* (1988) 'Influence of organ environment on the growth, selection, and metastasis of human colon carcinoma cells in nude mice', *Cancer Res.*, **48**(23), pp. 6863–6871.

Myers, J. N., Holsinger, F. C., Jasser, S. A., Bekele, B. N. and Fidler, I. J. (2002) 'An orthotopic nude mouse model of oral tongue squamous cell carcinoma', *Clin. Cancer Res.*, **8**(1), pp. 293–298.

Nagy, J. A., Masse, E. M., Herzberg, K. T., Meyers, M. S. *et al.* (1995) 'Pathogenesis of ascites tumor growth: vascular permeability factor, vascular hyperpermeability, and ascites fluid accumulation', *Cancer Res.*, **55**(2), pp. 360–368.

Onn, A., Isobe, T., Itasaka, S., Wu, W. *et al.* (2003) 'Development of an orthotopic model to study the biology and therapy of primary human lung cancer in nude mice', *Clin. Cancer Res.*, **9**(15), pp. 5532–5539.

Onn, A., Isobe, T., Wu, W., Itasaka, S. *et al.* (2004) 'Epidermal growth factor receptor tyrosine kinase inhibitor does not improve paclitaxel effect in an orthotopic mouse model of lung cancer', *Clin. Cancer Res.*, **10**(24), pp. 8613–8619.

Paget, S. (1989) 'The distribution of secondary growths in cancer of the breast', *Cancer Metastasis Rev.*, **8**(2), pp. 98–101.

Pardanaud, L., Altmann, C., Kitos, P., Dieterlen-Lievre, F. and Buck, C. A. (1987) 'Vasculogenesis in the early quail blastodisc as studied with a monoclonal antibody recognizing endothelial cells', *Development*, **100**(2), pp. 339–349.

Peron, J. M., Couderc, B., Rochaix, P., Douin-Echinard, V. *et al.* (2004) 'Treatment of murine hepatocellular carcinoma using genetically modified cells to express interleukin-12', *J. Gastroenterol. Hepatol.*, **19**(4), pp. 388–396.

Pettaway, C. A., Pathak, S., Greene, G., Ramirez, E. *et al.* (1996) 'Selection of highly metastatic variants of different human prostatic carcinomas using orthotopic implantation in nude mice', *Clin. Cancer Res.*, **2**(9), pp. 1627–1636.

Pietras, K. and Hanahan, D. (2005) 'A multitargeted, metronomic, and maximum-tolerated dose "chemo-switch" regimen is antiangiogenic, producing objective responses and survival benefit in a mouse model of cancer', *J. Clin Oncol*, **23**(5), pp. 939–952.

Presta, M., Rusnati, M., Dell'Era, P., Tanghetti, E. *et al.* (2000) 'Examining new models for the study of autocrine and paracrine mechanisms of angiogenesis through FGF2-transfected endothelial and tumour cells', *Adv. Exp. Med. Biol.*, **476**, no. 7–34.

Qian, D. Z., Wang, X., Kachhap, S. K., Kato, Y. *et al.* (2004) 'The histone deacetylase inhibitor NVP-LAQ824 inhibits angiogenesis and has a greater antitumor effect in combination with the vascular endothelial growth factor receptor tyrosine kinase inhibitor PTK787/ZK222584', *Cancer Res.*, **64**(18), pp. 6626–6634.

Reinmuth, N., Fan, F., Liu, W., Parikh, A. A. *et al.* (2002) 'Impact of insulin-like growth factor receptor-I function on angiogenesis, growth, and metastasis of colon cancer', *Lab. Invest.*, **82**(10), pp. 1377–1389.

Reinmuth, N., Liu, W., Ahmad, S. A., Fan, F. *et al.* (2003) 'Alphavbeta3 integrin antagonist S247 decreases colon cancer metastasis and angiogenesis and improves survival in mice', *Cancer Res.*, **63**(9), pp. 2079–2087.

Ryschich, E., Schmidt, J., Hammerling, G. J., Klar, E. and Ganss, R. (2002) 'Transformation of the microvascular system during multistage tumorigenesis', *Int. J. Cancer*, **97**(6), pp. 719–725.

Saadoun, S., Papadopoulos, M. C., Hara-Chikuma, M. and Verkman, A. S. (2005) 'Impairment of angiogenesis and cell migration by targeted aquaporin-1 gene disruption', *Nature*, **434**(7034), pp. 786–792.

Sacco, M. G., Gribaldo, L., Barbieri, O., Turchi, G. *et al.* (1998) 'Establishment and characterization of a new mammary adenocarcinoma cell line derived from MMTV neu transgenic mice', *Breast Cancer Res. Treat.*, **47**(2), pp. 171–180.

Sacco, M. G., Caniatti, M., Cato, E. M., Frattini, A. *et al.* (2000) 'Liposome-delivered angiostatin strongly inhibits tumor growth and metastatization in a transgenic model of spontaneous breast cancer', *Cancer Res.*, **60**(10), pp. 2660–2665.

Sacco, M. G., Soldati, S., Indraccolo, S., Cato, E. M. *et al.* (2003) 'Combined antiestrogen, antiangiogenic and anti-invasion therapy inhibits primary and metastatic tumor growth in the MMTVneu model of breast cancer', *Gene Ther*, **10**(22), pp. 1903–1909.

Samant, R. S., Debies, M. T., Hurst, D. R., Moore, B. P. *et al.* (2005) 'Suppression of murine mammary carcinoma metastasis by the murine ortholog of breast cancer metastasis suppressor 1 (Brms1)', *Cancer Lett.*, **235**(2), pp. 260–265.

Sasaki, A., Boyce, B. F., Story, B., Wright, K. R. *et al.* (1995) 'Bisphosphonate risedronate reduces metastatic human breast cancer burden in bone in nude mice', *Cancer Res.*, **55**(16), pp. 3551–3557.

Schmid, G., Guba, M., Papyan, A., Ischenko, I. *et al.* (2005) 'FTY720 inhibits tumor growth and angiogenesis', *Transplant Proc.*, **37**(1), pp. 110–111.

Schmidt, J., Ryschich, E., Daniel, V., Herzog, L. *et al.* (2000) 'Vascular structure and microcirculation of experimental pancreatic carcinoma in rats', *Eur. J. Surg*, **166**(4), pp. 328–335.

Schniewind, B., Christgen, M., Kurdow, R., Haye, S. *et al.* (2004) 'Resistance of pancreatic cancer to gemcitabine treatment is dependent on mitochondria-mediated apoptosis', *Int. J. Cancer*, **109**(2), pp. 182–188.

Senger, D. R., Galli, S. J., Dvorak, A. M., Perruzzi, C. A. *et al.* (1983) 'Tumor cells secrete a vascular permeability factor that promotes accumulation of ascites fluid', *Science*, **219**(4587), pp. 983–985.

Sikder, H., Huso, D. L., Zhang, H., Wang, B. *et al.* (2003) 'Disruption of Id1 reveals major differences in angiogenesis between transplanted and autochthonous tumors', *Cancer Cell*, **4**(4), pp. 291–299.

Singh, R. K. and Fidler, I. J. (1996) 'Regulation of tumor angiogenesis by organ-specific cytokines', *Curr. Top. Microbiol. Immunol.*, **213**(2), pp. 1–11.

Sipos, B., Moser, S., Kalthoff, H., Torok, V. *et al.* (2003) 'A comprehensive characterization of pancreatic ductal carcinoma cell lines: towards the establishment of an in vitro research platform', *Virchows Arch.*, **442**(5), pp. 444–452.

Sippola-Thiele, M., Hanahan, D. and Howley, P. M. (1989) 'Cell-heritable stages of tumor progression in transgenic mice harboring the bovine papillomavirus type 1 genome', *Mol. Cell Biol.*, **9**(3), pp. 925–934.

Solorzano, C. C., Baker, C. H., Bruns, C. J., Killion, J. J. *et al.* (2001) 'Inhibition of growth and metastasis of human pancreatic cancer growing in nude mice by PTK 787/ZK222584, an inhibitor of the vascular endothelial growth factor receptor tyrosine kinases', *Cancer Biother. Radiopharm.*, **16**(5), pp. 359–370.

Staroselsky, A. N., Fan, D., O'Brian, C. A., Bucana, C. D. *et al.* (1990) 'Site-dependent differences in response of the UV-2237 murine fibrosarcoma to systemic therapy with adriamycin', *Cancer Res.*, **50**(24), pp. 7775–7780.

Staton, C. A., Stribbling, S. M., Tazzyman, S., Hughes, R. *et al.* (2004) 'Current methods for assaying angiogenesis in vitro and in vivo', *Int. J. Exp. Pathol.*, **85**(5), pp. 233–248.

Stephenson, R. A., Dinney, C. P., Gohji, K., Ordonez, N. G. *et al.* (1992) 'Metastatic model for human prostate cancer using orthotopic implantation in nude mice', *J. Natl. Cancer Inst.*, **84**(12), pp. 951–957.

Stoeltzing, O., Ahmad, S. A., Liu, W., McCarty, M. F. *et al.* (2002) 'Angiopoietin-1 inhibits tumour growth and ascites formation in a murine model of peritoneal carcinomatosis', *Br. J. Cancer*, **87**(10), pp. 1182–1187.

Stoeltzing, O., Ahmad, S. A., Liu, W., McCarty, M. F. *et al.* (2003a) 'Angiopoietin-1 inhibits vascular permeability, angiogenesis, and growth of hepatic colon cancer tumors', *Cancer Res.*, **63**(12), pp. 3370–3377.

Stoeltzing, O., Liu, W., Reinmuth, N., Fan, F. *et al.* (2003b) 'Regulation of hypoxia-inducible factor-1alpha, vascular endothelial growth factor, and angiogenesis by an insulin-like growth factor-I receptor autocrine loop in human pancreatic cancer', *Am. J. Pathol.*, **163**(3), pp. 1001–1011.

Stoeltzing, O., Liu, W., Reinmuth, N., Fan, F. *et al.* (2003c) 'Inhibition of integrin alpha5beta1 function with a small peptide (ATN-161) plus continuous 5-FU infusion reduces colorectal liver metastases and improves survival in mice', *Int. J. Cancer*, **104**(4), pp. 496–503.

Stoeltzing, O., McCarty, M. F., Wey, J. S., Fan, F. *et al.* (2004) 'Role of hypoxia-inducible factor 1alpha in gastric cancer cell growth, angiogenesis, and vessel maturation', *J. Natl. Cancer Inst.*, **96**(12), pp. 946–956.

Takahashi, Y., Kitadai, Y., Bucana, C. D., Cleary, K. R. and Ellis, L. M. (1995) 'Expression of vascular endothelial growth factor and its receptor, KDR, correlates with vascularity, metastasis, and proliferation of human colon cancer', *Cancer Res.*, **55**(18), pp. 3964–3968.

Takahashi, Y., Cleary, K. R., Mai, M., Kitadai, Y. *et al.* (1996) 'Significance of vessel count and vascular endothelial growth factor and its receptor (KDR) in intestinal-type gastric cancer', *Clin. Cancer Res.*, **2**(10), pp. 1679–1684.

Tang, Z. Y., Sun, F. X., Tian, J., Ye, S. L. *et al.* (2001) 'Metastatic human hepatocellular carcinoma models in nude mice and cell line with metastatic potential', *World J. Gastroenterol.*, **7**(5), pp. 597–601.

Terayama, N., Terada, T. and Nakanuma, Y. (1996) 'An immunohistochemical study of tumour vessels in metastatic liver cancers and the surrounding liver tissue', *Histopathology*, **29**(1), pp. 37–43.

Thaker, P. H., Yokoi, K., Jennings, N. B., Li, Y. *et al.* (2005) 'Inhibition of experimental colon cancer metastasis by the GABA-receptor agonist Nembutal', *Cancer Biol. Ther.*, **4**(7), pp. 753–758.

Wang, C. Y. and Chang, Y. W. (1997a) 'A model for osseous metastasis of human breast cancer established by intrafemur injection of the MDA-MB-435 cells in nude mice', *Anticancer Res.*, **17**(4A), pp. 2471–2474.

Wang, H. Y., Ross, H. M., Ng, B. and Burt, M. E. (1997b) 'Establishment of an experimental intrapulmonary tumor nodule model', *Ann. Thorac. Surg.*, **64**(1), pp. 216–219.

Weidner, N., Semple, J. P., Welch, W. R. and Folkman, J. (1991) 'Tumor angiogenesis and metastasis-correlation in invasive breast carcinoma', *N. Engl. J. Med.*, **324**(1), pp. 1–8.

Wesseling, P., Schlingemann, R. O., Rietveld, F. J., Link, M. *et al.* (1995) 'Early and extensive contribution of pericytes/vascular smooth muscle cells to microvascular proliferation in glioblastoma multiforme: an immuno-light and immuno-electron microscopic study', *J. Neuropathol. Exp. Neurol.*, **54**(3), pp. 304–310.

Wilmanns, C., Fan, D., O'Brian, C. A., Bucana, C. D. and Fidler, I. J. (1992) 'Orthotopic and ectopic organ environments differentially influence the sensitivity of murine colon carcinoma cells to doxorubicin and 5-fluorouracil', *Int. J. Cancer*, **52**(1), pp. 98–104.

Winkler, F., Kozin, S. V., Tong, R. T., Chae, S. S. *et al.* (2004) 'Kinetics of vascular normalization by VEGFR2 blockade governs brain tumor response to radiation: role of oxygenation, angiopoietin-1, and matrix metalloproteinases', *Cancer Cell*, **6**(6), pp. 553–563.

Yamamura, K., Kibbey, M. C., Jun, S. H. and Kleinman, H. K. (1993) 'Effect of Matrigel and laminin peptide YIGSR on tumor growth and metastasis', *Semin. Cancer Biol.*, **4**(4), pp. 259–265.

Yashiro, M., Chung, Y. S., Nishimura, S., Inoue, T. and Sowa, M. (1996) 'Peritoneal metastatic model for human scirrhous gastric carcinoma in nude mice', *Clin. Exp. Metastasis*, **14**(1), pp. 43–54.

Yokoi, K., Thaker, P. H., Yazici, S., Rebhun, R. R. *et al.* (2005) 'Dual inhibition of epidermal growth factor receptor and vascular endothelial growth factor receptor phosphorylation by AEE788 reduces growth and metastasis of human colon carcinoma in an orthotopic nude mouse model', *Cancer Res.*, **65**(9), pp. 3716–3725.

Yoneda, T., Williams, P. J., Hiraga, T., Niewolna, M. and Nishimura, R. (2001) 'A bone-seeking clone exhibits different biological properties from the MDA-MB-231 parental human breast cancer cells and a brain-seeking clone in vivo and in vitro', *J. Bone Miner. Res.*, **16**(8), pp. 1486–1495.

15
Angiomouse: imageable models of angiogenesis

Robert M. Hoffman

Abstract

Human tumours labelled with the *Aequorea victoria* green fluorescent protein (GFP) for grafting into nude mice enable the non-luminous induced capillaries to be clearly visible against the very bright tumour fluorescence examined either intravitally or by whole-body luminance in real time. Opening a reversible skin flap in the light path markedly reduces signal attenuation, increasing detection sensitivity many-fold. The observable depth of tissue is thereby greatly increased and many tumours vessels that were previously hidden are now clearly observable. Dual-colour fluorescence imaging effected by using red fluorescent protein (RFP)-expressing tumours growing in GFP-expressing transgenic mice shows with great clarity the details of the tumour–stroma interaction, especially tumour-induced angiogenesis. The GFP-expressing tumour vasculature, both nascent and mature, can be readily distinguished interacting with the RFP-expressing tumour cells.

Keywords

GFP; RFP; imaging, blood vessels

15.1   Introduction

Previous models used to visualize angiogenesis

The discovery and evaluation of anti-angiogenic substances initially relied on methods such as the chorioallantoic membrane assay (Auerbach *et al.*, 1974; Crum *et al.*, 1985), the monkey iris neovascularization model (Miller *et al.*, 1993), the

Angiogenesis Assays   Edited by Carolyn A. Staton, Claire Lewis and Roy Bicknell
© 2006 John Wiley & Sons, Ltd

disc angiogenesis assay (Passaniti *et al.*, 1992), and various models that use the cornea to assess blood vessel growth (Alessandri *et al.*, 1983; Deutsch and Hughes, 1979; Korey *et al.*, 1977; Mahoney and Waterbury, 1985; Li *et al.*, 1991; Epstein *et al.*, 1990). Although they are important for understanding the mechanisms of blood vessel induction, these models do not represent tumour angiogenesis and are poorly suited to drug discovery.

Subcutaneous tumour xenograft mouse models have been developed to study tumour angiogenesis, but these require cumbersome pathological examination procedures such as histology and immunohistochemistry. Measurements require animal sacrifice and therefore preclude ongoing angiogenesis studies in individual, live, tumour-bearing animals. Moreover, subcutaneous tumour xenografts are not representative models of human disease.

Tumours transplanted in the cornea of the rodents (see Chapter 11, Gimbrone *et al.*, 1974; Fournier *et al.*, 1981; Muthukkaruppan and Auerbach, 1979) and rodent skin-fold window chambers (see Chapter 13) have also been used for angiogenesis studies (Papenfuss *et al.*, 1979; Gross *et al.*, 1982; Dewhirst *et al.*, 1984; Fukumura *et al.*, 1998; Li *et al.*, 2000; Al-Mehdi *et al.*, 2000; Huang *et al.*, 1999). The cornea and skin-fold chamber models provide a means for studying tumour angiogenesis in living animals. However, quantification requires specialized procedures, and the sites do not represent natural environments for tumour growth. The cornea and skin-fold window chamber tumour models do not allow metastatic angiogenesis to occur, which may involve mechanisms of angiogenesis (Cowen *et al.*, 1995) that are qualitatively different from those occurring in ectopic models.

15.2   Fluorescent proteins to image angiogenesis

Fluorescent proteins have been shown to be very useful for imaging in tumours including the formation of nascent vessels (Hoffman, 2002). We have developed unique mouse models to image tumour angiogenesis with fluorescent proteins, which are described in this review.

Orthotopic tumour models expressing fluorescent proteins to visualize tumour angiogenesis

We have developed surgical orthotopic implantation (SOI) metastatic models of human cancer (Hoffman, 2002). These models place tumours in natural microenvironments and replicate clinical tumour behaviour more closely than do ectopic implantation models. The orthotopically-growing tumours, in contrast to most other models, give rise to spontaneous metastases that resemble, both in target tissues and in frequency of occurrence, the clinical behaviour of the original human tumour (Hoffman, 1999). The tumours implanted in the orthotopic model

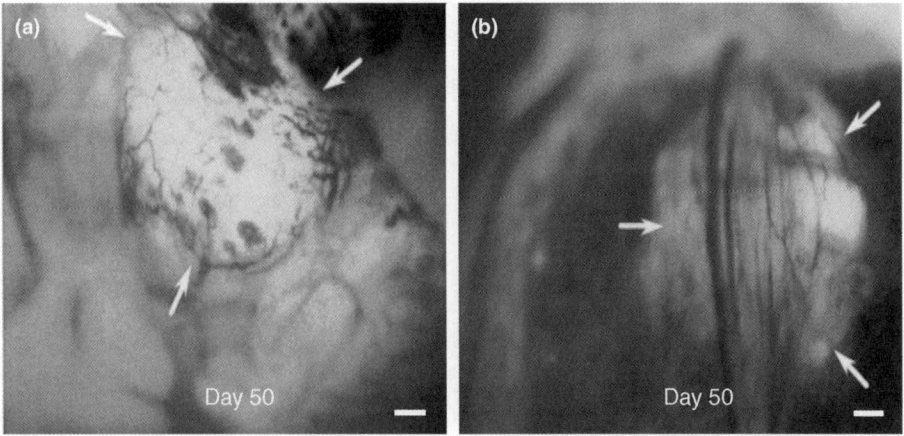

Figure 15.1 Intravital fluorescence imaging of Bx-PC-3-GFP human pancreas cancer angiogenesis. (a) Orthotopic tumour. (b) Metastatic lesion in spleen. The GFP-expressing human tumour was transplanted to nude mice by surgical orthotopic implantation (SOI) and intravitally imaged 50 days later (Bar = 200 μm) (Yang et al., 2001). (A colour reproduction of this figure can be viewed in the colour section towards the centre of the book).

have been transduced and selected strongly to express GFP *in vivo* (Hoffman, 2002). Orthotopically-implanted GFP-labelled tumours enable the visualization of the role of angiogenesis in metastasis. As Li *et al.* (2000) point out; angiogenesis initiation in metastatic tumours may be very different from that of primary tumours and may require different interventions. Moreover, the extreme detection sensitivity afforded by the strong GFP fluorescence allows imaging of very early events in blood vessel induction.

GFP expression in primary tumours and in their metastases in the mouse models can be detected by an intense fluorescence seen by intravital or by whole-body imaging. The non-luminous angiogenic blood vessels appear as sharply defined dark networks against this bright background (Figure 15.1). The high image resolution permits quantitative measurements of total vessel length. These genetically fluorescent tumour models thereby allow quantitative optical imaging of angiogenesis *in vivo*. Tumour growth, vascularization and metastasis can then be followed in real time (Yang *et al.*, 2001).

Intravital imaging of angiogenesis in orthotopic cancer models

The clarity of angiogenic blood vessel imaging was illustrated by intravital examination of the orthotopic growth of a Bx-PC-3-GFP pancreatic tumour. The non-luminous blood vessels were clearly visible against the GFP fluorescence of the primary tumour. Angiogenesis associated with metastatic growths was readily imaged by intravital examination (Yang *et al.*, 2001) (Figure 15.2). Because angiogenesis could be measured without animal sacrifice, it was possible to

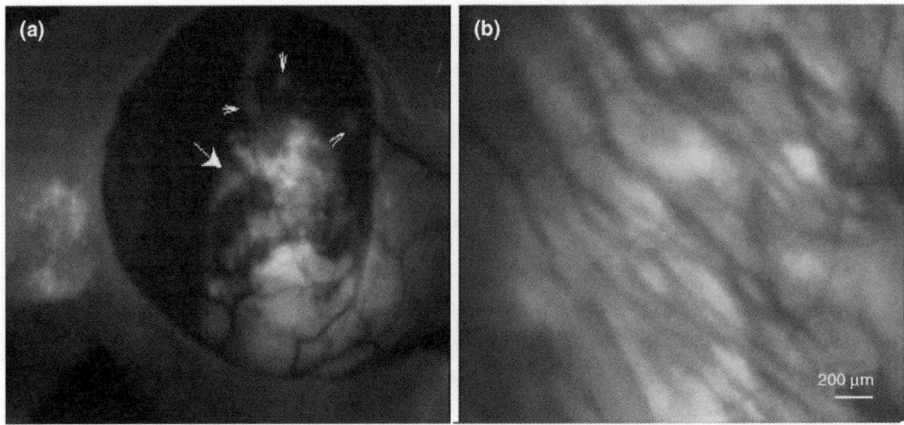

Figure 15.2 Direct view of microvessels of orthotopically growing pancreatic cancer (a) The human Bx PC-3-GFP pancreatic tumour, microvessels, and metastasis to the spleen (fine arrows) were directly viewed via a skin flap window over the abdominal wall of a nude mouse at day 58 after SOI. (b) Microvessels were directly viewed and highly resolved through the skin flap window at higher magnification (Bar = 200 μm) (Yang et al., 2002). (A colour reproduction of this figure can be viewed in the colour section towards the centre of the book).

determine a time course for individual animals. Sequential intravital images of angiogenesis for the PC-3 human prostate tumour expressing GFP and growing orthotopically in a single nude mouse were acquired. The tumour-associated blood vessels were clearly visible by day 7 and continued to increase at least until day 20 (Yang *et al.*, 2001).

Whole-body imaging of angiogenesis in orthotopic breast cancer

We have demonstrated whole-body images and quantitation of the time course of angiogenesis of the MDA-MB-435-GFP human breast cancer growing orthotopically in the mammary fat pad in a nude mouse. The development of the tumour and its angiogenesis could be imaged in a completely non-invasive manner (Yang *et al.*, 2001). The mouse mammary fat pad is the orthotopic environment for the implanted MDA-MB-435-GFP breast cancer and allows non-invasive, whole-body imaging of tumour angiogenesis. The quantitative angiogenesis data show that microvessel density increased over 20 weeks. Thus, tumours in their natural microenvironment, growing orthotopically in sites such as the fat pad, can be whole-body imaged for quantitative angiogenesis studies (Yang *et al.*, 2001). The advantages of visualizing blood vessels by their contrast to GFP expression of tumour are the simplicity of the method and the possibility of whole-body or external imaging. The disadvantage is that very small capillaries may not be visible.

Skin flaps enable ultra-high resolution external imaging of tumour angiogenesis

Opening a reversible skin flap in the light path markedly reduced signal attenuation, increasing detection sensitivity many-fold. The observable depth of tissue is thereby greatly increased (Yang *et al.*, 2002). The brilliance of the tumour GFP fluorescence, facilitated by the reduced absorption through the skin flap window, allowed imaging of the induced microvessels by their contrast against a bright background. The orthotopically growing BxPC-3-GFP human pancreatic tumour was externally visualized under fluorescence microscopy to be surrounded by its microvessels visible by their dark contrast (Yang *et al.*, 2002). Although the skin flap approach may increase resolution of visualizing vessels in GFP expressing tumour, visualization of very small capillaries may still prove difficult.

15.3   Dual-colour tumour host models

Okabe *et al.* (1997) produced transgenic mice with GFP under the control of a chicken beta-actin promoter and cytomegalovirus enhancer. All of the tissues from these transgenic mice, with the exception of erythrocytes and hair, fluoresce green.

Tumour cells to be transplanted in the GFP mouse were made visible by transforming them with the red fluorescent protein (RFP) (Hoffman, 2002). In order to gain further insight into tumour–host interaction in the living state, including tumour angiogenesis, we have visualized RFP-expressing tumours transplanted in the GFP-expressing transgenic mice under dual-colour fluorescence microscopy. The dual-colour fluorescence made it possible to visualize the tumour growth in the host by whole-body imaging as well as visibly to distinguish interacting tumour and host cells in fresh tissue. The dual-colour approach affords a powerful means of both visualizing and distinguishing the components of the host–tumour interaction (Yang *et al.*, 2003).

Dual-colour images of early events in tumour angiogenesis induced by a B16F10 mouse melanoma in the transgenic GFP-expressing mouse were acquired in fresh tissue preparations. Host-derived GFP-expressing fibroblast cells and endothelial cells from nascent blood vessels were visualized clearly against the red fluorescent background of the RFP-expressing mouse melanoma. Host-derived GFP-expressing mature blood vessels within the RFP-expressing mouse melanoma also became visible. The images were acquired 3 weeks after subcutaneous injection of B16F10-RFP melanoma cells in the GFP mouse (Yang *et al.*, 2003) (Figure 15.3).

The advantages of the dual-colour model is the great resolution it affords to visualize very fine vessels. The disadvantage is that for the highest resolution, tissue preparations may be needed.

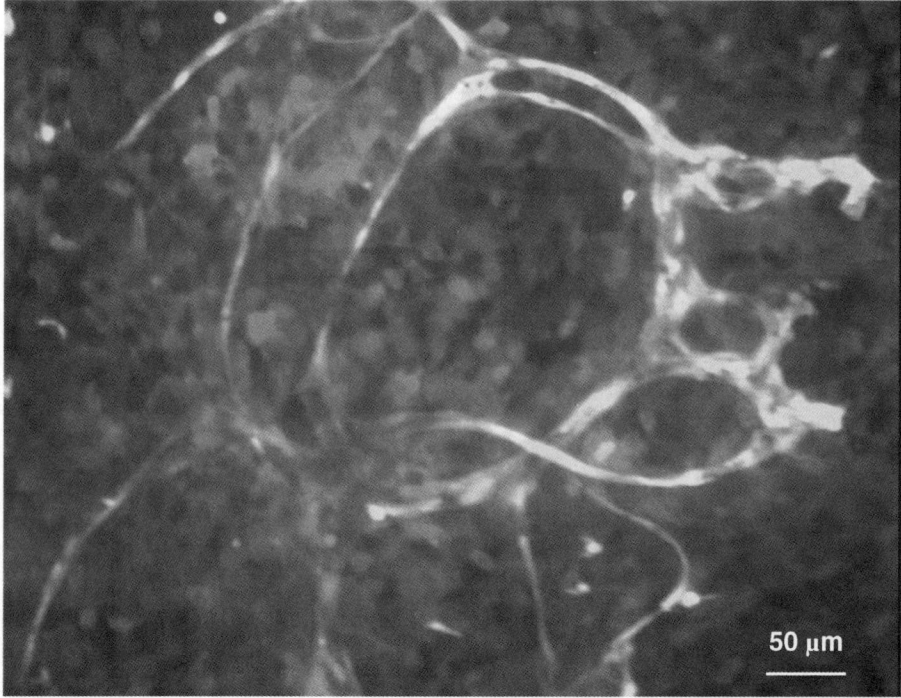

Figure 15.3 Visualization of angiogenesis in live tumour tissue 3 weeks after s.c. injection of B16F10-RFP melanoma cells in the transgenic GFP mouse. Well developed, host-derived GFP-expressing blood vessels are visualized in the RFP-expressing mouse melanoma (Scale bars, 50 μm) (Yang *et al.*, 2003). (A colour reproduction of this figure can be viewed in the colour section towards the centre of the book).

Development and description of the GFP nude mouse

In crosses between *nu/nu* GFP male mice and *nu/+* GFP female mice cells, the resultant embryos were green, apparently at the single-cell stage. Newborn mice and adult mice were very bright green and expressed GFP in essentially all cells and tissues (Yang *et al.*, 2004).

In the adult mice, the organs all brightly expressed GFP, including the heart, lungs, spleen, pancreas, oesophagous, stomach and duodenum. The entire digestive system was dissected out from tongue to anus and could be seen to fluoresce brilliant green upon blue-light excitation. The male and female reproductive systems were dissected out and all components fluoresced bright green upon blue-light excitation. The dissected brain and spinal cord also had brilliant GFP fluorescence. The dissected-out circulatory system, including the heart and major arteries and veins, had a brilliant green fluorescence. The skinned skeleton highly expressed GFP. Pancreatic islets were GFP fluorescent. The spleen cells were also GFP positive (Yang *et al.*, 2004).

The GFP nude mice appear to have a similar lifespan as non-GFP nude mice, such that long-term tumour growth and metastasis studies can be carried out. The GFP nude mouse has a critical advantage over the GFP C57-B6 immunocompetent mouse in that human tumours can grow in the GFP nude mouse. In addition, the lack of hair in the GFP nude mouse makes imaging more facile (Yang *et al.*, 2004).

Transplantation and whole-body imaging of red fluorescent protein-expressing human tumours in the GFP nude mouse

Whole-body images visualized the tumour–host interaction of RFP-expressing human tumours in the GFP nude mice, including orthotopic growth of the MDP-435-GFP mammary tumour; orthotopic growth of PC-3-RFP prostate tumour; orthotopic growth of the HCT-116-RFP colon cancer; and growth in the tibia of the HT1080-RFP fibrosarcoma (Yang *et al.*, 2004).

The GFP nude mouse enables visualization of human tumour angiogenesis in live tissue. Both the tumour and the host blood vessels are uniquely identified by their fluorescent colour: RFP for the tumour, GFP for the host blood vessels (Yang *et al.*, 2004).

Recently, Duda *et al.* transplanted non-fluorescent mouse tumour cells growing in transgenic immunocompetent mice with GFP-expressing stromal cells to non-fluorescent mouse hosts. They found that the fluorescent stroma cells, especially blood vessels, continued to grow in the non-fluorescent transplanted host mice. Our approach, using RFP-expressing human tumour cells growing in GFP nude mice, can allow simultaneous visualization of the growing human tumour cells with the surviving blood vessels to give further information on the development of tumour angiogenesis during growth as well as transplantation (Yang *et al.*, 2004).

The introduction of the nude mouse to cancer research led to a paradigm change in cancer biology enabling human tumours to be consistently grown in a mouse model. The GFP nude mouse should lead to another paradigm change enabling the visualization of the interaction of the human tumour and host in the living mouse (Yang *et al.*, 2004).

The greatest resolution of the two-colour model may be in tissue preparations. However, newer instrumentation such as the Olympus IV100 whole mouse scanning laser imaging system as well as the spectral resolution system described below may allow whole-body imaging at very high resolution.

Dual-colour imaging with spectral resolution with ultra-high resolution whole-body imaging of angiogenesis

The RFP-expressing MDA-MB-435 breast tumours growing in the GFP-mice were visualized using excitation centred at 470 nm and appropriate >500 nm emission filters. Using a spectral imaging system based on liquid crystal tunable filters, we were able to perform whole-body imaging yielding high-resolution spectral information at each pixel of the resulting image. Analysis algorithms allow the separation (unmixing) of individual spectral species on a pixel-by-pixel basis. Such techniques non-invasively the presence of host GFP-expressing stroma within the RFP-labelled tumour. Moreover, fluorescence spectra emitted in the far-red allow the whole-body imaging of tumour angiogenesis. This new differential dual-coloured fluorescence imaging tumour–host model, along with spectral unmixing, can non-invasively visualize in real-time the onset and progression of angiogenesis in a tumour. Other host cells and structures in the tumour may also be visualized by whole-body spectral imaging (Levenson *et al.* 2004). This method depends on having appropriate models to visualize.

15.4   ND-GFP mouse model

We have shown that the neural-stem cell marker nestin is expressed in hair follicle stem cells (Li *et al.* 2003) and the blood vessel network interconnecting hair follicles in the skin of transgenic mice with nestin regulatory element-driven green fluorescent protein (ND-GFP) (Amoh *et al.*, 2004). The hair follicles were shown to give rise to the nestin-expressing blood vessels in the skin. We visualized tumour angiogenesis by dual-colour fluorescence imaging in ND-GFP transgenic mice after transplantation of the murine melanoma cell line B16F10 expressing red fluorescent protein. ND-GFP was highly expressed in proliferating endothelial cells and nascent blood vessels in the growing tumour. Results of immunohistochemical staining showed that the blood vessel-specific antigen CD31 was expressed in ND-GFP-expressing nascent blood vessels. ND-GFP expression was diminished in the vessels with increased blood flow. Progressive angiogenesis during tumour growth was readily visualized by GFP expression. Doxorubicin inhibited the nascent tumour angiogenesis as well as tumour growth in the ND-GFP mice transplanted with B16F10-RFP (Amoh *et al.*, 2005a).

   We have now developed a transgenic ND-GFP nude mouse for the visualization of human tumour angiogenesis. The ND-GFP gene was crossed with nude mice on the C57/B6 background to obtain ND-GFP nude mice. ND-GFP was expressed in the brain, spinal cord, pancreas, stomach, oesophagus, heart, lung, blood vessels of glomeruli, blood vessels of skeletal muscle, testes, hair follicles and blood vessel network in the skin of ND-GFP nude mice. Human lung cancer, pancreatic cancer and colon cancer cell lines, as well as a murine melanoma cell line and breast cancer tumour cell line expressing RFP, were implanted orthotopically, and an RFP-expressing human fibrosarcoma was implanted subcutaneously in the ND-GFP nude mice. These tumours grew extensively in the ND-GFP mice. ND-GFP was highly expressed in proliferating endothelial cells and nascent blood vessels in the growing tumours, visualized by dual-colour fluorescence imaging. The ND-GFP transgenic nude mouse model enables the visualization of nascent angiogenesis in human and mouse tumour progression. All of these models are useful for the imaging of the angiogenesis of human as well as rodent tumours and visualization of the efficacy of angiogenetic inhibitors (Amoh *et al.* 2005b).

ND-GFP-labelled vessel network interconnects hair follicles

In ND-GFP transgenic mice, nestin-expressing hair follicles are interconnected by an ND-GFP-labelled dermal vascular network. Immunohistochemical staining showed that the network vessels display CD31 antigen and VWF, indicating that they are blood vessels (Amoh *et al.*, 2004) (Figure 15.4).

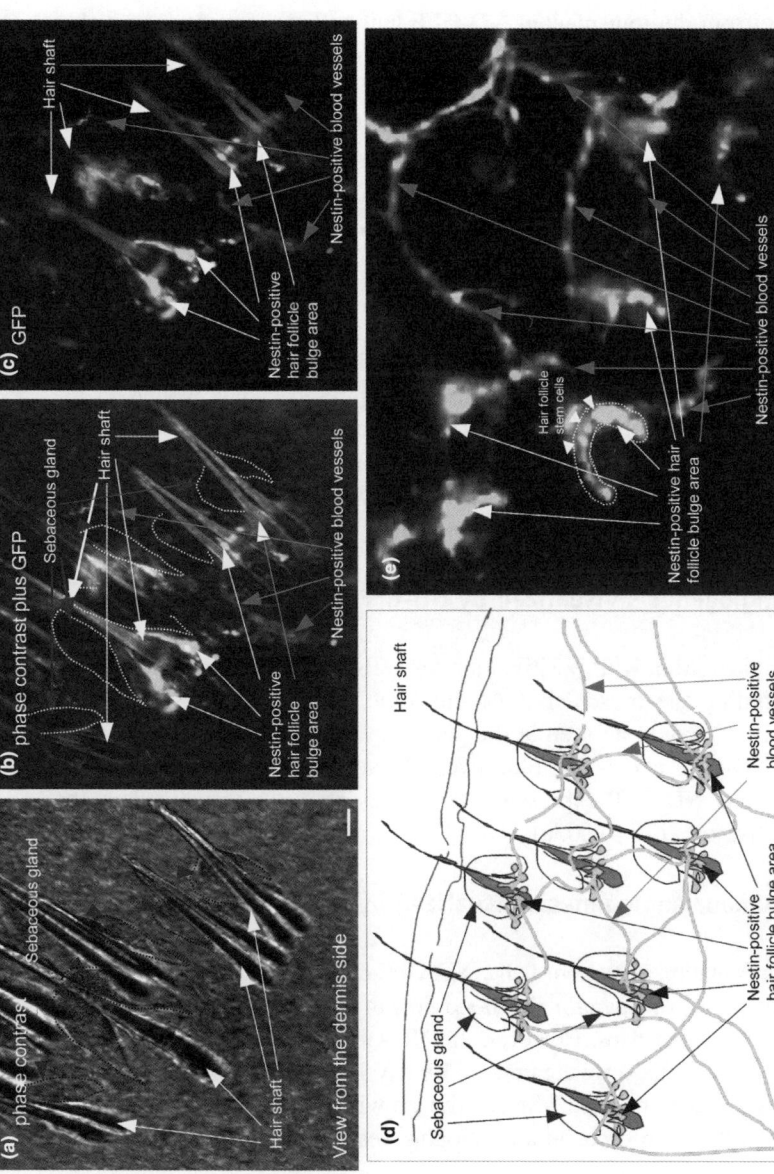

Figure 15.4 View from the dermis side of the dorsal skin in ND-GFP transgenic mice (a) Phase-contrast microscopy. Sebaceous glands (blue arrows) are located around the hair shaft (yellow arrows). (b) Phase-contrast microscopy plus GFP fluorescence. ND-GFP cells are visualized in the follicular bulge area (white arrows) and blood vessels (red arrows). The follicular bulge area is located beneath the sebaceous gland. (c) GFP fluorescence. The ND-GFP blood vessels (red arrows) are connected to ND-GFP hair follicles (white arrows). (d) Schematic of telogen hair follicle showing position of ND-GFP hair-follicle bulge areas (black arrows) and blood vessel network (red arrows). (e) GFP fluorescence. The ND-GFP blood vessels (red arrows) are associated with ND-GFP hair-follicle bulge areas (white arrows) (Scale bars, 100 µm) (Amoh *et al.*, 2004). (A colour reproduction of this figure can be viewed in the colour section towards the centre of the book).

Transplanted ND-GFP vibrissa follicles give rise to blood vessels

Vibrissa follicles expressing ND-GFP were isolated from transgenic mice and transplanted into the s.c. tissues of unlabelled nude mice. ND-GFP vessels were detected growing from the transplanted ND-GFP hair follicle in nude-mouse skin by day 3. By day 28, the nestin-GFP-expressing vessels had developed into an extensively branched network and appeared to anastomose with existing vessels in the recipient nude mice. Immunohistochemical staining showed that CD31 antigen and GFP fluorescence co-localized in nascent vessels (Amoh *et al.*, 2004).

Enhanced growth of ND-GFP vessels from transplanted hair follicles in wounded skin

Wounded skin samples containing transplanted ND-GFP vibrissa hair follicles were harvested for fluorescence microscopy. The images showed that ND-GFP vessels grew from the hair follicles toward the wound. The presence of a wound in the vicinity of the transplanted follicle significantly enhanced vessel outgrowth. Apparently, vessels originating in the follicle responded to angiogenic signals arising from the wound vicinity. Immunohistochemical staining showed that CD31 was expressed in the ND-GFP-expressing vessels growing into the wound (Amoh *et al.*, 2004).

Early tumour angiogenesis visualized by ND-GFP

At day 1 after implantation, RFP-positive tumour cells were visualized growing in the subcutis including the muscularis. At this time, ND-GFP-expressing blood vessels of the subcutis could be seen at the periphery of the RFP-positive tumour cells. By day 2, ND-GFP-expressing blood vessels were growing into the tumour. By day 5, numerous ND-GFP-expressing blood vessels were visualized in the growing tumour (Amoh *et al.*, 2005a).

Intermediate tumour angiogenesis visualized by ND-GFP

By day 7 after implantation of tumour cells, ND-GFP-expressing blood vessels were forming a network in the growing tumour. Blood flow in the pre-existing blood vessels could be visualized via skin flaps. However, the ND-GFP-expressing vessels did not seem to have blood flow in the growing tumour. By day 14, ND-GFP-expressing blood vessels with vessel dilatation could seen in the growing tumour. Vessel dilatation occurs before blood flow can occur in a developing vessel (Amoh *et al.*, 2005a).

Late-stage angiogenesis visualized by ND-GFP

By day 28 after implantation of tumour cells, some ND-GFP-expressing blood vessels were beginning to have apparent blood flow. However, ND-GFP expression was diminished in the blood vessels with apparent blood flow. By day 35 after implantation of tumour cells, large vessels had apparent blood flow in the growing tumour. ND-GFP expression was diminished in the blood vessels with significant blood flow. ND-GFP expression was maintained in the peripheral area of the tumour. In contrast to the skin, ND-GFP-expressing blood vessels were not visualized in the intestinal lymph node metastasis of the B16F10-RFP melanoma (Amoh et al., 2005a).

Immunohistochemical staining showed that CD31 and nestin were co-localized in the blood vessels in the growing tumour. A frozen section showing the ND-GFP blood vessels and RFP-expressing B16F10 melanoma was compared with a sister section stained for CD31 demonstrating co-localization of ND-GFP and CD-31 (Amoh et al., 2005a).

Visualization of intravasated tumour cells

By day 7, RFP-expressing B16F10 melanoma cells could be seen inside the ND-GFP blood vessels. These results suggest that the ND-GFP blood vessels can be used by the tumour cells for haematogenous metastasis. The dual-colour model shows that haematogenous metastasis is readily imaged (Amoh et al., 2005a).

Effects of doxorubicin on tumour growth and angiogenesis

Mice were given daily i.p. injections of 5 μg/g of doxorubicin at days 0, 1 and 2 after implantation of the B16F10-RFP melanoma cells. This protocol was used in order to minimize doxorubicin toxicity. At day 10 after implantation, only the marginal area of the tumour had ND-GFP-expressing blood vessels in the treated animals. ND-GFP-expressing blood vessels were not observed in the central area of the tumour. The number of ND-GFP-expressing blood vessels was much less in the doxorubicin-treated animals than in NaCl-injected control mice. Tumour volume was determined on days 10, 14 and 21 after implantation of tumour cells. Treatment with doxorubicin significantly decreased tumour volume as well as nascent blood vessel formation. By day 10 after implantation of tumour cells, treatment with doxorubicin significantly decreased the mean nascent blood vessel length per tumour volume (*$P < 0.05$ vs. NaCl solution-injected mice). These results show the utility of the dual-colour ND-GFP mouse-RFP-tumour model to visualize and quantitate angiogenesis and to screen for angiogenesis inhibitors (Amoh et al., 2005a).

Nascent angiogenesis of human tumours implanted in the ND-GFP nude mouse

HT1080 human fibrosarcoma cells, expressing histone H2B-GFP in the nucleus and RFP in the cytoplasm (Alessandri et al., 1983), were implanted into the subcutis of the ND-GFP nude mice. On day 14 after implantation of tumour cells, ND-GFP-expressing nascent blood vessels were visualized growing into the dual-colour tumour mass. The dual-colour tumour cells polarized toward the ND-GFP-expressing nascent blood vessels (Amoh *et al.*, 2005b).

ND-GFP-expressing nascent blood vessels vascularized the orthotopically transplanted RFP-expressing Bx-PC3 and MiaPaCa human pancreatic tumours. The endothelial-cell marker CD31 and ND-GFP fluorescence were both expressed in the newly formed blood vessels growing into the pancreatic tumour (Amoh *et al.*, 2005b). ND-GFP-expressing nascent blood vessels vascularized human colon tumour HCT116-RFP. CD31 and ND-GFP fluorescence were co-expressed in the newly formed ND-GFP-expressing blood vessels in the colon tumour (Amoh *et al.*, 2005b). Extensive vascularization by ND-GFP-expressing blood vessels of the orthotopically-implanted U87-RFP human glioma was visualized. Many RFP-expressing tumour cells appeared to grow closely associated with the ND-GFP vessels after implantation (Amoh *et al.*, 2005b) (Figure 15.5).

The ND-GFP nude mouse model should be useful for the visualization of human tumour angiogenesis and evaluation of angiogenetic inhibitors especially in the most important early stage of tumour growth and metastasis (Amoh *et al.*, 2005b). The greatest resolution of the two-colour model may be in tissue preparations. However, newer instrumentation such as the Olympus IV100 whole mouse scanning laser imaging system as well as the spectral resolution system described below may allow whole-body imaging at very high resolution.

Figure 15.5 Fluorescence imaging of tumour angiogenesis in transgenic ND-GFP nude mice. (a), RFP-expressing mouse B16F10 melanoma growing in a nestin-GFP transgenic nude mouse. Host-derived ND-GFP-expressing blood vessels were visualized in the RFP-expressing mouse melanoma on day 10 after s.c. injection of B16F10-RFP cells in the transgenic ND-GFP nude mouse. Bar, 100 µm. (b), numerous host-derived ND-GFP-expressing blood vessels were visualized in the RFP-expressing mouse mammary tumour on day 14 after orthotopic inoculation of MTT-060562-RFP cells. Bar, 100 µm. (c), RFP-expressing U87 human glioma growing in the ND-GFP transgenic nude mouse. ND-GFP-expressing blood vessels were visualized in the RFP-expressing human glioma on day 14 after s.c. injection of U87-RFP cells. Bar, 100 µm. (d), human HT1080 fibrosarcoma on day 14 after injection. Dual-colour tumour cells expressing GFP in the nucleus and RFP in the cytoplasm are polarized towards ND-GFP-expressing blood vessels (white arrows). Bar, 100 µm. (e), RFP-expressing Bx-PC-3 human pancreatic tumour vascularized with ND-GFP vessels on day 14 after orthotopic implantation. Bar, 100 µm. (f), RFP-expressing human HCT-116 colon tumour vascularized with ND-GFP vessels on day 14 after orthotopic implantation. Bar, 100 µm. (g), extensive ND-GFP-expressing blood vessels were visualized in the RFP-expressing human fibrosarcoma 8 days after injection of HT1080 cells. Only ND-GFP vessels are visualized. Bar, 100 mm. (h), extensive inhibition of ND-GFP-expressing blood vessel formation in the RFP-expressing HT-1080 human fibrosarcoma by 5 µg/g doxorubicin (i. p.) on days 0, 1 and 2. Bar, 100 mm (Amoh et al., 2005b). (A colour reproduction of this figure can be viewed in the colour section towards the centre of the book).

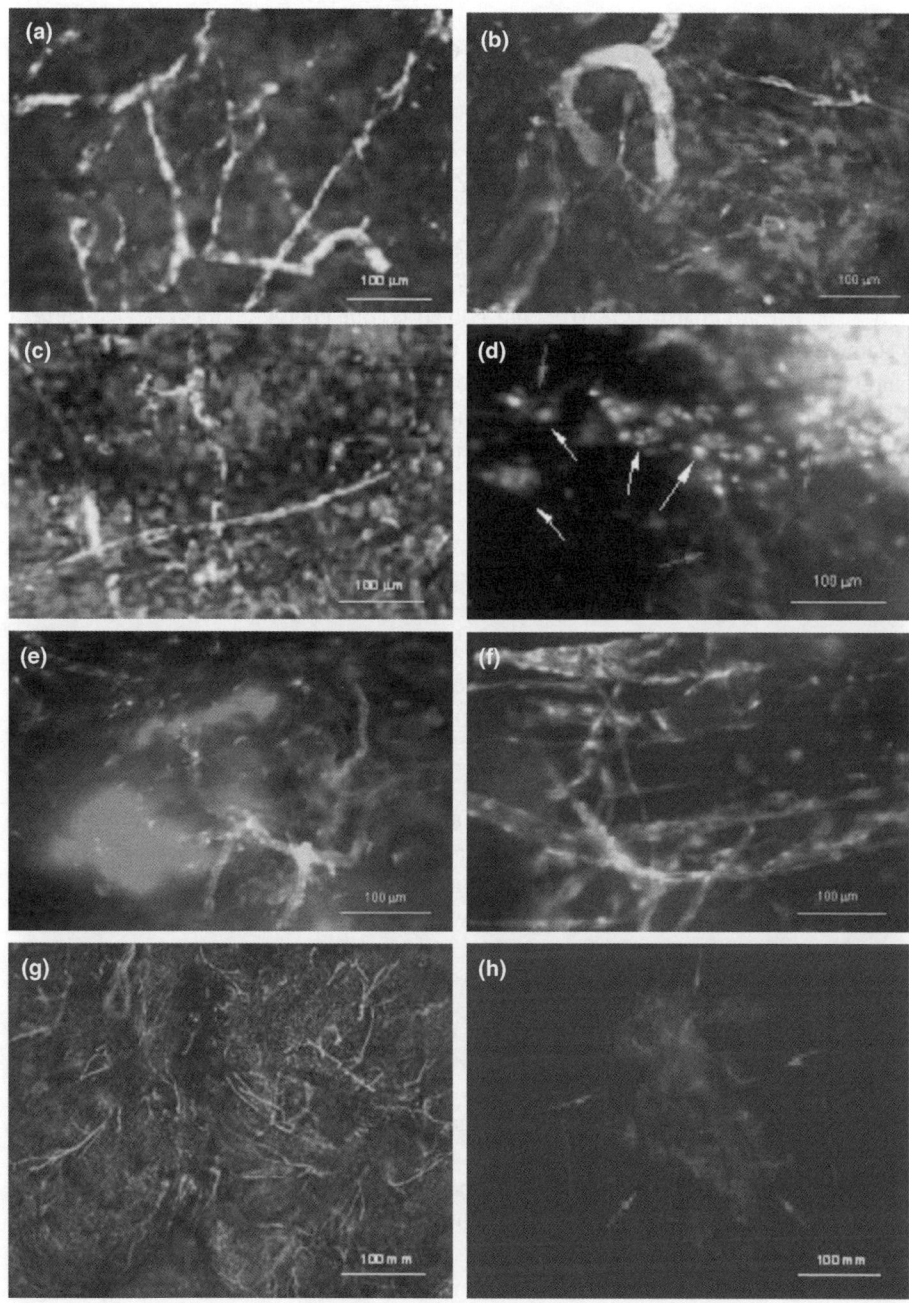

15.5   Methods of angiogenesis analysis in GFP models

Fluorescence optical imaging (Yang *et al.*, 2001)

Whole-body imaging was performed in a fluorescent light box illuminated by fibreoptic lighting at 470 nm (Lightools Research, Encinitas, CA, USA). A Leica fluorescence stereo microscope model LZ12 equipped with a mercury lamp and a 50 W power supply was used. Selective excitation of GFP was produced through a D425y60 band-pass filter and a 470 DCXR dichroic mirror. Emitted fluorescence was collected through a long-pass filter (GG475; Chroma Technology, Brattleboro, VT, USA) on a Hamamatsu C5810 3-chip cooled colour charge-coupled device camera (Hamamatsu Photonics, Bridgewater, NJ, USA). Images were processed for contrast and brightness and analysed with the use of IMAGE PRO PLUS 3.1 software (Media Cybernetics, Silver Spring, MD, USA). High-resolution images of 1024×724 pixels were captured directly on an IBM PC or continuously through video output on a high-resolution Sony VCR (model SLVR1000; Sony, Tokyo, Japan).

Quantitative analysis of angiogenesis (Yang *et al.*, 2001)

Periodically, the tumour-bearing mice were examined by intravital or whole-body fluorescence imaging. The extent of blood vessel development in a tumour was evaluated based on the total length of blood vessels (L) in chosen areas: areas containing the highest number of vessels were identified by scanning the tumours by intravital or whole-body imaging. To compare the level of vascularization during tumour growth, the 'hot' areas with the maximum development of vessels per unit area were then quantitated for L expressed in pixels. Captured images were corrected for unevenness in illumination. Then the total number of pixels derived from the blood vessels was quantified with IMAGE PRO PLUS software.

Spectral resolution (Levenson *et al.*, 2004)

Spectral imaging is the generation of images containing a high-resolution optical spectrum at every pixel, to 'unmix' the autofluorescence signal from that of the fluorescent protein-labelled tumours and bacteria. The standard GFP imaging system (long-pass emission filter) is modified by replacing the usual colour camera with a cooled monochrome camera and a liquid crystal tunable filter (CRI, Inc., Woburn, MA, USA) positioned in front of a conventional macro-lens. Typically, a series of images is taken every 10 nm from 500 to 650 nm and assembled automatically in memory into a spectral 'stack'. Using pre-defined GFP or RFP and autofluorescence spectra, the image can be resolved into different images using a linear combination chemometrics-based algorithm that generates images containing only the autofluorescence signals or only the GFP or RFP signals, now visible against essentially a black background. Using spectral autofluorescence subtraction, sensitivity is enhanced due to improvements in signal-to-noise ratio.

Skin flap windows (Yang *et al.*, 2002)

Tumour cells on the various internal organs were visualized through the body wall through a skin flap windows over the abdomen. The animals were anaesthetized with a ketamine mixture. An arc-shaped incision was made in the skin, and s.c. connective tissue was separated to free the skin flap. The skin flap could be opened repeatedly to image tumour cells on the internal organs through the nearly transparent mouse body wall and simply closed with an 6–0 suture. This procedure greatly reduced the scatter of fluorescent photons.

15.6 Conclusions

We have adapted the surgical orthotopic implantation (SOI) mouse tumour models to angiogenesis measurement by using human tumours labelled with GFP for grafting into nude mice. The non-luminous induced capillaries are clearly visible against the very bright tumour fluorescence examined either intravitally or by whole-body luminance in real time. The orthotopic implantation model of human cancer has been well characterized, and fluorescence shadowing replaces the laborious histological techniques for determining blood vessel density. Whole-body and intravital optical imaging of tumour angiogenesis were demonstrated (Yang *et al.*, 2001).

Opening a reversible skin flap in the light path markedly reduces signal attenuation, increasing detection sensitivity many-fold. The observable depth of tissue is thereby greatly increased and many tumours that were previously hidden are now clearly observable. Pancreatic tumours and their angiogenic microvessels were imaged by means of a peritoneal wall skin flap (Yang *et al.*, 2002).

We have also developed a simple yet powerful technique for delineating the morphological events of tumour-induced angiogenesis and other tumour-induced host processes with dual-colour fluorescence. The method clearly images implanted tumours and adjacent stroma, distinguishing unambiguously the host and tumour-specific components of the malignancy. The dual-colour fluorescence imaging is effected by using RFP-expressing tumours growing in GFP-expressing transgenic mice. This model shows with great clarity the details of the tumour–stroma interaction, especially tumour-induced angiogenesis. The GFP-expressing tumour vasculature, both nascent and mature, could be readily distinguished interacting with the RFP-expressing tumour cells (Yang *et al.*, 2003; Yang *et al.*, 2004; Amoh *et al.*, 2005a, Amoh *et al.*, 2005b).

The advantages of visualizing blood vessels by their contrast to GFP expressing tumours are the simplicity of the method and the possibility of whole-body or external imaging. The disadvantage is that very small capillaries may not be visible.

The advantage of the dual-colour model is the great resolution it affords to visualize very fine vessels. The disadvantage is that for highest resolution, tissue preparations may be needed. The fluorescent models of angiogenesis will enable the process to be visualized at unprecedented resolution in real time.

Acknowledgement

These studies were funded in part by National Cancer Institute grant numbers CA099258 and CA103563 and CA101600.

References

Alessandri, G., Raju, F. and Gullino, P. M. (1983) 'Mobilization of capillary endothelium in vitro induced by effectors of angiogenesis in vivo', *Cancer Res.*, **43**, pp. 1790–1797.

Al-Mehdi, A. B., Tozawa, K., Fisher, A. B., Shientag, L. *et al.* (2000) 'Intravascular origin of metastasis from the proliferation of endothelium-attached tumour cells: a new model for metastasis', *Nat. Med.*, **6**, pp. 100–102.

Amoh, Y., Li, L., Yang, M., Moossa, A. R. *et al.* (2004) 'Nascent blood vessels in the skin arise from nestin-expressing hair follicle cells', *Proc. Natl. Acad. Sci. USA*, **101**, pp. 13291–13295.

Amoh, Y., Li, L., Yang, M., Jiang, P. *et al.* (2005a) 'Hair follicle-derived blood vessels vascularize tumours in skin and are inhibited by doxorubicin', *Cancer Res.*, **65**, pp. 2337–2343.

Amoh, Y., Yang, M., Li, L., Reynoso, J. *et al.* (2005b) 'Nestin-linked-green fluorescent protein transgenic nude mice for imaging human tumour angiogenesis', *Cancer Res.*, **65**, pp. 5352–5357.

Auerbach, R., Kubai, L., Knighton, D. and Folkman, J. (1974) 'A simple procedure for the long-term cultivation of chicken embryos', *Dev. Biol,.* 41, pp. 391–394.

Cowen, S. E., Bibby, M. C. and Double, J. A. (1995) 'Characterisation of the vasculature within a murine adenocarcinoma growing in different sites to evaluate the potential of vascular therapies', *Acta Oncol.*, **34**, pp. 357–360.

Crum, R., Szabo, S. and Folkman, J. (1985) 'A new class of steroids inhibits angiogenesis in the presence of heparin or a heparin fragment', *Science*, **230**, pp. 1375–1378.

Deutsch, T. A. and Hughes, W. F. (1979) 'Suppressive effects of indomethacin on thermally induced neovascularization of rabbit corneas', *Am. J. Ophthalmol.*, **87**, pp. 536–540.

Dewhirst, M., Gross, J., Sim, D., Arnold, P. and Boyer, D. (1984) 'The effect of rate of heating or cooling prior to heating on tumour and normal tissue microcirculatory blood flow', *Biorheology*, **21**, pp. 539–558.

Epstein, R. J., Hendricks, R. L. and Stulting, R. D. (1990) 'Interleukin-2 induces corneal neovascularization in A/J mice', *Cornea*, **9**, pp. 318–323.

Fournier, G. A., Lutty, G. A., Watt, S., Fenselau, A. and Patz, A. (1981) 'A corneal micropocket assay for angiogenesis in the rat eye', *Invest. Ophthalmol. Vis. Sci.*, **21**, pp. 351–354.

Fukumura, D., Xavier, R., Sugiura, T., Chen, Y. *et al.* (1998) 'Tumour induction of VEGF promoter activity in stromal cells', *Cell*, **94**, pp. 715–725.

Gimbrone, M. A., Cotran, R. S., Leapman, S. B. and Folkman, J. (1974) 'Tumour growth and neovascularization: an experimental model using the rabbit cornea', *J. Natl. Cancer Inst.*, **52**, pp. 413–427.

Gross, J., Roemer, R., Dewhirst, M. and Meyer, T. (1982) *Int. J. Heat Mass Transfer*, **25**, 1313–1320.

Hoffman, R. M. (1999) 'Orthotopic metastatic mouse models for anticancer drug discovery and evaluation: a bridge to the clinic', *Investigational New Drugs*, **17**, pp. 343–359.

Hoffman, R. M. (2002) 'Green fluorescent protein imaging of tumour growth, metastasis, and angiogenesis in mouse models', *Lancet Oncol.*, 3, pp. 546–556.

Huang, Q., Shan, S., Braun, R. D., Lanzen, J. *et al.* (1999) 'Noninvasive visualization of tumours in rodent dorsal skin window chambers', *Nat. Biotechnol.*, **17**, pp. 1033–1035.

Korey, M., Peyman, G. A. and Berkowitz, R. (1977) 'The effect of hypertonic ointments on corneal alkali burns', *Ann. Ophthalmol.*, **9**, pp. 1383–1387.

Levenson, R., Yang, M. and Hoffman, R. M. (2004) 'Whole-body dual-color differential fluorescence imaging of tumour angiogenesis enhanced by spectral unmixing', *Proc. Am. Assoc. Cancer Res.*, **45**, pg. 46 (Abst #202).

Li, C. Y., Shan, S., Huang, Q., Braun, R. D. *et al.* (2000) 'Initial stages of tumour cell-induced angiogenesis: evaluation via skin window chambers in rodent models', *J. Natl. Cancer Inst.*, **92**, pp. 143–147.

Li, L., Mignone, J., Yang, M., Matic, M. *et al.* (2003) 'Nestin expression in hair follicle sheath progenitor cells', *Proc. Natl. Acad. Sci. USA*, **100**, pp. 9958–9961.

Li, W. W., Grayson, G., Folkman, J. and D'Amore, P. A. (1991) 'Sustained-release endotoxin. A model for inducing corneal neovascularization', *Invest. Ophthalmol. Vis. Sci,.* **32**, pp. 2906–2911.

Mahoney, J. M. and Waterbury, L. D. (1985) 'Drug effects on the neovascularization response to silver nitrate cauterization of the rat cornea', *Curr. Eye Res.*, **4**, pp. 531–535.

Miller, J. W., Stinson, W. G. and Folkman, J. (1993) 'Regression of experimental iris neovascularization with systemic alpha-interferon', *Ophthalmology*, **100**, pp. 9–14.

Muthukkaruppan, V. and Auerbach, R. (1979) 'Angiogenesis in the mouse cornea', *Science*, **205**, pp. 1416–1418.

Okabe, M., Ikawa, M., Kominami, K., Nakanishi, T. and Nishimune, T. (1997) '"Green mice" as a source of ubiquitous green cells', *FEBS Letters*, **407**, pp. 313–319.

Papenfuss, H. D., Gross, J. F., Intaglietta, M. and Treese, F. A. (1979) 'A transparent access chamber for the rat dorsal skin fold', *Microvasc. Res.* **18**, pp. 311–318.

Passaniti, A., Taylor, R. M., Pili, R., Guo, Y. *et al.* (1992) 'A simple, quantitative method for assessing angiogenesis and antiangiogenic agents using reconstituted basement membrane, heparin, and fibroblast growth factor', *Lab. Invest.*, **67**, pp. 519–528.

Yang, M., Baranov, E., Li, X-M., Wang, J.-W. *et al.* (2001) 'Whole-body and intravital optical imaging of angiogenesis in orthotopically implanted tumours', *Proc. Natl. Acad. Sci. USA*, **98**, pp. 2616–2621.

Yang, M., Baranov, E., Wang, J.-W., Jiang, P. *et al.* (2002) 'Direct external imaging of nascent cancer, tumour progression, angiogenesis, and metastasis on internal organs in the fluorescent orthotopic model', *Proc. Natl. Acad. Sci. USA*, **99**, pp. 3824–3829.

Yang, M., Li, L., Jiang, P., Moossa, A. R. *et al.* (2003) 'Dual-color fluorescence imaging distinguishes tumour cells from induced host angiogenic vessels and stromal cells', *Proc. Natl. Acad. Sci. USA*, **100**, pp. 14259–14262.

Yang, M., Reynoso, J., Jiang, P., Li, L. *et al.* (2004) 'Transgenic nude mouse with ubiquitous green fluorescent protein expression as a host for human tumours', *Cancer Res.*, 64, pp. 8651–8656.

16

Techniques and advances in vascular imaging in *Danio rerio*

Kenna R. Mills Shaw and Brant. M. Weinstein

Abstract

Danio rerio, commonly referred to as zebrafish, represent a powerful genetic and developmental model organism for the study of cardiovascular development. The genetic and optical tractability of zebrafish make it uniquely suited for *in vivo* characterization of vertebrate vascular ontogeny. An extensive repertoire of techniques is available for studying the vascular endothelium at any stage of zebrafish development, from the earliest differentiation of angioblasts to the established pattern in the adult organism. This review discusses the advantages and disadvantages of each of these techniques and how they are being employed in the pursuit of discovering novel genetic and small-molecule regulators of blood vessel development.

Keywords

cardiovascular; imaging; zebrafish; *Danio rerio*; digital motion analysis; microangiography

16.1   Introduction

The development of the vascular tree is a complex process involving the specification of angioblasts, migration of these cells to designated locations in the embryo, the coalescence of the angioblasts into tubes (vasculogenesis) and subsequent branching and remodelling of existing tubes fully to perfuse the entire organism with a blood supply (angiogenesis). Proper vascular development is

---

Angiogenesis Assays   Edited by Carolyn A. Staton, Claire Lewis and Roy Bicknell
© 2006 John Wiley & Sons, Ltd

required for viability. Moreover, aberrant vessel growth can result in or exacerbate a number of human diseases, including cancer. While multiple systems have been employed to examine endothelial cell behavior *in vitro*, fewer models are available to analyse this complex organ system *in vivo*. Zebrafish offer a variety of unique advantages making them an ideal system to utilize for the study of vertebrate vascular development.

Zebrafish (*Danio rerio*) are teleosts that, under optimal conditions, yield hundreds of embryos per mating. These large numbers and the optical clarity of developing embryos facilitate the study of embryonic development in general. Vascular transgenes and other vascular-specific methods discussed in this review make it possible directly to visualize and functionally manipulate blood vessels in the living animal. Furthermore, because of their small size zebrafish embryos receive enough oxygen by passive diffusion to survive many days without a functional cardiovascular system, a fact that enables scientists to conduct large-scale genetic screens for mutations affecting vascular development.

This review will cover how different techniques are utilized to image the developing vasculature in both fixed and living zebrafish embryos. Moreover, we will discuss how genetic screens and knockdown approaches combined with a variety of vascular imaging techniques have identified novel factors involved in proper angiogenesis. Additionally, we will address the more recent use of chemical screens to isolate compounds capable of perturbing vascular development or rescuing vascular defects.

16.2   Visualizing the developing vasculature

A variety of different methods can be employed to visualize the cardiovascular system in zebrafish. These methods range from techniques only suitable in fixed embryonic tissue to those that permit imaging of the vasculature *in vivo* throughout adulthood. Summarized in Table 16.1, each method offers its own advantages and constraints. Lineage tracing experiments offer valuable insight into the developmental origins of different features of the vascular system. Dye injection and alkaline phosphatase (AP) staining both enable visualization of the vasculature in fixed tissue. AP staining is useful for visualizing the early vasculature, as patent vessels are not a prerequisite for proper staining, while dye injection can be used to visualize patent vessels in older larvae or even adult animals. One of the most effective methods for examining the patterning of the adult vasculature is resin casting, a method that produces a plastic mould of the blood vessels. Confocal microangiography is currently the gold standard for short-term examination of the vasculature in living embryos and larvae; vessels with patent lumens can be imaged using this method with very high resolution. Knowledge of the genetics of vascular development has also improved imaging of this organ system, with the development of vascular-specific molecular markers and transgenic animals that permit dynamic imaging of blood vessels *in vivo*.

Table 16.1 Summary of techniques for imaging the zebrafish vasculature

Method	Advantages	Disadvantages
Lineage tracing	Rapid and inexpensive; coupled with *in situ* hybridization, can follow earliest stages of vascular differentiation	Detection of injected markers fade over time
Alkaline phosphatase staining	Amenable for high throughput analysis	Requires fixation; not dynamic; not completely specific for endothelial cells; useful over a limited temporal window
Dye injection	Rapid and inexpensive	Small dyes can leak from vasculature
Resin casting	Suitable for adult vasculature	Not suitable for embryonic vasculature
Digital motion analysis	Can perform over multiple days in the same animal; provides data on dynamics of blood flow; time-lapse/dynamic imaging is possible	Requires blood cells transiting vessels to be imaged; not directly assessing blood vessel development
Confocal microangiography	Does not require flow; detects lumenized vessels	Must perform rapidly, not amenable for high throughput
In situ hybridization	High throughput; provides specific data on differentiation	Requires fixation; difficult to perform on older embryos
GFP-transgenic lines	High throughput; live animal imaging, time-lapse/dynamic imaging is possible; does not require flow	Requires several months to generate a new stable transgenic line (existing lines are readily available)

Lineage tracing experiments

Lineage tracing experiments have been utilized in many developmental systems to follow the development of a subset of embryonic cells, as they form a differentiated tissue. These protocols have proven valuable to determine the cellular origins of multiple organ systems (Stern and Fraser, 2001; Mills *et al.*, 1999) Zhong *et al.* (2001) injected photoactivatable 'caged' fluorescein into developing zebrafish at the one-cell stage and then later laser-activated fluorescence in selected cells in the lateral plate mesoderm (LPM). They found that the progeny of labelled cells were incorporated into either trunk dorsal aorta or posterior cardinal vein, but not both, suggesting that arterial–venous identity may be specified as angioblasts arise in the LPM. In a subsequent study using similar methods, Childs *et al.* (2002) showed that the LPM also gives rise to the intersomitic vessels, a set of angiogenic vessels that form between the somites of all vertebrates. In fact, LPM progenitors were incorporated into all the major vascular structures of the animal, indicating that the LPM is a major source of vascular progenitors in the zebrafish, as it is in other vertebrates (Poole and Coffin, 1989). Using mosaic transgene expression Childs *et al.* (2002) also found that each intersegmental vessel (SeV) was initially composed of only three cells, each of which occupies a different characteristic position in the vessel. Lineage tracing can be utilized for as long as the dye used to label the tissue of interest is sufficiently bright to be imaged. Many non-fluorescent dyes are diluted after multiple rounds of cell division, making them difficult to detect in later stage embryos or adults. The cre-lox system has been used durably to mark endothelial cells for lineage studies in mice, but this technology has not yet been exploited in the zebrafish (Kisanuki *et al.*, 2001; Soriano, 1999).

Alkaline phosphatase activity

Alkaline phosphatase (AP) staining of developing embryos is another method that has been used to visualize the vasculature. This method relies on the fact that endothelial cells exhibit relatively high levels of endogenous AP activity and are thus preferentially stained by this protocol. Two different groups recently utilized this method to identify mutations in the vascular system in large-scale mutant screens (Childs *et al.*, 2002; Habeck *et al.*, 2002). Although AP staining is a simple and well-accepted method for defining the endothelial compartment in zebrafish, this method cannot be used to visualize vessels in living animals since staining can only be performed on fixed embryos. Furthermore, this method is most useful for only a narrow developmental window around approximately 3 days, since AP activity in vessels is not apparent prior to approximately 2 days post-fertilization and thereafter background staining in surrounding tissues gradually obscures the vascular staining. The method is most useful for visualizing more superficial vessels such as the sub-intestinal vessels on the zebrafish yolk sac.

Dye injection

Intravascular injection of dyes such as Berlin-Blue or India ink can also be used to visualize the cardiovascular system in zebrafish. The chosen dye is injected through the sinus venosus to label the bloodstream and then the animals are fixed and the vasculature analysed by microscopy. Many small molecular weight dyes leak from the vasculature in early embryos as endothelial cell junctions are quite permeable, limiting use of these dyes to older animals where this problem is less severe. India ink and large particulate dyes can be used at these earlier stages, however. Like other methods that simply label the luminal spaces of vessels connected to the active circulation, dye injection is not suitable for studying the earliest steps of vascular development when endothelial progenitor cells or angioblasts assemble together into vascular cords.

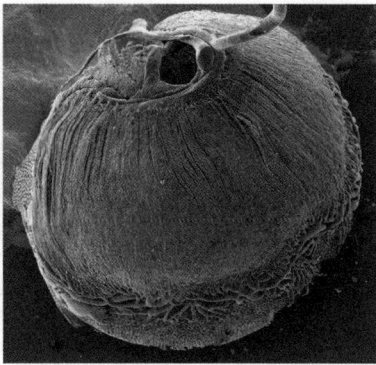

Figure 16.1   Scanning electron micrograph of a Mercox resin casting of the vasculature of the adult zebrafish eye *(image courtesy of S. Isogai).*

Resin casting

Large calibre and adult vessels in multiple vertebrates, including rodents and humans, have been visualized via corrosion resin casting methods (Hossler and Douglas, 2001; Murakami, 1972). Later, this method was combined with scanning electron microscopy to facilitate its use for examination of the microvasculature. Scanning EM of corrosion casts has been effectively used to examine the adult zebrafish vasculature (S. Isogai, personal communication, see Figure 16.1). For corrosion casting, Mercox plastic resin is injected into the vasculature at a stage of interest. After the plastic hardens inside the animal, the tissues of the animal are slowly removed by soaking in potassium hydroxide, hot water and acid. Corrosive resin casting, while not amenable to large-scale screening techniques, remains a valuable option for viewing the vasculature in adult animals where tissue opacity and thickness makes viewing of the deep vasculature essentially impossible by other methods. Moreover, it can yield highly quantitative and reproducible data on vessel diameter, cardiac volume and alterations in vessel architecture.

Digital motion analysis

Digital motion analysis (DMA) is a simple non-invasive method that has been used to assess the morphology and activity of the cardiovascular system in zebrafish (Schwerte and Pelster, 2000). DMA measures changes in dynamic pixels created by the flow of erythrocytes though the vessels. DMA is carried out by visualizing the dynamic pixels in video images collected of circulation through zebrafish blood vessels. Pairs of sequential images separated by 20 ms intervals are digitally compared and the stable (unchanged) pixels are subtracted. This leaves behind only the dynamic pixels and shifting vectors caused by the movement of

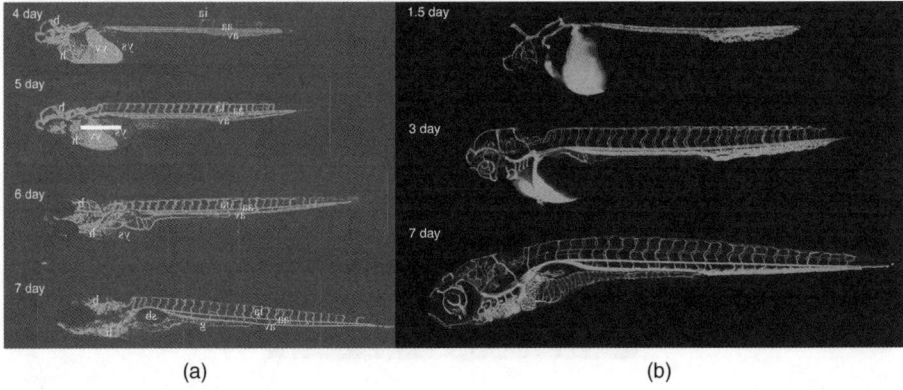

Figure 16.2   Digital motion analysis and microangiopathy. (a) False-colored image collected via digital motion analysis, a method that indirectly visualizes the vasculature by measuring erythrocyte movement in four to seven day old zebrafish larvae. *(Image reprinted with permission from J Exp Biol 2000 203: 1659–1669).* (b) Confocal microangiography of zebrafish embryos at 1.5, three and seven days post-fertilization. (A colour reproduction of this figure can be viewed in the colour section towards the centre of the book).

erythrocytes through the vessels, providing a digital 'cast' of the functional vasculature (Figure 16.2A). These shifting vectors can also be added together to calculate erythrocyte velocity and other flow parameters (Schwerte and Pelster, 2000). This method is the only one that provides useful information on flow in the vessels. Since the technique is entirely non-invasive, it can also be used repeatedly or even continuously to image the same animal over extended periods of time. However, since DMA relies on the transit of erythrocytes through a vessel segment, it is more limited in its ability to image vessels compared with other methods for imaging the luminal space such as dye injection or microangiography (described below). Many newly formed or small calibre vessels have few or no erythrocytes passing though them, making them invisible to DMA. These same vessels can be clearly visualized by microangiography, which requires only that a vascular segment be lumenized and at least partially connected to the active circulation. For example, the extent of SeV development appears delayed when these vessels are visualized with DMA as compared with microangiography (compare Figure 16.2A with Figure 16.2B).

Confocal microangiography

Confocal microangiography has emerged as one of the most powerful methods for visualizing the vasculature in developing zebrafish (Weinstein *et al.*, 1995). Briefly, fluorescent microspheres are injected into the vasculature of living animals, and optical sections are collected of the fluorescent blood vessels using confocal or multiphoton microscopy. Optical sections are then digitally reconstructed to yield three-dimensional renderings of the vasculature. Because this method relies on optical imaging, it provides very high (sub-micron) resolution images of the vasculature, while the small (0.02 μm) fluorescent microspheres employed penetrate even the smallest lumenized vascular spaces. The small size and optical clarity of zebrafish embryos permits imaging through the entire depth of the organism, making possible three-dimensional (3D) rendering of every portion of the developing vasculature (Figure 16.2B). Indeed, confocal microangiography was used to prepare a complete staged atlas of the vasculature of the developing zebrafish between days one and seven post-fertilization (Isogai *et al.*, 2001), also available online at http://mgchd1.nichd.nih.gov:8000/zfatlas/Intro%20Page/intro1. html. This atlas describes in detail the anatomy and wiring of all of the major blood vessels throughout early development, providing an invaluable standardized resource for further genetic and experimental studies of the vasculature in this model organism. This atlas has also supported the validity of extrapolating zebrafish data to other vertebrate species, since most of the major vascular structures identified in zebrafish have orthologues in other developing vertebrates.

Confocal microangiography is a very useful method for visualizing the vasculature within living zebrafish embryos and larvae, but it does have several limitations. First, the method is not amendable to time-lapse imaging. Following injection the fluorescent microspheres are rapidly taken up by endothelial cells lining the vessels. They are cleared from the circulation within about 30 min, necessitating rapid imaging and limiting the number of images collected. Development of macromolecular fluorescent labels that are not taken up by endothelium might eliminate this problem and allow longer-term imaging. As described below, long-term dynamic imaging of blood vessels can be carried out instead using transgenic zebrafish lines expressing a fluorescent reporter gene under the control of a vascular promoter. A second limitation of microangiography is that, like other methods labeling the luminal spaces of the vasculature, it is only compatible with imaging vessels connected to the active circulation. Early steps of vasculogenesis and growing but not yet lumenized angiogenic blood vessels cannot be visualized by this method.

Immunohistochemical staining of the zebrafish vasculature

Antibodies specific to the developing vasculature have also been important tools in the study of cardiovascular development. A novel monoclonal antibody, Phy-V002, labels the developing vasculature (Seng *et al.*, 2004). Moreover, direct treatment of the vascular endothelium with Phy-V002 via intravascular injection resulted in a potent disruption of angiogenesis, suggesting that in some cases antibodies that reveal vascular specific expression can also elicit inhibitory effects in this organ system. Several commercial antibody producers, like R&D Systems, have recently begun to design antibodies specifically for detecting zebrafish proteins *in situ*. For example, detection of Tie-2 in whole mount zebrafish embryo can be observed online at www.RnDSystems.com (immunohistochemistry image using R&D Systems goat anti-zebrafish Tie-2 antibody, catalog #AF928). To date, however, few other cardiovascular-specific antibodies have been used with success in zebrafish. However, due to the genetic conservation of genes that regulate cardiovascular development between zebrafish and other organisms, immunohistochemistry should be a viable option for future studies.

16.3   *In situ* hybridization reveals vascular specific expression

The methods described above offer a variety of means of imaging the anatomical form of the vasculature in both living and fixed animals, and at a variety of different stages of development or in adults. However, work in other vertebrates, in particular the mouse, has also identified specific molecular markers of vascular cells, the expression of which can be visualized by *in situ* hybridization (ISH) and immunohistochemical staining. In addition to providing useful reagents for highlighting the vasculature, these markers also provide important windows into vascular function, since many of them define key molecular players in vascular specification, differentiation and morphogenesis. Zebrafish orthologues have been uncovered for most of the critical vascular-specific genes identified in other species, and *in situ* hybridization and functional analysis of these genes has confirmed that genetic mechanisms of vessel formation are well conserved between zebrafish and other vertebrates (Table 16.2).

It has become clear that homologous pathways regulate cardiovascular development in zebrafish, chick, mouse and human. For example, vascular endothelial growth factor-A (VEGF-A) acts as an endothelial selective mitogen in each of these systems. Two predominant isoforms, $VEGF_{165}$ and $VEGF_{121}$ are expressed as secreted ligands in regions neighbouring the developing zebrafish vasculature (Liang *et al.*, 1998; Liang *et al.*, 2001). The VEGF receptor tyrosine kinase (RTK) genes flk1 and flt4 exhibit expression in the lateral plate mesoderm before becoming limited to angioblasts and the vascular endothelium. At 24 h, when

Table 16.2   Vascular endothelial *in situ* hybridization markers in zebrafish

Expression Pattern	Gene	Reference
Pan-endothelial	Fli-1	(Brown *et al.*, 2000)
	Flk-1	(Sumoy *et al.*, 1997, Fouquet *et al.*, 1997)
	Tie1 and 2	(Lyons *et al.*, 1998, Sato *et al.*, 1995)
	VE-cadherin	(Larson *et al.*, 2004)
	PlexinD1	(Torres-Vasquez *et al.*, 2004)
	flt4 (early)	(Lawson *et al.*, 2001)
Artery restricted	EphrinB2	(Lawson *et al.*, 2001)
	Grl	(Zhong *et al.*, 2001)
	notch5	(Lawson *et al.*, 2001)
Vein restricted	EphB4	(Lawson *et al.*, 2001)
	Flt4 (late)	(Lawson *et al.*, 2001)

flk1 expression is observed in the axial vessels and the sprouting intersomitic vessels, the VEGF transcript is found in the adjacent somitic tissue (Fouquet *et al.*, 1997; Liang *et al.*, 2001). Thorough investigation of the role of VEGF-A in murine vascular development has been hampered by the strict dose-dependent requirement for VEGF. VEGF-A is one of the few genes that demonstrates haploinsufficiency in mice; with loss of a single copy of the VEGF gene resulting in embryonic lethality (Carmeliet *et al.*, 1996; Ferrara, 1996). A genetic mutant in VEGF-A has not yet been isolated in zebrafish, but investigators have employed a potent anti-sense gene knockdown method to examine the loss of VEGF function in zebrafish embryos. Using morpholino-modified oligonucleotides (MO), Nasevicius *et al.* (2000) have successfully decreased VEGF-A expression in the zebrafish and have observed concomitant deficiencies in vessel development. The most severely affected morphant embryos lack all cardiovascular structures except those around the yolk and the heart itself (Nasevicius *et al.*, 2000). However, lower doses appear only to affect the sprouting of the SeV and leave the expression of early vascular markers essentially normal in the axial vessels (Nasevicius *et al.*, 2000). Comparable phenotypes were observed in analogous experiments in mice (Haigh *et al.*, 2000; Carmeliet *et al.*, 1996).

Tie1 and Tie2 are two additional vascular specific RTKs. Again, the zebrafish Tie1 and Tie2 genes exhibit expression patterns similar to those of their mouse counterparts (Lyons *et al.*, 1998; Sato *et al.*, 1995). Zebrafish Tie-1 *in situ* hybridization reveals a pattern indistinguishable from the VEGF receptor flk1, while Tie-2 expression is somewhat more restricted. Zebrafish tie-2 is not expressed in the intersegmental vessels (Lyons *et al.*, 1998). Like VEGF-A, the ligands for Tie-2, the angiopoietins, are not expressed in the vasculature but are instead found in adjacent tissues (Pham *et al.*, 2001).

PlexinD1 is an additional receptor with vascular-endothelial restricted expression. Plexin protein family members act as receptors for semaphorins, which serve primarily as guidance cues in the nervous system (Deutsch, 2004; Fiore and Puschel, 2003). Recently, however, this family of receptors was demonstrated to coordinate trunk intersegmental vessel guidance in zebrafish through a plexinD1–semaphorin interaction (Torres-Vazquez *et al.*, 2004). Elimination of vascular plexinD1 expression causes disruption of the normally stereotypic guidance of intersomitic vessels that occurs along somite boundaries. Work by Gitler *et al.* (2004) and Gu *et al.* (2005) revealed similar intersegmental vessels guidance defects in mice with targeted disruption of the murine plexinD1gene (Gu *et al.*, 2005), further demonstrating the effectiveness of the zebrafish as a model for mammalian vascular development.

Pan-endothelial expression of vascular-endothelial cadherin (VE-cadherin) is also conserved across species (Larson *et al.*, 2004; Breier *et al.*, 1996). The cadherin protein family consists of a group of molecules that mediate calcium dependent cell–cell adhesion. VE-cadherin (also known as Cdh5) expression is restricted to the vasculature in mouse (Breier *et al.*, 1996). Moreover, Gory *et al.* (1999) have demonstrated that 2.5 kb of 5′ sequence is sufficient to drive vascular-specific expression in the mouse (Gory *et al.*, 1999). A single, highly conserved VE-cadherin homologue has been isolated from zebrafish (Larson *et al.*, 2004) and expression of zebrafish VE-cadherin is also restricted to the vascular system.

Finally, the ets family transcription factor fli-1 exhibits vascular expression in developing zebrafish, as it does in mice. Originally identified as the locus disrupted in Ewing's sarcoma (Hromas and Klemsz, 1994), fli-1 expression initially overlaps during early zebrafish embryogenesis with the gata-2 positive haematopoietic compartment, suggesting the possibility of a bi-potential precursor population for both blood and endothelial cells (Brown *et al.*, 2000). However, fli-1 and gata-2 soon become restricted to distinct endothelial and haematopoietic compartments, respectively. It is important to note that other ets domain-containing genes found in zebrafish have similar vascular restricted expression patterns (V. Pham and BMW, unpublished observations).

Some genes are expressed in specific regions of the vasculature, as can be seen in the axial vessels. Prior to the 30 somites stage, flt4 exhibits pan-endothelial expression. However, after this point flt4 expression becomes restricted to the vein (Lawson *et al.*, 2001). Other markers of specific endothelial compartments include ephrinB2 and ephB4, which are differentially expressed in the dorsal aorta and posterior cardinal vein, respectively (Lawson *et al.*, 2001). These markers show similar differential arterial–venous expression in the mouse (Wang *et al.*, 1998).

16.4   Vascular specific fluorescent reporter lines

While *in situ* hybridization has been a potent method for addressing cardiovascular patterning, the advent of real-time, *in vivo* imaging of vascular gene expression via transgenic zebrafish lines has stimulated the study of embryonic vascular development. The zebrafish is amenable to real-time imaging via transgenic lines expressing green fluorescent protein (GFP), or one of the additional colour variants, under the control of a tissue-specific promoter. As described in more detail below, a number of different transgenic zebrafish lines have been developed to express GFP in the developing vasculature (Motoike *et al.*, 2000; Lawson and Weinstein, 2002; Cross *et al.*, 2003). These lines permit very high optical resolution imaging of the vasculature in developing animals over extended periods of time.

Fli

Lawson and Weinstein (2002) demonstrated the successful isolation of a 15 kb region surrounding the transcriptional start site of the zebrafish Fli-1 gene. This promoter was used to drive enhanced-GFP (eGFP) and generate a stable transgenic line with remarkably strong fluorescent protein expression throughout the entirety of the developing vasculature, as well as in cranial neural crest derivatives

(Lawson and Weinstein, 2002). Invaluable tools for numerous studies, Fli-eGFP transgenics have been utilized in an ENU chemical mutagenesis screen designed to isolate novel vascular mutations (see below). Use of transgenic zebrafish for genetic screening offers a number of advantages over AP staining. Zebrafish are alive when screened and the same animals can be examined at multiple time points. No experimental procedures are required for the screening process. A much wider range of vessels can be effectively visualized, and the developmental window for screening vessels in Fli-eGFP transgenics is essentially unlimited.

Tie2

Tie2 expression is tightly restricted to the vascular endothelium (Dumont *et al.*, 1992; Sato *et al.*, 1993), and the murine promoter has been shown to drive uniform expression of GFP in the vasculature of transgenic mice. This same murine construct efficiently elicited vascular specific expression in transgenic zebrafish (Motoike *et al.*, 2000). Expression levels in transgenic mice were sufficient to use fluorescence activated cell sorting (FACS) specifically to isolate GFP-positive endothelial cells from murine tissue, but the analogous experiment was not performed in Tie2 transgenic zebrafish (Motoike *et al.*, 2000). The expression of GFP in mTie2 transgenic zebrafish was not as strong as that observed in Fli-eGFP transgenic line, and expression was noted in the hindbrain and neural tube.

Flk1

Another, Flk promoter-driven GFP transgenic zebrafish vascular reporter line has also been derived (Cross *et al.*, 2003). This line was used in a study of vascular tubulogenesis in the zebrafish (Parker *et al.*, 2004). Although like the mTie2-GFP transgenic, these lines exhibit a weaker expression of GFP than the Fli-eGFP transgenic zebrafish, they are expressed only in endothelium and lack the cranial neural crest expression of Fli-eGFP animals that can obscure visualization of vessels in the head, particularly the aortic arch vessels. Using specific lines such as this, however, the possibility exists to isolate a pure population of zebrafish endothelial cells by flow cytometry for subsequent biochemical or cell biological experiments.

16.5 Chemical mutagenesis screens reveal novel genes involved in cardiovascular development

The genetic accessibility of the fish, physical accessibility and optical clarity of zebrafish embryos and larvae, highly stereotypic nature of zebrafish cardiovascular development and advent of stable vascular-specific fluorescent reporter lines have all facilitated the identification of novel genes involved in vascular development through forward-genetic analysis. Forward-genetic screens have proven to be an effective strategy for unbiased identification of genes important for zebrafish embryonic development (*Development*, vol. 123, 1996), including the development of the vasculature (Weinstein *et al.*, 1995). The majority of these genetic screens have employed the alkylating agent N-ethyl-N-nitrosylurea (ENU) as mutagen. Generally causing single base-pair alterations, ENU mutagenesis has yielded multiple cardiovascular-specific mutations, including *gridlock, (grl), plcg1^{yll}, violet beauregard (vbg), heart of glass (heg), out of bounds (obd)* and *cloche (clo)* (Lawson *et al.*, 2003; Childs *et al.*, 2002; Mably *et al.*, 2003; Stainier *et al.*, 1995; Weinstein *et al.*, 1995; Roman *et al.*, 2002). These and other mutants have vascular defects ranging from misguided SeV *(obd)* to nearly complete loss of the endothelial and haematopoeitic compartments *(clo)*. The cloning and identification of the genes responsible cardiovascular zebrafish mutant phenotypes has led to the elucidation of multiple genetic pathways that regulate aspects of vascular development. For example, the *plcg1^{yll}* mutant was found to be due to a mutation in phospholipase C gamma-1, a gene subsequently placed downstream of sonic hedgehog and VEGF in arterial–venous differentiation (Lawson *et al.*, 2003). The *obd* mutant was shown to have a mutation in plexinD1, a vascular specific receptor for type 3 semaphorin ligands. These results suggest that plexin-semaphorin signalling, shown previously to regulate axon guidance, also regulates guidance in the vascular system (Torres-Vazquez *et al.*, 2004).

Small molecule and peptide screens

In addition to the advantages of zebrafish for studying the vasculature, zebrafish are also an informative model for whole-organism chemical screens. Small-molecule screens have recently become widely used for the identification of novel chemicals and peptides that either mediate or inhibit specific cellular processes. Several investigators have demonstrated that zebrafish are a reasonable *in vivo* tool to assess small-molecule activity. As aberrant cardiovascular development is linked to the progression of numerous diseases, coupling chemical screens with visualization of the zebrafish cardiovascular system may prove a valuable method for identifying novel drugs for the treatment of cardiovascular abnormalities.

   Peterson *et al.* (2004) performed a small-molecule based equivalent of a genetic suppressor screen. Zebrafish *gridlock (grl)* mutants have defects in the aorta caused by lack of an orthologue of mammalian hey2, and lack all circulation in the trunk (Zhong *et al.*, 2001). *grl* mutant embryos were exposed to a panel of 5000 compounds and rescue of the circulatory defects was assessed by presence of circulation after 48 h. Two chemicals successfully suppressed the gridlock phenotype, and subsequent analysis revealed that these compounds elicit increased VEGF receptor signalling (Peterson *et al.*, 2004). While screens such as this remain hampered by the difficulty in identifying the cellular target of specific small molecules, the molecules obtained often demonstrate signs of therapeutic potential. These chemicals are often more efficient at suppression of phenotypic abnormalities than their genetic counterparts. Secondly, small molecules identified in the zebrafish can elicit comparable effects in mammalian cells (Peterson *et al.*, 2004). Thus, the coupling of vascular imaging techniques in zebrafish with the advent of such 'chemical genetic' approaches has the potential of identifying and validating novel pharmacological agents that may play a role of treating human cardiovascular disease.

16.6 Conclusions

Since George Streisinger began working with zebrafish as a genetic organism in the 1970s, *Danio rerio* has become a widely used research tool for studying the genetics of vertebrate organogenesis. Moreover, as neither *Drosophila* nor *C. elegans* have blood vessels, and large-scale forward genetic screens are generally impractical to perform in mice, zebrafish offers the only vertebrate system to perform a high-throughput mutagenesis screen aimed at the identification of novel genes required for vascular development. Moreover, aspects of both vasculogenesis and angiogenesis can be successfully monitored in the zebrafish system. Specifically, vasculogenesis can be assessed by examining the differentiation and subsequent migration of the pool of angioblasts that form the axial vessels along the trunk. The branching of the SeVs from the primary axial vessels is considered angiogenesis.

In the future, the completion of the zebrafish genome sequence should prove to be a valuable asset in hastening the identification of the genes responsible for vascular phenotypes found in these mutant screens. To date, multiple genes have been characterized for their involvement in zebrafish angiogenesis and many of them have functional orthologues in mammalian systems. Moreover, the current screens have yet to reach saturation suggesting that there are many more novel genes yet to discover.

The ability rapidly to generate transgenic zebrafish lines with fluorescent reporter proteins expressed under vascular-restricted promoters has further facilitated the identification of mutants as well as highly rigorous and detailed examination of the steps involved in vascular development. In the future, promoters driving expression in a specific subset of vascular cells (for example, ephB4 in venous endothelial cells), might be beneficial for the study of certain aspects of vascular dynamics. Many genes, like delta-c, are expressed in both the endothelium and other tissues (Smithers *et al.*, 2000). In order to address the role of such genes in the vasculature specifically, it is possible to over-express genes in the endothelium by placing them under the control of a vascular specific driver. Such experiments may help to delineate the roles of different genes in different cellular subsets.

Finally, the ability to couple visualization of the vasculature through transgenic reporter zebrafish lines or *in situ* hybridization of vascular-specific probes, with assays that screen thousands of chemicals for phenotypic effects will also prove to profoundly affect the identification of compounds with therapeutic potential. In short, the genetic tractability of zebrafish makes it a profoundly powerful system to study vertebrate organogenesis. Moreover, the conservation of genes between zebrafish and mammals imparts credibility to *Danio rerio* as a model with broad applicability. As investigators continue to develop vascular-specific transgenic lines and isolate novel genetic regulators of vascular development, zebrafish will continue to serve as a compelling model system with which to study cardiovascular development.

References

Breier, G., Breviario, F., Caveda, L., Berthier, R. *et al.* (1996) 'Molecular cloning and expression of murine vascular endothelial-cadherin in early stage development of cardiovascular system', *Blood*, **87**, 630–641.

Brown, L. A., Rodaway, A. R., Schilling, T. F., Jowett, T. *et al.* (2000) 'Insights into early vasculogenesis revealed by expression of the ETS-domain transcription factor Fli-1 in wild-type and mutant zebrafish embryos', *Mech. Dev.*, **90**, 237–252.

Carmeliet, P., Ferreira, V., Breier, G., Pollefeyt, S. *et al.* (1996) 'Abnormal blood vessel development and lethality in embryos lacking a single VEGF allele', *Nature*, **380**, 435–439.

Childs, S., Chen, J. N., Garrity, D. M. and Fishman, M. C. (2002) 'Patterning of angiogenesis in the zebrafish embryo', *Development*, **129**, 973–982.

Cross, L. M., Cook, M. A., Lin, S., Chen, J. N. and Rubinstein, A. L. (2003) 'Rapid analysis of angiogenesis drugs in a live fluorescent zebrafish assay', *Arterioscler. Thromb. Vasc. Biol.*, **23**, 911–912.

Deutsch, U. (2004) 'Semaphorins guide PerPlexeD endothelial cells', *Dev. Cell*, **7**, 1–2.

Dumont, D. J., Yamaguchi, T. P., Conlon, R. A., Rossant, J. and Breitman, M. L. (1992) 'tek, a novel tyrosine kinase gene located on mouse chromosome 4, is expressed in endothelial cells and their presumptive precursors', *Oncogene*, **7**, 1471–1480.

Ferrara, N. (1996) 'Vascular endothelial growth factor', *Eur. J. Cancer*, **32A**, 2413– 2422.

Fiore, R. and Puschel, A. W. (2003) 'The function of semaphorins during nervous system development', *Front Biosci.*, **8**, s484–499.

Fouquet, B., Weinstein, B. M., Serluca, F. C. and Fishman, M. C. (1997) 'Vessel patterning in the embryo of the zebrafish: guidance by notochord', *Dev. Biol.*, **183**, 37–48.

Gory, S., Vernet, M., Laurent, M., Dejana, E. *et al.* (1999) 'The vascular endothelial-cadherin promoter directs endothelial-specific expression in transgenic mice', *Blood*, **93**, 184–192.

Habeck, H., Odenthal, J., Walderich, B., Maischein, H. and Schulte-Merker, S. (2002) 'Analysis of a zebrafish VEGF receptor mutant reveals specific disruption of angiogenesis', *Curr. Biol.*, **12**, 1405–1412.

Haigh, J. J., Gerber, H. P., Ferrara, N. and Wagner, E. F. (2000) 'Conditional inactivation of VEGF-A in areas of collagen2a1 expression results in embryonic lethality in the heterozygous state', *Development*, **127**, 1445–1453.

Hossler, F. E. and Douglas, J. E. (2001) 'Vascular corrosion casting: review of advantages and limitations in the application of some simple quantitative methods', *Microsc. Microanal.*, **7**, 253–264.

Hromas, R. and Klemsz, M. (1994) 'The ETS oncogene family in development, proliferation and neoplasia', *Int. J. Hematol*, **59**, 257–265.

Isogai, S., Horiguchi, M. and Weinstein, B. M. (2001) 'The vascular anatomy of the developing zebrafish: an atlas of embryonic and early larval development', *Dev. Biol.*, **230**, 278–301.

Kisanuki, Y. Y., Hammer, R. E., Miyazaki, J., Williams, S. C. *et al.* (2001) 'Tie2-Cre transgenic mice: a new model for endothelial cell-lineage analysis in vivo', *Dev. Biol.*, **230**, 230–242.

Larson, J. D., Wadman, S. A., Chen, E., Kerley, L. *et al.* (2004) 'Expression of VE-cadherin in zebrafish embryos: a new tool to evaluate vascular development', *Dev. Dyn.*, **231**, 204–213.

Lawson, N. D., Mugford, J. W., Diamond, B. A. and Weinstein, B. M. (2003) 'Phospholipase C gamma-1 is required downstream of vascular endothelial growth factor during arterial development', *Genes Dev.*, **17**, 1346–1351.

Lawson, N. D., Scheer, N., Pham, V. N., Kim, C. H. *et al.* (2001) 'Notch signaling is required for arterial-venous differentiation during embryonic vascular development', *Development*, **128**, 3675–3683.

Lawson, N. D. and Weinstein, B. M. (2002) 'In vivo imaging of embryonic vascular development using transgenic zebrafish', *Dev. Biol.*, **248**, 307–318.

Liang, D., Chang, J. R., Chin, A. J., Smith, A. *et al.* (2001) 'The role of vascular endothelial growth factor (VEGF) in vasculogenesis, angiogenesis, and hematopoiesis in zebrafish development', *Mech. Dev.*, **108**, 29–43.

Liang, D., Xu, X., Chin, A. J., Balasubramaniyan, N. V. *et al.* (1998) 'Cloning and characterization of vascular endothelial growth factor (VEGF) from zebrafish, Danio rerio', *Biochim. Biophys. Acta*, **1397**, 14–20.

Lyons, M. S., Bell, B., Stainier, D. and Peters, K. G. (1998) 'Isolation of the zebrafish homologues for the tie-1 and tie-2 endothelium-specific receptor tyrosine kinases', *Dev. Dyn.*, **212**, 133–140.

Mably, J. D., Mohideen, M. A., Burns, C. G., Chen, J. N. and Fishman, M. C. (2003) 'Heart of glass regulates the concentric growth of the heart in zebrafish', *Curr. Biol.*, **13**, 2138–2147.

Mills, K. R., Kruep, D. and Saha, M. S. (1999) 'Elucidating the origins of the vascular system: a fate map of the vascular endothelial and red blood cell lineages in Xenopus laevis', *Dev. Biol.*, **209**, 352–368.

Motoike, T., Loughna, S., Perens, E., Roman, B. L. *et al.* (2000) 'Universal GFP reporter for the study of vascular development', *Genesis*, **28**, 75–81.

Murakami, T. (1972) 'Vascular arrangement of the rat renal glomerulus. A scanning electron microscope study of corrosion casts', *Arch. Histol. Jpn.*, **34**, 87–107.

Nasevicius, A., Larson, J. and Ekker, S. C. (2000) 'Distinct requirements for zebrafish angiogenesis revealed by a VEGF-A morphant', *Yeast*, **17**, 294–301.

Parker, L. H., Schmidt, M., Jin, S. W., Gray, A. M. *et al.* (2004) 'The endothelial-cell-derived secreted factor Egfl7 regulates vascular tube formation', *Nature*, **428**, 754–758.

Peterson, R. T., Shaw, S. Y., Peterson, T. A., Milan, D. J. *et al.* (2004) 'Chemical suppression of a genetic mutation in a zebrafish model of aortic coarctation', *Nat. Biotechnol.*, **22**, 595–599.

Pham, V. N., Roman, B. L. and Weinstein, B. M. (2001) 'Isolation and expression analysis of three zebrafish angiopoietin genes', *Dev. Dyn.*, **221**, 470–474.

Poole, T. J. and Coffin, J. D. (1989) 'Vasculogenesis and angiogenesis: two distinct morphogenetic mechanisms establish embryonic vascular pattern', *J. Exp. Zool.*, **251**, 224–231.

Roman, B. L., Pham, V. N., Lawson, N. D., Kulik, M. *et al.* (2002) 'Disruption of acvrl1 increases endothelial cell number in zebrafish cranial vessels', *Development*, **129**, 3009–3019.

Sato, T. N., Qin, Y., Kozak, C. A. and Audus, K. L. (1993) 'Tie-1 and tie-2 define another class of putative receptor tyrosine kinase genes expressed in early embryonic vascular system', *Proc. Natl. Acad. Sci. USA*, **90**, 9355–9358.

Sato, T. N., Tozawa, Y., Deutsch, U., Wolburg-Buchholz, K. *et al.* (1995) 'Distinct roles of the receptor tyrosine kinases Tie-1 and Tie-2 in blood vessel formation', *Nature*, **376**, 70–74.

Schwerte, T. and Pelster, B. (2000) 'Digital motion analysis as a tool for analysing the shape and performance of the circulatory system in transparent animals', *J. Exp. Biol.*, **203**, 1659–1669.

Seng, W. L., Eng, K., Lee, J. and McGrath, P. (2004) 'Use of a monoclonal antibody specific for activated endothelial cells to quantitate angiogenesis in vivo in zebrafish after drug treatment', *Angiogenesis*, **7**, 243–253.

Smithers, L., Haddon, C., Jiang, Y. J. and Lewis, J. (2000) 'Sequence and embryonic expression of deltaC in the zebrafish', *Mech. Dev.*, **90**, 119–123.

Soriano, P. (1999) 'Generalized lacZ expression with the ROSA26 Cre reporter strain', *Nat. Genet.*, **21**, 70–71.

Stainier, D. Y., Weinstein, B. M., Detrich, H. W., 3rd, Zon, L. I. and Fishman, M. C. (1995) 'Cloche, an early acting zebrafish gene, is required by both the endothelial and hematopoietic lineages', *Development*, **121**, 3141–3150.

Stern, C. D. and Fraser, S. E. (2001) 'Tracing the lineage of tracing cell lineages', *Nat. Cell Biol.*, **3**, E216–218.

Sumoy, L., Keasey, J. B., Dittman, T. D. and Kimelman, D. (1997) 'A role for notochord in axial vascular development revealed by analysis of phenotype and the expression of VEGR-2 in zebrafish flh and ntl mutant embryos', *Mech. Dev.*, **63**, 15–27.

Torres-Vazquez, J., Gitler, A. D., Fraser, S. D., Berk, J. D. *et al.* (2004) 'Semaphorin-plexin signaling guides patterning of the developing vasculature', *Dev. Cell*, **7**, 117–123.

Wang, H. U., Chen, Z. F. and Anderson, D. J. (1998) 'Molecular distinction and angiogenic interaction between embryonic arteries and veins revealed by ephrin-B2 and its receptor Eph-B4', *Cell*, **93**, 741–753.

Weinstein, B. M., Stemple, D. L., Driever, W. and Fishman, M. C. (1995) 'Gridlock, a localized heritable vascular patterning defect in the zebrafish', *Nat. Med.*, **1**, 1143–1147.

Zhong, T. P., Childs, S., Leu, J. P. and Fishman, M. C. (2001) 'Gridlock signalling pathway fashions the first embryonic artery', *Nature*, **414**, 216–220.

17

Biological and clinical implications of recruitment of stem cells into angiogenesis

Gianluigi Castoldi, **Antonio Cuneo** and **Gian Matteo Rigolin**

Abstract

Emerging evidence suggests that bone marrow may represent a reservoir of endothelial progenitor cells (EPCs), which could contribute to post-natal tissue vasculogenesis. These cells can be mobilized to the circulation through the action of specific angiogenic factors and contribute to neoangiogenic processes. The level of circulating EPCs has been proposed as a surrogate biological marker for vascular function suggesting that EPCs may have a role in maintenance and reparative processes and in tumour development. The phenotypic characterization of EPCs remains controversial because of the lack of specific endothelial markers and functional assays. The most promising areas of current research focus on the investigation and understanding of the role EPCs in cardiovascular disorders and tumours in order to develop therapeutic strategies that can modulate EPC trafficking and function.

Keywords

stem cells; endothelial progenitor cells; phenotype; ageing

17.1   Introduction

Angiogenesis and neovascularization are increasingly recognized as having an important role in a wide array of diseases. For many years it was believed that vasculogenesis, the *in situ* differentiation of primitive angioblasts into endothelial cells, was a phenomenon restricted to the developing embryo while angiogenesis,

Angiogenesis Assays   Edited by Carolyn A. Staton, Claire Lewis and Roy Bicknell
© 2006 John Wiley & Sons, Ltd

the vascular growth in which a blood vessel network expands through divisions of existing cells within the vascular network, was considered a distinct mechanism during growth, tissue ischaemia, and tumour proliferation and metastasis (Khakoo and Finkel, 2005). In recent years, emerging evidence suggests that bone marrow may represent a reservoir of endothelial progenitor cells (EPCs), which could contribute to post-natal tissue vasculogenesis. These cells can be mobilized to the circulation through the action of specific angiogenic factors and contribute to neoangiogenic processes. The co-recruitment of angiocompetent haematopoietic cells delivering specific angiogenic factors facilitates incorporation of endothelial progenitor cells into newly sprouting blood vessels (Rafii and Lyden, 2003).

17.2   Phenotypic and functional characterization of EPCS

Accumulating evidence indicates that the peripheral blood of adults contains a population of circulating bone marrow-derived EPCs with properties similar to those of embryonal angioblasts (Asahara *et al.*, 1997). EPCs are capable of proliferation and readily migrate from the bone marrow, circulate, differentiate into mature endothelial cells (ECs) and contribute to tissue revascularization. EPCs are extremely rare in normal peripheral blood, representing somewhere between 0.01 per cent and 0.0001 per cent of peripheral mononuclear cells.

The identification and direct quantification of EPCs from blood samples is usually performed by flow cytometry measurement of the percentage of positive cells using specific combinations of monoclonal antibodies (Khan *et al.*, 2005). Different cell surface markers have been used to detect these cells (Table 17.1;

Table 17.1   Cell surface antigens present on endothelial cells

Antigen	Other name	Expression by other cells
CD31	PECAM-1	Platelet, monocytes, neutrophils, T-cell subsets
CD34		Haematopoietic stem cells
CD62e	E-selectin	Activated skin/synovial endothelium
CD54	ICAM-1	Monocytes, activated B and T lymphocytes
CD105	Endoglin	Activated monocytes, tissue macrophages, erythroid marrow precursors
CD106	VCAM-1	Activated endothelium, stromal cells
CD133	AC133	Stem cells, EPCs
CD141	Thrombomodulin	Keratinocytes, platelets, monocytes, neutrophils
CD144	VE-cadherin	Endothelial adherens junction
CD146	P1H12, S-endo-1	Activated T lymphocytes
CD202b	Tie-2	Haematopoietic stem cells
VEGFR-2	KDR, Flk-1	Haematopoietic stem cells

PECAM-1: platelet endothelium cell adhesion molecule 1; ICAM-1: intracellular cell adhesion molecule-1; VCAM-1: vascular cell adhesion molecule-1; VE: vascular endothelium; VEGFR-2: vascular endothelium growth factor receptor 2; KDR: kinase insert domain receptor. Flk-1: fetal liver kinase 1.

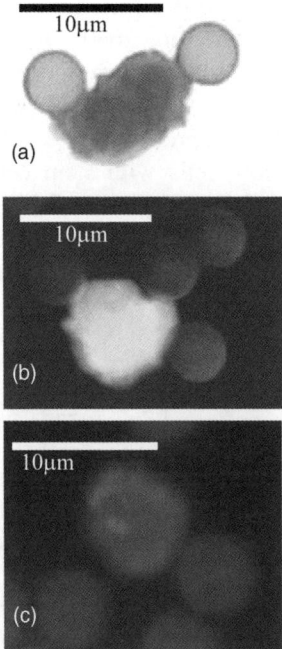

Figure 17.1 Immunophenotypic characterization of immunomagnetic sorted EPCs express von Willebrand factor (a), UEA-1 (b) and CD133 (c). Note that several Dynabeads are attached to the cells. (A colour reproduction of this figure can be viewed in the colour section towards the centre of the book).

Figure 17.1) but the phenotypic characterization of EPCs remains controversial because of the lack of specific endothelial markers and functional assays to distinguish these cells from mature circulating ECs (CECs). To date, there is no clear definition as to when an endothelial progenitor cell turns into a mature, fully differentiated endothelial cell *in vivo* (Hristov *et al.*, 2003). In addition, identification of EPCs and mature CECs may also be complicated by the fact that EPCs and mature CECs share several markers with haematopoietic cells including CD34, CD31 (PECAM), von Willebrand factor and VEGFR1. Some haematopoietic cells also incorporate acetylated low density lipoprotein and bind lectins including Bandeiraea Simplicifolia (BS-1) and Ulex Europeus-1 (UEA-1), which were once considered endothelial-specific. Finally, subsets of myelomonocytic cells have been misrepresented as EPCs because of the expression of endothelial-specific antigens thereby making difficult the interpretation of the nature of cellular contribution to neoangiogenesis (Raffi and Lyden, 2003).

The stem cell marker CD133 (AC133) could represent an important antigen in the identification of EPCs. CD133 is a five-transmembrane glycoprotein whose function is not well understood and whose expression appears to define a population of cells with pluripotent haematopoietic stem cell properties. CD133 is also expressed on EPCs but not on mature ECs or on myelomonocytic cells and therefore could be useful to distinguish EPCs from mature CECs and from cells of myelomonocytic origin (Peichev *et al.*, 2000). An intriguing hypothesis is that CD133+/CD34+/VEGFR2+ cells could represent more primitive EPCs with high proliferative potential (Figure 17.2), which then give rise to CD133-/CD34+/VEGFR2+ EPCs with more limited proliferative capacity. Very recently it has been proposed that CD14+/CD34low cells with stem cell phenotype and functional features (Romagnani *et al.*, 2005) and CD14+/VEGFR2+ monocytes (Elsheikh *et al.*, 2005) might represent the major source of circulating endothelial cells.

The expression of CD45, the leukocyte common antigen, on EPCs has also been reported by several groups to be positive or negative (Asahara *et al.*, 1997; Hur *et al.*, 2004): the low levels of expression of this molecule by EPCs is likely to be the

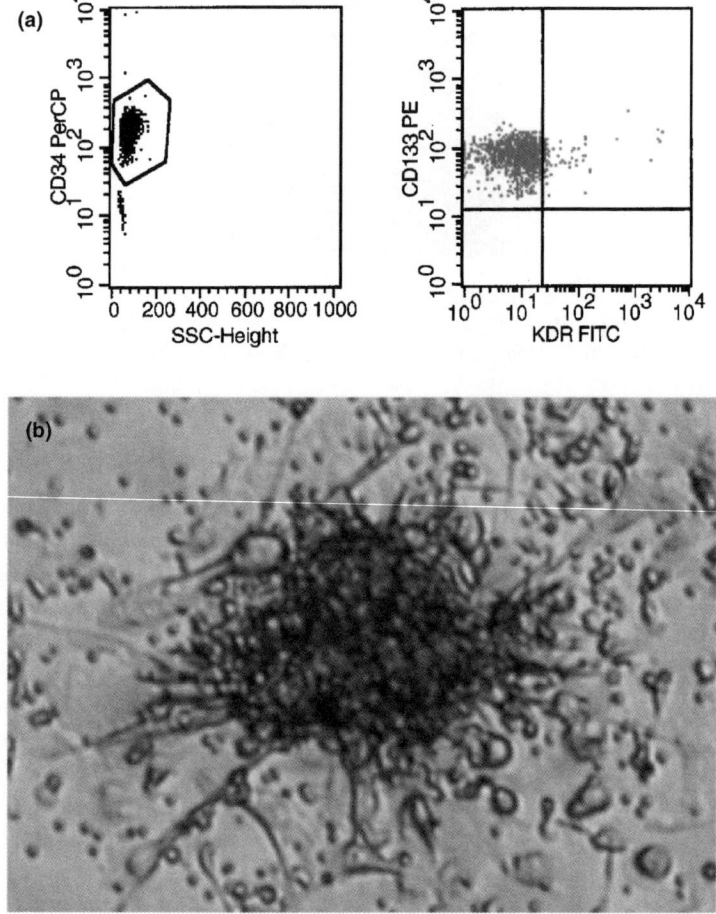

Figure 17.2   EPC enumeration in (a), flow cytometric characterization of PB EPC: after gating on CD34[+] cells exhibiting low side scatter (SSC) properties (left dot plot), EPC were identified as CD34[+] cells co-expressing CD133 and KDR/VEGFR-2 (right dot plot). In (b), representative phase contrast micrograph of an EPC-colony, characterized by multiple thin, flat cells emanating from a central cluster of rounded cells (magnified at x32).

cause of this discrepant interpretation (Khan *et al.*, 2005). Mature ECs are unequivocally CD45 negative.

Quantification of EPCs can also be performed by using colony-forming assays. One common method for culturing EPCs from peripheral blood is to plate the low-density mononuclear cells on fibronectin-coated plates using commercially available tissue-culture media with added endothelial growth supplements. The adherent cells are then examined for expression of several characteristics of endothelial cells. Using this approach, it is possible to characterize two types of EPCs with different morphology, proliferation rates, survival behaviour and gene expression profiles leading to different functions. Early EPCs, with spindle shape, show peak growth at 2 to 3 weeks, die at 4 weeks, and secrete angiogenic cytokines (VEGF and IL8). Late EPCs with cobblestone shape appear at 2 to 3 weeks, show exponential growth at 4–8 weeks and live up to 12 weeks. Late EPCs produce more nitric oxide, incorporate more readily into human umbilical vein EC monolayers and form capillary tubes (Hur *et al.*, 2004). Furthermore, it was shown that patients who had received a gender-mismatched bone marrow transplant gave rise in culture to two different types of EC colonies. EC colonies appearing within 9 days are predominantly of recipient origin suggesting that these cells may represent CECs from the vascular endothelium of the recipient, while colonies that appear later and exhibit tremendous proliferative capacity are of donor origin and therefore bone marrow-derived (Lin *et al.*, 2000). However, the peripheral blood contains several cell types that possess the ability to differentiate into cells with endothelial-like characteristics including haematopoietic stem cells with angioblastic potential, mononuclear phagocytes and sloughed endothelial cells (Ingram *et al.*, 2005). To this extent, it was shown that mononuclear phagocytes (monocytes and macrophages) represent an extremely flexible cell population in their phenotype and functional properties in response to changing microenvironmental states (Stout and Suttles, 2004). These observations led to the hypothesis that mononuclear phagocytes could represent a unique form of macrophage differentiation (angiogenic macrophage) or that alternatively the EPCs may be derived from an earlier common myeloid progenitor (Ingram *et al.*, 2005). In an attempt to avoid possible contamination of EPC cultures with monocytes, haematopoietic progenitors and CECs, an alternative culture method uses a pre-plating step of 24–48 h and then replates the non-adherent cells onto fibronectin-coated dishes. After 7 days of culture, the plates are scanned for the presence of EPC colonies (Figure 17.2b) and the number of colonies are counted and characterized (Hill *et al.*, 2003). These colonies disappear by day 10 to 14 in culture leaving behind spindle-shaped adherent cells expressing the endothelial antigens. The proliferative potential of these cells has not been well examined (Ingram *et al.*, 2005).

The functional characterization of EPCs includes *in vitro* and *in vivo* studies. In some of these studies, EPC functionality was evaluated on a purified population of cells isolated by immunomagnetic sorting with microbeads against surface markers like CD133 and CD34. *In vitro* analyses may evaluate different aspects including gene expression of angiogenesis-related genes (growth factors and

receptors, chemokines, cytokines, adhesion molecules, matrix proteases and inhibitors), measurement of cytokine production, Matrigel capillarogenesis and chick
embryo chorioallontoic membrane assay (Staton *et al.*, 2004). Animal models of
ischaemia (including hind-limb ischaemia and myocardial infarction) or tumour
angiogenesis (for example, xenograft tumour models in immunodeficient mice)
may be used to study *in vivo* angiogenesis. *In vivo* analyses have shown that infusion
of distinct cell types, either from bone marrow or by *ex vivo* cultivation, augmented
capillary density and neovascularization. Using immunological or genetic markers
(for example, human antigens in animal models, or the Y chromosome in sex-
mismatched models) and mutant mice, it was possible to determine the contribution
of the different cell populations to ischaemic or tumour angiogenesis. However, the
use of different animal models, cell numbers and species limited the comparability
and the efficiency of distinct cell populations. In most of these studies, the
functional improvement appeared similar, suggesting that the functional activity
of the different subsets of stem/progenitor cells to augment neovascularization is
rather dependent on the type of EPCs used. The functional capacity of EPCs to
augment blood flow may also be dependent on the co-recruitment of haematopoietic
cells (Rafii and Lyden, 2003) including monocytic cells providing angiogenic and
chemoattractant factors that play a crucial role in the process of neovascularization.
In addition, other as yet undefined functional characteristics might contribute to
EPC-mediated augmentation of blood flow after ischaemia or in tumour angiogenesis, including the expression of chemokine or integrin receptors mediating homing
(Urbich and Dimmeler, 2004).

Finally, a multipotent adult progenitor cell (MAPC, CD34−,VE-cadherin−,
AC133+, and Flk1+) has been characterized (Jiang *et al.*, 2002), which represents
a human bone marrow-derived adult progenitor cell that co-purifies with mesench-
ymal stem cells and has the ability to differentiate *in vitro* into cells that express
EC markers, and function as mature ECs as they contain von Willebrand Factor
stored in Weibel-Palade bodies, take up LDL, form vascular tubes when plated on
Matrigel or extracellular matrix, respond to hypoxia by upregulating expression of
VEGF, VEGFR2 and upregulate expression of HLA antigens and cell adhesion
ligands in response to inflammatory cytokines. Transplantation of MAPC-derived
ECs to an animal model for tumour and wound angiogenesis has shown that these
cells may also contribute to neoangiogenesis *in vivo* as demonstrated by the
observation that 35–40 per cent of ECs in tumour and wound vessels were positive
for human antigens (Reyes *et al.*, 2002). MAPCs hold great promise because they
could represent an important source of ECs for the treatment of degenerative or
inherited diseases.

17.3   Mobilization, homing and differentiation of EPCS

In adult life, rapid revascularization is essential for the survival of growing, injured
and ischaemic tissues in order to restore and maintain organ function (Khakoo and

Finkel, 2005). The release and mobilization of EPCs from the bone marrow is a complex process regulated by a variety of growth factors, enzymes, ligands and surface receptors. The physiological mechanisms and the signals that regulate mobilization, homing and differentiation of EPCs into mature ECs in the injured area have not yet been completely elucidated. The initial step of this process involves the activation of matrix metalloproteinase-9, which promotes the transformation of membrane-bound Kit ligand to soluble Kit ligand. Subsequently, cKit-positive stem and progenitor cells, including also a common haematopoietic and haemangioblast precursor cell, move to the vascular zone of the bone marrow microenvironment and enter into the circulation. Mobilized EPCs are then attracted to ischaemic areas by locally elevated VEGF or stromal-derived factor-1 (SDF-1).

VEGF appears to be one of the most important mediators of EPC mobilization into peripheral blood circulation. VEGF expression is markedly increased in hypoxic tissues and tumours mainly because of the effects of hypoxia-inducible factors (HIF-1) on VEGF transcription. It has been shown that vascular trauma induces a rapid but transient mobilization of VEGFR2+/AC133+ EPCs on to the peripheral circulation and that this mobilization is promoted by and correlates with the release of chemokines such as VEGF (Gill *et al.*, 2001). In patients with acute myocardial infarction (MI), one week after MI, a significant increase of both circulating CD34+ cells and EPCs as assessed by flow cytometry and cell culture assay, respectively, was demonstrated, which closely mirrors the level of VEGF (Shintani *et al.*, 2001). However, the exact kinetics of mobilization and the actual contribution of these cells to the repair of ischaemic tissues is still a matter of research. To this extent, it was shown, by flow cytometry and *in vitro* cultures, that a spontaneous mobilization of haematopoietic stem cells (HSCs) and EPCs occurs within a few hours from the onset of MI and is detectable until 2 months, suggesting that, in tissue repair, a cooperation between EPCs and other HSCs is possibly involved (Massa *et al.*, 2005).

SDF-1 is a chemokine of the CXC family that binds to the chemokine receptor CXCR4. SDF-1 is a mediator of stem cell homing to the bone marrow. EPCs express CXCR4 and systemic SDF-1 over-expression induces mobilization of endothelial and haemopoietic cells (Yamaguchi *et al.*, 2003). Transplantation of genetically altered cardiac fibroblasts constitutively expressing SDF-1 combined with stem-cell mobilization by G-CSF could restore depressed myocardial function (Askari *et al.*, 2003). The combined application of SDF-1 and G-CSF increased mobilization and homing of CD117+ cells to the infarct and enhances the formation of new blood vessels within the infarct zone. SDF-1 appears to play a critical role in regulating EPC mobilization and homing that could be exploited to direct stem cell engraftment into injured tissues, although this therapeutic strategy must be initiated within days of MI to obtain an efficient stem cell mobilization and homing.

In the infarct zone, bone marrow-derived cells produce new blood vessels via localized recruitment, proliferation and differentiation of circulating EPCs. The formation of new vessels involves several steps most of which are still a matter of active research. EPC migration from the bone marrow to the systemic circulation in response to angiogenic factors represents the starting point of EPC differentiation. Hypoxic conditions alter the vascular endothelium, causing circulating EPC arrest and egress into ischaemic tissues. Once in the interstitium, EPCs form cellular clusters and proliferate to increase the pool of cells available for neovascularization. Gradients of ischaemia drive the formation of vascular cords in the direction of hypoxic regions. Vascular cords then tubulize and unite with existing vasculature leading to tissue perfusion (Tepper *et al.*, 2005).

EPCs are also involved in tumour neoangiogenesis. Transplantation models have shown that tumour neoangiogenesis is mostly derived from bone marrow-derived EPCs (Lyden *et al.*, 2001). By using the angiogenic defective, tumour resistant Id-mutant mice it was shown that transplantation of wild-type BM or vascular endothelial growth factor (VEGF)-mobilized stem cells restored tumour angiogenesis and growth suggesting that recruitment of VEGF-responsive BM-derived precursors is necessary and sufficient for tumour angiogenesis. Bone marrow may contribute to tumour expansion in two different ways: first by supplying EPCs that directly incorporate into the vascular endothelium of the tumour and secondly by contributing cells that home to perivascular sites within the tumour and secrete pro-angiogenic growth factors (Khakoo and Finkel, 2005). The contribution of bone marrow-derived EPCs to tumour vascular growth in human neoplasms is not completely understood. Recent data suggest that only a minor percentage of blood vessels were derived from bone marrow EPCs and point to a substantial difference with mouse models which may have important implications for the translation of experimental anti-angiogenic therapies to the clinical applications (Peters *et al.*, 2005). Recent data also showed that microvascular ECs in B-cell lymphomas were in part tumour-related, reflecting a novel aspect of tumour angiogenesis (Streubel *et al.*, 2004; Rigolin *et al.*, 2006). All together these observations suggest that tumours can elicit the sprouting of new vessels from existing capillaries through the secretion of angiogenic factors (Carmeliet, 2000) and that, in some cases, cancer cells can also mimic the activities of ECs by participating in the formation of vascular-like networks.

Tracking of EPCs represents a further critical issue in the study of the recruitment of these cells into areas of neovascularization. During MI, an increased homing of radiolabelled human EPCs to the damaged heart of athymic nude mice, has been described using a radioactive technology (Aicher *et al.*, 2003). By magnetic resonance imaging and confocal microscopy, it was also possible to demonstrate that, in a mice model of xenografted glioma, human CD34+/CD133+ EPCs labelled with ferumoxides-protamine sulfate complexes migrate and incorporate into tumour neovasculature of implanted flank tumours (Arbab *et al.*, 2005). In mouse, cytokine-mediated mobilization of bone marrow cells is beneficial in terms of improving heart function and survival during MI (Orlic *et al.*, 2001). However, in a primate model of acute MI, the combined use of stem cell factor and granulocyte-colony stimulating factor (G-CSF) improved myocardial neovascularization but failed to improve myocardial function (Norol *et al.*, 2003).

17.4   Ageing

The maintenance of an intact endothelium is critical for preservation of the integrity of the vascular system (Khakoo and Finkel, 2005). Accumulating evidence suggests that age may have a critical role in the mobilization and functional activity of stem cells and EPCs (Dimmeler and Vasa-Nicotera, 2003).

Age may interfere with neovascularization at various steps. There is evidence that increasing age may lead to 1) a limited capacity of response towards EPC mobilizing stimuli after critical ischaemia (Geiger and Van Zant, 2002), 2) a reduced functional activity of stem and progenitor cells (Edelberg et al., 2002), 3) a significantly reduced migratory capacity of peripheral blood-derived EPCs (Vasa et al., 2001b), 4) a progressive progenitor cell reduction that may contribute to the development of atherosclerosis (Rauscher et al., 2003), 5) a decrease of VEGF production with a limited mobilization and survival/differentiation of EPCs (Scheubel et al., 2003), and 6) an increased turnover rate with increased susceptibility to apoptosis (Dimmeler and Vasa-Nicotera, 2003).

A possible intriguing hypothesis is, therefore, that in some patients genetic and environmental factors may contribute, in the presence of a continuous endothelial damage, to a lower number of poorly functioning EPCs as a consequence of an anticipated exhaustion of bone marrow-derived EPCs. It is also well known that age leads to telomere shortening and dysfunction, which are implicated in senescence and apoptosis. To this extent it has been observed that accelerated telomere shortening, an indicator of a process of senescence, contributes to the rise in mortality rates from heart disease (Cawthon et al., 2003).

17.5   Therapeutic applications

The most important areas of current clinical research focus on the role EPCs in cardiovascular disease and tumours. Concerning cardiovascular disorders, several pre-clinical and clinical studies have exploited the possibility of using bone marrow-derived cells containing EPCs for the neovascularization and repair of ischaemic tissues (Mathur and Martin, 2004). Stem cells when directly injected into necrotic areas have been shown to improve ventricular function both in experimental models and in humans. To this extent it has been proposed that CD34+ cells may contribute to maintenance of the vasculature, not only as a pool of EPCs but also as the source of angiogenic and growth factors (Rafii and Lyden, 2003).

The interpretation of these results and the effective contribution of EPCs to the observed benefit is difficult to establish due to the heterogeneity of the cells injected: many have in fact used unfractionated bone marrow as the therapeutic agent, whereas others have used different types of purified cell populations with various degrees of EPC-like features (Khakoo and Finkel, 2001).

The molecular interactions between EPCs, bone marrow stromal cells and other factors in the bone marrow microenvironment that control EPC release into the circulation represent the targets of new therapeutic strategies in the treatment of several diseases. Several studies have described the influence of pathological conditions on the number of EPCs *in vivo*. A reduced number of circulating EPCs have been associated with ischaemic cardiovascular disease and Framingham cardiovascular risk factor score (Hill *et al.*, 2003; Vasa *et al.*, 2001b), type II diabetes mellitus (Tepper *et al.*, 2002), smoking (Kondo *et al.*, 2004), inadequate coronary collateral development (Lambiase *et al.*, 2004), congestive heart failure (Valgimigli *et al.*, 2004) and chronic renal failure (Choi *et al.*, 2004). Different mechanisms, either separated or in combination, have been implicated: a decreased mobilization from the bone marrow, an increased consumption of EPCs at sites of vascular injury and a reduced half-life of circulating EPC (Hristov *et al.*, 2003).

Efforts to increase mobilization of EPCs and promote neovascularization at sites of vascular injury have become one of the most promising areas of research in ischaemic heart diseases. Pharmacological agents have been shown to enhance EPC mobilization and could therefore represent promising therapies in cardiovascular diseases. Some studies have suggested that statins may mobilize and increase the number of EPCs in patients with stable coronary heart disease (Vasa *et al.*, 2001a) and that oestrogens may mediate re-endothelialization after injury by increasing EPC mobilization (Iwakura *et al.*, 2003).

Cytokines promoting granulocyte proliferation and peripheral mobilization have also been implicated in EPC mobilization (Takahashi *et al.*, 1999). In animal models, GM-CSF and G-CSF promoted myocardial infarct reperfusion (Norol *et al.*, 2003) and repair (Orlic *et al.*, 2001). In clinical trials, G-CSF-treated patients have shown some improvement (Mathur and Martin, 2004) although safety concern was raised in some studies (Kang *et al.*, 2004) but not in others (Valgimigli *et al.*, 2005). Recent findings suggest that the beneficial effect of G-CSF after myocardial infarction is mediated by an activation of the Jak-Stat pathway in cardiomyocytes thus promoting cardiac myocyte survival and preventing left ventricular remodelling (Harada *et al.*, 2005). Erythropoietin is also a potent physiological stimulus for EPC mobilization (Heeschen *et al.*, 2003) and it has been shown that in humans may markedly mobilize functionally active EPCs (Bahlmann *et al.*, 2004) thus opening new therapeutic strategies in patients with impaired recruitment capacity.

Very interestingly, it has also been shown that, compared with adult peripheral blood, umbilical cord blood contains higher amounts of EPCs that are also characterized by a greater proliferative capacity (Peichev *et al.*, 2000; Murohara *et al.*, 2000), an observation which may suggest new potential clinical applications for umbilical cord blood EPCs.

Factors that influence EPC mobilization have been shown to interfere with tumour growth by regulating tumour angiogenesis. To this extent, the possibility of regulating the mobilization and homing of EPCs to the developing vascular network within neoplastic tissues by using angiostatin that inhibits the contribution of EPCs to angiogenesis by altering the growth of mature endothelial cells (Ito *et al.*, 1999), endostatin that inhibits the mobilization and clonogenic potential of EPCs (Capillo *et al.*, 2003) or synthetic metalloproteinase inhibitors that determine and regulate EPC mobilization from the bone marrow (Heissig *et al.*, 2002) was investigated.

17.6 Conclusions

The level of circulating EPCs has been proposed as a surrogate biological marker for vascular function suggesting that EPCs may have a role not only in tumour development but also in the maintenance and reparative processes, including neovascularization, during tissue and organ ischaemia. As shown in this review, different biological approaches (phenotype, cultures) may represent just some facets of the angiogenetic differentiation process in specific organs, possibly endowed with prognostic features. Further work is, however, needed to better understand EPC biology before the enormous expectations suggested by animal studies and preliminary clinical trials became real therapeutic options.

References

Aicher, A., Brenner, W., Zuhayra, M., Badorff, C. *et al.* (2003) 'Assessment of the tissue distribution of transplanted human endothelial progenitor cells by radioactive labeling', *Circulation*, **107**(16), pp. 2134–2139.

Arbab, A. S., Pandit, S. D., Anderson, S. A., Yocum, G. T. *et al.* (2005) 'MRI and confocal microscopy studies of magnetically labeled endothelial progenitor cells trafficking to sites of tumor angiogenesis', *Stem Cells*, Sept 22, Epub ahead of print.

Asahara, T., Murohara, T., Sullivan, A., Silver, M. *et al.* (1997) 'Isolation of putative progenitor endothelial cells for angiogenesis', *Science*, **275**(5302), pp. 964–967.

Askari, A. T., Unzek, S., Popovic, Z. B., Goldman, C. K. *et al.* (2003) 'Effect of stromal-cell-derived factor 1 on stem-cell homing and tissue regeneration in ischaemic cardiomyopathy', *Lancet*, **362**(9385), pp. 697–703.

Bahlmann, F. H., De Groot, K., Spandau, J. M., Landry, A. L. *et al.* (2004) 'Erythropoietin regulates endothelial progenitor cells', *Blood*, **103**(3), pp. 921–926.

Capillo, M., Mancuso, P., Gobbi, A., Monestiroli, S. *et al.* (2003) 'Continuous infusion of endostatin inhibits differentiation, mobilization, and clonogenic potential of endothelial cell progenitors', *Clin. Cancer Res.*, **9**(1), pp. 377–382.

Carmeliet, P. (2000) 'Mechanisms of angiogenesis and arteriogenesis', *Nature Med.*, **6**(4), pp. 389–395.

Cawthon, R. M., Smith, K. R., O'Brien, E., Sivatchenko, A. and Kerber, R. A. (2003) 'Association between telomere length in blood and mortality in people aged 60 years or older', *Lancet*, **361**(9355), pp. 393–395.

Choi, J. H., Kim, K. L., Huh, W., Kim, B. *et al.* (2004) 'Decreased number and impaired angiogenic function of endothelial progenitor cells in patients with chronic renal failure', *Arteriosclerosis, Thromb. Vasc. Biol.*, **24**(7), pp. 1246–1252.

Dimmeler, S. and Vasa-Nicotera, M. (2003) 'Aging of progenitor cells: limitation for regenerative capacity', *J. Am. Coll. Cardiol.*, **42**(12), pp. 2081–2082.

Edelberg, J. M., Tang, L., Hattori, K., Lyden, D. and Rafii, S. (2002) 'Young adult bone marrow-derived endothelial precursor cells restore aging-impaired cardiac angiogenic function', *Circ. Res.*, **90**(10), pp. E89–93.

Elsheikh, E., Uzunel, M., He, Z., Holgersson, J. *et al.* (2005) 'Only a specific subset of human peripheral-blood monocytes has endothelial-like functional capacity', *Blood*, **106**(7), pp. 2347–2355.

Geiger, H. and Van Zant, G. (2002) 'The aging of lympho-hematopoietic stem cells', *Nature Immunol.*, **3**(4), pp. 329–333.

Gill, M., Dias, S., Hattori, K., Rivera, M. L. *et al.* (2001) 'Vascular trauma induces rapid but transient mobilization of VEGFR2$^+$AC133$^+$ endothelial precursor cells', *Circ. Res.*, **88**(2), pp. 167–174.

Harada, M., Qin, Y., Takano, H., Minamino, T. *et al.* (2005) 'G-CSF prevents cardiac remodeling after myocardial infarction by activating the Jak-Stat pathway in cardiomyocytes', *Nature Med.*, **11**(3), pp. 305–311.

Heeschen, C., Aicher, A., Lehmann, R., Fichtlscherer, S. *et al.* (2003) 'Erythropoietin is a potent physiologic stimulus for endothelial progenitor cell mobilization', *Blood*, **102**(4), pp. 1340–1346.

Heissig, B., Hattori, K., Dias, S., Friedrich, M. *et al.* (2002) 'Recruitment of stem and progenitor cells from the bone marrow niche requires MMP-9 mediated release of kit-ligand', *Cell*, **109**(5), pp. 625–637.

Hill, J. M., Zalos, G., Halcox, J. P. I., Schenke, W. H. *et al.* (2003) 'Circulating endothelial progenitor cells, vascular function and cardiovascular risk', *N. Eng. J. Med.*, **348**(7), pp. 593–600.

Hristov, M., Erl, W. and Weber, P. C. (2003) 'Endothelial progenitor cells. Mobilization, differentiation, and homing', *Arteriosclerosis, Thromb. Vasc. Biol.*, **23**(7), pp. 1185–1189.

Hur, J., Yoon, C. H., Kim, H. S., Choi, J. H. *et al.* (2004) 'Characterization of two types of endothelial progenitor cells and their different contributions to neovasculogenesis', *Arteriosclerosis, Thromb. Vasc. Biol.*, **24**(2), pp. 288–293.

Ingram, D. A., Caplice, N. M, and Yoder, M. C. (2005) 'Unresolved questions, changing definitions, and novel paradigms for defining endothelial progenitor cells', *Blood*, **106**(5), pp. 1525–1531.

Ito, H., Rovira, I. I., Bloom, M. L., Takeda, K. *et al.* (1999) 'Endothelial progenitor cells as putative targets for angiostatin', *Cancer Res.*, **59**(23), pp. 5875–5877.

Iwakura, A., Luedemann, C., Shastry, S., Hanley, A. *et al.* (2003) 'Estrogen-mediated, endothelial nitric oxide synthase-dependent mobilization of bone marrow-derived endothelial progenitor cells contributes to reendothelialization after arterial injury', *Circulation*, **108**(25), pp. 3115–3121.

Jiang, Y., Jahagirdar, B. N., Reinhardt, R. L., Schwartz, R. E. *et al.* (2002) 'Pluripotency of mesenchymal stem cells derived from adult marrow', *Nature*, **418**(6893), pp. 41–49.

Kang, H. J., Kim, H. S., Zhang, S. Y., Park, K. W. *et al.* (2004) 'Effects of intracoronary infusion of peripheral blood stem-cells mobilised with granulocyte-colony stimulating factor on left ventricular systolic function and restenosis after coronary stenting in myocardial infarction: the MAGIC cell randomised clinical trial', *Lancet*, **363**(9411), pp. 751–756.

Khakoo, A. Y. and Finkel, T. (2005) 'Endothelial progenitor cells', *Ann. Rev. Med.*, **56**, pp. 79–101.

Khan, S. S., Solomon, M. A. and McCoy Jr, J. P. (2005) 'Detection of circulating endothelial cells and endothelial progenitor cells by flow cytometry', *Cytometry*, **64B**(1), pp. 1–8.

Kondo, T., Hayashi, M., Takeshita, K., Numaguchi, Y. *et al.* (2004) 'Smoking cessation rapidly increases circulating progenitor cells in peripheral blood in chronic smokers', *Arteriosclerosis, Thromb. Vasc. Biol.*, **24**(8), pp. 1442–1447.

Lambiase, P. D., Edwards, R. J., Anthopoulos, P., Rahman, S. *et al.* (2004) 'Circulating humoral factors and endothelial progenitor cells in patients with differing coronary collateral support', *Circulation*, **109**(24), pp. 2986–2992.

Lin, Y., Weisdorf, D. J., Solovey, and A., Hebbel, R. P. (2000) 'Origins of circulating endothelial cells and endothelial outgrowth from blood', *J. Clin. Invest.*, **105**(1), pp. 71–77.

Lyden, D., Hattori, K., Dias, S., Costa, C. *et al.* (2001) 'Impaired recruitment of bone-marrow-derived endothelial and hematopoietic precursor cells blocks tumor angiogenesis and growth', *Nature Med.*, **7**(11), pp. 1194–1201.

Massa, M., Rosti, V., Ferrario, M., Campanelli, R. *et al.* (2005) 'Increased circulating hematopoietic and endothelial progenitor cells in the early phase of acute myocardial infarction', *Blood*, **105**(1), pp. 199–206.

Mathur, A. and Martin, J. F. (2004) 'Stem cells and repair of heart', *Lancet*, **364**(9429), pp. 183–192.

Murohara, T., Ikeda, H., Duan, J., Shintani, S., *et al.* (2000) 'Transplanted cord blood-derived endothelial progenitor cells augment postnatal neovascularization', *J. Clin. Invest.*, **105**(11), pp. 1527–1536.

Norol, F., Merlet, P., Isnard, R., Sebillon, P. *et al.* (2003) 'Influence of mobilized stem cells on myocardial infarct repair in a nonhuman primate model', *Blood*, **102**(13), pp. 4361–4368.

Orlic, D., Kajstura, J., Chimenti, S., Limana, F. *et al.* (2001) 'Mobilized bone marrow cells repair the infarcted heart, improving function and survival', *Proc. Natl. Acad. Sci. USA*, **98**(18), pp. 10344–10349.

Peichev, M., Naiyer, A. J., Pereira, D., Zhu, Z. *et al.* (2000) 'Expression of VEGFR-2 and AC133 by circulating human CD34$^+$ cells identifies a population of functional endothelial precursors', *Blood*, **95**(3), pp. 952–958.

Peters, B. A., Diaz, L. A., Polyak, K., Meszler, L. *et al.* (2005) 'Contribution of bone marrow-derived endothelial cells to human tumor vasculature', *Nature Med.*, **11**(3), pp. 261–262.

Rafii, S. and Lyden, D. (2003) 'Therapeutic stem and progenitor cell transplantation fro organ vascularization and regeneration', *Nature Med.*, **9**(6), pp. 702–712.

Rauscher, F. M., Goldschmidt-Clermont, P. J., Davis, B. H., Wang, T. *et al.* (2003) 'Aging, progenitor cell exhaustion, and atherosclerosis', *Circulation*, **108**(4), pp. 457–463.

Reyes, M., Dudek, A., Jahagirdar, B., Koodie, L. *et al.* (2002) 'Origin of endothelial progenitors in human postnatal bone marrow', *J. Clin. Invest.*, **109**(3), pp. 337–346.

Rigolin, G. M., Fraulini, C., Ciccone, M., Mauro, E. *et al.* (2006) 'Neoplastic circulating endothelial cells in multiple mieloma with 13q14 deletion', *Blood*, **107**, pp. 2531–2535.

Romagnani, P., Annunziato, F., Liotta, F., Lazzeri, E. *et al.* (2005) 'CD14+CD34low cells with stem cell phenotypic and functional features are the major source of circulating endothelial progenitors', *Circ. Res.*, **97**(4), pp. 314–322.

Scheubel, R. J., Zorn, H., Silber, R. E., Kuss, O. *et al.* (2003) 'Age-dependent depression in circulating endothelial progenitor cells in patients undergoing coronary artery bypass grafting', *J. Am. Coll. Cardiol.*, **42**(12), pp. 2073–2080.

Shintani, S., Murohara, T., Ikeda, H., Ueno, T. *et al.* (2001) 'Mobilization of endothelial progenitor cells in patients with acute myocardial infarction', *Circulation*, **103**(23), pp. 2776–2779.

Staton, C. A., Stribbling, S. M., Tazzyman, S., Hughes, R. *et al.* (2004) 'Current methods for assaying angiogenesis *in vitro* and *in vivo*', *Int. J. Exp. Pathol.*, **85**(5), pp. 233–248.

Stout, R. D. and Suttles, J. (2004) 'Functional plasticity of macrophages: reversible adaptation to changing microenvironments', *J. Leukocyte Biol.*, **76**(3), pp. 509–513.

Streubel, B., Chott, A., Huber, D., Exner, M. *et al.* (2004) 'Lymphoma-specific genetic aberrations in microvascular endothelial cells in B-cell lymphomas', *New England J. Med.*, **351**(3), pp. 250–259.

Takahashi, T., Kalka, C., Masuda, H., Chen, D. *et al.* (1999) 'Ischemia- and cytokine-induced mobilization of bone marrow-derived endothelial progenitor cells for neovascularization', *Nature Med.*, **5**(4), pp. 434–438.

Tepper, O. M., Galiano, R. D., Capla, J. M., Kalka, C. *et al.* (2002) 'Human endothelial cells from type II diabetics exhibit impaired proliferation, adhesion, and incorporation into vascular structures', *Circulation*, **106**(22), pp. 2781–2786.

Tepper, O. M., Capla, J. M., Galiano, R. D., Ceradini, D. J. *et al.* (2005) 'Adult vasculogenesis occurs through in situ recruitment, proliferation, and tubulization of circulating bone marrow-derived cells', *Blood*, **105**(3) 3, pp. 1068–1077.

Urbich, C. and Dimmeler, S. (2004) 'Endothelial progenitor cells: characterization and role in vascular biology', *Circulation Res.*, **95**(4), pp. 343–353.

Valgimigli, M., Rigolin, G. M., Fucili, A., Porta, M. D. *et al.* (2004) 'CD34+ and endothelial progenitor cells in patients with various degrees of congestive heart failure', *Circulation*, **110**(10), pp. 1209–1212.

Valgimigli, M., Rigolin, G. M., Cittanti, C., Malagutti, P. *et al.* (2005) 'Use of granulocyte-colony stimulating factor during acute myocardial infarction to enhance bone marrow stem cell mobilization in humans: clinical and angiographic safety profile', *Eur. Heart J.*, **26**(18), pp. 1838–1845.

Vasa, M., Fichtlscherer, S., Adler, K., Aicher, A. *et al.* (2001a) 'Increase in circulating endothelial progenitor cells by statin therapy in patients with stable coronary artery disease', *Circulation*, **103**(24), pp. 2885–2890.

Vasa, M., Fichtlscherer, S., Aicher, A., Urbich, C. *et al.* (2001b) 'Number and migratory activity of circulating endothelial progenitor cells inversely correlate with risk factors for coronary artery disease', *Circulation Res.*, **89**(1), pp. E1–7.

Yamaguchi, J., Kusano, K. F., Masuo, O., Kawamoto, A. *et al.* (2003) 'Stromal cell-derived factor-1 effects on ex vivo expanded endothelial progenitor cell recruitment for ischaemic neovascularization', *Circulation*, **107**(9), pp. 1322–1328.

18

Methods for monitoring of the anti-angiogenic activity of agents in patients: novel trial design

Shannon Smiley, Michael K. K. Wong and Shaker A. Mousa

Abstract

The field of angiogenesis modulation is at a major crossroad. Tremendous advancement in basic science in this field is providing several novel targets and strategies, which is in contrast to the limited clinical support, resulting in a large gap between experimental and clinical data. This gap might be helped through the selection of clinically relevant surrogate biomarkers that correlate with clinical outcome. Additionally, in the design of clinical trials, hard endpoints (such as mortality, quality of life, and cost-effectiveness of the anti-angiogenesis strategy) in addition to surrogate biomarkers and non-invasive imaging of tumour and its micro-vascular environment should be used. Furthermore, adjunct therapies that maximize the efficacy and minimize complications associated with cancer and its current standard therapy should also be used.

Keywords

angiogenesis; clinical trial; anti-angiogenesis; cancer; vascular disorder; imaging; biomarkers; adjunct therapy

18.1   Introduction

Angiogenesis, the process of new blood vessel formation, is considered critical to the growth of tumours. Judah Folkman (1971) first proposed the idea that tumour angiogenesis could serve as a potential target for anticancer therapy. Numerous studies (Folkman, 1989; Moehler *et al.*, 2003) have established that angiogenesis is

Angiogenesis Assays   Edited by Carolyn A. Staton, Claire Lewis and Roy Bicknell a necessary component for solid tumour growth and may also contribute to haematologic malignancies. This has led to the development and clinical testing of numerous novel anti-angiogenic agents with potentially broad applicability for different tumour types.

Angiogenesis inhibition represents a new paradigm in the treatment of cancer. Unlike traditional chemotherapy, angiogenesis inhibitors specifically target blood vessels. In addition, angiogenesis inhibitors have the potential to treat a wide variety of tumours because the requirement for neovascularization is nearly universal across different tumour types (Collins and Hurwitz, 2005). Numerous vascular endothelial growth factor (VEGF)-based and non-VEGF-based anti-angiogenic therapies are being evaluated in clinical trials, including high-grade glioma, renal cell carcinoma, lung cancer, metastatic colorectal cancer, Kaposi's sarcoma, multiple myeloma, Waldenstrom's macroglobulinemia, and agnogenic myeloid metaplasia.

Anti-angiogenesis trials are trials that test molecules that specifically inhibit newly formed vessels. At one time this was taken to mean those molecules that target a specific endothelial pathway, for example endostatin or angiostatin. This has since been broadened to include strategies that have a higher therapeutic efficacy against the vascular compartment compared with other host tissues. Examples of this include antibodies against VEGF, cytotoxic drugs to which endothelial cells may be more sensitive (such as Combretastatin), or different administration schedules of cytotoxic drugs (such as paclitaxel) designed to preferentially expose the vascular compartment.

Angiogenesis is controlled by both positive and negative influences (Carmeliet, 2003; Ferrara and Alitalo, 1999; Gale and Yancopoulos, 1999; Kyriakides et al., 1999; O'Reilly, 1997; O'Reilly et al., 1997; Panetti et al., 1997; Streit et al., 1999; Volpert et al., 1995). The existence of these two classes of regulatory molecules has led to two separate strategies for inhibiting pathological angiogenesis. One strategy is the inhibition of positively acting agents (for example, small molecule VEGF receptor inhibitors) and antibodies directed against VEGF, its receptor, or endothelial cell surface proteins (Yan et al., 2003). A second strategy is the exogenous administration of agents that possess endogenous negative regulatory influence on angiogenesis (e.g. angiostatin, endostatin, fumagillin [TNP-470], matrix metalloproteinase inhibitors) (Hidalgo and Eckhardt, 2001). Promising results have been obtained in using monoclonal anti-VEGF antibodies in colorectal and renal cell cancer, providing rigorous proof of concept in randomized trials.

There are a multitude of anti-angiogenic strategies. It is not possible to discuss all these in detail; however, some strategies have had a significant impact on the development of this field. Prototypical molecules are discussed to illustrate each of these strategies.

Bevacizumab is a humanized monoclonal antibody directed against VEGF that recognizes isoforms of VEGF-A. It has an extremely long half-life of 17–21 days after intravenous (i.v.) infusion. Bevacizumab has been evaluated in several phase II and phase III studies in solid tumours (such as colorectal cancer, advanced non-small cell lung cancer, renal cell cancer, and breast cancer), as well as in relapsed and resistant acute myeloid leukemia. It was approved by the USFDA on February 26, 2004, for use in combination with i.v. 5-FU for first-line treatment of metastatic colorectal cancer. No molecular markers are known to predict efficacy.

Endostatin is the 20 kD C-terminal fragment of collagen XVIII and is thus a naturally occurring non-VEGF-based angiogenesis inhibitor. The preclinical data on endostatin showed that it appeared to possess all the desired attributes of a pure anti-angiogenic agent. It is potent and specific for tumour vasculature; tumour 'resistance' did not occur and there is a striking lack of toxicity. Thus it was not surprising that endostatin appeared on the clinical scene with great fanfare and expectations (Beecken *et al.*, 2001; Katzenstein *et al.*, 2001; O'Reilly, 1997; O'Reilly *et al.*, 1997; Prox *et al.*, 2003; Zhang *et al.*, 2002).

Eder *et al.* (2002) administered human recombinant endostatin (rh-Endo) as a 20 min intravenous infusion to 15 patients with advanced and treatment-refractory solid tumours. The dose was escalated from 15 to 240 mg/m^2 in successive cohorts of patients. No clinically significant treatment-related toxicities were seen; however, there were no objective responses as measured by conventional criteria. The use of the 20 min daily infusion meant that the systemic exposure to endostatin at the 240 mg/m^2 dose was 50 per cent lower than that provided by the dosing regimen that afforded maximum growth inhibition in preclinical studies (Daly *et al.*, 2003). Although endostatin's clinical activity did not live up to its promise, the development and testing of this drug broke new ground in trial design and the assessment of clinical efficacy for anti-angiogenic drugs. Interest in endostatin has waned considerably; however, it remains the prototype of the 'pure' anti-angiogenic drug.

Thalidomide possesses some of the most significant clinical activity of the non-VEGF-based angiogenesis inhibitors. It exhibits clinical activity in several malignant disorders that have been linked to abnormal angiogenesis, including refractory myeloma, idiopathic myeloid metaplasia and Kaposi's sarcoma. In addition to its anti-angiogenic properties, thalidomide also exhibits potent immunomodulatory effects. It is unclear whether the efficacy of thalidomide in these disorders is derived solely from inhibition of tumour vasculature because the anti-angiogenic properties remain undefined (Raje and Anderson, 1999).

18.2 Tumour micro-environmental factors

Haemostasis and angiogenesis

In recent years it has become apparent that the haemostatic system has a role to play in regulating angiogenesis and tumour biology. This is achieved by components of the coagulation and fibrinolytic systems regulating proteolysis in the milieu and context of the extracellular matrix and the fibrin clot. There is also an emerging paradigm suggesting that cryptic fragments released by proteolysis from components of the coagulation and fibrinolytic systems have a role to play in regulating angiogenesis.

Fibrin and angiogenesis

Documentation shows that capillary sprouting results in the dissolution of fibrin clots (fibrinolysis) at the tips of these sprouts (Clark and Clark, 1939). As capillary sprouts mature, the fibrin is replaced by other components of ECM. Thus, fibrin provides a temporary matrix scaffold along which endothelial cells can migrate and has a fundamental role to play in angiogenesis. Fibrinogen, the precursor of fibrin, is a large glycoprotein found in the blood plasma of all vertebrates. Thrombin is the main enzyme involved in generating insoluble fibrin from its soluble precursor fibrinogen. The central enzyme in fibrin proteolysis is plasmin. The activity of thrombin and plasmin are regulated by a variety of molecules.

The coagulation cascade is initiated by 1) either the intrinsic (contact factor) pathway, or 2) the extrinsic (tissue factor) pathway. The main pathway for initiation of the coagulation cascade is the extrinsic pathway. The intrinsic pathway serves to augment or amplify the coagulation cascade. Cleavage of HK by plasma kallikrein results in the generation of a two-chain form of HK (HKa). The light chain of HKa contains two domains, D5 and D6. One of these domains, D5, binds to anionic surfaces, including heparin and phospholipids as well as zinc (Colman *et al.*, 2000).

Intrinsic system (contact activation system) and angiogenesis

Two-chain high molecular weight kininogen (HKa) is reported to bind to endothelial cells via the urokinase receptor (Colman *et al.*, 2000), and this interaction has been proposed as the basis for the anti-endothelial effect of HKa (Colman *et al.*, 2000; Colman *et al.*, 2003). Recent experiments demonstrate that D5, derived from human kininogen, can inhibit angiogenesis (Colman *et al.*, 2003). Thus, the protein components involved in the initiation of the intrinsic pathway are also involved in the regulation of angiogenesis.

HKa D5 (kininostatin) has been shown to inhibit endothelial cell migration toward vitronectin, endothelial cell proliferation and angiogenesis in the chicken chorioallantoic membrane assay (Colman *et al.*, 2000). Structure–function studies demonstrate that D5 contains a peptide that inhibits only migration, as well as another peptide that inhibits only proliferation (Colman *et al.*, 2000). Data on a related inhibitory monoclonal antibody confirmed the role of kininogen in angiogenesis modulation (Colman *et al.*, 2003).

Extrinsic system (tissue factor/factor VIIa) and angiogenesis

Tissue factor is a membrane-bound protein that is found wherever it may play a protective role, initiating coagulation upon blood vessel damage. Upon exposure to blood, tissue factor activates factor VII to factor VIIa (with the aid of calcium and phospholipids), and this complex in turn activates factors IX and X. Tissue factor is found on the surfaces of tumour cells, as well as on tumour-associated macrophages and tumour-associated endothelium (Rickles *et al.*, 2001). In fact, some would argue that tissue factor expression is a hallmark of cancer progression and metastases that are mediated by pathways independent of blood coagulation (Bromberg *et al.*, 1995). The importance of tissue factor in tumour angiogenesis is highlighted by the observation that dramatic reductions in experimental tumour angiogenesis can be induced by inhibiting tissue factor activity via the use of specific antibodies (Fernandez and Rickles, 2002). Furthermore, experiments involving the transfection of low-VEGF-producing tumour cells with full-length clones of tissue factor restore VEGF production in these cells, whereas transfection of these same cells with a clone of tissue factor lacking the cytoplasmic serine residues does not restore VEGF production. These results suggest that the cytoplasmic tail of tissue factor is, in some way, required for full expression of VEGF (Abe *et al.*, 1999). Thus tissue factor has a role to play in tumour angiogenesis as well as in coagulation.

Tissue factor pathway inhibitor

Tissue factor pathway inhibitor (TFPI) is a multi-domain protein with three Kunitz-type proteinase inhibitor domains. Recent reports suggest that TFPI can inhibit the proliferation of basic fibroblast growth factor (bFGF)-driven endothelial cells (Hembrough *et al.*, 2001). Thus, the entire cascade leading to coagulation and fibrinolysis contains proteins that regulate angiogenesis (Pirie-Shepherd, 2003; Rickles *et al.*, 2001; Thompson *et al.*, 1991). In some cases this is simply by leading to the deposition or degradation of fibrin, which is inherently angiogenic due to its ability to sequester growth factors and guide the migration of endothelial cells. Other proteins appear to have direct effects on endothelial cells, either in their native form or after proteolysis leads to the exposure of cryptic domains with pro- or anti-angiogenic activity.

It is of particular interest that many of the cryptic fragments are or can be generated by proteinases that already have a function in generating the active form of a given protein. It appears that many proteins have a more active role than we have discovered thus far. For example, plasminogen is activated to plasmin, which can be inactivated by further proteolysis, such as the proteolytic removal of microplasmin from the cell surface. This reaction leads to the rapid inhibition of microplasmin by alpha 2 antiplasmin and results in the generation of angiostatin as a byproduct (Eriksson *et al.*, 2003). Maximal angiogenic activity appeared to be associated with proteolysis events. Thus, a healing wound offers a clear example of the gross coordination of these events. Given the rapid discovery of anti-angiogenic fragments of proteins within the coagulation cascade in recent years, we should expect to find more fragments and domains in the near future. Not only will this allow us to expand our arsenal of anti-angiogenic drugs, but it will also allow us to understand much more about the process of wound healing and the role that coagulation and fibrinolysis play in angiogenesis.

The role of the coagulation system in angiogenesis

The processes of blood coagulation and the generation of new blood vessels both play crucial roles in wound healing. Platelets, for example, are the first line of defence during vascular injury and contain at least a dozen promoters of angiogenesis, which they may be induced to secrete into the surrounding vasculature upon activation by thrombin. Targeting both the coagulation and angiogenesis pathways may provide more potent anti-tumour effect than targeting either pathway alone. Elucidation of the tissue factor signalling pathway using tumour cells as a model system should provide new insights into the cellular biology of tissue factor that might be applied to signalling in endothelial cells, smooth muscle cells, and fibroblasts. Additionally, because new classes of anticoagulant molecules have been developed over the past several years that selectively target tissue factor and/or tissue factor-VIIa complex, an understanding of this pathway might provide the rational basis for the development of new agents to prevent and/ or reduce angiogenesis-related disorders, tumour-associated thrombosis, and the positive feedback loop between thrombosis and cancer (Mousa, 2002).

Activation of the blood coagulation system stimulates the growth and dissemination of cancer cells through multiple mechanisms, and anticoagulant drugs inhibit the progression of certain cancers. Laboratory data on the effects of anticoagulants in various tumours suggest that this treatment approach has considerable potential in some cancers but not others. For example, renal cell cancer is one of a small number of human tumour types in which the tumour cell contains an intact coagulation pathway, leading to thrombin generation and conversion of fibrinogen to fibrin immediately adjacent to viable tumour cells (Wojtukiewicz *et al.*, 1990). Similar observations have been made in melanoma, ovarian, and small cell lung cancers but not in breast, colorectal, and non-small cell lung cancers (Zacharski *et al.*, 1992). This is of considerable relevance to the finding that growth of melanoma and small cell lung cancer is inhibited by anticoagulants, but that no such effect has been observed in those other tumour types.

Coagulation and cancer

The association between coagulation system activation and systemic thrombosis in human cancers has been recognized for over a century since Trousseau's original description of migratory thrombophlebitis complicating gastrointestinal malignancy (Trousseau, 1865). An improved appreciation in recent years of the interdependency of the coagulation system and malignant behaviour has led to an understanding of how an activated coagulation system in turn may enhance cancer cell growth (Mousa, 2002; Zacharski and Meehan, 1993). While this does not establish causality or even a biological association, it is of interest that a recent Danish study showed that patients with cancer who developed venous thrombosis during the course of their disease had significantly shorter cancer-related survival than similar patients who remained thrombosis-free (Sorensen *et al.*, 2000). In addition, several studies (including randomised clinical trials) have documented improved cancer-related survival in patients treated with anticoagulants compared with those not receiving anticoagulants (Lebeau *et al.*, 1994; Thornes, 1983; Zacharski *et al.*, 1984).

Platelets and angiogenesis

Platelets are involved in the promotion of angiogenesis due to the release of several pro-angiogenesis growth factors upon platelet interaction with the cancer cell and its activation. Additionally, platelets facilitate fibrin-tumour binding and metastasis. These data suggest the potential of antiplatelet and perhaps antiplatelet plus anticoagulant as an adjunct with the standard cancer therapies (El-Naggar and Mousa, 2005; Mousa, 2005).

Anticoagulants and angiogenesis

Many cancer patients reportedly have a hypercoagulable state, with recurrent thrombosis due to the impact of cancer cells and chemotherapy on the coagulation cascade (Mousa, 2002). Analysis of biomarkers of the coagulation cascade and of vessel wall activation was performed and showed significant increases in thrombin generation and endothelial cell perturbation in a treatment cycle-dependent manner when combining angiogenesis inhibitors and chemotherapeutic agents (Bennett *et al.*, 2002; Kuenen *et al.*, 2002). The incidence of thromboembolic events, possibly related to the particular regimen tested in this study (SU015 and chemotherapeutic agents), discourages further investigation of this regimen (Kuenen *et al.*, 2002). This investigation, along with the increased incidences of deep vein thrombosis in multiple myeloma patients receiving thalidomide and che- motherapeutic agents, suggests the potential advantages of using an anticoagulant such as heparin or low molecular weight heparin (LMWH). Studies have also demonstrated that unfractionated heparin (UFH) or LMWH interfere with processes involved in tumour growth and metastasis. Clinical trials have indicated a clinically relevant effect of LMWH, as compared with UFH, on the survival of cancer patients with deep vein thrombosis. Mechanism and efficacy of the LMWH and its *in vivo* releasable TFPI on the regulation of angiogenesis and tumour growth was documented (Mousa, 2002; Mousa and Mohamed, 1999). Heparin, steroids, and heparin/steroid combinations have been used in a variety of *in vitro* models and *in vivo* in animal models as effective inhibitors of angiogenesis (Jung *et al.*,

2001; Mousa and Mohamed, 1999). Additionally, platelet-tumour cell interactions could play a significant role in tumour metastasis (McCarty *et al.*, 2000).

Anti-angiogenic agents and thrombosis

Anti-angiogenic agents might mediate endothelial cell dysfunction, which could be associated with increased incidence of thrombosis (Bennett *et al.*, 2002; Kaushal *et al.*, 2002; Kuenen *et al.*, 2002; Tosi and Tura, 2001). The effect of anti-angiogenic mechanisms on the haemostatic system needs further exploration.

18.3   Possible mechanisms of acquired resistance to anti-angiogenic drugs

The ultimate target of anti-angiogenic drugs is the genetically stable, activated endothelial cell of a newly forming tumour blood vessel, rather than the genetically unstable tumour cell population per se. This leads to the notion that acquired resistance to such drugs may not develop as readily, if at all. While there is some evidence that this lack of resistance development may be the case for some direct-acting angiogenesis inhibitors, it is becoming apparent that resistance can develop over time to many types of angiogenesis inhibitors, possibly including some direct inhibitors, especially when used as monotherapies (Wojtukiewicz *et al.*, 1990). Possible mechanisms for such acquired or induced resistance include the following: 1) redundancy of pro-angiogenic growth factors when the drug used targets a single such growth factor or its cognate endothelial cell-associated receptor tyrosine kinase; 2) the anti-apoptotic/pro-survival function of growth factors such as VEGF, which, in high local concentrations, can antagonize the pro-apoptotic effects of various angiogenesis inhibitors; 3) epigenetic, transient upregulation, or induction of various anti-apoptotic effects or molecules in host endothelial cells; and 4) heterogeneous vascular dependence of tumour cell populations. It is suggested that long-term disease control with anti-angiogenic drugs can be best achieved by judicious combination therapy. In this regard, the great molecular diversity of anti-angiogenic drug targets, in contrast to chemotherapy, makes this a particularly attractive therapeutic option, especially when used in combinations (Kerbel *et al.*, 2001).

18.4   Standard chemotherapy versus angiogenesis inhibitors

Several differences between standard chemotherapy and anti-angiogenesis therapy result from the fact that angiogenesis inhibitors target dividing endothelial cells rather than tumour cells. Anti-angiogenic drugs are not likely to cause bone marrow suppression, gastrointestinal symptoms, or hair loss. Additionally, because anti-angiogenic drugs may not necessarily kill tumours, but rather hold them in check indefinitely, the endpoint of early clinical trials may be different than for standard therapies. Rather than looking only for tumour response, it may be appropriate to evaluate increases in survival and/or time to disease progression. Drug resistance is a major problem with chemotherapy agents; this occurs because most cancer cells are genetically unstable, are more prone to mutations, and are therefore likely to produce drug-resistant cells. Because anti-angiogenic drugs target normal endothelial cells, which are not genetically unstable, drug resistances may not develop. So far, resistance has not been a major problem in long-term animal studies or in clinical trials. Finally, anti-angiogenic therapy may prove useful in combination with therapy directly aimed at tumour cells. Because each therapy is aimed at a different cellular target, the hope is that the combination will prove more effective.

18.5 Monitoring clinical trials

Traditional/conventional measures of clinical efficacy

Phase I trials are designed to determine the appropriate dose and schedule for further evaluation and describe the pharmacological behaviour of the drug and its toxic effects. The goal of phase II trials is to screen agents for evidence of anti-tumour activity. Anti-angiogenesis trials present unique challenges to these traditional paradigms. One of the areas of greatest activity in the design of such trials concerns the means to measure the study's endpoint. In other words, how do you know your drug has anti-tumour activity?

The tradition of utilizing toxicity as the phase I endpoint is most useful for cytotoxic drug studies because there is a direct relationship between the dose of a drug and its effects on both tumour and toxicity. This means that the propensity and severity of the drug's toxicity correlates directly with drug dose. Thus, the dose-limiting toxicity is the primary determinant of the dose that is chosen for further study in subsequent trials because it represents the highest possible dose.

The goal of phase II trials is to screen agents for their potential efficacy. Traditionally this has been measured by objective tumour regression, as described using standard criteria (e.g. World Health Organization). Tumour shrinkage has proven to be useful as a phase II endpoint because it has allowed us to select drugs earlier in the drug development process. It is important to point out that response per se is not synonymous with efficacy because the gold standard remains: improved cure rates, survival, or quality of life.

Novel trial design and evaluation

Anti-angiogenic therapeutic trials have successfully challenged traditional paradigms of success. One of the major reasons for this has been the contention that agents that possess high selectivity for a specific target (in this case, the tumour vasculature) will, by definition, possess little cross-over toxicity to normal host tissue. This concept has fuelled much of the excitement and enthusiasm for such drugs. Indeed, the first generation of anti-angiogenesis trials suggests that this prediction may hold true. For example, over a dose range from 100 to 1000 mg/day, thalidomide possesses almost none of the grade III and IV (NCI toxicity grading system) toxicity that is commonly seen with cytotoxic drugs. Perhaps the most striking example is endostatin. As described above, there were very few grade II or higher toxicities with this drug over a log-range of doses. Bevacizumab (Avastin®), when administered with 5-FU, was recently FDA approved for the treatment of metastatic colon cancer. The sum of the information from these and other Bevacizumab clinical trials show a striking absence of the severe toxicities commonly seen in cytotoxic clinical trials. One practical implication of this is that selection of the subsequent phase II or III drug dose based on the observed phase I toxicities may be very difficult.

One of the novel features of anti-angiogenesis clinical trials is that this lack of toxicity makes these drugs ideal companions for combining with other therapeutic modalities. Thus, combinations with other cytotoxic drugs become possible, and indeed it is the combination of Bevacizumab with 5-FU that was proven to provide significant increased longevity in colon cancer patients. Combination with other biological response modifiers is also possible (for example, in combination with interferon). This was done in melanoma with endostatin and in renal cell carcinoma with Bevacizumab. Combinations with radiation therapy have also been carried out. New combinations become possible in ways that have not been traditionally possible with cytotoxic drugs.

Traditional measures of therapeutic efficacy have come under scrutiny as it becomes increasingly obvious that anti-angiogenic drugs may be effective in ways that would not trigger tumour regression. There is a solid body of experimental evidence that successful inhibition of tumour angiogenesis results in a state of tumour dormancy. This means that a new state of tumour–host equilibrium is induced that prevents further tumour growth and metastasis, but frank tumour regression or shrinkage does not occur. Thus traditional radiological imaging would not show any tumour change; in the current evaluation systems, this would not be scored as a success, or, worse, it would be scored as a therapeutic failure.

18.6 Tumour dormancy and tumour progression

There are early hints from angiogenesis clinical trials that a state of dormancy may indeed occur. In addition, new methods of imaging tumours show that blood flow and tumour activity decrease as anti-angiogenesis therapies take hold. This has led to re-examining how success is measured in clinical trials and to devising new measures to capture this information. One way is to directly measure the impact of anti-angiogenesis therapy on tumour blood flow. Imaging methods such as dynamic MRI imaging and PET scanning are particularly useful if serially carried out over the course of therapy.

Success in a trial is being redefined. The category of 'stable disease' as radiologically measured has traditionally been associated with either therapeutic inactivity or failure. Because tumour dormancy may manifest as 'stable disease' and because the length of time a tumour is held dormant is an important measure of anti-angiogenic efficacy, then parameters such as stable disease, minor response, and time to tumour progression (TTP) are now part of the evaluation process in therapeutic trials. In addition, molecular surrogate markers of efficacy become even more important in this setting.

What this means for trial design is that the definition of tumour progression has now been tightened to mean definitive increase in tumour size and no longer includes stable disease. The accuracy of TTP is directly tied to the interval between efficacy assessments. Therefore, the difference between success and failure may rest on when exactly a patient's tumour increased in size. Thus, many anti-angiogenesis trials have begun to shorten the interval between tumour assessments to no more than 8 weeks and sometimes even less.

18.7   Tumour markers

Tumour markers have the advantage that they can be molecularly characterized and easily quantified. A substantial number of clinical trials using anti-angiogenic therapies are ongoing worldwide. How to achieve maximum benefit from these therapies and how to monitor patient responses are of paramount concern to investigators. There are currently no markers of net angiogenic activity of a tumour available to aid investigators in the design of anti-angiogenic treatment schemes. No marker used alone has been validated as a phase II endpoint (Eisenhauer, 1998).

Quantification of various aspects of tumour vasculature might provide an indication of angiogenic activity. Angiogenesis can be assessed directly, by counting new blood vessels, or indirectly, through the measurement of putative angiogenic factors and their receptors. Selected examples are discussed below.

18.8   Microvessel density

One often-quantified aspect of tumour vasculature is microvessel density (MVD). Studies over the last decade have demonstrated the value of using tumour MVD as a prognostic indicator for a wide range of cancers (Gasparini and Harris, 1999). MVD is the measure of the number of vessels per high-power (microscope) field and, as such, reflects the number of capillaries and intercapillary distance. The central concept is that this would decrease as anti-angiogenesis therapy takes hold, but there are several important caveats to this.

MVD can vary widely. Angiogenesis is a focal event, and microvascular 'hotspots' occur in tumours (Weidner *et al.*, 1992). Thus, it is not surprising that the minute sampling cores obtained even with the widest biopsy needles in clinical use can miss these hot spots. In addition, there is a significant degree of inter-observer variability, further confounding these issues.

The utility of MVD in clinical trials is variable. A higher MVD predicts for more extensive cancer stage at diagnosis and a worse prognosis in cancers of the lung, breast, and colon (Takebayashi *et al.*, 1996). The lesson from thalidomide in myeloma patients is that the propensity of a patient to respond to therapy does not correlate with pre-treatment MVD and that responders did not show a concomitant decrease in MVD. In contrast, the colon cancer data from Bevacizumab suggests that MVD predicted for response. Current knowledge supports the contention that clinical measurements of MVD by itself does not accurately and reliably reflect angiogenic activity or angiogenic dependence of a tumour and, so far, has not been shown to be a robust measure to guide or evaluate anti-angiogenic treatment (Hlatky *et al.*, 2002). Trial design has shifted to the discovery and validation of blood markers that can supplement or supplant MVD.

18.9 Surrogate endpoints

Survival remains the gold standard for therapeutic clinical trials in cancer. However, this often does not become evident until enough events have occurred, and this eventuality can take years to happen. Indeed, in adjuvant clinical trials, where the event rate is low, survival data may not achieve robust significance for close to a decade (Fisher, 1999). Because of this long timeline, other means have evolved to evaluate drug efficacy.

Surrogate endpoints are loosely referred to as endpoints that can be used in lieu of traditional endpoints in the evaluation of experimental treatments or other interventions. The need to evaluate treatment benefits as fast as possible on easily measured endpoints has always been a preoccupation in clinical research. Whereas dose selection is still primarily guided by pharmacokinetics and safety considera-tions, several early phase clinical trials of angiogenesis inhibitors have tried to incorporate biomarkers to help define optimal dosing.

One surrogate endpoint is to assess directly the intended drug target for the intended effects and to project forward to the hypothesis that hitting the target will correlate with the intended clinical response. For example, in trials in which the vascular endothelial growth factor receptor (VEGFR) is being targeted, the tumour would be sampled to determine if VEGFR was indeed being obliterated.

In angiogenesis trials, the most direct measure is to assess the degree to which tumour angiogenesis changes. This is usually reflected in the MVD, as discussed above. This requires that serial biopsies be obtained in the patient undergoing treatment. Although this may be easily accessible in superficial tumour deposits that may occur with certain malignancies such as melanoma, sampling deep visceral tumours may be more problematic and riskier for the patient. In trials where serial sampling is at the discretion of medical judgement, there is often an over-representation of superficial tumours. This would not matter if the biology of metastases was site-independent; however, there is molecular evidence that deep visceral metastatic deposits possess biology distinctive from its more superficial counterpart.

Investigators have examined the association between angiogenic factors that are either produced or shed by the tumour or have examined ancillary proteins induced as a result of angiogenesis. Angiogenic factors are an example of the former. The most commonly studied angiogenic factors are bFGF and VEGF. VEGF mRNA and protein are markedly upregulated in the vast majority of human tumours; in some, VEGF over-expression is associated with a poor prognosis and reduced survival (Duff et al., 2003; Paley et al., 1997). In addition, there is evidence that elevated bFGF levels correlate with more aggressive cancers and translate into decreased overall survival (Nguyen et al., 1994).

These angiogenic factors are measured in the fluids of the body, including blood, urine, cerebral spinal fluid, and pleural effusions. Because these are more readily accessible than tumour biopsies, the evaluation of these surrogate markers was commonly carried out in angiogenesis-based clinical trials.

However, the evaluation of VEGF and FGF levels in this setting is fraught with difficulties. One major issue is that there is wide intra- and inter-patient variability in these levels. This is likely a consequence of multiple factors, including tissue and blood sequestration, affinity for heparin-like molecules and other matrix molecules, and the fact that blood may not accurately reflect the tumour micro-environment. Thus it is not surprising that angiogenic factor levels have not been found to track faithfully with tumour response. To date, all biomarkers for this field have limitations and none has been validated (Collins and Hurwitz, 2005).

Coagulation factors as surrogate markers

Interestingly, elements that have traditionally been associated with the coagulation pathway have emerged as promising biomarkers. For example, plasminogen activator inhibitor-1 (PAI-1) is a member of the serine protease inhibitor family that can modulate tumour angiogenesis in experimental systems (McMahon et al., 2001; Stefansson et al., 2003). PAI-1 has emerged as a marker of poor prognosis in breast cancer (Duffy, 2002). The reason for this remains unclear, but presumably it relates to tissue turnover at the tumour–host interface.

As discussed earlier in this chapter, increased pro-thrombotic states, as well as decreased natural anticoagulant factors, are common in cancer patients with different tumour types and at different stages (Mousa, 2005). Haemostasis activation, increased pro-inflammatory biomarkers, and pro-angiogenesis biomarkers in cancer patients have been shown to correlate with the clinical outcome (Alizadeh et al., 2005a; Alizadeh et al., 2005b; Alizadeh et al., in press. Mousa, 2005).

Circulating endothelial cells

Normal vasculature is a highly organized structure that possesses a very low mitotic index. In contradistinction, angiogenic vessels are disorganized, poorly functional, and actively growing. This chaotic state results in the shedding of vascular endothelial cells into the circulation; thus capturing and quantifying these cells will serve as an index of endothelial activity and, by inference, of the angiogenic effect of these drugs.

18.10   Molecular imaging

Dynamic contrast enhanced MRI (DCE-MRI) has been increasingly used to evaluate novel anti-angiogenic agents, both preclinically and in phase I/II trials. DCE-MRI can be a pharmacodynamic indicator of biologic therapy for anti-angiogenic agents by demonstrating changes in tumour vasculature through parameters reflecting tumour perfusion and permeability. For example, SU5416 is an i.v. inhibitor of the tyrosine kinase of Flk-1 that was administered to patients with metastatic melanoma involving the liver. In this study, DCE-MRI was able to detect a statistically significant decrease in the ratio of peak contrast uptake in the tumour versus normal liver at 8 weeks when compared with pre-treatment. In addition, the changes in tumour permeability and vascularity seen with DCE-MRI have also been shown to mirror clinical response (Morgan et al., 2003; Peterson et al., 2004).

18.11   New insights into trial design

One of the ways in which trial design has evolved to adapt to the advent of biological therapy and, in particular, angiogenesis therapy is to move to a rapid titration trial design. This is designed to escalate the dose to rapidly the maximum planned dose. The rationale for this is that agents that are not known to exert dose-related toxicities (unlike the dose-related organ dysfunction caused by conventional cytotoxic chemotherapy agents) and that are specifically targeted to a biological molecule (i.e. biological therapy such as endogenous proteins like endostatin or antibodies) can be given at the maximum planned dose with a minimum of toxicity. This contention is supported by the ever-growing data from antibody-based therapies currently under review or approved for use in cancer. Studies on antibody-based therapeutic agents such as Herceptin (target: Her2/neu), Rituxan (target: CD20), Erbitux (target: EGFR) and most recently Bevacizumab (target: VEGF) are notable for the lack of acute, dose-limiting toxicity. These studies also show a dose-related therapeutic effect, with higher efficacy generally at the higher doses. Likewise, a maximum tolerated dose was not established in treatments using the endogenous anti-angiogenic molecule, endostatin, even when dosing schemas spanned several logs in dose.

There is also growing concern that traditional phase I designs doom the first patients that come onto trial to receive sub-therapeutic doses of the agent under study and furthermore locks them at these dose levels. In addition, the stepwise escalation scheme over-stacks patients at the lower dose levels such that the majority of patients on trial are treated at the lower doses.

One solution is to use a rapid titration clinical trial design (Simon *et al.*, 1997). This is a design that allows for the rapid escalation of the drug dose. The danger in a rapid titration design is that toxicities may begin to appear at the higher doses and that it will instead stack patients up at these higher doses. At Roswell Park Cancer Institute, we have developed a modified titration trial design that balances these concerns by using a rapid titration design to escalate an anti-angiogenic drug in the *same patient*, so long as there is no grade 2 or greater drug-related toxicities. Subsequent patients will enter onto the study at the next higher dose level. If significant toxicity is encountered, then that patient and two others will be treated at the dose level where the toxicity occurred. Entry of patients onto the study is staggered so that the next patient will only go onto treatment if the patient ahead of him/her remains free of toxicity. Our mathematical modelling shows that such a design retains the statistical power to detect significant toxicities while reducing the number of patients required.

18.12 Concluding remarks

The unique attributes of anti-angiogenic agents have challenged the traditional design and evaluation of clinical trials. Time to tumour progression and stable disease are becoming important parameters of efficacy. Surrogate markers of efficacy become increasingly important in this arena. Tissue measures of MVD remain the most direct measure of an anti-angiogenesis effect. Plasma factors such as VEGF and FGF, as well as newer measures (such as PAI-1, coagulation factors, and tissue factor), are currently under study. This area is evolving rapidly as newer and more effective anti-angiogenesis agents with multiple targets and mechanisms. Additionally, the use of adjunct therapies that maximize the efficacy of standard cancer therapy and minimize cancer-associated complications would be introduced in upcoming trials.

References

Abe, K., Shoji, M., Chen, J. Bierhaus, A. *et al.* (1999) 'Regulation of vascular endothelial growth factor production and angiogenesis by the cytoplasmic tail of tissue factor', *Proc. Natl. Acad. Sci. USA.*, **96**(15), pp. 8663–8668.

Alizadeh, H., Al-Tajer, S. and Mousa, S. A. (2005a) 'Gastric, colorectal, and pancreatic carcinoma: the relationship between hemostasis and cancer prognostic markers', *Int. J. Cancer Prevention*, **2**, pp.157–168

Alizadeh, H., Al-Tajer, S. and Mousa, S. A (2005b), 'Hemostatic state in female patients with breast and ovarian cancer', *Int. J. Cancer Prevention*, **2**, pp. 77–86.

Alizadeh, H., Al-Tajer, S. and Mousa, S. A. (in press), 'Hemostatic state in lung cancer patients: pilot study', *Int. J. Cancer Prevention*.

Beecken, W. D., Fernandez, A., Joussen, A. M. *et al.* (2001) 'Effect of antiangiogenic therapy on slowly growing, poorly vascularized tumors in mice', *J. Natl. Cancer Inst.*, **93**(5), pp. 382–387.

Bennett, C. L., Schumock, G. T., Desai, A. A. *et al.* (2002) 'Thalidomide-associated deep vein thrombosis and pulmonary embolism', *Am J. Med.*, **113**(7), pp. 603–606.

Bromberg, M. E., Konigsberg, W. H. and Madison, J. F. (1995) 'Tissue factor promotes metastasis by a pathway independent of blood coagulation', *Proc. Natl. Acad. Sci. USA*, **92**, pp. 8205–8209.

Carmeliet, P. (2003) 'Angiogenesis in health and disease', *Nature Med.*, **9**(6), pp. 653–660.

Clark, E. R. and Clark, E. L. (1939) 'Microscopic observations on the growth of blood capillaries in the living', *Am. J. Anat.*, **64**, pp. 251–301.

Collins, T. S. and Hurwitz, H. I. (2005) 'Targeting vascular endothelial growth factor and angiogenesis for the treatment of colorectal cancer', *Sem. Oncol.*, **32**(1), pp. 61–68.

Colman, R. W., Jameson, B. A., Lin, Y. *et al.* (2000) 'Domain 5 of high molecular weight kininogen (kininostatin) down-regulates endothelial cell proliferation and migration and inhibits angiogenesis', *Blood* **95**, pp. 543–550.

Colman, R. W., Pixley, R. A., Sainz, I. M., Song, J. S. *et al.* (2003) 'Inhibition of angiogenesis by antibody blocking the action of proangiogenic high-molecular-weight kininogen', *J. Thrombosis Haemostasis*, **1**(1), pp. 164–173.

Daly, M. E., Makris, A., Reed, M. and Lewis, C. E. (2003) 'Hemostatic regulators of tumor angiogenesis: a source of antiangiogenic agents for cancer treatment?', *J. Natl. Cancer Inst.*, **95**(22), pp. 1660–1673.

Duff, S. E., Li, C., Jeziorska, M., Kumar, S. *et al.* (2003) 'Vascular endothelial growth factors C and D and lymphangiogenesis in gastrointestinal tract malignancy', *Br. J. Cancer*, **89**(3), pp. 426–430.

Duffy, M. J. (2002) 'Urokinase plasminogen activator and its inhibitor, PAI-1, as prognostic markers in breast cancer: from pilot to level 1 evidence studies', *Clin. Chem.*, **48**(8), pp. 1194–1197.

Eder, J. P. Jr., Supko, J. G., Clark, J. W., Puchalski, T. A. *et al.* (2002) 'Phase I clinical trial of recombinant human endostatin administered as a short intravenous infusion repeated daily', *J. Clin. Oncol.*, **20**(18), pp. 3772–3784.

Eisenhauer, E. A. (1998) 'Phase I and II trials of novel anti-cancer agents: endpoints, efficacy and existentialism. (The Michel Clavel Lecture, held at the 10th NCI-EORTC Conference on New Drugs in Cancer Therapy, Amsterdam, 16–19 June 1998)', *Ann. Oncol.*, **9**(10), pp. 1047–1052.

El-Naggar, M. M. and Mousa, S. A. (2005) Patent No. US 6,908,907 B2, issued June 21.

Eriksson, K., Magnusson, P., Dixelius, J., Claesson-Welsh, L. and Cross, M. J. (2003) 'Angiostatin and endostatin inhibit endothelial cell migration in response to FGF and VEGF without interfering with specific intracellular signal transduction pathways', *FEBS Letters*, **536**(1–3), pp. 19–24.

Fernandez, P. M. and Rickles, F. R. (2002) 'Tissue factor and angiogenesis in cancer', *Curr. Opin. Hematol.*, **9**(5), pp. 401–406.

Ferrara, N. and Alitalo, K. (1999) 'Clinical applications of angiogenic growth factors and their inhibitors', *Nature Med.*, **5**(12), pp. 1359–1364.

Fisher, B. (1999) 'Highlights from recent National Surgical Adjuvant Breast and Bowel Project studies in the treatment and prevention of breast cancer', *CA: A Cancer J. for Clinicians*, **49**(3), pp. 159–177.

Folkman, J. (1971) 'Tumor angiogenesis: therapeutic implications', *New Eng. J. Med.*, **285**(21), pp. 1182–1186.

Folkman, J. (1989) 'What is the evidence that tumors are angiogenesis dependent?' *J. Natl. Cancer Inst.*, **82**(1), pp. 4–6.

Gale, N. W. and Yancopoulos, G. D. (1999) 'Growth factors acting via endothelial cell-specific receptor tyrosine kinases: VEGFs, angiopoietins, and ephrins in vascular development', *Genes Dev.*, **13**(9), pp. 1055–1066.

Gasparini, G. and Harris, A. (1999) 'Prognostic significance of tumor vascularity', in *Antiangiogenic Agents in Cancer Therapy*, B. A. Teicher, Ed., Totowa, NJ: Humana Press, pp. 317–399.

Hembrough, T. A., Ruiz, J. F., Papathanassiu, A. E., Green, S. J. and Strickland, D. K. (2001) 'Tissue factor pathway inhibitor inhibits endothelial cell proliferation via association with the very low-density lipoprotein receptor', *J. Biol. Chem.*, **276**, pp. 12241–12248.

Hidalgo, M. and Eckhardt, S. G. (2001) 'Development of matrix metalloproteinase inhibitors in cancer therapy', *J. Natl. Cancer Inst.*, **93**(3), pp. 178–193.

Hlatky, L., Hahnfeldt, P. and Folkman, J. (2002) 'Clinical application of antiangiogenic therapy: microvessel density, what it does and doesn't tell us,' *J. Natl. Cancer Inst.*, **94**(12), pp. 883–893.

Jung, S. P., Siegrist, B., Wade, M. R., Anthony, C. T. and Woltering, E. A. (2001) 'Inhibition of human angiogenesis with heparin and hydrocortisone', *Angiogenesis*, **4**(3), pp. 175–186.

Katzenstein, H. M., Salwen, H. R., Nguyen, N. N., Meitar, D. and Cohn, S. L. (2001) 'Antiangiogenic therapy inhibits human neuroblastoma growth', *Med. Pediatric Oncol.*, **36**(1), pp. 190–193.

Kaushal, V., Kohli, M., Zangari, M., Fink, L. and Mehta, P. (2002) 'Endothelial dysfunction in antiangiogenesis-associated thrombosis', *J. Clin. Oncol.*, **20**(13), p. 3042.

Kerbel, R. S., Yu, J., Tran, J., Man, S. *et al.* (2001) 'Possible mechanisms of acquired resistance to anti-angiogenic drugs: implications for the use of combination therapy approaches', *Cancer Metastasis Rev.*, **20**(1–2), pp. 79–86.

Kuenen, B. C., Rosen, L., Smit, E. F., Parson, M. R. *et al.* (2002) 'Dose-finding and pharmacokinetic study of cisplatin, gemcitabine, and SU5416 in patients with solid tumors', *J. Clin. Oncol.*, **20**(6), pp. 1657–1667.

Kyriakides, T. R., Leach, K. J., Hoffman, A. S., Ratner, B. D. and Bornstein, P. (1999) 'Mice that lack the angiogenesis inhibitor, thrombospondin 2, mount an altered foreign body reaction characterized by increased vascularity', *Proc. Natl. Acad. Sci. USA*, **96**(8), pp. 4449–4454.

Lebeau, B., Chastang, C., Brechot, J.-M. *et al.* (1994) 'Subcutaneous heparin treatment increases survival in small cell lung cancer', *Cancer*, **74**, pp. 38–45.

McCarty, O. J. T., Mousa, S. A., Bray, P. and Konstantopoulos, K. (2000) 'Immobilized platelets support human colon carcinoma cell tethering, rolling and firm adhesion under dynamic flow conditions', *Blood*, **96**(5), pp. 1789–1797.

McMahon, G. A., Petitclerc, E., Stefansson, S., Smith, E. *et al.* (2001) 'Plasminogen activator inhibitor-1 regulates tumor growth and angiogenesis', *J. Biol. Chem.*, **276**(36), pp. 33964–33968.

Moehler, T. M., Ho, A. D., Goldschmidt, H. and Barlogie, B. (2003) 'Angiogenesis in hematologic malignancies', *Crit. Rev. Oncol./Hematol.*, **45**(3), pp. 227–244.

Morgan, B., Thomas, A. L., Drevs, J., Hennig, J. *et al.* (2003) 'Dynamic contrast-enhanced magnetic resonance imaging as a biomarker for the pharmacological response of PTK787/ZK 222584, an inhibitor of the vascular endothelial growth factor receptor tyrosine kinases, in patients with advanced colorectal cancer and liver metastases: results from two phase I studies', *J. Clin. Oncol.*, **21**(21), pp. 3955–3964.

Mousa, S. A. (2002) 'Anticoagulants in thrombosis and cancer: the missing link', *Sem. Thromb. Haem.*, **28**(1), pp. 45–52.

Mousa, S. A. (2005) 'Emerging links between thrombosis, inflammation and cancer: role of heparin', *Acta Chirurgica Belgica*, **105**(3), pp. 237–248.

Mousa, S. A. and Mohamed, S. (1999) 'Anti-angiogenesis efficacy of the LMWH, tinzaparin and TFPI', *Blood*, **94**(10), pp. 22a, 82.

Nguyen, M., Watanabe, H., Budson, A. E., Richie, J. P. *et al.* (1994) 'Elevated levels of an angiogenic peptide, basic fibroblast growth factor, in the urine of patients with a wide spectrum of cancers', *J. Natl. Cancer Inst.*, **86**, pp. 356–361.

O'Reilly, M. S. (1997) 'The preclinical evaluation of angiogenesis inhibitors', *Invest. New Drugs*, **15**(1), pp. 5–13.

O'Reilly, M. S., Boehm, T., Shing, Y., Fukai, N. *et al.* (1997) 'Endostatin: an endogenous inhibitor of angiogenesis and tumor growth', *Cell*, **88**(2), pp. 277–285.

Paley, P. J., Staskus, K. A., Gebhard, K., Mohanraj, D. *et al.* (1997) 'Vascular endothelial growth factor expression in early stage ovarian carcinoma', *Cancer*, **80**(1), pp. 98–106.

Panetti, T. S., Chen, H., Misenheimer, T. M., Getzler, S. B. and Mosher, D. F. (1997) 'Endothelial cell mitogenesis induced by LPA: inhibition by thrombospondin-1 and thrombospondin-2', *J. Lab. Clinl Med.*. **129**(2), pp. 208–216.

Peterson, A. C., Swiger, S., Stadler, W. M., Medved, M. *et al.* (2004) 'Phase II study of the Flk-1 tyrosine kinase inhibitor SU5416 in advanced melanoma', *Clin. Cancer Res.*, **10**(12 Pt 1), pp. 4048–4054.

Pirie-Shepherd, S. R. (2003) 'Regulation of angiogenesis by the haemostatic system', *Front. Biosci.*, **8**, pp. 286–293.

Prox, D., Becker, C., Pirie-Shepherd, S. R., Celik, I. *et al.* (2003) 'Treatment of human pancreatic cancer in mice with angiogenic inhibitors', *World J. Surg.*, **27**(4), pp. 405–411.

Raje, N. and Anderson, K. (1999) 'Thalidomide – a revival story', *New Eng. J. Med.*, **341**(21), pp. 1606–1609.

Rickles, F. R., Shoji M. and Abe, K. (2001) 'The role of the haemostatic system in tumor growth, metastasis, and angiogenesis: tissue factor is a bifunctional molecule capable of inducing both fibrin deposition and angiogenesis in cancer', *Int. J. Hematol.*, **73**, pp. 145–150.

Simon, R., Freidlin, B., Rubinstein, L., Arbuck, S. G. *et al.* (1997) 'Accelerated titration designs for phase I clinical trials in oncology', *J. Natl. Cancer Inst.*, **89**(15), pp. 1138–1147.

Sorensen, H. T., Mellemkjaer, L., Olsen, J. H. and Baron, J. A. (2000) 'Prognosis of cancers associated with venous thromboembolism', *New Eng. J. Med.*, **343**(25), pp. 1846–1850.

Stefansson, S., McMahon, G. A., Petitclerc, E. and Lawrence, D. A. (2003) 'Plasminogen activator inhibitor-1 in tumor growth, angiogenesis and vascular remodeling', *Curr. Pharmaceutical Design*, **9**(19), pp. 1545–1564.

Streit, M., Riccardi, L., Velasco, P., Brown, L. F. *et al.* (1999) 'Thrombospondin-2: a potent endogenous inhibitor of tumor growth and angiogenesis', *Proc. Natl. Acad. Sci. USA*, **96**(26), pp. 14888–14893.

Takebayashi, Y., Aklyama, S., Yamada, K., Akiba, S. and Aikou, T. (1996) 'Angiogenesis as an unfavorable prognostic factor in human colorectal carcinoma', *Cancer*, **78**(2), pp. 226–231.

Thompson, W. D., Harvey, J. A., Kazmi, M. A. and Stout A. J. (1991) 'Fibrinolysis and angiogenesis in wound healing', *J. Pathol.*, **165**, pp. 311–318.

Thornes, R. (1983) 'Coumarins, melanoma and cellular immunity', in *Protective Agents in Cancer*, McBrien, D. and Slator, T., eds, London: Academic Press, pp. 43–56.

Tosi, P. and Tura, S. (2001) 'Antiangiogenic therapy in multiple myeloma,' *Acta Haematologica*, **106**(4), pp. 208–213.

Trousseau, A. (1865) *Phlegmasia alba dolens: clinique medicale de l'hotel-dieu de Paris*, London: New Sydenham Society, ch. 3, p. 94.

Volpert, O. V., Tolsma, S. S., Pellerin, S., Feige, J. J. *et al.* (1995) 'Inhibition of angiogenesis by thrombospondin-2', *Biochem. Biophys. Res. Comm.*, **217**(1), pp. 326–332.

Weidner, N., Folkman, J., Pozza, F., Bevilacqua, P. *et al.* (1992) 'Tumor angiogenesis: a new significant and independent prognostic indicator in early-stage breast carcinoma' [see comments], *J. Natl. Cancer Inst.*, **84**(24), pp. 1875–1887.

Wojtukiewicz, M., Zacharski, L. and Memoli, V. (1990) 'Fibrinogen–fibrin transformation in situ in renal cell carcinoma', *Anticancer Res.*, **10**, pp. 579–582.

Yan, X., Lin, Y., Yang, D., Shen, Y. *et al.* (2003) 'A novel anti-CD146 monoclonal antibody, AA98, inhibits angiogenesis and tumor growth', *Blood*, **102**(1), pp. 184–191.

Zacharski, L. and Meehan, K. (1993) 'Anticoagulants and cancer therapy', *Cancer J.*, **6**, pp. 16–20.

Zacharski, L., Henderson, W., Rickles, F. *et al.* (1984) 'Effect of warfarin anticoagulation on survival in carcinoma of the lung, colon, head and neck and prostate', *Cancer*, **53**, pp. 2046–2052.

Zacharski, L., Wojtukiewicz, M., Costantini, V. *et al.* (1992) 'Pathways of coagulation/ fibrinolysis activation in malignancy,' *Seminars Thromb. Haem.*, **18**, pp. 104–116.

Zhang, G. F., Wang, Y. H., Zhang, M. A., Wang, Q. *et al.* (2002) '[Inhibition of growth and metastases of human colon cancer xenograft in nude mice by angiogenesis inhibitor endostatin]', *Ai Zheng*, **21**(1), pp. 50–53.

19

An overview of current angiogenesis assays: Choice of assay, precautions in interpretation, future requirements and directions

Robert Auerbach

Abstract

This chapter places the various angiogenesis assays into a broader picture of what we can do now and what we cannot yet do, by discussing the limitations imposed by each of the specific assays described in the various chapters. It also outlines some of the conditions which still need to be fulfilled before there can be a reliable translation from *in vitro* to *in vivo* and from *in vivo* to the real world of patient care. The importance of considering heterogeneity is discussed, not only because of differences among endothelial cells themselves, but also because of the substantive differences present in the microenvironment at different organ sites. Progress in methodology utilizing computer-based imaging systems and mathematical modelling can be anticipated given the paramount need for improved quantitation and rapid assessment of angiogenic responses.

Keywords

angiogenesis; assays; overview; *in vivo*; *in vitro*

19.1 Introduction

In this volume we have discussed three types of assays: *in vitro* assays, *in vivo* assays and assays applied in clinical research. Within each chapter the benefits and

Angiogenesis Assays Edited by Carolyn A. Staton, Claire Lewis and Roy Bicknell
© 2006 John Wiley & Sons, Ltd

problems of particular methods have been delineated. This chapter will try to put the various assays into a broader picture of what we can do now and what we cannot yet do, what limitations are imposed by the three broad assay groups, as well as by each of the specific assays described in the various chapters, and, finally, to outline some of the conditions which still need to be fulfilled before there can be a reliable translation from *in vitro* to *in vivo* and from *in vivo* to the real world of patient care.

19.2 *In vitro* assays

It is hard to underestimate the importance that *in vitro* studies have played in furthering our understanding of angiogenesis. Most have focused on endothelial cells in isolation – their proliferation (Chapter 2), their migration (Chapter 3) and their reorganization into three-dimensional structures (Chapter 4). These studies have led to the identification of angiogenic growth factors such as VEGF, angiogenic inhibitors such as thrombospondin, and regulators of vascular morphogenesis such as the angiopoietins. However, we must keep the results obtained with these assays in clear perspective, and can best do this by always asking ourselves a number of critical questions (Auerbach *et al.*, 2000; Auerbach *et al.*, 2003).

Perhaps foremost of these is the question of endothelial cell heterogeneity. All endothelial cells are not alike (Auerbach and Joseph, 1983; Joseph *et al.*, 1983; Auerbach, 1983; Alby and Auerbach, 1984; Auerbach *et al.*, 1985; Gumkowski *et al.*, 1987; Auerbach *et al.*, 1987; Auerbach; 1988; Kaminski and Auerbach, 1988; Auerbach, 1991; Plendl *et al.*, 1996; Fajardo, 1989; Zetter, 1990; Zetter and Blood, 1990; Ruoslahti and Rajotte, 2000; Aird, 2003; Gebb and Stevens, 2004; Shaked *et al.*, 2005; Voskas *et al.*, 2005). For years, the most prevalent studies were carried out on bovine aortic endothelial cells (BAECs) or human umbilical vein endothelial cells (HUVECs), yet these cells are poor surrogates for endothelial cells located in blood vessels *in vivo* (for that matter, all BAECs as well as all HUVECs are not alike). Most work was done on long-term established cell lines, long after they had become adapted to their isolated survival in the incubator. Their adaptation belied their normal quiescence, failed to consider their karyotypic alterations, and had, of necessity, lost their dependence on their normal microenvironment. Even when dealing with 'primary cultures', there has been a silent acceptance of the optimistic idea that cells can be passaged at least a few times before they lose their 'normal' physiological properties. This is simply not so.

One of the early demonstrations of the exquisite nature of endothelial cell specificity is shown by the selective affinity of different tumour cell types for specific endothelial cells (Figure 19.1a–c). Thus glioma cells showed preferential adhesion to brain-derived endothelial cells whereas teratocarcinoma cells adhered most avidly to ovary-derived endothelial cells. The two tumour types showed no differential adhesion when plated onto 'neutral' fibroblasts. Similarly, hepatoma cells showed preferential adhesion to liver-derived endothelial cells. It may be

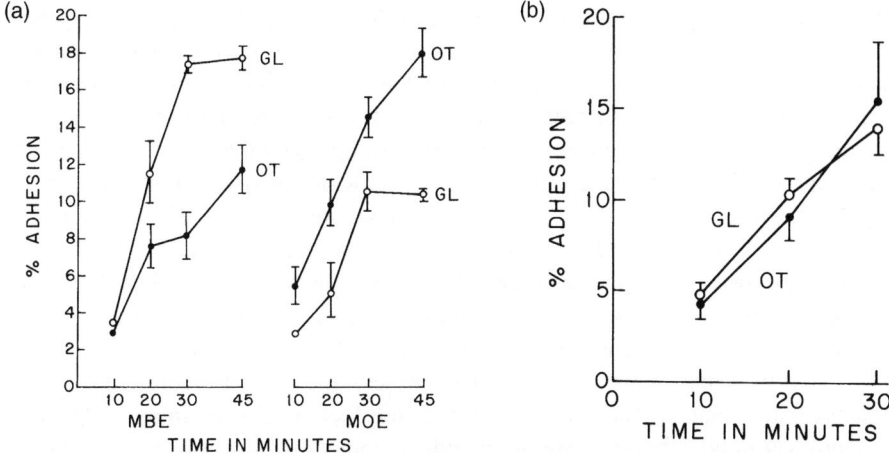

Figure 19.1 Organ-selective adhesion between murine tumour and endothelial cells (a) Mouse glioma (GL-26) or teratocarcinoma (OTT6050) cells were seeded onto a monolayer of murine brain-derived endothelial cells (MBE), murine ovary-derived (MOE) endothelial cells. Cultures were placed on a slowly rotating platform and adhesion was measured at 10-min intervals (Alby and Auerbach, 1983). Organ preference is readily apparent. (b) Similar assay of GL26 and OTT6050 cells seeded on murine fibroblasts (L929). The pattern of adhesion is similar for these tumour lines when plated on a 'neutral' cell layer.

highly significant that mammary adenocarcinoma cells were found to show strong affinity for lymphatic endothelial cells as well as aortic endothelial cells. Only recently has lymphangiogenesis gained prominence in vascular biology, and it has yet to be considered in the development of new angiogenesis assays.

Much of the work involving organ-specific endothelial cells has used the mouse as the model organism. Murine endothelial cells are available from a large number of organs, but the dominant karyotype, even within two to three passages, is likely to be near-tetraploid. True primary cultures are virtually useless because the number of cells obtained in an initial isolate is low. The availability of organ-specific human endothelial cells in early passage is limited to a few sites and the isolates vary from batch to batch. Moreover, human cells, although not as sensitive to karyotypic changes *in vitro* as are mouse cells, nonetheless rapidly undergo modifications in the expression of cell surface specificities as they adapt to (or are selected for) rapid proliferation *in vitro*.

This leads to another question: Does the adaptation or selection for rapid proliferation alter endothelial cells in a way that obscures their natural (*in vivo*) reactivity? One needs to bear in mind that most endothelial cells, with the exception of those activated in processes such as reproduction, regeneration or wound healing, are quiescent (cf. Chapter 1). This makes for a potential problem for studies which use cultured endothelial cells to assess pro-angiogenic factors, inasmuch as these cells already express an angiogenic phenotype. This might, however, not be a problem in studies designed to inhibit endothelial cell responses following angiogenic stimulation, such as might be seen during tumour development. But it is premature to conclude that all angiogenic responses are alike. Indeed, the concept of 'differential angiogenesis' was proposed many years ago (Auerbach, 1987), but is only now beginning to be recognized as a reality which must be considered when selecting specific angiogenesis assays. For example, the angiogenic response evoked by activated lymphocytes appears different from that evoked by tumour cells (Muthukkaruppan and Auerbach, 1979).

This introduces yet a third question: How important is the physical conformation and tissue environment in which endothelial cells are maintained? Agents which inhibit proliferation of endothelial cells in monolayer-cultured endothelial cells may actively promote proliferation when these same cells are cultured in three-dimensional explants. One could argue that endothelial cells do, in a way, grow as a monolayer *in vivo*, but they are in immediate contact with accessory cells such as pericytes and smooth muscle cells (cf. Chapter 7). In addition, they form three-dimensional structures subject to shear stress and differential microenvironmental cues (cf. Chapter 5). Efforts to simulate these types of environmental conditions *in vitro* have revealed many interesting perturbations not seen when endothelial cells are maintained under standard monolayer culture conditions (cf. Chapter 6). The numerous studies using Matrigel or collagen scaffolds, the employment of culture vessels that allow the monitoring of reactions to hydrodynamic forces, and the application of microchamber or hollow fibre technology designed to permit controlled variation in oxygen concentrations and pressure almost universally have yielded insight into the complexity of the factors regulating endothelial cell responses.

The idea that the three-dimensional structure of blood vessels plays an important role in angiogenesis is but an extension of what has been recognized by cell/tissue culturists since the 1920s when pioneers such as Alexis Carrel, Honor Fell and Alexander Maximow carried out their studies on organ cultures in plasma clots (see Greene, 1943 for an exhaustive bibliography). A good example was seen in the development of methods to obtain immune reactions *in vitro*. Organ cultures of spleen fragments could undergo the classic transitions from IgM to IgG (Globerson and Auerbach, 1965), a transition which could not be observed when spleen cells were grown in cell suspension cultures (Mishell and Dutton, 1966). When spleen cell suspensions re-aggregated into three-dimensional arrays, however, their ability to transition from IgM to IgG was restored (Figure 19.2a–c).

None of these questions – and there are others such as the complications introduced by transient or permanent adaptation to culture conditions such as serum, hormones, growth factors, or gas environment – imply that *in vitro* studies cannot be valuable, but rather that these questions must be kept in mind when drawing conclusions about what a particular agent or condition is likely to be able to achieve *in vivo*. It would, for example, be unfortunate if failure to observe an effect of a particular drug *in vitro* automatically eliminates that drug from consideration as a potential agent *in vivo*, and *vice versa*. At the very least, one

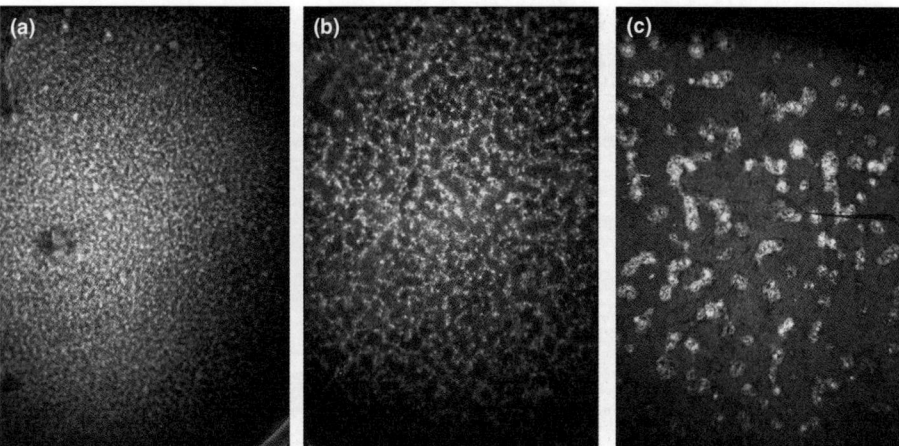

Figure 19.2   Three-dimensional reorganization of dissociated spleen cells *in vitro* mouse adult spleen cells were plated onto a sheep red blood cell-coated plastic culture dish (a). Reorganization began within 24 h (b) and three-dimensional structures formed within the next two days (c). These aggregates were capable of responding to the sheep red cells by first producing IgM and subsequently producing IgG as well, as tested in a plaque-forming assay.

would want to use multiple *in vitro* assays, both three-dimensional and monolayer, with or without accessory cells, and employ more than one endothelial cell source.

Perhaps one of the most important aspects of *in vitro* assays is their ability to act as correlates of *in vivo* activity. We will return to this point when we discuss clinical trials later in this chapter.

19.3   *In vivo* assays

The arguments for using *in vivo* assays are many, but again one must use caution in interpreting the results, especially when using these results to design clinical studies. One must ask how well a particular *in vivo* test reflects reality, where reality is defined in terms of applicability to patient treatments and evaluation. This is especially true because in our efforts to simplify assay procedures and make them subject to quantitation, we have had to make many compromises with that reality.

Consider, for example, the use of the chicken chorioallantoic membrane (CAM) as an assay system (Chapter 10). Leaving aside the questions of species differences (cf. Chapter 16) or even strain differences (Chan *et al.*, 2004) as well as whether the extra-embryonic membrane of a chick embryo is really an 'in vivo assay' (Auerbach *et al.*, 2003), one needs to recognize that the growth of new blood vessels on the CAM is almost two-dimensional, that the tissue environment is highly specialized in a manner quite distinct from what is found in the animal itself, that the cells responding are embryonic and already in an 'activated' state,

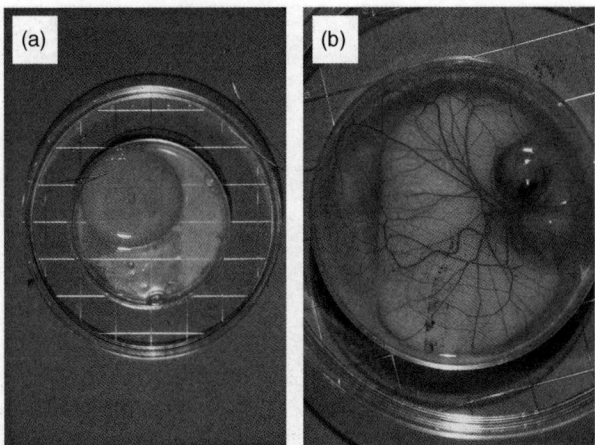

Figure 19.3   Chick embryo cultures (a) The egg contents of embryonated eggs were transferred to petri dishes after 60–70 h of incubation. (b) After an additional 5–6 days of culture the chorio-allantoic membrane covers the entire surface of the culture (from Auerbach *et al.*, 1974). (A colour reproduction of this figure can be viewed in the colour section towards the centre of the book).

and that the diffusion of drugs and growth factors is compromised by the membrane's peripheral location. Even a superficial examination of the CAM of a specific stage of chick embryo development shows the dramatic heterogeneity of the vasculature (Figure 19.3a). There are dramatic changes in these blood vessels from day to day as they develop. To illustrate: Tumour grafts are typically inserted into a slit made through the CAM, thus exposing the graft to the underlying mesenchymal cells and sub-membrane fluids, whereas filter or methylcellulose discs are placed loosely on top. Similarly, some investigators use very young (6–7 day) embryos so that the entire assay period measures responses of highly proliferative endothelial cells in an expanding membrane, while others elect to use older embryos where endothelial cells have stopped dividing and the membrane is no longer expanding. Incubation temperatures have varied also, as has the length of the assay period. Some have even used the 'CAM' before this membrane had actually formed, confusing the yolk sac (Figure 19.3a) for the CAM (Figure 19.3b) This does not mean the assay is not valuable – on the contrary, it has led to many significant discoveries. But the results from CAM assays need to be kept in perspective.

Perhaps the assay systems that have gained the most prominence in the last few years are those employing sponges, Matrigel, or synthetic polymer implants (Chapter 9). Here, too, there are questions that need to be considered. Perhaps most important is the site of implantation: Given the increasing evidence that orthotopic grafts respond differently from heterotopic ones (Killion *et al.*, 1998–1999), one must wonder, for example, how relevant a response induced by a subcutaneously placed sponge or polymer (or even a tumour, Chapter 14) is to the response induced by a tumour growing in a specific organ. It is imperative to question whether the finding that a particular agent can inhibit such a response in a heterotypic site is likely to be relevant when that agent is tested for therapeutic efficacy in the clinic.

Many times, the corneal assay system (Chapter 11) has been cited as the 'gold standard' assay system, because it is a neovascular reaction in an ostensibly neutral site and its progress can be monitored throughout the course of the reaction. However, the concept of 'neutral' is paradoxical: normally, blood vessels do not grow in an avascular site such as the cornea. Moreover, the cornea is a 'privileged site', protected from circulating cells, rapid immune reactions and many serum constituents. Again, the assay has proved most valuable in studies of angiogenesis, but it will be important to keep the limitations of the assay in mind.

Interestingly, the anterior eye chamber of the mouse or rat eye has served as a site for tissue transplantation for more than 60 years (Greene, 1943), but has yet to be exploited for angiogenesis assays. This site, in contrast to the cornea, provides ample room for implantation of tissue, tumours or sponges, and grafts can be positioned for direct contact with the vascular choroid layer. But this site, too, is considered a privileged site, with nutrients provided by the anterior eye chamber fluid rather than serum, and with only a delayed immunological response by the host.

Tumour models have played a key role in many of the *in vivo* studies of angiogenesis (Chapter 12–14), and certainly they would appear to be most relevant when attempting to develop anti-angiogenic therapy protocols. Here, one might have relatively fewer reservations, but several do exist. For example, most studies have been carried out with tumour cell lines or transplantable tumours, long removed from the autochthonous tumours one would encounter in patients. Studies done with murine tumours avoid the problem of histoincompatibility by using syngeneic hosts. Mouse tumours, however, are almost universally very fast-growing and highly aneuploid, and thus quite different from the kind of tumours generally found in the clinic. On the other hand, when human tumours are implanted into immunocompromised mice, the vessels induced by the grafted tumour cells will be murine ones, not human ones, and the host reactions mediated by immune cells, such as NK cells, will be murine, not human. Assays employing genetically modified animals remove some of these problems, but create others by the very nature of these modifications (cf. Chapter 15).

In studying transplanted tumours, one must be aware of the importance of regional differences in the vascular system (Auerbach and Auerbach, 1982). In the trunk of the mouse, even a few millimetres are enough to lead to major differences in tumour growth or wound healing. As shown in Figure 19.4a–c, tumours transplanted subcutaneously in the flank of syngeneic mice showed up to a fourfold difference in growth rates depending on their proximo-distal placement and this has been associated with variation in vascular density (Auerbach *et al.*, 1978a; Auerbach *et al.*, 1978b). A dorso-ventral gradient was also observed (Auerbach *et al.*, 1978c). Figure 19.5a–c illustrates a similar finding for wound healing and subsequent hair growth.

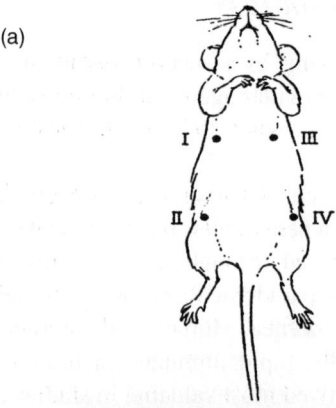

(a)

I   III

II   IV

Diagram showing sites of inoculation used in this study.

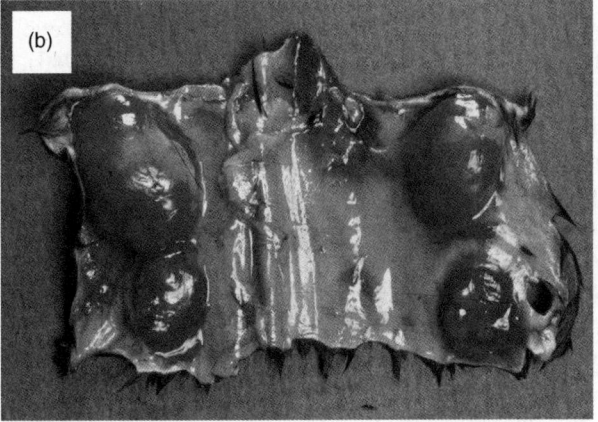

(b)

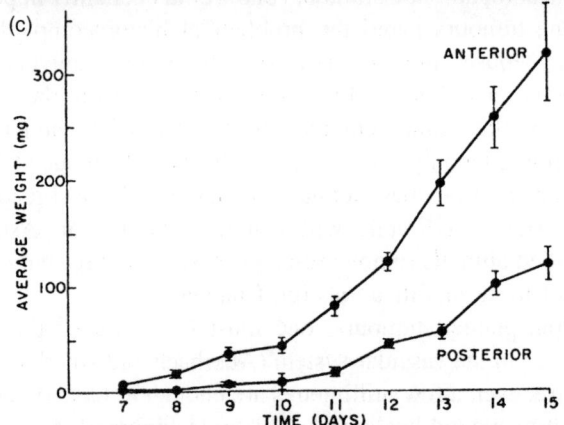

(c)

ANTERIOR

POSTERIOR

Chart 2. Average tumor weight determined 7 to 15 days following inoculation of 5 × 10⁴ C755 mammary tumor cells into anterior (Sites I and III) or
posterior (Sites II and IV) regions of the trunk skin. *Bars*, S.E.

Figure 19.4 Regional differences in tumour growth (a) Diagram showing the location of anterior and posterior injection sites in the flank of adult mice. (b) Characteristic growth differences seen 15 days after 5×10^6 mouse C755 mammary adenocarcinoma tumours were inoculated anteriorly vs. posteriorly into syngeneic C57BL/6Au mice. (c) Shows the average tumour weights obtained following transplantation of C755 tumour cells (n = 58 for anterior, 60 for posterior). Similar results were obtained for murine lymphoma, sarcoma, mastocytoma and teratocarcinoma cells (from Auerbach *et al.*, 1987).

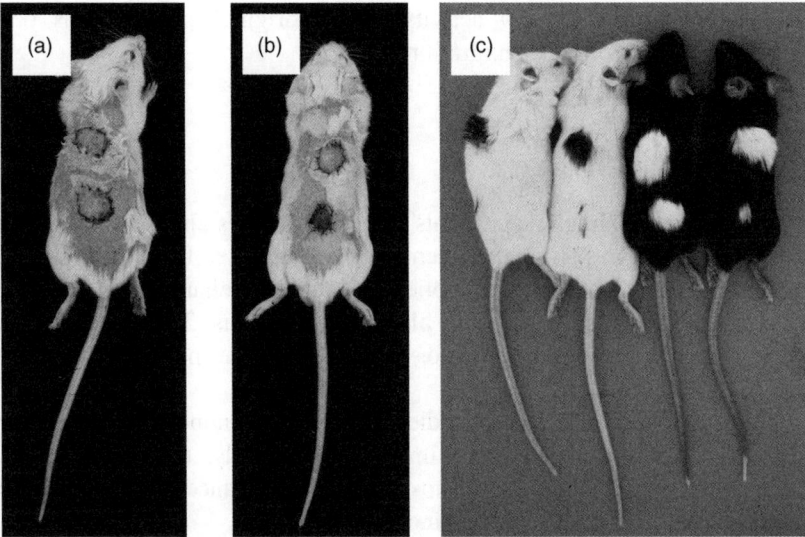

Figure 19.5 Regional differences in the healing in and growth of syngeneic skin transplants black (C57BL/6J) and white (C57BL/6c2J) mice differed only in that the white mice carried mutant albino genes (cc). (a) Shows grafts after one day. (b) Shows same grafts after one week. (c) Shows typical appearance of reciprocal grafts between C57BL6J and C57BL/6c2J mice after 6 weeks. (from Kubai and Auerbach, 1980).

One of the exciting new approaches in angiogenesis studies is the generation of new vessels from embryonic or post-natal stem cells (Chapter 17). It is now possible to generate pure (or near-pure) endothelial cells from primate embryonic (ES) cells, and incorporate these cells into the neovasculature (Kaufman *et al.*, 2004). This is still a long way removed from what is likely to be accomplished clinically. As yet we do not have knowledge about the long-term effect, growth regulation, genetic stability, or immunological problems of transplanted stem cells even in syngeneic systems. Our knowledge of circulating endothelial cell precursors or bone marrow derived adult stem cells is even more fragmentary at this time, but this area of study shows great potential for clinical applications.

The monitoring of clinical studies involving anti-angiogenic agents is at this time at a critical stage (Chapter 18). It is here where, perhaps, many of the angiogenesis assays described in this book can be useful. It is essential for the reagents used in the clinic to be active, and activity has generally been measured in several angiogenesis assays. The same assays should therefore be applied to validate the efficacy of the agents under test. For example, if large-batch preparations of endostatin are administered to patients, these preparations should show the previously defined reactivity in one or more of the animal assay systems (Mundhenke *et al.*, 2001; Thomas *et al.*, 2003). Similarly, if serum levels of an anti-angiogenic reagent serve to establish half-life, then the reagent, isolated from the patients' serum, should show activity in these original assay systems, be they CAM, corneal micropocket or migration assays.

19.4   Clinical trials

The transition from animal experiments to patient studies almost always requires compromise, for the invasive experimental techniques used to study angiogenesis in test animals cannot be directly applied to patients. Reliance thus falls on two major approaches: radioimaging and blood flow analysis. There are, of course, potential surrogate markers in serum or plasma, but these markers have had only limited utility to date.

The major emphasis in clinical studies has been on endpoints such as tumour progression, morbidity or survival times. Unfortunately, these studies almost always are carried out on patients whose disease is advanced to the point where the response is least likely to be obtained.

Lessons learned from *in vitro* and *in vivo* studies, however, should be heeded in designing clinical studies. If, for example, an agent affects wound healing in animals, then skin biopsies would be a reasonable adaptation. If an agent affects circulating levels of FGF-2 or VEGF, then measuring these growth factor levels can prove valuable. If an agent causes mobilization of endothelial cell precursors (EPCs) from the bone marrow, then the determination of EPC numbers in the blood can become a biomarker for efficacy of treatment.

One of the most important aspects of clinical trials is the need to validate the reagents being administered to patients. For this, the specific *in vitro* and *in vivo* small animal tests used in the development of these reagents serve an important function. These tests can be used to determine whether the scale-up of reagents necessary for clinical testing has permitted the reagents to retain their expected level of activity and stability. These tests can also be used to determine half-time values of reagent levels remaining in the circulation. In the absence of these determinations, doubts about the validity of toxicity studies would be justified. More importantly, an unwarranted pessimism about the potential of an anti-angiogenic therapeutic agent may cause the abortion of clinical trials that could, when appropriately designed, lead to successful results in the clinic (Herbst, 2003; Thomas *et al.*, 2003).

19.5   Future directions

It is always hazardous to predict what scientific efforts will lead to. The goals are well-defined: better quantitation, more rapid assessments, applicability to clinical practice, reproducibility and ease of execution. These goals were defined a few years ago in a review published by Rakesh Jain and his colleagues (1997), and they have been reiterated in numerous publications as well as in many of the chapters in this volume.

In the past – and presumably in the future – assays will primarily be developed as needs arise in specific research studies. For example, the need to administer short-lived drugs led to the utilization of slow-release formulations such as ELVAX and HYDRON, and slow-release devices such as ALZET osmotic pumps. The need for making multiple observations led to chamber assays and *in vitro* CAM protocols. Multiple sample analysis was implemented by developing 96-well assays for cell migration and cell proliferation. Semi-solid substrates were created to permit three-dimensional formation of vessels. Various approaches to quantitation, especially using computer-based image analysis of vessel structure, patterning and density, as well as mathematical models such as fractals, were made in order to provide better resolution not only for *in vitro* tests but for observations made in animal models and patients.

A major new technology centres around the use of micro-imaging methods as MRI and CT scans of small animals such as mice are becoming more widely available. There is an expanding laser-mediated technology used both for confocal and fine structure imaging, for sorting of rare cell subpopulations and for single cell isolations. Methods for isolating and expanding embryonic and organ-specific stem cells are already being applied in studies of angiogenesis, and these methods are rapidly being refined through the use of controlled growth factors, serum-independent culture media, and viral integration methods to produce stable cell lines. There is likely to be an increasing translation of information obtained from model organisms such as zebrafish or Drosophila. The generation of specific disease models such as diabetes idiopathic pulmonary fibrosis or macular degeneration, may be achieved by transgene protocols or selective gene targeting, and these will almost certainly lead to new diagnostic assays that incorporate angiogenic responses.

19.6 Concluding comments

In this chapter we have discussed some of the limitations of current assay procedures and a number of potential pitfalls in design and interpretation. We have suggested some of the directions which are likely to be taken in the near future. Perhaps the most important thing to remember will be that the biological system we are studying, angiogenesis, is a highly complex one – involving cell differentiation, intra- and inter-cellular signalling, diffusible mediators, genetic modulations, phenotypic variability – in short, all the vagaries inherent in the study of individual organisms, including humans. Angiogenesis research has moved to the forefront of biology, due in no small part to the development of many creative, useful assays. The ever more rapid movement of angiogenesis research into the area of patient care (Liu *et al.*, 2005; Rugo *et al.*, 2005) will promote modifications of existing protocols and will lead to the development of new methods for correlating basic laboratory findings with clinical studies.

Acknowledgement

I am indebted to Wanda Auerbach for extensive editorial assistance.

References

Aird, W. C. (2003) 'Endothelial cell heterogeneity', *Crit. Care Med.,* **31**(4), Suppl., pp. S221–230.

Alby, L. and Auerbach, R. (1984) 'Differential adhesion of tumor cells to capillary endothelial cells *in vitro*', *Proc. Natl. Acad. Sci. USA*, **81**(18), pp. 5739–5743.

Auerbach, R. (1987) 'Differential angiogenesis', in *Angiogenesis. Mechanisms and Pathology*, Rifkin, D. and Klagsbrun, M., eds, *Current Communications in Molecular Biology*, Cold Spring Harbor Laboratory, pp. 131–133.

Auerbach R. (1988) 'Patterns of tumor metastasis: Organ selectivity in the spread of cancer cells', *Lab. Invest.*, **58**(4), pp.361–364.

Auerbach, R. (1991) 'Vascular endothelial cell differentiation: organ-specificity and selective affinities as the basis for developing anti-cancer strategies'. *Int. J. Rad. Biol.*, **60**(1/2), pp.1–10.

Auerbach, R. and Auerbach, W. (1982) 'Regional differences in the growth of normal and neoplastic cells', *Science*, **215**(4529), pp. 127–134.

Auerbach, R. and Joseph, J. (1983) 'Cell surface markers on endothelial cells: A developmental perspective', in *The Biology of Endothelial Cells*, Jaffe, E. A., Ed., Martinus Nijhoff, The Hague, pp. 393–400.

Auerbach, R., Morrissey, L. W. and Sidky, Y. A. (1978a) 'Regional differences in the incidence and growth of mouse tumors following intradermal or subcutaneous inoculation', *Cancer Res.*, **38**(6), pp. 1739–1744.

Auerbach, R., Morrissey, L. W., Kubai, L. and Sidky, Y. A. (1978b) 'Regional differences in tumor growth: Studies of the vascular system', *Int. J. Cancer*, **22**(1), pp. 40–46.

Auerbach, R., Morrissey, L. W. and Sidky, Y. A. (1978c) 'Gradients in tumour growth', *Nature*, **274**(5672), pp. 697–699.

Auerbach, R., Alby, L., Morrissey, L. W., Tu, M. and Joseph, J. (1985) 'Organ-specificity of capillary endothelial cells', Microvasc. Res., **29**(3), pp. 401–411.

Auerbach, R., Lu, W. C., Pardon, E., Gumkowski, F. *et al.* (1987) 'Specificity of adhesion between tumor cells and capillary endothelium: an *in vitro* correlate of preferential metastasis *in vivo*', *Cancer Res.*, **47**(6), pp. 1492–1496.

Auerbach, R., Auerbach, W. and Polakowski, I. (1991) 'Assays for angiogenesis: A review', *Pharmacol. Therapeutics*, **51**(1), pp. 1–11.

Auerbach, R., Akhtar, N., Lewis, R. and Shinners, B. L. (2000) 'Angiogenesis assays: problems and pitfalls', *Cancer Metastasis Rev.*, **19**(1/2), pp. 167–172.

Auerbach R., Lewis, R., Shinners, B., Kubai, L. and Akhtar, N. (2003) Angiogenesis assays: A critical overview', *Clin. Chem.*, **49**(1), pp. 132–140.

Auerbach, R., Kubai, L., Knighton, D. and Folkman, J. 'A simple procedure for the long term cultivation of chicken embryos', *Dev. Biol.*, **41**(2), pp. 391–39.

Chan, C. K., Pham, L. N., Chinn, C.; Spee, C. *et al.* (2004) 'Mouse strain-dependent heterogeneity of resting limbal vasculature', *Invest. Ophthalmol. Visual Sci.*, **45**(2), pp. 441–447.

Fajardo, L. F. (1989) 'The complexity of endothelial cells. A review', *Am. J. Clin. Pathol.*, **92**(2), pp. 241–250.

Gebb, S. and Stevens, T. (2004) 'On lung endothelial cell heterogeneity', *Microvasc. Res.*, **68**(1), pp. 1–12.

Globerson, A. and Auerbach, R. (1965) 'Primary immune reactions in organ cultures', *Science*, **149**(687), pp. 991–993.

Greene, H. S. N. (1943) 'The heterologous transplantation of embryonic mammalian tissues', *Cancer Res.*, **3**, pp. 809–822.

Gumkowski, F., Kaminska, G., Kaminski, M., Morrissey, L. W. and Auerbach, R. (1987) 'Heterogeneity of mouse vascular endothelium: *in vitro* studies of lymphatic, large blood vessel and microvascular endothelial cells', *Blood Vessels*, **24**(1/2), pp. 11–23.

Herbst, R. S., Mullani, N. A., Davis, D. W., Hess, K. R. *et al.* (2003) 'Development of biologic markers of response and assessment of antiangiogenic activity in a clinical trial of human recombinant endostatin', *J. Clin. Oncol.*, **20**(18), pp. 3804–3814.

Jain R. K., Schlenger, K., Hoeckel, M. and Yuan, F. (1997) 'Quantitative angiogenesis assays: Progress and problems', *Nature Med.*, **3**(11), pp. 1203–1208.

Joseph, J., Tu, M., Alby, L., Grieves, J. *et al.* 'Immunological probes for the study of endothelial cell diversity', in *The Endothelial Cell - A Pluripotent Control Cell of the Vessel Wall*, Thilo-Korner, D. G. S. and Freshney, R. I., eds, Karger, Basel, pp. 55–66.

Kaminski, M. and Auerbach, R. (1988) 'Differences among endothelial cells: Their relation to tumor growth and metastasis', in *Tumor Progression and Metastasis*, Nicholson, G. L. and Fidler, I. J., eds, Alan R. Liss, New York, pp. 161–166.

Kaufman, D. S., Lewis, R. L., Hanson, E. T., Auerbach, R. *et al.* (2004) 'Functional endothelial cells derived from rhesus monkey embryonic stem cells'. *Blood*, **103**(4), pp. 1325–1332. (electronic edition published Nov. 2003).

Killion, J. J., Radinsky. R. and Fidler, I. J. (1998/1999) 'Orthotopic models are necessary to predict therapy of transplantable tumors in mice', *Cancer Metastasis Rev.*, **17**(3), pp. 279–284.

Kubai, L. and Auerbach, R. (1980) 'Regional differences in the growth of skin transplants', *Transplantation* **30**(2), pp. 128–131.

Lelkes, P. I. (1991) 'New aspects of endothelial cell biology', *J. Cellular Biochem.*, **45**(3), pp. 242–244.

Liu, G., Rugo, H. S., Wilding, G., McShane, T. M. *et al.* (2005) 'Dynamic contrast-enhanced magnetic resonance imaging as a pharmacodynamic measure of response after acute dosing of AG-013736, an oral angiogenesis inhibitor, in patients with advanced solid tumors: results from a phase I study', *J. Clin. Oncol.*, **23**(24), pp. 5464–5473.

Mishell, R. I. and Dutton, R. W. (1966) 'Immunization of normal mouse spleen cell suspensions *in vitro*', *Science* **153**(739), pp. 1004–1006.

Mundenke, C., Thomas, J. P., Wilding, G., Lee, F. T. *et al.* (2001) 'Tissue examination to monitor antiangiogenic therapy: a phase I clinical trial with endostatin', *Clin. Cancer Res.*, **7**(11), pp. 3366–3374.

Muthukkaruppan, V. R. and Auerbach, R. (1979) 'Angiogenesis in the mouse cornea', *Science*, **205**(4413), pp. 1416–1418.

Plendl, J, Neumüller, C, Vollmar, A, Auerbach, R. and Sinowatz, F. (1996) 'Isolation and characterization of endothelial cells from different organs of fetal pigs', *Anat. Embryol.*, **194**(5), pp. 445–456.

Rugo, H. S., Herbst, R. S., Liu, G., Park, J. W. *et al.* (2005) 'Phase I trial of the oral antiangiogenesis agent AG-013736 in patients with advanced solid tumors: pharmaco-kinetic and clinical results', *J. Clin. Oncol.* **23**(24), pp. 547–583.

Ruoslahti, E, and Rajotte, D. (2000) 'An address system in the vasculature of normal tissues and tumors', *Ann. Rev. Immunol.*, **18**, pp. 813–827.

Shaked, Y, Bertolini, F. Man, S., Rogers, M. S. *et al.* (2005) 'Genetic heterogeneity of the vasculogenic phenotype parallels angiogenesis; Implications for cellular surrogate marker analysis of antiangiogenesis', *Cancer Cell*, **7**(1), pp. 101–111.

Thomas, J. P., Arzoomanian, R. Z., Albert, D., Marnocha, R. *et al.* (2003) 'Phase I pharmacokinetic and pharmacodynamic study of recombinant human endostatin in patients with advanced solid tumors', *J. Clin. Oncol.*, **21**(2), pp. 223–231.

Voskas, D., Dumont, D. J., Ben-David, Y., Lawler, J. *et al.* (2005) *Cancer Cell*, **7**(1), pp. 101–111.

Zetter, B. R. (1990) 'The cellular basis of site-specific tumor metastasis', *New Eng. J. Med.*, **322**(9), pp. 603–612.

Zetter, B. R. and Blood, C. H. (1990) 'Tumor interactions with the vasculature: angiogenesis and tumor metastasis', *Biochim. Biophys. Acta*, **1032**(1), pp. 89–118.

Index

Note: page numbers in *italics* refer to figures and tables.

Angiogenesis Assays   Edited by Carolyn A. Staton, Claire Lewis and Roy Bicknell
© 2006 John Wiley & Sons, Ltd